Waste Regulation Law

Waste Regulation Law

2nd Edition

Dr Louise Smail

Bloomsbury Professional

Bloomsbury Professional Limited, Maxwelton House, 41–43 Boltro Road, Haywards Heath, West Sussex RH16 1BJ

Bloomsbury Professional is an imprint of Bloomsbury Publishing plc

ISBN: 978 1 84592 137 8

Typeset by Phoenix Photosetting, Chatham, Kent, ME4 4TZ Printed and bound in the United Kingdom by CPI Group (UK) Ltd, Croydon, CR0 4YY

Preface

This book has been written over a period when British waste management law has been subject to large changes. Following the Brexit vote, an already changing environmental law landscape is now entering a new phase and may change in ways which until now had not been considered.

England, Ireland, Scotland and Wales have been enacting their legislation to ensure that European Directives are complied with. Although these Acts have not always been exactly aligned – they have had the same set of Directives to follow. How the future will change no one yet knows – but this book at least sets out where we currently are as a basis to understand where we may move to next. To this end, the legislation of England, Ireland, Scotland and Wales are each looked at in terms of how different aspects of waste legislation have been enacted.

The book has been written with three audiences in mind: lawyers, the regulatory community and industry. For lawyers, this book offers in-depth evaluation of the legislation and also of the relevant case law and hopes also to broaden the lawyer's understanding of the context within which the law operates by relating the legislation to the manner by which waste regulation is conducted in Britain.

The second audience for which the book was written is the regulatory community. Given its complexity, many regulators require a detailed guide to the law, particularly quick and easy access to key legal points, coupled with cross-referencing to appropriate cases. This is essential given the nature of the statutory provisions and the many amendments made to the legislation.

The third audience is the 'regulated', in other words the waste management industry and those industrial activities which themselves generate waste. Operators of waste management facilities work within a tightly regulated sector. Any manager of a waste management facility now requires a detailed knowledge of the relevant legislation and of the wording of the licence or registration under which the facility is operated. The nature of the legislation may mean that legal restraints affect the business on a daily basis. This high profile interface means that, like regulatory officials, each manager has to develop expertise in this area. Following the changes to sentencing for environmental offences in 2014 there have been some very large fines so it is essential that organisations have an understanding of the legal framework of environmental controls on business.

I would like to acknowledge the work that Duncan Laurence did on the previous edition which provided a detailed and useful framework.

Louise Smail
Manchester
01 December 2016

About the Author

Dr Louise Smail is the Corporate Risk Specialist with Fisher Scoggins Waters LLP, where she offers clients assistance with the preparation and validation of Risk Management Plans. Louise Smail is a specialist risk and safety consultant with many years' experience dealing with issues of Corporate Responsibility and has extensive experience within a wide range of organisations including utilities, waste recycling, local government, police, manufacturing and many train operating companies. Between 1994 and 1996, Louise acted as Advanced Works Safety Manager for Union Railways, (a government owned company charged with the task of deciding the route for the Channel Tunnel Rail Link project). Following the takeover of the rail link project by the successful consortium, Louise went to work for Bechtel based in Warrington, looking at issues related to construction risk for clients in the utilities sector and compliance with legislation concerning construction and environmental issues. Louise has worked as principal consultant and director of two international consulting groups and now works on high profile projects looking at business risk evaluation and performance at board level and the impact of changes in legislation in both UK and Europe and is especially interested on the changes that may occur following the Brexit vote.

Louise is a chartered member of the Institute of Occupational Safety and Health, member of the Health and Safety Lawyers' Association and member of the British Computing Society. She is also a Chartered Information Technology Practitioner.

Louise writes and lectures extensively and is an editor and contributor to Corporate Liability: Work Related Deaths and Criminal Prosecutions (published by Bloomsbury Professional).

Contents

Table of Statutes

[*All references are to paragraph numbers*]

Table of Statutory Instruments

[*All references are to paragraph numbers*]

Table of European and Other Legislation

[*All references are to paragraph numbers*]

DIRECTIVES

Table of Cases

[*All references are to paragraph numbers*]

P

R

S

T

V

Z

Chapter 1

Introduction

INTRODUCTION

1.1 Following the vote to leave the European Union there has been a lot of discussion about the effect that this may have on UK legislation in particular, that which covers the environment. This book looks at the current situation that we have in the UK which is based upon European Directives. Although there are differences between England, Ireland, Scotland and Wales – they are still in the main using the same EU Directives on which to base their law. Leaving Europe does not immediately change anything. The law doesn't cease to have force just because the UK have voted to leave the EU – what it may do in future is mean that a new standard on which to base UK law could be created instead of the EU Directives. Much of the original Directive on Waste was based on UK law – so the effect of no longer having an EU Directive may not make such large changes.

What is certain is that most human activities produce waste. Some wastes may be environmentally hazardous in themselves. Others may present significant hazards only if deposited together in large quantities. Inappropriately landfilled household waste, for example, may present a significant threat to water supplies and a gas hazard to neighbouring properties. Conversely, large amounts of other wastes are biologically and environmentally inert. A particular feature of the environmental control of wastes is that, unlike other air or water pollutants, they are infinitely transportable and may even be traded as commodities. The result is that they may pose a threat to the environment far away from their source. They need to be disposed of or recycled somewhere, often at significant cost and in spite of the fact that wastes are universal, no one wants to live near a waste management facility of any kind.

It follows then, that the transport and disposal of wastes must be subject to legal controls. As wastes are global, this legislation impacts on every commercial and industrial organisation in the UK. This is from a small builder to a large construction company, an owner-driver of a skip vehicle or a stock market quoted waste disposal company, the local corner shop or the largest manufacturing plant in the country.

The fact that wastes are global means that waste law is complex. This complexity needs to match a waste disposal and reclamation system whose subtleties and interactions are often little appreciated by those less familiar with the waste management industry. This is even more true now that there is an emphasis on processes involving waste minimisation, recycling and reclamation.

In statistics produced by Defra[1] the England 'waste from households' recycling rate reached 44.8% in 2014. Which is 44.2% higher than in 2013 and is an increase of almost 4% since 2010, the earliest year for which data for this specific measure are available. There is an EU target for the UK to recycle at least 50% of waste generated by households by 2020. This target is steep and leaving the EU may mean that the UK makes a different target. The tonnage of organic recycling (including food waste and other organics) increased by 9.6% in 2014 compared to 2013, and dry recycling increased by 2.3% compared to 2013. In the most recent quarterly data from January to March 2015, the rolling 12 month 'waste from households' recycling rate to end March 2015 was 44.7%. In 2014, the total 'waste from households' amounted to 413kg per person, up from 402kg per person in 2013 (2.9%) and very similar to 2012 levels. Total waste from households amounted to 22.4 million tonnes in 2014, an increase of 3.7% on 2013.

Looking at the management of all Local Authority collected waste for the financial year 2014/15 – which includes waste from households plus that from street bins, street sweepings, parks and grounds as well as other non-household waste collected by local authorities, the total local authority waste disposed to landfill has fallen by 71% since 2000/01 to 6.4 million tonnes in 2014/15, with a marked 20% fall in the last 12 months compared to 2013/14. There is an increase in the amount of waste sent to incineration with energy recovery of 25% in 2014/15 compared to 2013/14 with a more than three-fold increase since 2000/01. This is the first time that the amount of waste sent to incineration with energy recovery exceeded the amount of waste disposed to landfill, at 7.8 million tonnes and 6.4 million tonnes respectively accounting for 30% and 25% of the total local authority waste. In combination, the waste disposed to landfill and sent to incineration with energy recovery has remained very similar to 2013/14 even though the total collected increased by 0.7% (25.8 million tonnes). Local authority managed waste recycled (including waste prepared for reuse and composted) in 2014/15 was 11.1 million, and accounted for 43% of all local authority managed waste. This has increased more than three-fold since 2000/01 although increases in more recent years, especially since 2011/12, have been modest.

1 Defra statistics on waste managed by local authorities in England 2014–15.

This chapter puts into context some of the history of waste legislation prior to the more detailed discussion of the legislation, which follows in later chapters. This chapter sets out not only to summarise the statutory provisions and to describe the regulatory bodies responsible for them; it also attempts to introduce the system of British waste management in order to provide an understanding of the context within which the waste management legislation operates. It will demonstrate the fact that waste regulation law affects all industrial sectors and therefore requires compliance by a very large number of different bodies and individuals. It also provides a perspective on the size and significance of the regulatory task facing the agencies which need to enforce it.

The Development of British Waste Regulation Law

1.2 The statutory controls contained in Part II of the Environmental Protection Act 1990 ('EPA 1990') represent a version of the UK regulatory system which specifically focused on waste management. Before 1972, waste regulation was put into effect only by general legislation, such as that from land use planning and by way of the Public Health Act 1936[2]. Neither of these proved to be particularly effective, as illustrated by the case of the Malkins Bank site in Cheshire. This site had been receiving an uncontrolled variety of toxic wastes since 1932, with the result that half of the site was covered to a depth of 15 metres with various chemicals. By 1970, the smell of 'rotten eggs and tomcats'[3] could be detected several miles away, with periodic minor explosions from deposited drums. A useful view of typical industrial waste disposal practices up until the mid 1970s can be found in the Department of the Environment's publication Co-operative Programme of Research on the Behaviour of Hazardous Wastes in Landfill Sites[4]. It would seem that Malkins Bank was not untypical of industrial waste landfills at that time.

The first statute, which was specifically aimed at controlling waste management practices, was the Deposit of Poisonous Waste Act 1972. This Act was a stop-gap measure, rushed through Parliament in a record 11 days in the face of widespread public concern about the fly-tipping of hazardous waste[5]. The Act for the first time provided for a notification procedure to track and control toxic waste movements and also made it an offence to deposit poisonous,

2 Technical Committee on the Disposal of Solid Toxic Wastes (1970) Disposal of Solid Toxic Wastes, the Ministry of Housing and Local Government and the Scottish Development Department, London, HMSO (generally known as the 'Key Report').

3 Wilson D C, Smith E T and Pearce K W (1981) 'Uncontrolled Hazardous Waste Sites: A perspective of the Problem in the UK' Chemistry and Industry, No 1, Jan 1981, pp 18–23.

4 Department of the Environment (1978) Co-operative Programme of Research on the Behaviour of Hazardous Wastes in Landfill Sites, HMSO, London.

5 For an interesting account of the campaign for controls, see Open University (1978) Environmental Control and Public Health Course Unit 10 Toxic Wastes, Appendix 1, Oxford University Press.

noxious or polluting wastes in a manner which gave rise to an environmental hazard[6]. The Act empowered local authorities to act as the enforcing agency.

More comprehensive legislation followed with the Deposit of Poisonous Waste Act 1972 in the form of the Control of Pollution Act 1974. Part I of the 1974 Act introduced a system of waste disposal licences for waste management facilities[7] and provided the backbone of waste regulatory controls until 1994. The toxic waste notification system under the Deposit of Poisonous Waste Act 1972 continued until 1981, when it was replaced by the Control of Pollution (Special Waste) Regulations 1980[8].

Although the Control of Pollution Act 1974 provided a good start to the formalisation and control of waste management activities, by the 1980s it became apparent that waste regulation was not working. A series of highly critical reports from a variety of Select Committees[9], along with the Royal Commission on Environmental Pollution[10] and the Hazardous Waste Inspectorate[11], were a feature of the environmental history of that decade. A reasonable summary of the problems is apparent from the first paragraph of one of the Environment Committee's reports[12]:

> 'Never, in any of our enquiries into environmental problems, have we experienced such consistent and universal criticism of existing legislation and of central and local government as we have during the course of this inquiry.'[13]

Eventually in 1994 – the Control of Pollution Act 1974 was replaced by Part II of the EPA 1990 and the associated regulations, which are the subject of this volume. Waste regulation was subsequently moved from local government to the Environment Agency and the Scottish Environment Protection Agency in 1996.

For many years, a formal national waste policy for Britain was absent. In December 1995, a White Paper, Making Waste Work: A Strategy of Sustainable

6 Department of the Environment (1972) Deposit of Poisonous Waste Act 1972, Circular 70/72, HMSO, London.
7 Waste disposal licensing was enacted in 1976: Control of Pollution Act 1974 (Commencement No 5) Order 1976, SI 1976/731.
8 SI 1980/1709: subsequently repealed by SI 1996/972.
9 House of Lords Select Committee on Science and Technology (1981) Hazardous Waste Disposal, Vols I–III, HMSO, London; House of Commons Environment Committee (1989) Toxic Waste Vols I–III, HMSO, London; House of Commons Welsh Affairs Committee (1990) Toxic Waste Disposal in Wales, HMSO, London.
10 Royal Commission on Environmental Pollution (1985) Managing Waste: The Duty of Care, HMSO, London.
11 Hazardous Waste Inspectorate (1985) First Report Hazardous Waste Management – An Overview, Department of the Environment, London; Hazardous Waste Inspectorate (1986) Second Report, Hazardous Waste Management '... Ramshackle & Antediluvian'?, Department of the Environment, London; Department of the Environment (1988) The Hazardous Waste Inspectorate Third Report, HMSO, London.
12 House of Commons Environment Committee (1989) Toxic Waste Vol I, HMSO, London, page xi, para 1.
13 See also Laurence D S (1987) 'The Department of the Environment: Ramshackle or Antediluvian?' Wastes Management, January 1987, pp 28–35.

Waste Management in England and Wales[14], was published. Making Waste Work set out to provide a framework in which the statutory controls worked with the waste planning roles of the Agency and local authorities and also with the landfill tax. The Strategy contained a series of targets for waste management up to the year 2005. Two primary targets were singled out; a 'landfill diversion target' and a waste recovery target. These required that the amount of controlled waste passing to landfill should be reduced from 70% to 60% and that 40% of household waste be recovered, both by 2005. It was expected that the primary driver for the delivery of these goals was to be the landfill tax. A further target was to be set in 1998, which was for the overall reduction of controlled waste, but this was postponed[15].

In June 1998, a further consultation paper was published – Less Waste, More Value[16]. Separate strategies were proposed for Scotland, Wales and Northern Ireland. The lack of progress in these matters since Making Waste Work was starkly highlighted in June 1998 by the House of Commons Select Committee on the Environment[17]. In a report whose tone is similar to the critical Reports of Select Committees in the late 1980s, the Committee expressed the view that[18]:

> '... waste management in this country is still characterised by inertia, careless administration and ad-hoc, rather than science-based, decisions. Lip service alone, in far too many instances, has been paid to the principles of reducing waste and diverting it from disposal. Central government has lacked the commitment, and local government the resources, to put a sustainable waste management strategy into practice.'

In July 1999, A Way with Waste – A Draft Waste Strategy for England and Wales was finally published[19] and was finalised in 2000.

A new system for the regulation of contaminated land[20] came in with provisions for a modern control regime to identify and clean-up contaminated land, which had been awaited for most of the 1990s and was delayed due to commercial implications. These provisions operate in parallel to the waste management licensing system under Part II of the EPA 1990, with boundaries being set to prevent overlap. For the reasons discussed in the Preface, a detailed discussion of these matters is beyond the scope of this book.

14 Department of the Environment (1995), Cm 3040, HMSO, London.
15 'Waste Targets in doubt as Waste Strategy put back to 1999' Ends Report 276, Jan 1998, p 32.
16 Department of the Environment, Transport and the Regions (1998) Less Waste, More Value, DETR Sales, London.
17 Department of the Environment, Transport and the Regions, London.
18 EPA 1990, s 78YB, as amended by the Environment Act 1995, s 57.
19 Hansard 27 April 1999 col 91.
20 EPA 1990, s 78A, as amended by the Environment Act 1995, s 57.

European Union Influences

1.3 Although the issues of the Union's influence over UK waste regulation may come into question after the vote to leave there is no doubt about how this has shaped Europe's waste regulation.

The Community Policy Context

1.4 In parallel to the British developments, which started in the 1970s and matured in the 1980s with the European Community's environmental policy, the first European Environment Programme was started in July 1973[21]. This was firstly, as a parallel to the Treaty of the European Community's emphasis on economic expansion, where there was a need to effect an improvement in the quality of life of Community citizens as well as their standards of living. The second reason was that Member States with high environmental standards might find themselves at a competitive disadvantage to countries with lower standards. Conversely, there was also a need to safeguard against a member state obstructing free trade by using its environmental legislation as a market barrier[22].

The international aspects of global environmental policy-making were also being recognised at this time. Discharges to major European rivers – such as the Rhine – and the whole issue of acid rain provided a context for an environment component in Community policy making. The large volume of residuals generated by Member States made Community waste management practices an obvious candidate for an initiative.

As shown in the figure below and the table, the quantity of waste produced by activity within the EU is vast[23].

	Total	Mining and quarrying	Manufacturing	Energy	Construction	Other economic activities	Households
EU-28	2 514 220	733 980	269 630	96 480	821 160	379 560	213 410
Belgium	67 630	115	17 736	1 314	24 570	18 891	5 004
Bulgaria	161 252	141 083	3 009	9 533	1 033	3 841	2 755
Czech Republic	23 171	167	4 376	1 063	8 593	5 739	3 233
Denmark	16 332	18	1 610	893	3 867	6 216	3 727
Germany	368 022	8 625	56 596	8 050	197 528	60 752	36 472
Estonia	21 992	9 355	4 121	6 258	657	1 165	436

21 Although until the Single European Act came into force in 1987, there was no mention of a Community environment policy within the EC Treaty.
22 That this is a genuine concern is apparent from the cases on waste definitions.
23 EUROSTAT Waste Statistics. Data extracted in September 2015.

	Total	Mining and quarrying	Manufacturing	Energy	Construction	Other economic activities	Households
Ireland	13 421	2 025	4 599	396	366	4 379	1 657
Greece	72 328	47 832	4 183	12 259	813	2 383	4 859
Spain	118 562	22 509	14 594	5 772	26 129	28 333	21 224
France	344 732	2 477	21 431	2 100	246 702	42 024	29 996
Croatia	3 379	5	425	108	682	968	1 191
Italy	162 765	720	34 142	3 616	52 966	41 708	29 613
Cyprus	2 086	218	98	2	965	353	451
Latvia	2 310	2	396	133	8	558	1 213
Lithuania	5 679	26	2 551	29	419	1 477	1 177
Luxembourg	8 397	131	509	2	7 079	426	249
Hungary	16 310	91	2 991	2 872	4 038	3 638	2 681
Malta	1 452	45	9	2	1 041	201	155
Netherlands	123 613	179	14 115	1 342	81 354	17 758	8 864
Austria	34 047	51	3 636	622	19 471	6 247	4 020
Poland	163 378	68 035	31 135	20 706	15 368	18 809	9 324
Portugal	14 184	243	3 188	422	928	4 672	4 731
Romania	266 976	223 293	6 029	9 043	1 325	22 638	4 647
Slovenia	4 547	14	1 345	1 069	535	941	641
Slovakia	8 425	311	2 516	1 046	806	2 090	1 657
Finland	91 824	52 880	14 531	1 011	16 034	5 635	1 734
Sweden	156 307	129 481	6 158	1 852	7 656	6 967	4 193
United Kingdom	241 101	24 044	13 596	4 965	100 230	70 759	27 506
Iceland	529	0	93	2	11	191	233
Liechtenstein	467	29	12	0	107	2	316
Norway	10 721	470	2 639	89	1 881	3 205	2 438
Montenegro	386	1	33	351	0	0	0
FYR of Macedonia	8 472	802	1 304	6	0	6 360	0
Serbia	55 003	47 896	760	5 744	364	238	0
Turkey	1 013 226	950 587	13 141	18 424	0	289	30 785
Bosnia and Herzegovina	4 457	72	1 213	3 171	0	0	0
Kosovo	1 167	177	80	151	0	268	490

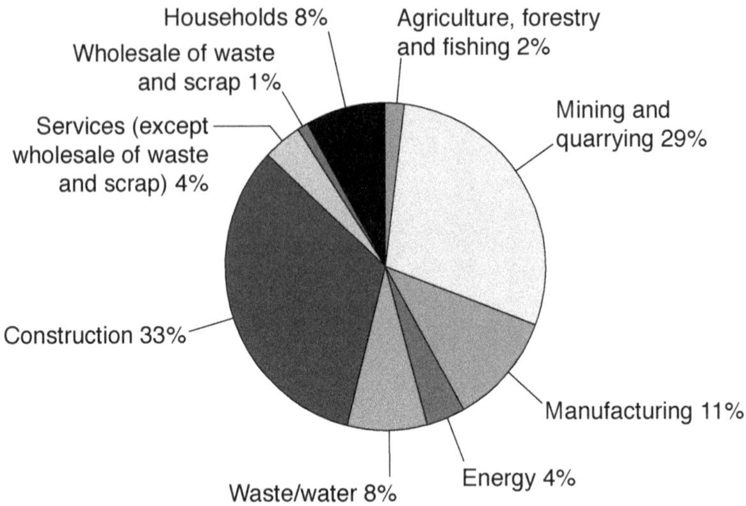

The extensive internal waste trade between Member States particularly required Community-wide action in this area.

Waste Cross Borders for EU Countries

1.5 The need for international controls on cross-border waste movements was highlighted by an incident in 1983 involving the temporary loss of 41 barrels of dioxin contaminated waste from a major industrial accident at Seveso in Italy[24].

Since the Single European Act came into force in 1987, Community waste management policy has been effectively a subset of the obligations on environmental protection provided for in the amended Treaty[25]. An indication of the direction of Community waste management policy is shown by five successive Environmental Action Programmes[26] and by the 1989 Community

24 Anon (1986) 'Carrier's Wife Paid to Reveal Seveso Dioxin in Plan Brokered by German Interior Minister', International Environment Reporter 9 (1) p 8.

25 Contained in Title XIX of the EC Treaty, particularly in Article 174 (previously Title XVI and Article 130(r)).

26 Programme of Action of the European Communities on the Environment ([1973] OJ C112/1 20.12.73); European Community Action Programme on the Environment (1977–1981) ([1977] OJ C139/3 13.6.77); EEC Third Environmental Action Programme (1983–1987) ([1983] OJ C46 17.2.83); EEC Fourth Environmental Action Programme (1987–1992) ([1987] OJ C328/1 7.12.87); Towards Sustainability – A European Community Programme of Policy and Action in relation to the Environment and Sustainable Development ([1993] OJ C138/5 17.5.93). The latter runs until the year 2000.

Strategy on Waste Management[27]. The latter was reviewed in 1996[28], being the subject of a Council resolution in 1997[29] which covers a further five-year period.

Community waste management policy at the outset had a dual objective: a resource conservation strategy and a disposal policy. This was explicitly reinforced by the Fifth Action Programme. Consequently, the Community Strategy on Waste Management[30] suggests a series of Strategic Guidelines:

(1) Waste Prevention by the use of 'clean' technologies and by waste minimisation;
(2) Recycling and Re-use by the effective re-utilisation of waste after it arises;
(3) Optimisation of Final Disposal Methods and Technologies by the expansion of physical, chemical and biological waste pre-treatment. Landfill should be a last resort;
(4) Effective Regulation of Waste Transportation;
(5) Remedial Action on Contaminated Dump Sites.

These guidelines are condensed into the form of a simple, three level hierarchy. This is composed – in order of priority – of waste prevention, recycling and final disposal. The centrality of the hierarchy in policy-making was re-confirmed in 1996 by the Communication from the Commission which reviewed the Waste Strategy[31]. However, like many individual countries, until recently the emphasis of Community waste management policy has been placed on attaining an effective regulatory framework aimed particularly at controlling waste disposal activities and waste movements. The Communication from the Commission was itself critical of the slowness of progress in respect of the upper elements of the hierarchy[32], noting in particular that waste generation rates across the Community are still increasing.

27 Commission of the European Communities (1989) A Community Strategy for Waste Management, SEC(89) 934 Final, 18 September 1989; Council Resolution of 7 May 1990 on Waste Policy (90C 122/02) ([1990] OJ C122/2 18.5.90).

28 Communication from the Commission on the Review of the Community Strategy for Waste Management draft Council Resolution on Waste Policy, Brussels 30.07.96, COM(96) 399 Final.

29 Council Resolution of 24 February 1997 on a Community Strategy for Waste Management ([1997] OJ C76 11 March 1997), p 1.

30 Commission of the European Communities (1989) A Community Strategy for Waste Management, SEC(89) 934 Final, 18 September 1989, sections IV to VIII.

31 Communication from the Commission on the Review of the Community Strategy for Waste Management draft Council Resolution on Waste Policy, Brussels 30.07.96, COM(96) 399 Final, para 20.

32 Communication from the Commission on the Review of the Community Strategy for Waste Management draft Council Resolution on Waste Policy, Brussels 30.07.96, COM(96) 399 Final, para 33.

EC Policy Implementation

1.6 The implementation of Community waste management initiatives started in the mid 1970s by way of a series of Directives which required the establishment of structures and responsibilities by Member States. The most significant of these initiatives is the Directive on Waste[33], which is sometimes referred to as the 'Waste Framework Directive'[34]. This was followed by Directives on Toxic and Dangerous Waste[35], PCBs[36] and waste oils[37].

Initially, the Directive on Waste had little influence on British practices, as from a 1990s perspective, the control system in place in Britain was ahead of many Member States and the Directive on Waste was partly based on British practice[38]. Indeed, the first time a Directive changed UK waste regulation practices was in 1981 when the Directive on Toxic and Dangerous Waste[39] caused the repeal of the Deposit of Poisonous Waste Act 1972 and its replacement by the Control of Pollution (Special Waste) Regulations 1980[40].

Since then, Community instigated environmental law has attained wide-ranging influence over the UK's system of environmental controls. Britain's early lead in these matters dropped rapidly in the 1980s. Indeed, the integration of many Community environmental provisions has been problematic ever since, despite virtually all of them being subject to the UK's earlier prior agreement at the Council of Ministers. For example, the implementation of the waste management licensing system under the EPA 1990 was delayed for a year due to technical problems involved with the matching of that Act to the provisions of the Directive on Waste.

33 Council Directive of 15 July 1975 on Waste (75/442/EEC) ([1975] OJ L194/39 25.7.75) (as amended by Council Directive of 18 March 1991 amending Directive 75/442/EEC on Waste (91/156/EEC) ([1991] OJ L78/32 26.3.91)) and by Commission Decision of 24 May 1996 adapting Annexes IIA and IIB to Council Directive 75/442/EEC on Waste (96/350/EEC) ([1996] OJ L135/32 6.6.96).

34 Note that, despite the Directive on Waste being substantially amended by Directive 91/156, it retains its original number (75/442/EEC): see, for example, the title of the subsequent decision to establish a Waste Catalogue: 'Commission Decision of 20 December 1993 establishing a list of waste pursuant to Article 1(a) of Council Directive 75/442/EC on waste'.

35 Council Directive of 20 March 1978 on Toxic and Dangerous Waste ([1978] OJ L84/43 31.3.78) (repealed by Council Directive of 12 December 1991 on Hazardous Waste (91/689/EEC) ([1991] OJ L377/20 31.12.91).

36 Council Directive of 6 April 1976 on the disposal of polychlorinated biphenyls and polychlorinated terphenyls (76/403/EEC) ([1976] OJ L108 26.4.76) (repealed by Council Directive of 16 September 1996 on the disposal of polychlorinated biphenyls and polychlorinated terphenyls (PCB/PCT) (96/59/EC) ([1996] OJ L243 p 31).

37 Council Directive of 16 June 1975 on the disposal of waste oils (75/439/EEC) ([1985] OJ L194/23 25.7.85) (subsequently amended by Council Directive of 22 December 1986 amending Directive 75/439/EEC on the disposal of waste oils (87/101/EEC) ([1987] OJ L42/43 12.2.87).

38 Haigh N (1998) Manual of European Environmental Policy: The EC and Britain Cartermill Publishing, London, paras 5.3-4 and 5.3-6.

39 Council Directive on Toxic and Dangerous Waste (78/319) ([1978] OJ L84/43 31.3.78).

40 SI 1980/1709.

Many of the original waste management Directives could be criticised for their vagueness. Key problems were around a lack of clear definitions of 'waste', 'toxic waste' and of the nature of acceptable disposal. Even after the wording of the Directive on Waste was tightened in 1991[41], the overall environmental objective for Member States' waste management controls, namely disposal or recovery '... without endangering human health and without using processes or methods which could harm the environment ...'[42] is clearly very open ended and hence likely to be subject to varied interpretation between countries.

There is a trend for the embellishment of the Community provisions, due to the need to respond to the lack of uniform transposition of Directives into national law by Member States. Generally, the later Community initiatives are characterised by their greater precision and leave Member States with proportionately less discretion. For example, since the late 1980s, a series of Directives emerged which have been specifically aimed at defining acceptable practices at particular types of waste management facility. There are now two Directives on municipal waste incineration[43], and a Directive on hazardous waste incineration[44]. There was also a succession of proposals for a Directive on landfill[45]. However, by the Autumn of 1996, the draft Landfill Directive had become so weakened and internally confused that it was not surprising that the European Parliament vetoed it by a majority of 445 votes to 18[46].

A significantly different proposal for a Landfill Directive emerged in the winter of 1996 and was formally published in 1997[47]. Given certain Member States' opposition to the earlier proposal, the new proposal is more stringent and wider ranging in its potential effect. For example, whilst the earlier version was controversial for the reason that it wished to restrict co-disposal[48] practices at

41 By way of Council Directive of 18 March 1991 amending Directive 75/442/EEC on Waste (91/156/EEC) ([1991] OJ L78/32 26.3.91).

42 Council Directive of 15 July 1975 on Waste (75/442/EEC) ([1975] OJ L194/39 25.7.75) (as amended by Council Directive of 18 March 1991 amending Directive 75/442/EEC on Waste (91/156/EEC) ([1991] OJ L78/32 26.3.91)) and by Commission Decision of 24 May 1996 adapting Annexes IIA and IIB to Council Directive 75/442/EEC on Waste (96/350/EEC) ([1996] OJ L135/32 6.6.96), Article 4.

43 Council Directive of June 1989 on the Prevention of Air Pollution from New Municipal Waste Incineration Plants (89/369/EEC); Council Directive of 21 June 1989 on the Reduction of Air Pollution from Existing Municipal Incineration Plans, (89/429/EEC) ([1989] OJ L203/50).

44 Council Directive 94/67/EC of 16 December 1994 on the Incineration of Hazardous Waste ([1994] OJ L365/34 31.12.94). It is proposed to replace this Directive – along with the two on municipal waste incincerators: Proposal for a Council Directive on the Incineration of Waste ([1998] OJ C372/11 2.12.98) and Proposal for a Council Directive amending Directive 94/67/EC on the Incineration of Hazardous Waste ([1998] OJ C13/6 17.1.98). Since these were published, a more recent proposal is to merge these draft Directives. There is also a proposed Council Directive on Port Reception Facilities for Ship-Generated Waste and Cargo Residues ([1998] OJ C271/79 31.8.98).

45 Proposal for a Council Directive on the Landfill of Waste, COM(91)102 ([1991] OJ C190/1 22.7.91). A number of other drafts were also circulated.

46 ENDS Report 256 May 1996, p 38.

47 Proposal for the Council Directive on the Landfill of Waste ([1997] OJ C156/10 24.5.97).

48 The mixing of particular types of hazardous waste with household and other similar wastes.

landfills, the revised version will extensively reduce the intake of biodegradable wastes at most landfills.

There have also been attempts to tighten up on the subject of regulation, by way of defining 'waste' and 'hazardous waste' in much greater detail. A new definition of waste was made in the 1991 amendments to the Directive on Waste[49], although it did not fully standardise waste definitions between Member States. The latter criticism is itself raised by the 1996 Communication on the Waste Strategy[50]. A redefinition of hazardous waste occurred in 1995 when the earlier Toxic and Dangerous Waste Directive[51] was repealed and replaced by a Directive on Hazardous Waste[52].

A final initiative of relevance is a Directive on Integrated Pollution Prevention and Control (IPPC)[53]. Unlike the other Directives already referred to, this was more general in its environmental protection requirements, perhaps reflecting the reality that Member States were not likely to agree a more stringent version. The IPPC Directive required the implementation of an integrated pollution control regime on a range of waste management facilities. These include:

- all hazardous waste disposal facilities in excess of 10 tonnes per day capacity[54];
- hazardous waste recovery facilities in excess of 10 tonnes per day capacity which involve the processing of waste solvents, acids and bases, 'components used for pollution abatement' and waste oil, along with processes which use waste to generate energy[55];
- non-hazardous waste landfills, other than those, which accept only inert wastes, of an input rate in excess of 10 tonnes per day or 25,000 tonnes total capacity[56];
- municipal waste incinerators of a capacity of over 3 tonnes per hour[57];
- disposal processes involving biological and physico-chemical waste treatment.

49 Council Directive of 18 March 1991 amending Directive 75/442/EEC on Waste (91/156/EEC) ([1991] OJ L78/32 26.3.91).
50 Communication from the Commission on the review of the Community Strategy for Waste Management. Draft Council Resolution on Waste Policy 30.7.1996, Com(96) 399 Final, para 13.
51 Council Directive of 20 March 1978 on Toxic and Dangerous Waste, ([1978] OJ L84/43 31.3.78).
52 Council Directive of 12 December 1991 on Hazardous Waste (91/689/EEC) ([1991] OJ L377/20 31.12.91); Council Directive 94/31/EC of 27 June 1994 amending Directive 91/689/EEC on Hazardous Waste ([1994] OJ L168/28 2.7.94).
53 Council Directive 96/61/EC of 24 September 1996 concerning Integrated Pollution Prevention and Control ([1996] OJ L257/26 10.10.96).
54 Directive 96/61, Annex 1, para 5.1.
55 Directive 96/61, Annex 1, para 5.1 as cross-referred to operations R1, R5, R6, R8 and R9 in Annex IIB of Directive 75/442.
56 Directive 96/61, Annex 1, para 5.4.
57 Directive 96/61, Annex 1, para 5.2.

Reducing Community Waste Generation

1.7 Most of the Community initiatives have focused on definitions and the control of particular waste management processes. Many such initiatives address waste management options, which are low down in the waste management hierarchy proposed in the Community Waste Strategy. Initiatives aimed at progressing activities towards the upper areas of the waste hierarchy – like waste minimisation and recycling – have taken longer. However, it seems likely that Community-led activity will increase in respect of this matter in the next decade. A start was made in 1994, by way of a Directive[58], which attempted to instigate more widespread recycling of the 25 million tonnes of packaging waste produced each year by Member States[59].

In the 1980s, a number of Member States started independent drafting of legislation to restrict the production of waste packaging and to encourage its re-use and recycling. Such initiatives were aided by a judgment of the European Court, which permitted Denmark unilaterally to ban metal drink cans and to require that all drinks containers be refillable[60]. Rather than allow such unilateral actions as these by individual Member States with the consequential trade barriers, the Commission attempted to regain some of the initiative by way of its own proposal for a Packaging Directive[61]. In 1996 the Directive replaced an earlier, and rather anodyne, Beverage Containers Directive[62].

The Packaging Directive contains ambitious recycling targets, requiring that between 50% and 65% of packaging waste be removed from the waste stream for recovery within five years of June 1996[63]. In addition, between 25% and 45% of such waste has to be recycled[64]. The distinction between 'recovery' and 'recycling' is principally that the former includes incineration with energy generation[65]. It would appear that this Directive is very much seen as the first in a series, which will set targets for the reduction and recovery of different waste types[66].

A prerequisite to any system, which attempts to establish national waste recovery targets, is a common statistical base to compare countries. While there remains no common and unified definition of waste across all Member States, a

58 The European Parliament and Council Directive 94/62 of 20 December 1994 on Packaging and Packaging Waste ([1994] OJ L365/10 31.12.94).
59 Figure extrapolated from Table 7.1B of OECD (1997) OECD Environmental Data Compendium 1997, OECD, Paris.
60 *Commission EC v Denmark*: 302/86 [1988] ECR 4607, [1990] 2 JEL No 1 p 89.
61 The European Parliament and Council Directive 94/62 of 20 December 1994 on Packaging and Packaging Waste ([1994] OJ L365/10 31.12.94).
62 Council Directive of 27 June 1985 on Containers of Liquids for Human Consumption (85/339) ([1985] OJ L176/18 6.7.85), which was repealed by Article 23 of Directive 94/62.
63 Directive 94/62, Article 6(1)(a).
64 Directive 94/62, Article 6(1)(b).
65 Directive 94/62, Article 3(6), (7) and (8).
66 Communication from the Commission on the review of the Community Strategy for Waste Management. Draft Council Resolution on Waste Policy 30.7.1996, Com(96) 399 Final, para 70.

European Waste Catalogue was finalised[67] in 1994 with the objective of setting down the foundations of Community-wide waste management statistics[68].

Transfrontier Waste Controls

1.8 A major policy problem for certain Member States was the need to adequately regulate cross-border movements of waste. In 1984, the extensive international waste trade became the specific subject of a Directive[69], which was a response to the considerable volume of waste moved internally within the Community and a reaction to the 41 wandering Seveso drums. When the Toxic and Dangerous Waste Directive was being drafted, it was felt that existing legislation and a number of ad-hoc conventions between Member States presented a sufficient level of control[70]. The Seveso drum incident therefore called into question the efficacy of Member States and the Community's existing regulatory framework.

However, the policy approach at Community level towards trans-frontier waste movements was markedly different in the 1980s to that observed in the 1990s. Until roughly late 1989, it was proposed that the free trade provisions concomitant with the creation of the Single Market should apply to international shipments of wastes as well as goods. This meant the creation of a type of 'Euro-Waste Free Market'. Eventually, practical realities prevailed. The 1989 Waste Strategy proposed[71] what has become known as the 'proximity principle': that wastes should be disposed of[72] as close to their origin as is possible.

67 Commission Decision of 20 December 1993 establishing a list of wastes pursuant to Article 1(a) of Council Directive 75/442/EEC on Waste (Decision 94/3) ([1994] OJ L5/15 7.1.94).

68 Communication from the Commission on the review of the Community Strategy for Waste Management. Draft Council Resolution on Waste Policy 30.7.1996, Com(96) 399 Final, para 83.

69 Council Directive of 6 December 1984 on the Supervision and Control within the European Community of the Transfrontier Shipment of Hazardous Waste ([1984] OJ L326/31 13.12.84), as amended by the Council Directive of 22 July 1985 adapting to technical progress Council Directive 84/631/EEC on the supervision and control within the European Community of the transfrontier shipment of hazardous waste (85/469/EEC) ([1985] OJ L272/1 12.10.85); Council Directive of 12 June 1986 amending Directive 84/631/EEC on the supervision and control within the European Community of the transfrontier shipment of hazardous waste (86/279/EEC) ([1986] OJ L181/13 4.7.86); Council Directive of 23 December 1986 adapting to technical progress for the second time Council Directive 84/631/EEC on the supervision and control within the European Community of the transfrontier shipment of hazardous waste, (87/112/EEC) ([1987] OJ L48/31 17.2.87).

70 Haigh N (1987) EEC Environmental Policy and Britain 2nd edn, Longman, p 147.

71 Commission of the European Communities (1989) A Community Strategy for Waste Management, SEC(89) 934 Final, 18 September 1989, p 23.

72 Typically, the proximity principle does not apply to waste movements destined for recycling: Article 5 of the Directive on Waste (as amended) refers only to 'disposal' activities, not recovery facilities. Similarly Article 4 of Council Regulation 259/93 allows Member States to prevent cross-border waste movements only in respect of disposal sites.

Although the proximity principle was enshrined in the Directive on Waste when it was amended in 1991[73], extra-Community institutions in the form of the United Nations Environment Programme and the OECD[74] mainly drove the enactment of the principle of cross-border waste movements. The Trans-frontier Directive[75] had to be scrapped entirely with the ratification by the Community of the Basel Convention[76]. All Member States are parties to the Basel Convention, which is aimed at protecting less developed countries from the indiscriminate dumping of 'developed' countries' waste. As explained in Chapter 5, the Basel Convention requires that European countries should be looking inwardly to provide the infrastructure to dispose of Europe's residuals. To accord with the Convention and the subsequent OECD Decision on wastes passing to recovery[77], the Community's cross border control system had to be expanded to address all trans-frontier waste movements, not only those involving hazardous wastes. In addition, a significant change was made to the tradition by which Community provisions were transposed[78] using a Regulation, rather than a Directive, as a vehicle.

THE WASTE MANAGEMENT INDUSTRY IN BRITAIN[79]

1.9 In parallel to the development of both domestic and Community legislation, the British waste industry has become increasingly sophisticated. Until the implementation of the EPA 1990, there were, in effect, three quite distinct and separate branches of the waste management industry. The first consisted of the private sector waste disposal companies, which ranged from owner-driver skip operators to major public limited companies. These dealt extensively in the carriage and disposal of commercial and industrial waste, also providing a limited household waste disposal infrastructure. The second sector of the waste industry centred upon the local authorities, which collected virtually all household waste and whose household waste disposal sites also

73 Council Directive of 15 July 1975 on Waste (75/442/EEC) ([1975] OJ L194/39 25.7.75) as amended by Council Directive of 18 March 1991 amending Directive 75/442/EEC on Waste (91/156/EEC) ([1991] OJ L78/32 26.3.91), Article 5.

74 Organisation for Economic Co-operation and Development.

75 Council Directive of 6 December 1984 on the Supervision and Control within the European Community of the Transfrontier Shipment of Hazardous Waste ([1984] OJ L326/31 13.12.84).

76 1989 Convention on the Control of Transboundary Movements of Hazardous Wastes and their Disposal (1 JEL No 2 (1989), pp 255–277).

77 OECD Decision on the Control of Transfrontier Movements of Waste Destined for Recovery Operations (OECD C(92)39/Final as amended by Decision C(93)74/Final 23.7.93, Decision C(94) 153 28.7.94 and Decision C(96)231 10.12.96: para 5.21ff).

78 Directives require transposition by a member state's own legislation, Regulations are directly applicable in all Member States and do not require transposition: EC Treaty Article 249 (formerly Article 189).

79 Council Regulation (EEC) No 259/93 of 1 February 1993 on the supervision and control of shipments of waste within, into and out of the European Community ([1993] OJ L30/1 6.2.93).

provided a significant outlet for commercial and industrial wastes gathered by the private sector. The final component was the recycling industry which covered the reclamation of scrap metal, vehicle dismantling, waste paper processing, plastics recovery and other similar activities.

By the 1990s the distinction between the public and private sector operators had become much less sharp, particularly in England and Wales. This decade also saw the start of the integration of some of the recycling industry within certain waste disposal companies.

The duality involving the private and public sectors was affected by two related government policy initiatives. First, the compulsory competitive tendering system established in the 1980s for local authority refuse collection introduced much more private sector involvement[80]. Secondly, under s 32 of the EPA 1990, local authorities in England and Wales were required to divest themselves of their waste disposal infrastructure, including landfills, household waste incinerators, transfer stations and civic amenity sites[81]. The legislation left the local authorities with what were essentially two options. The first was to form a Local Authority Waste Disposal Company (LAWDC) which, whilst wholly controlled by the local authority, was operated at 'arms length' to it, by way of an independent board of directors[82]. The alternative solution was to sell off the waste management assets to the private sector. Both of the options have been extensively utilised, and most LAWDCs have been subsequently sold by their local authority masters.

Besides the traditional differentiation between the public and private sector waste industry, activities involving recycling and reclamation did not, in the past, consider themselves to be part of the national waste management industry. The perception was that this sector dealt with 'secondary raw materials', not wastes[83]. Historic roots and ownership were often quite distinct and, until the late 1980s, much of the recycling industry was subject to considerably less formal regulation.

However, this distinction has also been subject to change, albeit that actual industrial restructuring appears by no means over. Structural change is encouraged by the recognition of the contribution of the recycling industry towards the delivery of the national waste strategy and to the Landfill Directive's biodegradable waste diversion targets. Given the increased emphasis on recycling and reclamation at the expense of the option of final

80 By way of the provisions contained in the Local Government Act 1988.

81 Transfer stations are facilities whereby waste is deposited by collection vehicles, being bulked up for long-distance transport into larger capacity vehicles. Civic amenity sites are facilities where the public can dispose of those wastes which are not gathered by the normal household waste collection rounds.

82 Department of the Environment (1991) Competition for Local Authority Waste Disposal Contracts and New Arrangements for Disposal Operations Circular 8/91, HMSO, London.

83 This distinction was perhaps the cause of the case of *R v Rotherham Metropolitan Borough Council, ex p Rankin* [1990] 1 PLR 93 where the planning permission of a solvent recovery facility was the subject of a judicial review due to a failure to advertise the application as 'bad neighbour' development under the Town and Country Planning Act 1971.

disposal, particularly disposal by way of landfill, commercial providers of waste management capacity are increasingly looking at the provision of integrated reclamation and disposal systems. Hence the traditional structural boundaries between the waste management and the recycling industry are breaking down. For example, in the 1990s a number of waste paper merchants were purchased by certain companies which had previously focused on landfill site management. Further integration would seem inevitable. Finally, the waste regulatory provisions of the EPA 1990 acknowledge that many recycling activities involve materials falling within the statutory meaning of controlled wastes. Many of these activities are also sources of pollution, particularly land contamination, and hence statutory controls are applicable to them.

The structure of the waste industry remains interesting. Firstly, a very large number of organisations are involved in different areas of the industry. Between 2009 and 2014[84], all sectors within the waste industry experienced increases in employment numbers except the wholesale of waste and scrap, with the largest increase being employment in the remediation activities (85%).

Mergers have been an extensive feature of this industrial sector since the late 1970s[85] and it seems inevitable that restructuring will continue. Water privatisation in the 1980s caused providers of water services to diversify. Increasing inward investment by French and US firms has also been a feature[86]. The Water Act, which comes into force fully in 2017, will allow businesses to choose their water supplier in much the same way as other utilities. How this will impact the water industry is not yet clear.

While not part of the bone fide waste management infrastructure, a small industry has grown up which attempts to exploit the financial benefits of unauthorised disposal activities. Although there does not appear to be a significant involvement in organised crime in British waste management outside certain major cities, a minority of small-scale operators are extensively involved in fly-tipping. Others are involved in illicit sorting and bulking-up activities as a way of avoiding spiralling landfill charges. Fortunately, the materials involved are often construction, demolition and less environmentally significant industrial wastes. The policing of unauthorised waste management activities occupies a critical amount of the Environment Agency's waste regulation staff time, and remediation is also a significant public expense.

84 DEFRA Digest of Waste and Resource Statistics – 2016 Edition (revised).

85 Department of Trade and Industry (1997) The Competitiveness of the UK Waste Management Industry, DTI, London, p 43.

86 Department of Trade and Industry (1997) The Competitiveness of the UK Waste Management Industry, DTI, London, para 2.5. also, for example, Miller-Bakewell M (1997) 'French take the lead' The Waste Manager, March 1997, pp 10–11; Owen D (1997) 'The French are coming' The Waste Manager, September 1997, pp 12–13.

The Environment Agency

1.10 On 1 April 1996, with the commencement of the relevant sections of the Environment Act 1995, the Environment Agency in England and Wales[87] was created, along with the Scottish Environment Protection Agency[88]. These two Agencies were formed out of a number of the earlier pollution control bodies whose powers, property and staff transferred over. In England, the single largest body to transfer was the National Rivers Authority[89]. The functions of this body transferred wholesale and these included[90] the policing of water quality by way of the Water Resources Act 1991, along with significant other powers in respect of flood defence[91].

Also transferred were the functions under Part I of the EPA 1990 of Her Majesty's Inspectorate of Pollution[92], along with those of the Inspectorate's duties under the Radioactive Substances Act 1993[93]. Part I of the EPA 1990 established a system of Integration Pollution Control (IPC) authorisations for major items of industrial plant. IPC authorisations cover much of the more significant controlled waste combustion and hazardous waste recovery infrastructure, involving hazardous waste incinerators, other incinerators of over one tonne per hour capacity[94], cement kilns, solvent recovery facilities and plants manufacturing fuel from waste[95].

The powers of the different agencies within the UK is discussed in Chapter 10.

The Environment Agency also gained certain new functions in respect of waste management. Section 92 of the Environment Act 1995 introduced a provision[96] which requires the Agency to provide advice to the Secretary of State on the production of a national waste strategy for England and Wales[97]. This function superseded the waste management planning provisions contained in s 50 of

87 By way of SI 1995/1983 and SI 1996/186.
88 By SI 1995/2649 and SI 1996/186.
89 In Scotland, the equivalent bodies were the River Purification Authorities: however flood defence functions were not transferred.
90 The following more significant NRA functions transferred (Environment Act 1995, s 2(1)(a)): Pts II (water resources management), III (pollution control) and IV (flood defence) of the Water Resources Act 1991; the functions under the Land Drainage Act 1991 (including functions under s 138(8) of the Water Act 1989); and functions under the Diseases of Fish Act 1937, the Sea Fisheries Regulation Act 1966 and the Salmon and Freshwater Fisheries Act 1975; functions as navigation authority, harbour authority or conservancy authority passed to the NRA by way of Pt III of the Water Act 1989.
91 Transferred by the Environment Act 1995, ss 2 and 3 and by SI 1996/234.
92 Environment Act 1995, s 2(1)(d), (f) and (g).
93 Environment Act 1995, s 2(1)(e).
94 Including all of the national household waste incinerator infrastructure and similar clinical waste disposal facilities.
95 SI 1991/972, Sch 1 as amended by SIs 1991/836, 1992/614, 1993/1749, 1993/2405, 1994/1271, 1994/1329, 1995/3247, 1996/2678 and 1998/767.
96 EPA 1990, s 44A.
97 EPA 1990, s 44A(6). By contrast SEPA itself produces the strategy for Scotland (EPA 1990, s 44B).

the EPA 1990, which the Environment Act 1995 repealed[98]. The Agency also became responsible for the oversight of the 'producer responsibility scheme' contained in s 93 of the Environment Act 1995. The scheme is enacted by way of secondary legislation[99] and at the time of writing, is mainly concerned with the implementation of the Packaging Directive[100]. The Agency will also have significant functions in respect of the Pollution Prevention and Control Act 1999[101], as well as the new contaminated land regime when it is implemented.

Up to the period when the EPA 1990 was brought into force, the actions of most regulatory authorities were informed by interpretation of the statutory remit. Even within the current British legislation, there is no explicit statutory duty on the Environment Agency to search out every possible potentially licensable waste management activity. Certainly, under the earlier Control of Pollution Act 1974, action could be taken if damage to the environment was implicated or if what was going on was a bona fide commercial scale waste management operation. If it was not, then it could be ignored. This was particularly the case with waste recovery activities of low potential environmental significance[102].

The tendency throughout the early 1990s to codify all conceivable permutations of waste transaction resulted in statutory provisions of considerable length and complexity. Examples include provisions, which exclude from the requirement to hold a waste management licence for the discharge of sewage from railway carriages[103] and the burial of a dead domestic cat in a private garden[104]. Whether it is strictly necessary to go this far is a moot point. It seems doubtful that litigation on this would be likely or well received in a court of law. It also does seem rather unlikely that a successful complaint for an infraction of the Directive on Waste would be upheld by the European Court of Justice on such a trivial matter.

Although a certain amount of complexity is inevitable in the regulation of a £3.5 billion industry, complexity is also a direct result of the manner by which the Community provisions are integrated with the EPA 1990 in the relevant legislation, principally the Waste Management Licensing Regulations

98 Environment Act 1995, s 120(1), Sch 22, para 78.

99 The Producer Responsibility Obligations (Packaging Waste) Regulations 1997, SI 1997/648, as amended by the Producer Responsibility Obligations (Packaging Waste) (Amendment) Regulations 1999, SI 1999/1361.

100 The European Parliament and Council Directive 94/62 of 29 December 1994 on Packaging and Packaging Waste ([1994] OJ L365/10 31.12.94).

101 EPA 1990, Part IIA as amended by the Environment Act 1995, s 57.

102 However, having removed much of the regulators' discretion on these matters, the guidance issued with the Waste Management Licensing Regulations 1994 and the Environment Act 1995 also appears to suggest that the Agency should utilise its discretion even more widely: for example, in Circular 11/94 regulators are cautioned about pursuing what are 'technical' offences, where the environment has not been affected (Circular 11/94, para 17 and para 12.34ff).

103 SI 1994/1056, Sch 3, para 31.

104 SI 1994/1056, Sch 3, para 37.

1994[105]. Despite the fact that the Bill which resulted in the EPA 1990 and the amendments to the Directive on Waste[106] were being drafted concurrently, this is not at all apparent from the wording of the EPA 1990. Indeed, it would appear that the EPA 1990 was drafted in isolation of what was being developed in the European dimension. Certainly, the British and European legislation do not sit particularly comfortably together, with the Directive being exclusively transposed in the form of 'bolt on' provisions contained mainly in the Schedules to the Waste Management Licensing Regulations 1994.

The result is that it is very difficult to grasp quickly the principles of the regulatory system, as it is not immediately apparent where the key parts of the legislation are to be located. For example, while the main powers to grant waste management licences are contained in ss 35 and 36 of the EPA 1990, further highly significant provisions are contained in the Waste Management Licensing Regulations 1994, particularly reg 15 in respect of groundwater protection and reg 19 – in conjunction with Sch 4 – in relation to the implementation of the 'relevant objectives' of the Directive on Waste.

An additional problem is the confusing terminology scattered around the EPA 1990 and Waste Management Licensing Regulations 1994. Again, this is a function of the less than satisfactory reconciliation of the EPA 1990 and the Directive on Waste. Certainly, there had been no attempt to rationalise terminology to a common base. For example, some sections refer to 'recycling', others to 'recovery'. The Waste Management Licensing Regulations 1994 set out to describe waste 'disposal' operations and include waste treatment under this heading[107]. But s 29(6) of the EPA 1990, which provides a list of key definitions in the primary legislation, makes a distinction between 'disposal' and 'treatment' operations. In addition, 'recycling and re-use' is viewed as a subset of 'treatment' in that section. By contrast, the Waste Management Licensing Regulations 1994 indicate that recovery and disposal operations are subject to quite distinct definitional approaches.

It should also be observed that, in certain instances, the transposition of the Community provisions could be described as somewhat half-hearted. The manner by which the requirements for registration of exempted activities is drafted, involving a maximum penalty of a £10 fine[108], is somewhat mocking of the requirements of the provisions on registration contained in the Directive on Waste. As is shown in Chapter 2, the revised definition of waste – particularly that involving 'Directive waste'[109] – is subject to an unclear and highly ambiguous definition. However, it follows the wording of the Directive exactly and hence results in transposition of EC law in its 'pure' form. That this change undermined a bundle of case law – much of which ironically already followed

105 SI 1994/1056 as amended by SIs 1994/1137, 1995/288, 1995/1950, 1996/593, 1996/634, 1996/972, 1996/1279, 1997/251, 1997/351, 1997/2203, 1998/606 and 1998/2746.
106 By way of Directive 91/156.
107 SI 1994/1056, Sch 4, Parts III and IV; Tables 2.5 and 2.6 in Chapter 2.
108 Besides this low level of fine, the provisions are drafted in a manner which makes them unlikely to be subject to proceedings.
109 SI 1994/1056, reg 1(3).

the tenet of the Directive – was obviously viewed as of much lesser significance by those responsible for the drafting of the legislation. Although Circular 11/94 sets out to clarify matters, this guidance is sometimes heavily qualified – and in places contradictory – as to be less than satisfactory. Furthermore, the guidance is of little application where it is most needed, in a court of law.

Finally, it also will be apparent from the forthcoming chapters that, although waste management is highly regulated, certain elements of the legislation have been subject to deliberate dilution, with the result that their application fails to serve fully the provisions' purpose or intent. Indeed, in the case of the duty of care there is a feeling that the dilution process has gone too far, particularly in respect, for example, of the transfer note system.

This book describes the waste regulations as they currently are – which has been to bring together the EU Directives in each part of the UK – interpreted into the law and although not exactly the same, following the same EU Directives. It is clear that what has happened has made the regulation clearer – with most of the previous law repealed. It remains to be seen what effect the vote to leave the EU will have and where the pressure to change will come from.

Chapter 2

Definition of Waste

WASTE

2.1 The simple definition of waste is:

> '…any substance or object which the holder discards or intends or is required to discard…'.

This definition has been in use for several decades and is in the 2008 Waste Framework Directive[1].

In simple terms it is an item that is thrown away because it is no-longer useful or required by its owner.

The definition of waste is important because the classification of substances as waste is the basis for the waste management policy and the application of regulatory controls to protect the environment and human health.

WASTE LEGISLATION

Waste Framework Directive (WFD)

2.2 The definition of waste is one of the key concepts of the WFD and it determines what falls under the Directive's scope. The WFD concept of waste affects the EU approach towards waste management. Only trans-frontier movements of waste as defined in the WFD are subject to the procedural requirements of the EU Waste Shipment Regulation[2]. It also helps in determining

1 Waste Framework Directive (2008/98/EC) ([2008] OJ L 312, 22.11.2008, pp 3–30), art 3(1).
2 EU Waste Shipment Regulation (EC) No 1013/2006 ([2006] OJ L 190, 12.7.2006, p 1–98).

the scope and application of other EU instruments directly concerning waste management[3].

The definition of 'waste' in the WFD is important because legal documents in other fields of environmental legislation refer to it. For example, Article 2(2) of the REACH Regulationstipulates that REACH does not apply to waste in the sense of the WFD. Waste in the sense of the WFD definition is explicitly and specifically excluded from REACH requirements. REACH is European Regulation (No 1907/2006) concerning chemicals and their safe use. It aims to improve the protection of human health and the environment through a system of Registration, Evaluation, Authorisation and Restriction of Chemicals.

The WFD was first adopted in 1975 as Directive 75/442/EEC and its focus was to ensure that there was safe disposal of waste. The original WFD allowed member states to adopt their own national definitions of waste. This Directive was then subject to a substantial amendment in 1991 by Directive 91/156/EEC which took into account the member states' experiences of implementing the original Directive. One of the changes made in 1991 was to extend the scope of the Directive to also include 'recovery' and this included recycling, re-use and reclamation of waste and the use of waste as a source of energy. This amended Directive also included a new EU-wide definition of waste because:

> '(3) Common terminology and a definition of waste are needed in order to improve efficiency of waste management within the community'.

Definition of waste in the 1991 Directive was:

> '"waste shall mean any substance or object in the categories set out in Annex I which the holder discards or intends or is required to discard"'.

In 2005 the European Commission published:

A Thematic Strategy on the prevention and recycling of waste ('the Waste Thematic Strategy')[4].
Proposals for associated legislation – revision of the WFD, the repeal of the Waste Oils Directive and the repeal and integration of the Hazardous Waste Directive into a revised WFD.

The definition of waste in the Thematic Strategy[5] is:

> 'The Waste Framework Directive defines waste as products or materials that are discarded. In the light of extensive stakeholder consultation the Commission has concluded that there is no need substantively to amend the definition of waste, but that it is necessary to clarify when a waste ceases to be a waste (and becomes a new or secondary raw material). Therefore, an amendment to the Directive is proposed which would establish waste-stream-based environmental criteria

3 Landfill Directive (1999/31/EC) ([1999] OJ L 182, 16.7.1999) and the Industrial Emissions Directive (2010/75/EC) ([2010] OJ L 334, 17.12.2010, p 17–119).

4 Report from the Commission to the European Parliament, the Council, the European Economic and Social Committee and eth Committee of the Regions on the Thematic Strategy on the Prevention and Recycling of Waste SEC(2011) 70 final COM/2011/0013 final.

5 Waste Thematic Strategy, Annex 1 Paragraph 1.

> to determine when a waste ceases to be a waste. This could both improve the environmental performance of recycled products, by encouraging businesses to produce recycled products that conform to these environmental criteria, and reduce unnecessary burdens for low-risk recycling activities.'

The terms 'substance' and 'object' are not to be understood in the sense of EU chemicals legislation, but as terms of waste legislation. Any substance or object is either waste or non-waste[6].

REACH Regulation (EC) 1907/2006 makes reference to waste in the sense of Directive 2006/12/EC. According to Article 41 and Annex V to the WFD, such references have to be understood as references to the WFD. In two key areas the concept of waste has been tightened up by including by-products (see para **2.46**) and of end-of-waste criteria (see para **2.57**). Both the concepts 'by-product' and 'end-of-waste' introduce a distinction between waste and non-waste.

The proposal was considered by the Council of Ministers and the European Parliament under the co-decision procedure. This was completed on 20 October 2008 when the Environment Council adopted the Common Position as amended by the European Parliament on 7 June 2008. The official Journal published the revised WFD on 19 November 2008 as the Directive 2008/98/EEC – the definition that we now use.

Transposition

2.3 The Directive 2008/98/EEC has been transposed into legislation in:

England and Wales

(i) The Waste (England and Wales) Regulations 2011,SI 2011/988[7];

(ii) In Wales, the Regulations at (i) are supplemented by the Waste (Miscellaneous Provisions) (Wales) Regulations 2011, SI 2011/971 (W 141)[8]. This makes a number of amendments to several Welsh Statutory Instruments and revokes a Wales-only instrument[9].

Scotland

The Waste (Scotland) Regulations 2012, SSI 2012/148[10].

Northern Ireland

The Waste Regulations (Northern Ireland) 2011, NISR 2011/127[11].

6 The definition of waste itself has not been modified compared to the previous Waste Framework Directive (2006/12/EC) ([2006] OJ L 396, 30.12.2006, p 1–849).
7 Available at www.legislation.gov.uk/uksi/2011/988/contents/made.
8 Available at www.legislation.gov.uk/wsi/2011/971/contents/made.
9 Environmental Protection (Duty of Care) (Amendment) (Wales) Regulations 2003, SI 2003/1720. Available at www.legislation.gov.uk/wsi/2003/1720/contents/made.
10 Available at www.legislation.gov.uk/ssi/2012/148/contents/made.
11 Available at www.legislation.gov.uk/nisr/2011/127/contents/made.

Why the Definition of Waste is Important

2.4 The definition of waste is one of the most complex and difficult elements of the regulatory regime. The difficulty mainly arises from defining substances that are going to be recovered and substances, which are 'discarded'. The term discarded has a special meaning which is different to the one in common usage. For materials which are sent to a disposal facility, the fact that they are being 'discarded' means they are defined as wastes. This chapter looks at the definitions of waste and at those substances which are borderline between 'wastes' and 'economic goods'. This becomes increasingly relevant when more waste is taken to recovery facilities.

Virtually every country has struggled with the drafting of a comprehensive legal definition of waste and it is perhaps correct to say that none have produced a wording which is entirely satisfactory.

The principal difficulty is that the dividing line between what are often termed 'goods' and 'wastes' is never clear-cut and may even overlap. In the absence of legislative intervention, economic circumstances would often dictate where that line is, but these circumstances – and hence the dividing line itself – will vary considerably. For example waste paper, a commodity whose price is unstable and highly cyclical. High prices may be paid by secondary users of waste paper at certain times while at other times holders of waste paper may end up paying a third party to take the material away. A second example is bone meal, a by-product of the rendering of animal carcasses. Until the BSE crisis, this material was sold extensively as an animal feed supplement. Since concerns were raised that bone meal may harbour the agent responsible for BSE, by the winter of 1998, 340,000 tonnes of bone meal were in storage awaiting disposal in Britain[12]. Bone meal became regarded as a waste in the middle of 1996 for the reason that the material became surplus, due to the collapse of its market.

The difficulties in defining waste comes from the realities of waste management and in particular the variety and complexity of commercial transactions involving substances and objects which may or may not be defined as 'wastes'. This is a challenge to a legislator attempting to devise a workable but comprehensive waste definition, and balancing two conflicting requirements. It is important that there is a definition, which covers substances which need to be the subject of environmental controls and produces a legal definition which inappropriately restricts the legitimate trade in economic goods through normal commercial transactions.

The value of wastes or raw materials is also not just a function of prices. There are other determining factors including transport costs, the quantities of the material involved, technological innovation and the availability of appropriate waste management infrastructure. All these aspects have a fundamental

12 'First Big Contract to Dispose of Waste from the BSE Cull' ENDS Report 285, October 1998, p 15; see also 'The Environmental Impact of the BSE Crisis' ENDS Report 268, May 1997, p 21.

influence on whether substances are to be seen as wastes or goods. Recently the increases in energy prices and the concentration of the Environment Agency on odour control has pushed up the cost of a lot of recycling.

In the past, there has been a tendency to view wastes as valueless articles. Although a lack of value may be a useful indicator that something is waste, focusing solely on value has become increasingly inappropriate for a number of reasons. First, if wastes are to be defined as valueless, the unstable value of some substances may cause them to switch in or out of regulatory controls. Clearly, it is untenable to construct a workable statutory definition – or indeed to centre a waste control system – on this basis.

Secondly, there is an observable trend of competition between the recycling and disposal infrastructure for the business of waste holders. Since the implementation of statutory controls in the 1970s, waste disposal prices have substantially increased. This trend will continue with further advances in the sophistication of the national waste management infrastructure, along with public opposition factors which restrict the supply of new disposal sites, legal restrictions such as the Landfill Directive[13] and economic mechanisms such as the landfill tax[14]. The result has stimulated interest in the possibility of reprocessing what were previously economically marginal substances. Examples include energy recovery from paint wastes as secondary fuel in cement manufacture, the chemical recovery of zinc-based wastes from galvanising and the composting of household waste.

Often materials such as these have a negative value when sent for re-processing: the holder still has to pay to have the materials taken away. In cases where a holder must pay for certain wastes to pass to recycling or recovery, it seems unarguable that such materials should be subject to statutory control. Paying to take these wastes away will always create an incentive on the holder to get rid of the material for free. A potentially attractive way of undertaking this would be indiscriminate unauthorised disposal. In addition, even if the holder receives some payment for the wastes, the financial incentive for uncontrolled dumping may still exist. Although the materials have a value, the value may be small. The value may also be totally dependent on the existence of the outlet, which is prepared to pay for the materials. Should that outlet close (or should the supply of wastes exceed the outlet's requirements), the waste materials will regain their negative value. The incentive for inappropriate disposal therefore continues.

While many countries acknowledge the desirability of the use of the recycling route at the expense of disposal[15], recycling and other recovery processes are by

13 Council Directive 1999/31/EC of 26 April 1999 ([1999] OJ L 182, 16.7.1999, p 1–19) on Landfill Waste.

14 Landfill Tax rates – HM Revenue and Customs update April 2015 Available at www.gov.uk/government/ publications/rates-and-allowances-landfill-tax/landfill-tax-rates-from-1-april-2013.

15 Waste Management Plan for England December 2013 Defra available at www.gov.uk/government/uploads/system/uploads/attachment_data/file/265810/pb14100-waste-management-plan-20131213.pdf.

no means environmentally benign. These operations can be highly polluting, whilst the residues resulting from the processes require careful handling and disposal. Clear evidence of this could be found at many UK car dismantlers and scrap yards, where the indiscriminate discharge of materials such as oil has made such sites one of the major sources of contaminated land in the country.

'Sham recycling' is also not unknown, particularly in countries which focus their controls solely on waste disposal installations. Materials are accepted at such plants, ostensibly for recovery, but where the amount of waste being recovered – if any – is very small. The residues are then disposed of in a way which avoids statutory control. The export of wastes from one country to another for sham recycling is a global problem. Hence EC controls on transfrontier waste movements[16] (see **Chapter 5**) allow an intended recipient country to formally object to such shipments on the grounds that the proportion of waste being recovered is very small, that the value of the recovered materials is low or that there is a disproportionate cost involved in the disposal of the non-recoverable residues[17].

All the above factors lead to the conclusion that the activities of the waste recycling industry requires statutory control. Equally, it should be clear that wastes with a positive value must be addressed by waste management law.

However, there comes a point where the value of the waste material is so significant that there is no financial incentive for the holder to consider unauthorised disposal. Many industries are based on the processing of intermediate products from other processes. The chemical industry in particular relies on the use of by-products – or more rightly 'co-products' – in a variety of ways. There are hundreds of possible examples, including alkali production. Caustic soda production by way of the electrolytic decomposition of salt solutions produces copious quantities of chlorine as a co-product. As both materials are produced together, an excessive demand for chlorine, particularly for PVC production, will mean that unwanted caustic soda will need to be stored[18]. The converse also applies, where an excessive demand for caustic soda may cause stocks of chlorine to build up. When the economics of the storage of the surplus materials become unfavourable, the production processes themselves may be curtailed, causing prices to increase. What is notable in this example is that the market provides controls. Regulatory intervention of an environmental nature has only a very indirect role (if at all) in this instance, relating primarily to general process emissions.

Innovation will always attempt to find uses for particular substances. The market may itself provide significant incentives for such innovation, but regulation also has a role in this process. Finding new uses for by-products

16 Communication from the Commission to the Council, the European Parliament, the European Economic and Social Committee and The Committee of the Regions ‚Äì Taking sustainable use of resources forward – A Thematic Strategy on the prevention and recycling of waste {SEC(2005) 1681} {SEC(2005) 1682} /* COM/2005/0666 final */.

17 Council Regulation 259/93, ([2006] OJ L 190, 12.7.2006), p 1–98, art 7(4)(a).

18 See Heaton C A, *The Chemical Industry* (Blackie, London, 1986) 132.

has been the basis of the chemical industry since its inception. Two examples of early chemical industry activities illustrate this point. In the 19th century, the Leblanc process was used to manufacture soda, an essential industrial feedstock. The process resulted in the production of vast quantities of hydrochloric acid, as a by-product of the action of sulphuric acid on sodium chloride. Originally, these acid gases were vented to the atmosphere causing significant environmental destruction to vegetation in proximity to any alkali works. Eventually, these releases were subject to control under the Alkali Works Act 1863. The development of water-based scrubbing systems to achieve the Alkali Works Act's 95% acid removal target resulted in the production of a by-product hydrochloric acid, for which end uses quickly developed. Likewise, the early production of town gas resulted in an obnoxious waste in the form of coal tar. Eventually, innovation resulted in the re-use of significant quantities; as one commentator enthused: '... black and apparently useless coal-tar was to prove to the organic chemist an almost unbelievably rich treasure house'[19]. Processed coal tar formed the basis for consumables as diverse as shampoos, disinfectants and creosote.

In 2012 Defra published guidance[20] on the legal definition of waste. This document was aimed at waste operators who already had a good knowledge and understanding of the waste management system in the UK. Defra also published another guide which is a practical guide for businesses and other organisations[21].

UK Waste Policy

2.5 Change is a constant feature of UK waste policy. Since the publication in 2000 of the Waste Strategy for England and Wales, changes, largely driven by EU waste laws, have been made as to how waste is produced and disposed of in the UK. Building on the 2000 policy and the subsequent 2007 Waste Strategy for England, Defra published in 2013 a new Waste Management Plan for England[22]. Wales, Scotland and Northern Ireland also have equivalent waste strategies.

19 Williams T I *The Chemical Industry: Past and Present*, (1952) (republished by E P Publishing, Wakefield, 1972), p 63.

20 Guidance on the Legal Definition of Waste. DEFRA, August 2012. www.gov.uk/government/uploads/system/ uploads/attachment_data/file/69590/pb13813-waste-legal-def-guide.pdf.

21 Guidance on the Legal Definition of Waste and its Application; A Practical Guide for Businesses and other Organisations, DEFRA, August 2012 www.gov.uk/government/uploads/system/uploads/attachment_data /file/69591/pb13813a-waste-business-guide.pdf.

22 Waste Management Plan for England, DEFRA, December 2013: www.gov.uk/government/uploads/system/uploads/attachment_data/file/265810/pb14100-waste-management-plan-20131213.pdf.

Waste Hierarchy

2.6 The EU uses a concept of waste hierarchy[23] – and the UK policies[24] on waste are built on this. The waste hierarchy has a number of stages – which requires that anyone managing waste to consider in order.

(a) prevention;
(b) preparing [25] for reuse[26];
(c) recycling[27];
(d) other methods of recovery[28] eg energy recovery;
(e) disposal[29].

Prevention, preparing for reuse and recycling should be given priority order in any waste legislation and policy. The WFD also encourages member states to provide options which deliver the best overall environmental outcome. It also acknowledges that this may require that specific waste streams depart from the hierarchy where this is justified by life-cycle thinking on the overall impacts of the generation and management of such waste.

Diversion of Waste from the Landfill

2.7 Using the waste hierarchy a government policy objective has been to reduce the level of waste going to landfill and to encourage more recycling. The Environmental Permitting (England and Wales) Regulations 2010, SI 2010/675 and the Landfill Allowances Scheme (Wales) 2004, SI 2004/1490, as amended, impose restrictions on the type and amount of waste that can be disposed of

23 WFD (2008/98/EC) ([2008] OJ L 312, 22.11.2008, p 3–30), Article 4.
24 The Waste (England and Wales) Regulations 2011, SI 2011/988, Sch 12.
25 The definition of 'preparing for re-use' (Article 3(16)) is: 'checking, cleaning or repairing recovery operations, by which products or components of products that have become waste are prepared so that they can be re-used without any other pre-processing.' By definition, preparing for re-use is a specific case of recovery.
26 WFD, Article 3(13) definition of 're-use' is: 'Any operation by which products or components that are not waste are used again for the same purpose for which they were conceived'.
27 WFD, Article 3(1): 'any recovery operation by which waste materials are reprocessed into products, materials or substances whether for the original or other purposes. It includes the reprocessing of organic material but does not include energy recovery and the reprocessing into materials that are to be used as fuels or for backfilling operations'.
28 Any waste treatment can only be either a recovery operation or a disposal operation; the CJEU has explicitly stated that no operation can be classified as disposal and recovery at the same time.
29 WFD, Article 3(19) the definition of 'disposal' is: 'any operation which is not recovery even where the operation has as a secondary consequence the reclamation of substances or energy'.

in landfills in England and Wales. Scotland[30] and Northern Ireland[31] also have similar landfill regulations.

Increase Recycling

2.8 A government objective is to make it easier for people and organisations to recycle more. Some measures have been put in place to encourage the general public to consider waste as a resource and adopt a recycle and reuse culture. The Welsh Government introduced a 5p charge for single use carrier bags in 2011 as did Northern Ireland[32]. Scotland[33] introduced the plastic bag charge in October 2014. In October 2015, the UK government introduced a 5p charge for single use plastic bags handed out by large retailers in England[34].

Reduction of Waste from the Economy

2.9 The UK waste policy aims to reduce the amount of waste produced by the commercial sector and policy changes have been made to get businesses and industries to reduce waste. For example, the Producer Responsibility Obligations (Packaging Waste) Regulations 2007[35] place responsibility on the producers of packaging waste to recover and recycle a certain amount of packaging. Products are also required to be designed to encourage easy dismantling and recycling at the end of the product's life cycle. Similar producer responsibility laws cover Waste Electrical and Electronic Equipment (WEEE), End of Life Vehicles and batteries by requiring producers to recover a certain amount.

30 Landfill (Scotland) Regulations 2003, SSI 2003/235 as amended by SSI 2003/343 and SSI 2013/222.
31 Landfill Regulations (Northern Ireland) 2003, NISR 2003/496.
32 Single Use Carrier Bags Act (Northern Ireland) 2011.
33 The Single Use Carrier Bags Charge (Scotland) Regulations 2014, SSI 2014/161.
34 The Single Use Carrier Bags Charges (England) Regulations 2015, SI 2015/776.
35 The Producer Responsibility Obligations (Packaging Waste) Regulations 2007, SI 2007/871 as amended.

EXCLUSIONS FROM THE DEFINITION OF WASTE

2.10 Not all substances fall within the definition of waste contained within the WFD. Exclusions are set out in Article 2(1), (2) and (3)[36]. This Article does not declassify substances as waste – but these provisions provide exclusions from the scope of the WFD for specified types of waste. So, if an article is not waste then it is not covered by the provisions of the WFD. If it is waste but is covered by the exclusions then it is still waste but not covered by the provision of the WFD.

These exclusions are transposed by means of the following provisions in the Regulations listed in para **2.10**. This effect is to provide that 'waste' in the transposition of the regulations means anything that is waste within the

36 WFD, Article 2:
Exclusions from the scope
1 The following shall be excluded from the scope of this Directive:
(a) gaseous effluents emitted into the atmosphere;
(b) land (in situ) including unexcavated contaminated soil and buildings permanently connected with land;
(c) uncontaminated soil and other naturally occurring material excavated in the course of construction activities where it is certain that the material will be used for the purposes of construction in its natural state on the site from which it was excavated;
(d) radioactive waste;
(e) decommissioned explosives;
(f) faecal matter, if not covered by paragraph 2(b), straw and other natural non-hazardous agricultural or forestry material used in farming, forestry or for the production of energy from such biomass through processes or methods which do not harm the environment or endanger human health.
2 The following shall be excluded from the scope of this Directive to the extent that they are covered by other Community legislation:
(a) waste waters;
(b) animal by-products including processed products covered by Regulation (EC) No 1774/2002, except those which are destined for incineration, landfilling or use in a biogas or composting plant;
(c) carcasses of animals that have died other than by being slaughtered, including animals killed to eradicate epizootic diseases, and that are disposed of in accordance with Regulation (EC) No 1774/2002;
(d) waste resulting from prospecting, extraction, treatment and storage of mineral resources and the working of quarries covered by Directive 2006/21/EC of the European Parliament and of the Council of 15 March 2006 on the management of waste from extractive industries (23).
3 Without prejudice to obligations under other relevant Community legislation, sediments relocated inside surface waters for the purpose of managing waters and waterways or of preventing floods or mitigating the effects of floods and droughts or land reclamation shall be excluded from the scope of this Directive if it is proved that the sediments are non-hazardous.
4 Specific rules for particular instances, or supplementing those of this Directive, on the management of particular categories of waste, may be laid down by means of individual Directives.

meaning of Article 3(1) of the WFD[37] read with Article 5(1)[38] on by-products and which is not excluded from the scope by Article 2(1), (2) and (3).

The Table below is a comparison between the Waste (England and Wales) Regulations 2011, SI 2011/988 and the Waste Regulations(Northern Ireland) 2011, SI 2011/127 showing where the regulations provide for the same element but at a different number.

The Waste (England and Wales) Regulations 2011, SI 2011/988	The Waste Regulations (Northern Ireland) 2011, SI 2011/127
Regulation 2	Regulation 9(2)
Regulation 47	Regulation 25
Schedule 2 Para 1	Regulation 46
Schedule 3 para 2, 4(2), 6 to 8, 13	Regulation 36(c) and (j)
Schedule 4 para 2 and 3(6)	

Three Types of Exclusion from the Scope of the WFD

2.11 **Type 1:** This exclusion is provided in Article 2(1) (a)–(e) and these are absolute exclusions as they are not dependent on the wastes being covered by other (Community) legislation.

Type 2: This exclusion is provided in Article 2(2)(a)–(d) and these wastes are only excluded if they are covered by other (Community) legislation.

Type 3: This is provided in Article 2(3) and is 'without prejudice to obligations under other relevant Community legislation'. These are waste waters provided that any sediment is non-hazardous. This type of exclusion is different from

37 WFD, Article 3(1), 'waste' means any substance or object which the holder discards or intends or is required to discard.

38 WFD, Article 5(1):

By-products

1 A substance or object, resulting from a production process, the primary aim of which is not the production of that item, may be regarded as not being waste referred to in point (1) of Article 3 but as being a by-product only if the following conditions are met:

(a) further use of the substance or object is certain;

(b) the substance or object can be used directly without any further processing other than normal industrial practice;

(c) the substance or object is produced as an integral part of a production process; and

(d) further use is lawful, i.e. the substance or object fulfills all relevant product, environmental and health protection requirements for the specific use and will not lead to overall adverse environmental or human health impacts.

those in Types 1 and 2 as they originate from an amendment to the Environment Council's Common Position proposed by the European Parliament (the exclusion was not proposed by the European Commission or the Member States).

Exclusion for Land (in situ), Including Unexcavated Contaminated Soil and Buildings Permanently Connected with Land

2.12 'in situ' means in the original position and the exclusion relates to land soils and buildings that have not been disturbed by excavation or demolition.

This exclusion was introduced into the WFD and follows from the '*van de Walle*' ruling (see para **2.67** where it was found that the holder of hydrocarbons which are accidentally spilled and which contaminate soil and groundwater, 'discards' this material. As a result, such hydrocarbons were found to be classified as waste; accordingly, contaminated soil became subject to the waste regime. This led to discussions on how to co-ordinate waste legislation with soil protection legislation. Under the current WFD, such contaminated soil would be excluded from the scope of the WFD and the soil protection measures would apply to it.

The term 'contaminated soil' is not defined in the WFD or in other legal acts at Community level. Soil is considered to be contaminated if it exhibits any of the 'properties of waste which render it hazardous' as per Annex III to the WFD. The term 'contaminated' can be clarified by comparing it to the term 'uncontaminated soil' in Article 2(1)(c) WFD[39]. From the wording of that provision 'uncontaminated soil and other naturally occurring material' it can be derived that uncontaminated soil essentially relates to virgin soil or soil that is equivalent to virgin soil. In the absence of EU standards, national soil legislation (where it exists) can be consulted to determine the type and level of trace contamination at which a soil might be considered equivalent to virgin soil.

Examples of exclusions under this provision are unexcavated soil beneath the forecourt of a service station; asbestos tiles on the roof of a building; green field site prior to construction.

Excavated Soil and Other Naturally Occurring Material

2.13 This exclusion relates to 'uncontaminated soil and naturally occurring material excavated in the course of construction activities where it is certain

39 WFD, Article 2(1)(c) uncontaminated soil and other naturally occurring material excavated in the course of construction activities where it is certain that the material will be used for the purposes of construction in its natural state on the site from which it was excavated.

that the material will be used for the purposes of construction in its natural state on the site from which it is excavated'.

This material has been excluded on the basis that a waste management regime was regarded as inappropriate for this kind of material, even if the definition of discarding is fulfilled.

For the material to be excluded form the scope it must be:

- uncontaminated;
- excavated during the construction activities;
- certain to be used in its natural state for construction purposes on the same site.

The waste management regime will apply to any material used in construction that does not meet these three criteria. Whether a material meets these criteria can be assessed by looking at by-products and end of waste (see para **2.46**).

'Uncontaminated soil' relates to virgin soil or soil that is equivalent to virgin soil. Other naturally occurring material means soil, stones, gravel, rock, etc. Man-made materials like concrete, or items that have been modified by man, eg wooden materials, are not excluded from the scope of the WFD.

In order to be excluded, the excavated material must be used in a construction activity on the site. Certainty of use could be inferred from, for example, construction plans or designs for the site in question. These may contain estimates of excavated amounts and whether there will be a surplus or deficit of such material. It could also be inferred from:

- Planning-permission conditions;
- Construction and demolition waste management plans, if required;
- For larger scale developments, the Environmental Impact Assessment (EIA).

A construction site will usually be defined in relation to the associated planning permission. Examples of what can be considered to be 'on the site' include[40]:

- A construction project for a 100 km motorway, where excavated material from one section of construction is used in its natural state in the same construction section.
- Soil or other such material temporarily taken from the site but returned later and used on the site for the purposes of construction (the transport operation as such is not relevant).

40 Guidelines on the interpretation of key provisions of Directive 2008/98/EC on Waste, available at: http://ec.europa.eu/environment/waste/framework/pdf/guidance_doc.pdf.

Exclusion for Agricultural and Forestry Material

2.14 Article 2(1)(f) WFD excludes:

> 'faecal matter, if not covered by paragraph 2(b), straw and other natural non-hazardous agricultural or forestry material used in farming, forestry or for the production of energy from such biomass through processes or methods which do not harm the environment or endanger human health'.

2.15 Faecal matter consists of faeces and urine excreted by animals in an agricultural or forestry setting. It does not include human faecal matter.

This exclusion applies to faecal matter only where the animal by-products (ABP) exclusion of Article 2(2)(b) WFD does not cover the material. The reason behind this was to link the exclusion unambiguously to a condition constituting an exception, not to create an exception from the exception. This means that ABP are in principle covered by the exclusion in Article 2(2)(b) WFD, except for faecal matter from farming, which is covered by the exclusion in Article 2(1)(f) of WFD.

Straw and Other 'natural non-hazardous' Material

2.16 The (non-) hazardousness of the material has to be assessed according to the criteria of Article 3(2)[41] and Annex III to WFD. Examples of materials from agriculture or forestry that could be considered natural non-hazardous materials are:

- straw from grain and other crops;
- cut grass;
- natural wood, wood off-cuts, wood chips, saw-dust, etc;
  other biomass.

Processes Which Do Not Harm the Environment or Endanger Human Health

2.17 The provision 'not harming the environment or endangering human health' applies to any use of the specified materials in agriculture or forestry, as well as their use to produce energy. The minimum standard for not harming the environment or endangering human health is compliance with the standards of EU environmental legislation.

41 WFD, Article 3(2), 'hazardous waste' means waste which displays one or more of the hazardous properties listed in Annex III.

Animal By-Products

2.18 Under Article 2(2)(b) WFD, 'animal by-products including processed products covered by Regulation (EC) No 1774/2002[42], except those which are destined for incineration, landfilling or use in a biogas or composting plant' are excluded from WFD to the extent that they are covered by that Regulation.

ABP Regulation (EC) 1069/2009 aims to exclude animals which have died (other than those slaughtered as fit for human consumption) and other condemned materials from the animal feed chain and to achieve safe processing and disposal of animal by-products produced in the Union. Under the ABP Regulation, only materials derived from animals declared fit for human consumption following veterinary inspection, subject to the restrictions set out in Article 11, may be used for the production of feed[43].

In order to avoid duplication of the rules, the WFD in regard to ABP that are subject to ABP Regulations, are excluded from the scope of the 'wherever they are intended for' uses that are not considered waste operations.

ABPs are included where they are destined for incineration, landfilling or use in a biogas or composting plant, since these are typical waste treatment operations with environmental risks, which have to be monitored under waste legislation. This means that to determine whether an ABP covered by Regulation (EC) No 1069/2009 is subject to waste legislation, will depend on what is going to happen to it.

The assessment of whether an ABP is considered under the ABP regime or under waste management legislation is sometimes complex. The outcome of this assessment may bring into question who the competent authority is within a member state.

The example given in the Guidelines[44] is on catering waste. ABP Regulation (EC) No 1069/2009 excludes catering waste from its scope, except under certain conditions. In all cases where these conditions are not met, material meeting the WFD definition of waste will be subject to waste management legislation and not the ABP Regulation.

The conditions under which catering waste is subject to the ABP Regulation are as follows:

- If catering waste originates from a means of transport operating internationally;

42 Note that Regulation (EC) 1774/2202 was repealed with effect from 4 March 2011 by ABP Regulation (EC) No 1069/2009. Article 54, para 2 of Regulation (EC) No 1069/2009 states that 'References to Regulation (EC) No 1774/2002 shall be construed as references to this Regulation and shall be read in accordance with the correlation table laid down in the Annex'.

43 Section 4 of the ABP Regulation 1069/2009 distinguishes between three categories of material (with material of Category 1 involving the most risks) and provides for a distinct system of treatment for each category.

44 Guidelines on the interpretation of key provisions of the Directive 2008/98/EC on waste.

- If catering waste is destined for feeding purposes;
- If catering waste is destined for processing by pressure sterilisation or other methods for processing animal by-products, in particular as regards the parameters to be applied for those processing methods, notably the time, temperature, pressure and size of particles; or for transformation into biogas or for composting.

As long as the material is destined for one of the treatment methods listed in Article 2(2)(b) WFD, it is subject to waste management legislation. For catering waste originating from a means of transport operating internationally, landfilling is indeed the only allowable treatment under the ABP Regulation (see Articles 8(f) and 12(d) ABP Regulation (EC) No 1069/2009). Thus, it is subject to the WFD.

Burning of ABPs and Derived Products as a Fuel

2.19 The ABP Regulation distinguishes between ABPs which are waste and ABPs which are used as a fuel. For example, whether the incineration of ABPs (including 'derived products' under ABP legislation, such as tallow) as a fuel is subject to waste legislation depends on whether the material meets the definition of waste or that of a 'by-product' under the WFD. If it is subject to waste legislation, the incineration operation must comply with the requirements of the Waste Incineration Directive 2000/76/EC.

The term 'incineration' is a typical waste treatment operation, thus including co-incineration even if not explicitly mentioned. Therefore, ABPs which are waste and intended for either incineration or for co-incineration are included in the scope of the WFD, and these operations should consequently be carried out in accordance with the Waste Incineration Directive 2000/76/EC.

Dredging Sediments Exclusion

2.20 This exclusion states:

> 'without prejudice to obligations under other relevant Community legislation' to 'sediments relocated inside surface waters for the purpose of managing waters and waterways or of preventing floods or mitigating the effects of floods and droughts or land reclamation (...) if it is proved that the sediments are non-hazardous'.

Obligations under other Community legislation, namely arising from the Water Framework Directive, are not affected by the exclusion.

Non-hazardous dredging sediment is excluded from the scope of the WFD only where it is put elsewhere within surface waters for the purposes (only) of: managing waters and waterways; preventing floods; mitigating the effects of floods and droughts; or land reclamation.

Excluded is sediment from dredging. This exclusion for the purpose of managing waters, etc applies not to the dredging but to the relocation of the sediment. If sediments are further used outside water bodies, the criteria for exclusion under Article 2(3) WFD are not fulfilled and waste management legislation applies. In all cases the sediments must be non- hazardous, taking into account the criteria of Article 3(2) WFD.

MEANING OF DISCARD

2.21 The definition in Article 3(1) of the WFD provides a general test for whether a substance or object is to be regarded as waste or not. The question which the European Court has addressed is whether the substance or object has been discarded by its holder. The reason for discard can be intentional or accidental, it may be voluntary or required and it may even occur without the holder's realising it. The 'discard' test is applied in a very wide range of circumstances but, in each case, the test must be interpreted so as to give effect to the aims of the WFD and related legislation.

There are three alternatives: These are any substance or object that the holder:

- Discards;
- Intends to discard;
- Is required to discard.

Discard has a Special Meaning

2.22 The question of 'discarding' can sometimes be complex and difficult with substances moving to recovery facilities. Firstly, the concept of discarding is itself less than clear in its scope and application. Secondly, the legislation does not make immediately clear the identity of the person taking the decision to discard. However, it seems vital to the wording of the definition of waste that such a person is correctly identified. The following sections will attempt to analyse these two, highly significant issues. What should become apparent is that the term 'discarding' must have a special meaning in the context of the legislation, which is different from its common language usage. Furthermore, the definition of waste is closely linked to the actions of the person who generates the waste at first instance. Of lesser significance will be the actions of other individuals who receive waste from the original generator.

The European Court has explained the term 'discard' as having a specially extended meaning in the WFD and includes the recovery of a substance or object as well as its disposal. One of the main aims of the WFD is to ensure that any recovery operations as well as disposal operations are carried out in a way which protects the environment and human health.

Often whether or not waste is discarded will be self-evident. This will particularly be the case when waste is passed to disposal facilities or is fly-tipped. Therefore only in the more involved cases – particularly those related to substances passing to recovery or re-use – is there a need to look at more interpretation.

The WFD makes it clear that a substance or object is to be classified as waste if the holder intends to discard it. The European Court recognised that this would make the application of the WFD too arbitrary and so the holder's intention is therefore inferred from their actions in the light of the aims of the WFD and regarding the factors set down in the European Court. The factual circumstances are considered with regard to the WFD and so the test is objective and not subjective.

Discarding and Waste Recovery

2.23 The case of *Vessoso and Zanetti* concerned a referral to the European Court of cases concerning waste carriers who were involved in the salvage of materials and the recovery of wastes. Cases C-206/88 and C-207/88 involved the collection and storage of general salvaged materials. In Case C-359/88, the materials in question were spent hydrochloric acids from galvanising, which were to be re-used in ferric chloride manufacture. The recovery plant did not have the benefit of an appropriate authorisation. In all the cases, the defendants claimed that a permit was not needed for the recovery operation as the materials being delivered did not constitute waste.

The judgment of Criminal Proceedings against *Tombesi* closely follows the earlier findings of the European Court described above. Hence it confirms that materials, which pass to recovery processes, can fall within the definition of waste in Article 1 of the Directive on Waste. *Tombesi* relates to a succession of cases concerning quite disparate circumstances, all of which involved Italian nationals. For example, Case C-304/94 pertained to the unauthorised disposal of waste from marble working, which appears to have been dumped or used for the construction of embankments; Case C-330/94 involved the handling of toxic wastes arising from the operation of electrostatic precipitators; Case C-342/94 addressed the unlicensed reprocessing by grinding of olive oil residues and Case C-224/95 related to the unauthorised transportation of copper cable and other metal scrap. All the cases were referred to the European Court due to the Italian Court wishing to seek guidance, under Article 177 of the Treaty establishing the European Community, on whether the relevant national legislation was compatible with the Directive on Waste.

Intention to Discard

2.24 The WFD makes it clear that a substance or object is to be classified as waste if the holder intends to discard it. So that this is not an arbitrary process, the holder's intention is to be inferred from their actions in the light of the aims of them and having regard to factors set down by the European Court. It is therefore an objective and not a subjective test.

Requirement or Obligation to Discard

2.25 It is not always necessary to infer the intention to discard. Under the WFD a substance or object, which is required to be discarded, is also to be classified as waste. If the substance or object is something which the holder is required to discard from either an obligation imposed by EU law, national law or even as a matter of contract then in each case the item will be waste.

Any oil containing PCBs above 50ppm must be discarded under the provision of EU PCB/PCT Directive 96/59/EC and is therefore considered to be waste.

An example is that of the EU Animal By-Products Regulations[45]. This categorises animal by-products into different categories. Category 1 material is defined under Article 8[46]. Under Article 12 the ways in which the material needs to be disposed of is described below.

Category 1 Material shall be:

(a) disposed of as waste by incineration:
   (i) directly without processing;

45 Available at: http://ec.europa.eu/food/food/biosafety/animalbyproducts/index_en.htm#Legislation.

46 Animal By-Products Regulations 1069/2009 ([2009] OJ L 300, 14.11.2009, p 1–33) and 142/2011 ([2011] OJ L 54, 26.2.2011, p 1–154) Article 8:
Category 1 material
Category 1 material shall comprise the following animal by-products:
(a) entire bodies and all body parts, including hides and skins, of the following animals:
   (i) animals suspected of being infected by a TSE in accordance with Regulation (EC) No 999/2001 or in which the presence of a TSE has been officially confirmed;
   (ii) animals killed in the context of TSE eradication measures;
   (iii) animals other than farmed and wild animals, including in particular pet animals, zoo animals and circus animals;
   (iv) animals used for experiments as defined by Article 2(d) of Directive 86/609/EEC without prejudice to Article 3(2) of Regulation (EC) No 1831/2003;
   (v) wild animals, when suspected of being infected with diseases communicable to humans or animals;
(b) the following material:
   (i) specified risk material;
   (ii) entire bodies or parts of dead animals containing specified risk material at the time of disposal;
(c) animal by-products derived from animals which have been submitted to illegal treatment as defined in Article 1(2)(d) of Directive 96/22/EC or Article 2(b) of Directive 96/23/EC;
(d) animal by-products containing residues of other substances and environmental contaminants listed in Group B(3) of Annex I to Directive 96/23/EC, if such residues exceed the permitted level laid down by Community legislation or, in the absence thereof, by national legislation;
(e) animal by-products collected during the treatment of waste water required by implementing rules adopted under point (c) of the first paragraph of Article 27:
   (i) from establishments or plants processing Category 1 material; or
   (ii) from other establishments or plants where specified risk material is being removed;
(f) catering waste from means of transport operating internationally;
(g) mixtures of Category 1 material with either Category 2 material or Category 3 material or both.

or where the material is that described in Article 8(f) catering waste from means of transport operating internationally;

...

(d) in the case of Category 1 material referred to in Article 8(f), disposed of by burial in an authorised landfill;

These requirements constitute an obligation to discard such material and it is therefore to be regarded as waste.

Another example of a requirement to discard is Article 2 of the Mercury Regulations[47].

> 'From 15 March 2011, the following shall be considered as waste and be disposed of in accordance with [the WFD] in a way that is safe for human health and the environment:
>
> (a) metallic mercury that is no longer used in the chlor-alkali industry;
> (b) metallic mercury gained from the cleaning of natural gas;
> (c) metallic mercury gained from non-ferrous mining and smelting operations; and
> (d) metallic mercury extracted from cinnabar ore in the Community as from 15 March 2011.'

Deliberate and Accidental Discards

2.26 Article 36(1) of the WFD requires

> 'Member States to take the necessary measures to prohibit the abandonment, dumping or uncontrolled management of waste'.

Member States are required under the WFD Article 36(1)[48] to take the necessary measures to prohibit abandonment, dumping or uncontrolled management of waste. In the UK there are a number of controls that have been put in place to ensure that the management of waste is properly controlled and to prohibit the dumping of waste. When a deliberate discarding of a substance or object takes place (eg by abandonment or dumping) then this is classified as waste. It is also possible that a substance or object that has been involuntarily discarded may also be waste. In the *Van de Walle* case (see para **2.67**) hydrocarbon were accidentally spilled which caused soil and grundwater contamination. This was held by the European Court to be waste even though no one was aware at the time that the spill was taking place.

In the UK a number of controls are in place to ensure that waste is properly managed and to prohibit the abandonment or dumping of waste. The deliberate

47 Regulation (EC) No 1102/2008, ([2008] OJ L 304, 14.11.2008, p 75–79) of the European Parliament and of the Council 22 Oct 2008 http://eur-lex.europa.eu/legal-content/EN/ALL/?uri=CELEX:32008R1102.

48 WFD, Article 36(1). Member States shall take the necessary measures to prohibit the abandonment, dumping or uncontrolled management of waste.

discarding of a substance or object (eg by abandonment or dumping) will result in its being classified as waste.

Oil Spills in the Marine Environment and Waste Legislation

2.27 After the Gulf of Mexico accident following off-shore drilling in April 2010 the Commission decided to review its existing legislation relating to oil spills and look at how waste legislation could apply.

The case is given in an example on the Guidance to the WFD[49]. Oil spills such as those in Brittany or in the Gulf of Mexico are considered waste under EU waste legislation. In its judgment (C-188/07 *Commune de Mesquer v Total*, Judgment of 24 June 2008) the CJEU found that hydrocarbons accidentally spilled at sea following a shipwreck, mixed with water and sediment and drifting along the coast of a Member State until they are washed up on that coast, constitute waste within the meaning of the WFD, where they are no longer capable of being exploited or marketed without prior processing.

Interpretation of the judgment means that any accidentally spilled hydrocarbons at sea, under circumstances where they are no longer capable of being exploited or marketed without prior processing, would have to be considered as waste. This would also apply to any oil spills from offshore drilling. Then any further treatment, storage or processing would have to satisfy the requirements of the waste legislation.

In addition, the characterisation of oil spills at sea as 'waste' can have direct consequences for the responsibility of the companies engaged in petroleum activities and on the extent of their liability.

Case-by-Case Approach

2.28 As there are difficulties in applying the discard test in a wide range of circumstances this has led the European Court to recognise a need for flexibility in adopting a case-by-case approach. The Court has cautioned Member States not to adopt modes of proof, such as statutory presumptions, which have the effect of restricting the scope of the Directive.

This means that whether or not a substance or object is waste should be considered in each case in the light of all the circumstances and in a way which does not undermine the effectiveness of the WFD. There are no decisive criteria within the Directive which allow the determination of the holder when the discard is a substance or object. There are however factors which the European Court consider to be relevant in the decision as to whether something is waste or not.

49 Guideline on the interpretation of key Directive 2008/98/EC on Waste 1.1.3 (Oil Spills in the marine environment and waste legislation).

CHARACTERISTICS OF THE SUBSTANCE OR OBJECT

2.29 The discard test relates to the conduct of the holder and not to the intrinsic characteristics of the substance or object. This is consistent with the idea that any substance or object can be waste if its holder decides to discard it. The consequence of this is that the characteristics of the substance or object cannot by themselves determine whether a substance or object is waste or not; this does not mean that the characteristics of the substance or object are not relevant at all.

Economic Value

2.30 The concept of waste does not exclude substances or objects even if they have a commercial value[50]. The WFD applies to the use of waste in recovery operations and in such circumstances the waste often has an economic value. Many wastes have some value in different parts of the world. If waste were to be defined in a way which confined it to items with no economic value that would undermine one of the WFD's main purposes (ie to control recovery as well as disposal operations and to ensure that both kinds of operation are carried out in a way which protects human health and the environment). There might also be an incentive in some cases to make nominal payments in order to avoid regulation.

Substances and objects which have been discarded remain waste even though they have value to their current holder who collected them on a commercial basis for recycling. Scrap metal is not recycled until it meets end-of-waste criteria adopted under Article 6 of the WFD or it is re-processed into a new product. In the *Mayer Parry* case[51], the court explained that:

> 'First, obstacles to trade could arise if different concepts of recycling applied in member states, so that the same material or product could be regarded as recycled in one Member State – and would accordingly have ceased to be classified as packaging waste and been freed form all waste-specific controls – while that would not be the case in another Member State.'

So someone who purchases scrap metal with the intention of it being re-processed into steel is taken to have the intention to discard that material even though they may regard it as a valuable secondary raw material. There is no distinction based on whether a substance or object is marketable or not.

50 On 28 March 1990 the European Court provided the following judgments in joined cases C-206/88 and C-207/88 (*Vessos and Zanetti* [1990] 2 LMELR 133) and CaseC-359/88 (*Zanetti and Others*).

51 On 19 June 2003 the European Court delivered its judgment on Case C-444/00 (*Mayer Parry Recycling Ltd*).

Polluting Characteristics of a Substance or Object

2.31 Neither the potential to pollute nor the harmlessness of the substance or object in question is a decisive criterion for determining what its holder intends to do with it and hence whether it is waste.

In the *Palin Granit* case[52], (see para **2.72**) leftover stone resulting from stone quarrying had very little potential to pollute as it was inert. However, it was stored for an indefinite length of time to await possible use and so was waste. The court found that the long-term stockpiling which was necessary, was a burden to the holder and also presented the type of nuisance which the WFD seeks to address.

There are also substances with hazardous properties, which are not waste even if they cause pollution. An ordinary fuel may be burned without becoming waste even if it causes pollution. However, waste which is recovered by burning as a fuel will be classified as waste and hence subject to regulation even if it presents a lower pollution risk when burned.

This latter point is quite naturally sometimes difficult to understand – particularly where the degree of regulation that turns on the application of the waste definition is significant. Where the level of regulation is accepted as proportionate then there is usually no debate about the definition of waste. But where there is little or no flexibility in the regulatory regime, attention is often focused on the definition of waste as the only means of avoiding regulation. The definition is seen as the problem even if the real argument concerns whether the controls applied are proportionate. The court recognised this in the *Arco Chemie* case[53] when it explained that the polluting characteristics are only relevant to the level of control to be applied:

> 'The fact that substances may be recovered as fuel in an environmentally responsible manner and without substantial treatment is, indeed, material to the question whether the use of that substance as fuel should be authorised or encouraged or to the decision as to the degree of control to be exercised.'

The differences can perhaps be explained by considering the differences in our approach to products and waste. There are many products, which present an environmental risk. The risks presented by such products are justified by the benefits of using the products. It is only when the risks are considered too high that products are banned (eg CFCs). Products will also be designed with particular uses in mind and will usually meet certain specifications. There will be instructions relating to their use and regulation in some form ensuring their safe use. Those who acquire the products will do so with the intention of using them in the way specified and will be careful to use them efficiently in order to keep costs down. However, the risks associated with the use of products are

52 On 18 April 2002 the European Court delivered its judgment on Case C-9/00 (*Palin Granit Oy and Vehmassalon kansanterveystyön kuntayhtymän hallitus*).

53 On 15 June 2000 the European Court delivered its judgement on Case C-418/97 and C-419/97 *ARCO Chemie Nederland Ltd etc.*

also increasingly being recognised and as this occurs, the distinction between waste regulation and product legislation becomes less marked. In this context, examples of product legislation include the Construction Products Directive[54] and the REACH Regulations[55].

Transfer to Another Person

2.32 Sometimes a person may not have a use for a substance or object but may know that some other person would have use for it. The questions in such cases are whether the fact that the first person wants to get rid of the substance or object means that it is waste or whether the fact that another person has a use for it means that it is not waste? The answer is that simply transferring a substance or object from one person to another does not in itself affect its classification as waste.

For example, the Freecycle NetworkTM66 is a scheme aimed at encouraging people to offer to others, free of charge, goods they no longer have a use for (eg a television (TV) or personal computer (PC)). In many ways, this is similar to someone taking clothes to a charity shop. In neither case is the substance or object waste. In both cases, the substance or object is being transferred with the intention that it should continue to be used for its original purpose (ie it is being re-used as non- waste).

However, a TV or PC is classified as waste if a householder decides to take their TV or PC to a local authority/district council 'civic amenity site'. This is because civic amenity sites are facilities, which local authorities/district councils, as waste disposal authorities, have a duty to provide as places 'at which persons resident in its area may deposit their household waste…'. The intention of the householder in taking the TV or PC to a civic amenity site is to ensure that it can be safely recovered/recycled or disposed of.

The difference between these two situations is the circumstances that surround the transfer. In the former, the person transferring the substance or object has made an assessment that someone else actually wants it. In the latter, the assumption is that no one else wants the TV or PC and they will therefore remain waste until they have been subject to a recovery operation.

Another context where the issue of transfer arises is where activities give rise to a residue for which the holder has no use; and if there is a demand for that residue the holder may argue that it is not waste. The European Court has addressed this type of situation through the case law on by-products; and whether the residue is waste or not will be assessed according to the principles set down in the case law. However, it is clear that the mere transfer again does not determine whether the residue is waste. For example, a construction activity may generate large amounts of excavated unpolluted natural soil. If

54 Council Directive 89/106/EC The Construction Products Directive, ECLI: EU: C:2007:200.
55 REACH Regulations http://echa.europa.eu/web/guest/regulations/reach/legislation.

the holder has no use for the soil then it may present a disposal problem that has to be addressed by the soil's classification as waste. However, if that soil is suitable for use without any treatment and is certain to be used for that purpose, it may be classified as a non-waste product, ie on the basis that it satisfies the by-products legal test.

It is the circumstances that surround the transfer that is the difference between these two cases. In the first case the person transferring the substance or object has made an assessment that someone else actually wants it. In the second, the assumption is that no one else wants the objects and they will therefore remain as waste until they have been subject to a recovery operation.

For example, a construction activity may generate large amounts of excavated unpolluted natural soil. If the holder has no use for the soil then it may present a disposal problem that has to be addressed by the soil's classification as waste. However, if the soil is suitable for use without treatment and is certain to be used for that purpose, it may be classified as a non-waste product, ie on the basis that it satisfies the by-products legal test.

SOME POSITIVE INDICATORS

2.33 The indicators below are positive indicators in favour of a substance or object being classified as waste, although no one indicator is conclusive.

The European Waste Catalogue

2.34 Article 1(2) of the WFD as codified in Directive 2006/12/EC required the Commission to draw up a list of wastes ('the European Waste Catalogue' (EWC)) belonging to the categories listed in Annex I to that Directive and to review and revise it from time to time. (See also Article 7 of the WFD as adopted in Directive 2008/98/EC). The first EWC was initially established by Commission Decision 94/3/EC and then replaced by Commission Decision 2000/532/EC.

The EWC is a hierarchical list of waste descriptions established by Commission Decision 2000/532/EC. It is divided into 20 main chapters each of which has a two-digit code between 01 and 20. Most of the chapters relate to industry but some are based on materials and processes. Individual wastes within each chapter are assigned a six-figure code.

Within the list hazardous wastes are signified by entries where the code is followed by an asterisk. Those that are 'Absolute Entries' (ie are always hazardous waste) are shown in red and the letter 'A' appears beside them. For some entries the judgment on whether they are hazardous depends on the threshold of any dangerous substance that is present. These entries are 'Mirror Entries', They are coloured blue and the letter 'M' appears beside them.

However, the EWC is not an exhaustive list of wastes. Indeed it would be difficult to list all wastes as new substances are produced and new uses for existing substances are found. A substance or object is therefore only to be classified as waste if the holder discards or intends or is required to discard it. It is also the case that an item not included in the EWC will be classified as waste if the holder discards or intends or is required to discard it. But the inclusion of a substance or object in the EWC is indicative that it may be waste.

Residues

2.35 Waste often arises from something that is left over from some other process or use. It is not the intended result and hence it is not designed for use in a particular manner. Or it can no longer be used for the purpose for which it was originally conceived. Its composition may be uncertain or it may contain impurities or contaminants from processing. Its economic value may be low. There are not the same incentives to treat it with the same care or the same guarantees that it will be used in the intended manner. The likelihood is that its holder will seek to get rid of it in some way, which has the potential to cause pollution. For these reasons, the starting point is that residues are waste – even if they can be used in some way. Many waste residues can be put to beneficial use.

Some residues may be viewed as by-products and classified as non-wastes. There is a large body of case law dealing with this and para **2.48** explains the criteria for deciding in what circumstances a residue is a non-waste by-product.

Negative Economic Value

2.36 Although the value of a substance or an object does not preclude it from being waste, its value may be a factor in deciding whether or not its holder discards or intends to discard it. If a substance or object has a negative value (so that the person disposing of it has to pay for it to be taken away) or depending on market conditions may at times have a negative value, that points towards it being waste since it is a burden on the producer or holder who therefore has an incentive to abandon, dump or otherwise dispose of it unlawfully.

Contaminating Substances

2.37 Waste may be contaminated by other substances and these may be a danger to human health or the environment. So, in the *Arco Chemie* case[56],

56 On 15 June 2000 the European Court delivered its judgment on Case C-418/97 and C-419/97 *ARCO Chemie Nederland Ltd etc.*

the European Court considered that waste in the form of wood, which was impregnated with toxic substances, did not lose its classification as waste when it was transformed into chips or those chips were reduced to wood powder since it did not purge the wood of the toxic substances.

Where a substance is contaminated by reason of its provenance, that may serve to indicate that it is waste. This is also true if its composition is uncertain. Substance or object is commonly regarded as waste.

If something is commonly regarded as waste then that may serve to indicate that it is in reality waste. Not too much store can be set by this – for instance in the *Saetti* case[57] there was an indication that the company concerned regarded the petcoke as waste but this did not affect the conclusion of the European Court that it was a product and not a production residue. The European Commission's guidance on by-products discounts this as a factor too. This may simply be a reflection that in most cases it is clear whether something is waste or not – but in difficult cases the way a substance is commonly regarded will not help very much.

SUBSTANCE OR OBJECT HAS BEEN CONSIGNED TO AN OPERATION WHICH IS A COMMON METHOD OF DISPOSING OR RECOVERING WASTE

2.38 When a substance or object has been consigned to an operation commonly regarded as a recovery operation (eg solvent regeneration) or a disposal operation (eg tipping underground), that will indicate the existence of waste. It is not always easy to tell, however, whether an operation is actually disposing or recovering waste. This is discussed below.

THE RELATIONSHIP BETWEEN THE CONCEPT OF WASTE AND DISPOSAL, RECOVERY, RECYCLING AND RE-USE

2.39 The point that the European Court always returns to when interpreting the definition of waste is that it must consider the objectives of the WFD. The designation of waste is therefore to ensure that the aims of the WFD are met.

The WFD has two main objectives relevant to the interpretation of the definition of waste and both of these objectives need to be met:

57 On 14 January 2004 the European Court delivered its judgment on Case C-235/02 *Mario Andonio Saetti and Andrea Frediani.*

- to ensure that waste is controlled until it has been safely disposed of or recovered (Article 13); and
- to reduce the overall impacts of resource use and to improve the efficiency of such use (Article 1). The WFD itself seeks to fulfil this objective by requiring Member States to apply a waste hierarchy as a priority order in waste prevention, waste management legislation and waste policy (Article 4).

There is a degree of tension between these two objectives because the WFD seeks both to encourage the use of waste as a resource and in the interests of the environment and human health, to regulate the use of waste. The question in many 'definition of waste' cases is whether a particular operation is to be regarded as the ordinary use of a substance or its submission to a waste management operation, which is subject to regulation. It is helpful therefore to consider the nature of these waste management operations.

Disposal

2.40 Disposal operations are primarily aimed at getting rid of waste. Any benefit that results as a secondary consequence will not affect the nature of the operation. There will usually be very little debate as to whether an operation, which looks like disposal, involves waste or not. If the ultimate aim of the operation is to get rid of the waste, there is a clear intent to discard.

Sometimes though, it will not necessarily be obvious. The use of slurry as a fertiliser does not amount to the disposal or recovery of waste where the use is part of a lawful practice of spreading and the spreading takes place on clearly identified parcels of land without prior processing. The reason for this is that animal faeces amount to a by-product of keeping animals and comprise a traditional fertiliser for which proprietary fertilisers may be regarded as a substitute. However, if slurry is simply spread as a means of disposal then it will be waste.

It should not be difficult in practice to distinguish between the use of slurry as a natural fertiliser and the spreading of slurry on land as a means of simply getting rid of the slurry. It should be obvious that slurry is being spread on land simply to get rid of it where such application is in quantities, which exceed those necessary to benefit the land. This can be judged from the point of view of good agricultural practice – and it will be clearer in cases where there is a breach of the Nitrates Directive (91/676/EC).

Recovery

2.41 Recovery is defined in Article 3(15) of the WFD as:

> '... any operation the principal result of which is waste serving a useful purpose by replacing other materials which would otherwise have been used to fulfil a particular function, or waste being prepared to fulfil that function, in the plant or in the wider economy'.

The principal objective of a recovery operation is to ensure that the waste serves a useful purpose by replacing other substances which would have otherwise been used for that purpose and thereby conserving natural resources. This is known as the 'substitution principle'.

Submission of a substance to a recovery operation will involve the discarding of the substance as any other interpretation would mean that recovery operations would not be subject to regulation. This is why it is important to distinguish between recovery and the ordinary use of substances or objects. Recovery operations are still to be promoted in order to achieve more efficient use of resources. However, the latter objective is not an argument to free from regulation what is properly regarded as a recovery operation. The effect of classifying something as waste is not to prevent it from being recovered by being used in some way – it is to prevent that operation taking place without the necessary precautions also being taken to protect the environment and human health. It is incumbent on legislators to promote recovery through the design of the regulatory regime. The UK has done this, in particular, by making use of its discretion under Article 24 of the WFD[58] to adopt a number of exemptions from the Directive's permit requirements for recovery operations.

Given that the primary objective of someone who carries out a recovery operation is to derive benefit from the substance (which they may have paid to obtain), it is unsurprising that there is often a great deal of difficulty distinguishing the recovery of waste from the ordinary use of substances. For instance, if a substance needs to be ground up in order to be used, is that to be regarded as ordinary processing or a recovery operation to enable use of the substance? The answer will depend on the circumstances as many factors will be relevant. The European Commission's WFD guidance on by-products gives one example where the grinding or crushing of blast furnace slag is not regarded as waste processing but is instead an ordinary part of the industrial process carried out by steel producers. But grinding up waste wood into powder so that it can be used as fuel in a power station is a recovery operation – the wood is to be recovered by burning and grinding it as a necessary step in the process.

The WFD includes a non-exhaustive list of recovery operations in Annex II to the Directive. These are intended to provide illustrations of the way these operations are carried out in practice. An operation may be a disposal or recovery operation within the meaning of the WFD even if it is not listed. Analogous operations also fall within the scope of the WFD. The Directive applies to new methods of disposing or recovering waste as well as existing methods, which are not listed.

58 WFD, Art 24.
Exemptions from permit requirements
Member States may exempt from the requirement laid down in Article 23(1) establishments or undertakings for the following operations:
(a) disposal of their own non-hazardous waste at the place of production; or
(b) recovery of waste.

A consequence of this is that the European Court has cautioned against concluding that a substance or object is waste solely on the basis that it undergoes an operation described in the Annexes to the Directive. Some of the descriptions in the Annexes are confined to operations which apply only to waste, but others are more abstractly expressed and could be applied to raw materials and other substances, which are not waste. Thus, for example, R1 (use principally as a fuel or other means to generate energy) could be read as applying to ordinary fuels such as oil, gas or kerosene, but those substances are not waste. Similarly, R10 (land treatment resulting in benefit to agriculture or ecological improvement) could be read as applying to the use of fertilisers on land, which similarly should not be regarded as waste.

This does not mean that there is no help in deciding whether a substance is waste or not can be derived from the way in which a substance is treated. The fact that a substance undergoes a process, which is a common method of recovering or disposing of waste, may be taken as evidence that it is being discarded and hence is waste. For example, a lot of waste burns well and it is a common way of getting rid of waste. If something is unwanted or cannot be used for its original purpose and there is a decision to burn it, that is a strong indication that it is waste.

Another point to note is that recovery may take place over several stages. This means that the recovery operation may not result in the substance or object being declassified as waste. If further recovery is necessary then it remains waste until it is the subject of a complete recovery operation and is fully recovered. Only in those examples of recovery operations that result in a final use of the substance (eg where waste is used as fuel to generate energy or as a fertiliser to benefit agriculture) will the waste always cease to be waste. In other cases, it will be necessary to assess whether the resulting substance needs to be controlled in order to meet the aims of the WFD. In the waste wood example, the grinding into powder is a necessary step in order to enable the waste to be used as fuel. In the *Arco Chemie* case[59], the resulting wood powder remained contaminated and so it was clear that it should still be controlled under the WFD and so continued to be classified as waste.

Recycling

2.42 Recycling is a form of recovery and as such pursues the same objective of saving natural resources. Article 3(17) of the WFD defines recycling as follows:

> '... any recovery operation by which waste materials are reprocessed into products, materials or substances whether for the original or other purposes. It includes the reprocessing of organic material but does not include energy recovery and the reprocessing into materials that are to be used as fuels or for backfilling operations'.

59 On 15 June 2000 the European Court delivered its judgment on Case C-418/97 and C-419/97 *ARCO Chemie Nederland Ltd etc.*

It is evident that a substance or object must first be classified as waste in order to be recycled within the terms of the WFD and related EU waste legislation (eg the Packaging Waste Directive). In other words, recycling can only ever be carried out on substances or objects that are classified as waste.

One of the key aims of the WFD and related EU legislation is to promote the better use of resources by recycling. In this context, it is important to emphasise that products made from recycled waste are not inherently inferior or more dangerous than those derived from virgin products. The 'stigma' of waste has been perceived as a problem, which limits the beneficial use of many wastes – regardless of any regulatory impact of waste controls. The definition of waste is characterised in this context as 'a barrier to recycling'. But anyone engaged in recycling is processing waste so it is misconceived to argue that a recycling operation cannot be carried out because something is regarded as waste. The underlying policy behind the WFD is that recycling is to be promoted as a virtuous activity and that waste is a useful resource –not simply inferior material that needs to be disposed of. This policy means the response to the perception of stigma is not to argue that only substances or objects that are disposed of are waste but is instead to acknowledge the waste origins of recycled products and to promote their use as a more sustainable use of resources.

The definition of recycling requires the waste to be re-processed so as to obtain a product, material or substance whether for the original or other purposes. This differs from recovery operations, which result merely in a change in the nature or composition of the waste (eg reducing waste wood into powder) so that the waste can serve a useful purpose. A new product, material or substance has been recovered and so all the material has re-entered the productive cycle. No further processing is required. A recycling operation is therefore different in nature to other recovery operations in that it will always result in the substance in question ceasing to be waste when it is transformed. The substance will cease to be waste before its final use.

Re-use

2.43 The WFD as adopted in Directive 2008/98/EC draws a clear distinction between 're-use' and 'preparing for re-use'. The distinction is that the former is an activity, which does not involve waste, and the latter is an activity which does involve waste. The WFD defines these terms as follows:

> '"re-use" means any operation by which products or components that are not waste are used again for the same purpose for which they were conceived;' and

> '"preparing for re-use" means checking, cleaning or repairing recovery operations, by which products or components of products that have become waste are prepared so that they can be re-used without any other pre- processing;'

A substance or object is not waste when it is being used for the purpose for which it was conceived – unless the circumstances are exceptional (eg the use is unlawful). An example of such exceptional circumstances is Article 5(1) of Directive 96/59/EC on the disposal of polychlorinated biphenyls and polychlorinated terphenyls (PCBs/PCTs) which provides that, 'By way of derogation from Article 3 of [the WFD] Member States shall prohibit the separation of PCBs from other substances for the purpose of reusing the PCBs'.

A Distinction Should be Made Between Re-use as a Waste Prevention Operation and Re-use Following a Recovery Operation

2.44 A substance or object may never become waste because it is designed or used in such a way as to enable re-use. An example is a milk bottle. In the other case, the object becomes waste but is then recovered in some way so that it can once again be used for the purpose for which it was created. However, used chemical drum containers are likely to require a recovery operation to remove contaminating material and make the drum ready for use again.

In *Inglenorth*[60] the High Court upheld a magistrates' decision to dismiss a waste prosecution on the basis that hardcore from a demolished building was not waste, even when the demolition was on one site and the intended re-use was on another. The Court considered the fact that the party producing the demolition material intended for it to be re-used as determinative. In reaching this conclusion, the Court focused on the status of the materials at the point where they were re-used, rather than the point at which they were produced and arguably this is the incorrect approach: If materials have already become waste, their subsequent re- use cannot 'cure' them of their waste status.

PRODUCTS, BY-PRODUCTS AND RESIDUES

2.45 Production processes produce a range of substances. Some of these substances will be sought (ie they will be products of the process) whereas others will not (ie they will be residues of the production process). If a residue can be regarded as a by-product then it does not become waste.

A product is something, which is manufactured or produced with the intention of using or marketing it. Short of some problem occurring (eg a defect in the product or a market collapse), it will not be waste when manufactured. Products are not manufactured with the intention of their being discarded.

A production process may be intended to produce several different products. Each will be a product as long as it is sought by the producer – that is it is intentionally produced as a result of a technical choice.

60 *Environment Agency v Inglenorth Ltd* Divisional Court 17 March 2009, [2009] EWHC 670 (Admin).

If a substance or object results automatically from the process (ie it is not intentionally produced), it may still be regarded as a product if it is like other products, which are produced by the same process, and is certain to be used. Some industries split raw materials into several different fractions which often have very similar end uses. In these cases it may be difficult to decide which fraction is the residue. The case law recognises that in such cases there is no practical difference between a residue that arises automatically and products, which are intended to be produced. The case of *Saetti* (refining of crude oil to produce fuels) is an example of this[61]. In that case the court commented that even if petroleum coke had automatically resulted from the refining process it would still be regarded as a petroleum product and not a waste residue if certain to be used mainly for the same purposes as the other substances produced in the refinery.

In *Commune de Mesquer v Total France SA, Total International Ltd*[62], the European Court confirmed that a marketable and useable fuel is not a waste, but if it is spilt and can no longer be used without having to be processed, it is waste. The question was whether heavy fuel oil, as the product of a refining process, meeting the user's specifications and intended by the producer to be sold as a combustible fuel, could be treated as waste within the meaning of the WFD. The European Court held that such a substance does not constitute waste where it is used or marketed on economically advantageous terms and is capable of actually being used as a fuel without requiring prior processing. The European Court also held that, conversely, such hydrocarbons accidentally spilled at sea following a shipwreck, mixed with water and sediment and washed up on the coast, constitute waste where they are no longer capable of being exploited or marketed without prior processing.

A residue is a substance or object, which results from a production process, which is not, in itself, sought for a subsequent use. The fact that a residue is unwanted when it is first produced is evidence that it is waste since unwanted objects are more likely to be discarded.

Note that this analysis applies to production residues. It does not apply to consumption residues. The circumstances when residues arise will influence their nature and composition and hence the likelihood that they are waste. It is felt that a general approach is possible in relation to production residues that cannot be applied to consumption residues. For example, scrap metal may arise in the context of a production process (eg offcuts) or following the use of products (eg from end-of-life vehicles). The former is likely to be less contaminated and more predictable and, if it complies with the by-product criteria, it will not be waste. That said – whether any substance is waste or not ultimately depends on the discard test.

61 On 14 January 2004 the European Court delivered its judgment on Case C-235/02 *Mario Andonio Saetti and Andrea Frediani.*

62 On 24 June 2008 the European Court delivered its judgment on Case C-188/07 *Commune de Mesquer v Total France SA, Total International Ltd.*

By-Products

2.46 It is recognised that there are circumstances where a genuine use may be found for a residue. In such circumstances the substance may not be regarded as waste but instead may be regarded as a by-product which the undertaking intends to exploit or market.

It is important to understand the distinction between a product and a by-product. The former is intended to be produced whereas the latter is not. If a substance is intentionally produced, it is a product and in principle is not waste. If it is not intentionally produced then it is necessary to assess whether it is a residue, which is being discarded, or a by-product, which is being used. The concept of a by-product is used to distinguish between residues, which are waste, and those, which are not. If it is considered to be a by-product then, at that point, it has the same legal status as a product – and so for instance would be subject to any applicable product legislation (eg the REACH Regulation).

Product – all material that is deliberately created in a production process. In many cases it is possible to identify one (or more) 'primary' products, this or these being the principal material(s) produced.

Production residue – a material that is not deliberately produced in a production process but may or may not be a waste.

A production residue that fulfils the conditions of WFD Article 5(1) is a by-product.

Is the Substance or Object a Product or a Residue?

2.47 It can be seen from the above that the first question is whether a particular substance is a product or a residue. If the substance is deliberately produced as a result of a technical choice then it will be a product and will not be waste.

The European Commission looks at some considerations, which are relevant to assessing this question. If the manufacturer could have produced the primary product without producing the other material, but chose to do so, then this is evidence that the material concerned is not a production residue. Other evidence that the production of the material concerned was a technical choice could include a modification of the production process in order to give the material concerned specific technical characteristics.

If the organisation seeks to limit the production of the substance concerned then that too might indicate that it should be considered a residue and not a product. If there is an attempt to limit the amount produced, that would suggest that there is an intention not to produce it. However, it may still be the case that it is certain to be used. This factor may be an indicator that the residue concerned might be waste.

The Commission's guidance includes some examples of substances, which are not waste. In the case of blast furnace slag for instance, there is a technical choice that governs what type of slag is produced, and that the material can be considered a by-product, falling outside the definition of waste.

The Environment Agency regards gypsum from flue gas desulphurisation (FGD) abatement which is produced to a European specification and which is certain to be used for the manufacture of plasterboard to be a product specifically sought and not a waste residue. In this case, additional steps have been introduced to the process to convert calcium sulphite to calcium sulphate and then refine it to a specification to be usable as gypsum raw material. There would seem to be no other reason for putting the additional steps in other than to produce gypsum and so it should be considered an intended product.

Condition for a production residue to be a by-product and not waste.

- Further use of the substance or object is certain;
- The substance or object can be used directly without any further processing other than normal industrial practice;
- The substance or object is produced as an integral part of a production process; and
- Future use is lawful, ie the substance or object fulfils all relevant product, environmental and health protection requirements for the specific use and will not lead to overall adverse environmental or human health impacts.

If the Substance or Object is not Intentionally Produced and hence is a Production Residue, in what Circumstances is it not Considered Waste?

2.48 Article 5(1) of the WFD provides that a substance or object, resulting from a production process, the primary aim of which is not the production of that item, may be regarded as a non-waste by-product – but only if all of the four conditions set out in Article 5(1)(a)–(d) are met. The provision was introduced on the initiative of Member States during the negotiation of Directive 2008/98/EC and was intended to reflect and place on the face of the Directive, existing case law by the European Court on the distinction between production residues as waste and non-waste by-products.

The decision as to whether or not a substance or object meets the four conditions set out in Article 5(1)(a)–(d) must be made on a case-by-case basis in the light of all the circumstances and in a way which does not undermine the effectiveness of the WFD. The conditions are that:

(a) further use of the substance or object is certain;
(b) the substance or object can be used directly without any further processing other than normal industrial practice;
(c) the substance or object is produced as an integral part of the production process; and

(d) further use is lawful, ie the substance or object fulfils all relevant product, environmental and health protection requirements for the specific use and will not lead to overall adverse environmental or human health impacts.

These conditions/criteria have come from principles developed in European Court case law which spell out when it is consistent with the WFD to regard the use of a production residue as a by-product which the undertaking wishes to exploit rather than a waste residue that is being discarded.

It is important to stress that the Article 5 conditions may not provide the whole picture. As the court explained in the *KVZ* case[63] (see para **2.79**):

> 'However, whether it is in fact "waste" within the meaning of [the WFD] must be determined in the light of all the circumstances, regard being had to the aim of that Directive and the need to ensure that its effectiveness is not undermined.'

Certainty of Further Use

2.49 'Further use is certain' means that it is not a mere possibility but a certainty; in other words, it is guaranteed that the material will be used. The purpose of this criterion is that if further use were not certain, there would be a risk of the material being disposed of as waste.

The main consideration here is to be satisfied that there is a genuine demand for the substance. If there is no demand, the undertaking which produces the substance or object, may have to store it for a significant period until a customer comes along. In those circumstances, the substance or object is to be regarded as a burden on the holder and hence waste.

In the *Palin Granit* case[64], the CJEU ruled that 'the holder of leftover stone resulting from stone quarrying which is stored for an indefinite length of time to await possible use discards or intends to discard that leftover stone, which is accordingly to be classified as waste'.

Circumstances may arise in which some of the residue which is produced, is classified as waste because it is discarded, but an identifiable proportion of the residue is not classified as waste because it is capable and certain of use without prior processing. This was made in relation to leftover rock in the *Avesta Polarit* case[65]. In that case, the European Court found that the holder of the leftover rock from a mining operation discarded it with the consequence that it was classified as waste – even where it was put to alternative use (because some prior processing was required in many of those cases). However, the Court also found that the leftover rock could be classified as a non-waste by-product

63 On 1 March 2007 the European Court delivered its judgment on Case C-176/05 *KVZ retec GmBH v Republik Österrich*.

64 On 18 April 2002 the European Court delivered its judgment on Case C-9/00 *Palin Granit Oy and Vehmassalon kansanwerveystyön kuntayhtymän hallitus*.

65 On 11 September 2003 the European Court delivered its judgment on Case C-114/01 *AvestaPolarit Chrome Oy*.

where the holder used it lawfully for the necessary backfilling of the mine from which it had been extracted and provided guarantees as to the identification and actual use of the leftover rock to be used for that purpose.

The way that certainty of further use may be demonstrated is through the existence of contracts – although contracts are not always necessary. The economic value of a residue can be relevant in assessing whether the likelihood of the proposed use occurring is sufficiently high to conclude that the substance will be used. This is not to say that economic value will by itself determine that the substance or object is not waste. Other circumstances will be relevant (eg whether it needs to be processed or whether the proposed use presents risks which the WFD seeks to control).

Certainty of further use can be difficult to prove in advance. However, indications of 'certainty of further use' may be:

- Existence of contracts between the material producer and subsequent user;
- A financial gain for the material producer;
- A solid market (sound supply and demand) existing for this further use;
- Evidence that the material fulfils the same specifications as other products on the market.

There are also indications that future use may be uncertain:

- There is no market for the material;
- Only part of the material is to be used, with the rest to be disposed of (should be initially treated as waste);
- The financial gain for the waste holder is nominal compared to the cost of waste treatment.

FURTHER PROCESSING PRIOR TO REUSE

2.50 All raw materials require some processing in order to be made into products. Processing as part of an ordinary production process is not ordinarily a waste management operation.

However, a substance or object may be classified as a production residue and as waste, if it is necessary to subject it to further processing in a pre-treatment or recovery process before it is suitable for use. Processing in this context is there to address the characteristics that render the substance waste.

Article 5 of the WFD does not preclude further processing which is 'normal industrial practice'. Whether further processing falls within this definition will be determined by reference to the overall circumstances. An example of the European Commission's approach is given in its guidance on by-products regarding blast furnace slag. Whilst this slag needs to be ground up so that it can be used for construction purposes, doing so is not to be regarded as further processing prior to re-use; and the Commission concludes that: 'This material can therefore be considered a by-product and fall outside the definition of waste.'

Produced as an Integral Part of a Production Process

2.51 Prior to the adoption of the WFD as Directive 2008/98/EC, there was some debate as to whether the residue had to be used in the production process, which gave rise to it being considered a by-product. In this context, the European Court considered what 'integral to the production process' meant. The requirement in their terms is not that the substance is used as an integral part of the production process but that it is made ready for further use as an integral part of the production process. This is consistent with the wording of Article 5(1)(c) of the WFD, which makes no reference to the substance being used in the process.

The production process may in principle be the original production process or a different one, run by the same or by different operators. A good example of these conditions applied in practice is spent grains at a brewery, which are used as animal feed or spent yeast used in MarmiteTM production. It would be misleading to describe the brewing process and the MarmiteTM production process as integral parts of the same production process. But it would be accurate to say that the spent grains or spent yeast are certain to be used in a subsequent process. As far as the brewery is concerned, spent yeast is as much a part of its business as beer production. It is certain to be used and its further use does not involve any processing outside a normal production process. An example of a by-product that can be marketed and used directly is uncontaminated sawdust from a sawmill that may be used without any further processing as animal bedding.

Further Use is Lawful

2.52 Article 5(1)(d) effectively defines this test by stating that the term means that 'the substance or object fulfils all relevant product, environmental and health protection requirements for the specific use and will not lead to overall environmental or human health impacts.'

Use of a Residue as a Fuel

2.53 If a residue is used as a fuel in place of normal fuel that will be evidence that the residue is waste. In a sense this is not specific to residues though the fact that the substance is a residue and is subject to a common form of waste recovery appears to combine to create a strong impression that the substance is waste. This is why it was significant in the *Saetti* case (Case C-235/02 *Saetti and Fredani* ECLI: EU: C:2004:26 at para **2.66**) that the petcoke was considered to be a product and not a production residue. It is more difficult to envisage circumstances where a residue which is burnt without prior processing is not waste.

Residue Can Only be Disposed of

2.54 Similarly if the proposed uses of a residue all involve its destruction or disappearance in some way then that too is evidence that anyone acquiring it is simply acquiring it in order to discard it. So a residue that can be used in some other production process (eg, in construction) is less likely to be considered waste than one which can only be burned.

Fitness for Purpose

2.55 If the residue is put to a use for which it is not suited, that may create the impression that it is in reality being discarded. This might be the case for a production residue, which is spread on land. There may be agricultural benefit from the spreading but it would be difficult to conclude that it is equivalent to traditional or proprietary fertilisers. It may also present a risk due to the presence of contaminants.

SPECIAL PRECAUTIONS DUE TO ENVIRONMENTAL HAZARDS

2.56 Many products present environmental risks and require special handling without being made subject to regulation as waste. The point about residues is that their provenance may often give rise to contamination of the type the WFD seeks to control. A residue, which automatically arises from a production process, will often present risks, which are quite different to a product designed to meet a particular specification and to be used for a particular purpose. If special precautions need to be taken in relation to the use of a residue then that is an indicator that it should be regarded as waste and subject to control.

END-OF-WASTE

2.57 The WFD incorporates the concept of end-of-waste (EoW) by setting out conditions whereby substances or objects which meet the waste definition can achieve, after undergoing a recovery operation (including recycling), a non-waste status and thus fall outside the scope of waste legislation.

The concept of end-of-waste follows the systematic approach of EU waste legislation. The specific obligations of waste producers and holders remain in force until a recovery operation has been completed in compliance with the aims of the WFD, therefore minimising the possible waste-related risks to health and the environment. In this respect, the concept of EoW is closely linked to the understanding of the completion of a recovery process, and the understanding of recovery itself which is now defined in Article 3(15) WFD as 'any operation the principal result of which is waste serving a useful purpose by substituting other materials, or waste being prepared to fulfil that function'. Note that Recital 22 WFD states that a recovery operation for the purposes of reaching EoW status may be as simple as just checking the waste to verify that it fulfils the EoW criteria.

The approach of the WFD with respect to setting criteria and taking decisions on EoW is twofold:

- For certain specified waste streams (eg, aggregates, paper, glass, metal, tyres and textiles), EoW criteria can be set at EU level by comitology procedure (Article 6(2)WFD), in accordance with (cumulative) conditions listed in Article 6(1) which are explained in detail below (see para **2.58**). Once EoW criteria are set at Community level, these are binding for Member States. If they have been set in an EU Regulation, they are also binding for private actors. Member States cannot apply different EoW provisions for the scope for which criteria have been set at EU level, with the exception of adopting more stringent protective measures, under the conditions set out in and pursuant to Article 193 TFEU.
- Where no such criteria have been set, Member States may decide whether certain waste has ceased to be waste, taking into account applicable case law.

Conditions for EoW Criteria to be Set at EU Level

2.58 The cumulative conditions for certain specific waste streams are laid down in WFD, Article 6(1)(a)–(d). These are:

- the substance or object is commonly used for specific purposes;
- a market or demand exists for such a substance or object.

These first two conditions are related. Compliance with these two criteria can be indicated by:

  - the existence of market conditions related to supply and demand;
  - a verifiable market price being paid for the material;
  - the existence of trading specifications or standards.

- The substance or object fulfils the technical requirements for the specific purpose and meets the existing legislation and standards applicable to products.
  - Compliance with this criterion can be indicated by compliance with established relevant technical specifications or technical standards that are used for virgin materials for the same purpose. The material should be ready for final use and no additional waste treatment steps should be needed.
- The use of the substance or object will not lead to overall adverse environmental or human-health impacts.

- Compliance with this criterion can be indicated by comparing the use of the material under the relevant product legislation with the use of the same material under waste legislation. The following questions are also relevant: Is the product legislation sufficient to adequately minimise the environmental or human health impacts? Would releasing the material from the waste regime lead to higher environmental or health risks?

An example given in the Guidance[66] is EoW for iron and steel scrap. On 31 March 2011, Council Regulation (EU) No 333/2011 was adopted containing EoW criteria for iron, steel and aluminium scrap. The document has been in force since 9 October 2011. The Regulation sets out criteria whereby iron and steel scrap and aluminium scrap (including aluminium alloy scrap) can cease to be waste. The requirements concern both the input and the output of the recovery process as well as treatment processes and techniques. The producer or importer of waste, which has ceased to be waste, is required to issue a statement of conformity in accordance with a model set out in Annex III to the Regulation.

One aim of setting EoW criteria at EU level is to create legal certainty and a level playing field throughout EU-28. Within Regulation 333/2011, this is emphasised by introducing responsibilities in relation to the role of Conformity Assessment Bodies and Environmental Verifiers (see Article 6(5) of that Regulation) as well as 'independent external verifiers' in third countries, as envisaged under Article 6(6).

Where no EoW Criteria Have Been Set at EU Level

2.59 In cases where EoW criteria have not been set at EU level (Article 6(4) of the WFD), Member States may decide at national level whether certain waste has ceased to be waste. This can relate to classes of materials recovered from waste or to single-case decisions. In their decisions, Member States (this means any level within the Member State entrusted with the task of developing such criteria under the national administrative structure) are bound by the applicable Directives and must take account of CJEU case law.

Member States have to observe the notification requirements in accordance with Directive 98/34/EC. Any draft technical regulations by Member States on EoW criteria have to be notified so that they can be compliance-checked by the Commission against Article 6(1) WFD and as regards their impacts on the functioning of the Internal Market. This includes de facto technical regulations, such as administrative provisions or voluntary agreements (for details, see Article 1 of Directive 98/34/EC). Single-case decisions do not have to be notified, even though they may be based on general administrative provisions for which notification is mandatory.

66 Guidelines on the interpretation of key provisions of Directive 2008/98/EC on Waste.

The Point when Waste which has been Subject to Recovery (including Recycling) Operations Ceases to be Waste

2.60 The WFD definition of recovery includes not only processes where a material is actually substituting other materials, but also processes preparing a waste material in such a way that it no longer involves waste-related risks and is ready to be used as a raw material in other processes. The point of completion of a recovery operation may be considered to be the moment where a useful input for further processing, not representing any waste-specific risks to health and the environment, becomes available.

Does End of Waste Status Need to be Demonstrated?

2.61 Article 6(1) WFD does not foresee a specific point in the chain where EoW has to be demonstrated. Measures adopted relating to Article 6(2) WFD may determine specific timelines and conditions. For example, in Council Regulation (EU) No 333/2011 on EoW criteria for scrap metal, the transfer of possession from one holder (the 'producer' of EoW material) to another holder is a legal condition for reaching the EoW status. Note that it is the material producer, ie the person who first transfers the material to another person as non-waste, who is responsible for providing evidence that EoW criteria have been fulfilled via the statement of conformity.

General Principles – Complete Recovery Operations

2.62 The European Court has considered the issue generally with reference to the concept of a complete recovery operation. In the *OSS* case[67] (see para **2.66**) this has not resulted in a clear distinction and much is left to national authorities to assess in relation to the cases that come before them.

A complete recovery operation is one which has the effect of transforming waste into a distinct product with characteristics that are the same as or analogous to a raw material, so that it may replace that raw material. The waste has not undergone a complete recovery operation if, after the operation has been carried out, the waste substance remains contaminated – for example because of its provenance. Waste controls are still needed where this is the case.

The European Court has only been prepared to say that when waste has undergone a complete recovery operation, it does not necessarily lose its classification as waste[68] – it is only one factor to be taken into consideration

67 *R (on the application of OSS Group Ltd) v Environment Agency and others* [2007] EWCA Civ 611.

68 On 15 June 2000 the European Court delivered its judgment on Case C-418/97 and C-419/97 *ARCO Chemie Nederland Ltd etc.*

in assessing the discard test. However, the European Court has not expressed what the other factors may be.

Examples of Recovery Operations that will not Result in 'End-of-Waste'

2.63 There have been some examples of recovery operations which have not resulted in waste ceasing to be waste. These are often referred to as 'pre-treatment' or 'pre-processing' operations.

In the *Arco Chemie* case (Case C-418/97 and C-419/97 *ARCO Chemie Nederland Ltd etc)* waste in the form of wood chips which were impregnated with toxic substances was ground into wood powder so that it could be burned in a power station. The European Court found that the waste was being subjected to pre-processing since these operations did not purge the wood of its toxic qualities or have the effect of transforming the waste into a product analogous to a raw material. The processing resulted in making the waste suitable for burning in a power station. This was therefore an example of pre-processing prior to recovery.

Similarly, in the *Castle Cement* case[69] (see para **2.83**), fuel produced by mixing waste solvents and in the *Scottish Power* case[70], fuel produced by drying and pelletising sewage sludge, were classified by the courts as waste throughout their recovery and final use since:

- the relevant operations did not alter the nature of the wastes from which they were derived;
- the fuels retained their environmentally threatening nature;
- the processes were therefore pre-treatment processes; and
- the objectives of the WFD were achieved only when the fuels had been safely burned and the energy generated had been utilised.

Examples of Recovery Operations that will Result in 'End of Waste'

2.64 The WFD definition of recovery includes not only processes where a material is actually substituting other materials, but also processes preparing a waste material in such a way that it no longer involves waste-related risks and is ready to be used as a raw material in other processes. Generally speaking, the point of completion of a recovery operation may be considered to be the moment where a useful input for further processing, not representing any waste-specific risks to health and the environment, becomes available. Specific

69 Case No: CO/2635/2000 *Castle Cement v Environment Agency* (22nd March 2001) [2001] EWHC Admin 224.

70 *Scottish Power Generation Ltd v Scottish Environment Protection Agency* 2005 SLT 641.

legislation on EoW criteria may determine a particular point where waste becomes non-waste.

There are two examples of where the recovery operations will result in substances and objects losing their waste status. The first is a recycling operation and the second is an operation which prepares discarded substances and objects for re-use.

Recovery operations, which make the waste suitable for re-use, are often very minor operations. A substance or object which becomes waste may not need very much to be done to it in order to rid it of its waste status (eg, a sorting operation may identify those substances and objects which can still be used for their original purpose). It will still be consistent with the WFD not to regulate the substance or object as waste – provided there is certainty that the substance or object can be used in the same way it was used before it became waste, it will have been recovered and should not be regarded as waste.

In the case of *Environment Agency v Thorn International UK Ltd*[71], the Divisional Court considered the case of electrical equipment collected by distributors of electrical equipment under new-for-old schemes, where the collected old equipment was then bought and refurbished for resale by a repair company. The Court held that the items bought by the repair company were not waste even though they needed repair before resale. The judgment turned on an assessment of the particular facts of the case and the Environment Agency does not propose to derive a broad principle from it; rather, the Agency intends to restrict the application of this judgment to cases where it can be shown that, where discarded electrical and electronic equipment is obtained for refurbishment, there is a very high degree of probability that all the items selected will be repaired and reused. The Agency's position is as follows:

- Waste electrical and electronic equipment (WEEE) remains waste until it has been completely recovered. Recovery will commonly involve recycling of materials or dismantling items to obtain parts, but can also occur, in limited circumstances, where WEEE is subject to a competent assessment and segregated for immediate reuse or for reuse following repair. If WEEE is segregated for reuse, the reusable WEEE will only cease to be waste at the time of segregation if the holders have an auditable system that ensures that almost all of the items will in fact be reused (following repair if needed).
- If an overseas refurbisher is contracted the home contractor must be able to clearly identify the refurbisher and have a mechanism for ensuring that the equipment will be refurbished and reused. In assessing whether repair is probable, the Agency will take account of the type and condition of equipment, the way it is contained and transported, and the basis for considering that there is a very high probability that almost all the segregated equipment will in fact be repaired and reused.

71 *Environment Agency v Thorn International* [2008] WLR (D) 219.

A recovery operation can also occur when a substance is recycled. 'Recycling' is defined in Article 3(17) of the WFD. A similar definition is used in the Packaging Waste Directive[72] as follows:

> 'the reprocessing in a production process of the waste materials for the original purpose or for other purposes including organic recycling but excluding energy recovery.'

The Packaging Waste Directive definition was analysed in some detail in the *Mayer Parry* case[73]. Although this case pre-dates the definition of recycling in Article 3(17) of the WFD and the provisions in Article 6(1) and (2) enabling the adoption of EU-wide end of waste criteria, the similarities between the two recycling definitions means that it is still of some use in considering what may be considered recycling. The reprocessing of the waste involves a transformation of the waste into a new material or product, which possesses characteristics comparable to those of the material from which the waste was derived. Recycling is seen as a process, which does more than change the nature or composition of the substance – it results in an entirely new material or product, which can no longer be regarded as waste. The environmental risks presented have been dealt with and the substance has been brought back into the productive cycle.

In the *Niselli* case[74], the court made clear that it is only in exceptional circumstances that recycled substances, like products, will be waste. It is not possible to distinguish products derived from recycled material from those made from primary raw materials:

> 'subject to the case where the products obtained are in their turn abandoned, the point at which the materials in question cease to be classified as "waste" cannot be fixed at an industrial or commercial stage subsequent to their reprocessing into steel products, because, from that point, they can hardly be distinguished from other steel products made from primary raw materials.'

It is therefore important to determine the point at which waste streams are considered to be recycled. In *Niselli*, the court held that secondary raw materials derived from scrap metal and used in steel making continued to be classified as waste until they had actually been recycled into steel products:

> 'In the earlier phases, they cannot yet be regarded as recycled, since the reprocessing has not yet been concluded.'

Does the 'End-of-Waste' Status Need to be Demonstrated?

2.65 Article 6(1) WFD does not foresee a specific point in the chain where EoW has to be demonstrated. Measures adopted pursuant to Article 6(2) WFD

72 Council Directive 94/62/EC of 20 December 1994 on packaging and packaging waste.
73 On 19 June 2003 the European Court delivered its judgment on Case C-444/00 *Mayer Parry Recycling Ltd.*
74 See Case C-457/02 *Niselli.*

may however determine specific timelines and conditions. For example, in Council Regulation (EU) No 333/2011 on EoW criteria for scrap metal, the transfer of possession from one holder (the 'producer' of EoW material) to another holder is a legal condition for reaching the EoW status. Note that it is the material producer, ie the person who first transfers the material to another person as non-waste, who is responsible for providing evidence that EoW criteria have been fulfilled via the statement of conformity.

CAN WASTE DERIVED FUEL CEASE TO BE WASTE PRIOR TO USE AS FUEL?

2.66 In the UK, the end-of-waste question that has been raised more than any other is whether a particular waste derived fuel has ceased to be waste before it is used as a fuel. The main reason why this is a particular issue is that the requirements of the Waste Incineration Directive (WID) mean that waste which is used as fuel is subject to stringent technical standards, which have financial implications for the businesses burning them. This affects not only those businesses that burn substances which are waste, but those that market waste derived fuels. If it is possible to process a waste in some way so that it is no longer subject to control under the WID, the economic value of the substance will be enhanced. Processing waste so that it loses its waste status therefore offers the prospect of a much cheaper option for those intending to recover energy from waste.

The question is whether this is consistent with the objectives of the WFD. The Directive ensures disposal and recovery operations are carried out safely by subjecting them to controls and it also seeks to ensure better use of resources by encouraging use of waste. If controls are dispensed with too easily, it is clear the Directive will be undermined.

There is also the risk that this would favour one disposal route over others, which may bring greater environmental benefits. The controls of the WID will result in benefits to the environment as emissions are strictly controlled. For waste streams that are well suited for use as a fuel, if it is possible to avoid those controls too easily, there will be less impetus to invest in WID-compliant plant. The overall environmental outcome will therefore be worse. If incineration is unregulated, energy recovery may be prioritised over alternative uses for the waste. Again this would undermine the WFD.

This is not to say that waste derived fuel will never cease to be waste before its use, but it is clear that something more than pre-treatment is needed to be certain that the aims of the WFD are not undermined if waste controls are allowed to be dispensed with before the substance is used.

The court in the *OSS* case suggested that the Government and the Environment Agency should provide practical guidance for those affected by this question. In response to the court's suggestion, the Environment Agency developed a national end-of-waste protocol for 'the production and use of processed fuel oil from waste lubricating oils'. The protocol, which was notified to the European Commission in compliance with Article 6(4) of the WFD and the Technical Standards Directive (98/34/EC), is available on the Environment Agency's website. However, this guidance is also part of the exercise to fulfil the court's suggestion.

Although the *OSS* case was about waste lubricating oil being used as a fuel, the competent authorities consider that the three-part test it sets out can be used to determine whether other wastes have ceased to be waste before being burned as fuel.

As to the application of the test, the first practical question is how to distinguish a distinct marketable product from a marketable waste. Many waste fuels are marketed as such – no-one would set out to produce them from primary materials and in that sense they are not regarded as products. They are burnt subject to waste legislation but are not subject to certain other legislation, for example, the REACH Regulation.

If the recovery process, which is carried out, is mere pre-treatment it will be clear that the substance remains waste. If the recovery process is not mere pre-treatment, the process must produce a distinct substance for which there is a market for use as a fuel. A distinct substance needs to be reclaimed from the waste to escape the suggestion that the recovery process is only completed when the energy is actually recovered by burning.

In the *Saetti* case, it was noted that various factors were not relevant in assessing whether petcoke was waste because the petcoke was an intended product of the production process. One such factor was that special measures were necessary to protect the environment when petcoke was used as a fuel. This is a reflection of the general principle that the environmental impact of a substance does not itself determine waste status.

However, where you start with a substance that has been discarded, the WFD requirement is to ensure that the waste is recovered in a manner which ensures protection of the environment and human health. The factors, which apply to residues but not to intended products, also apply to substances that result from recovery processes.

In the *Total France* case[75], it was argued that heavy fuel oil should be considered waste by virtue of the environmental risks its use presents. The Advocate General rejected this argument, explaining that environmental risks are indicative of an intention to discard but that other factors could take precedence where you have a 'wanted product' like heavy fuel oil. In its judgment the court concluded that 'heavy fuel oil sold as a combustible fuel, does not constitute waste within the meaning [of the WFD], where it is exploited or marketed on economically advantageous terms and is capable of actually being used

75 On 24 June 2008 the European Court delivered its judgment on Case C-188/07 *Commune de Mesquer v Total France SA, Total International Ltd.*

as a fuel without requiring further processing'. It is important that end-of-waste criteria recognise the need to show that, aside from its environmental characteristics, the fuel genuinely constitutes a 'wanted product' in the same way as any comparator.

Generally of course, comparisons by themselves do not provide a means to distinguish waste from non-waste fuels. As the European Court stated in the *Arco Chemie* case:

> 'An ordinary fuel may be burnt without regard to environmental standards without thereby becoming waste, whereas substances which are discarded may be recovered as fuel in an environmentally responsible manner and without substantial treatment yet still be classified as waste.'

Comparators need to be appropriate and relevant so they provide good evidence that the submission to a common form of treatment is not evidence of waste recovery.

For example, where a substance is returned to its original state, the relevant comparator is clear. It is usually a matter of removing contamination from the waste so that it is returned to its original state. In this situation, it is usually quite straightforward to decide whether the substance has been processed sufficiently to be regarded as an ordinary fuel. There will not be any features, which distinguish it from primary resources. It would not make sense to compare it to another less environmentally friendly fuel as a means to avoid having to process the waste to deal with the contaminants.

Another situation where the relevant comparator is clear is where a waste is transformed into an ordinary fuel such as gasoil. This may occur in the context of a regeneration process. Or it may be possible to generate biodiesel from animal by-products or waste cooking oil.

The product of a recovery process can be used for several different purposes, including use as a fuel. If it is not considered waste when used for some other purpose, can it at the same time be considered waste when used as a fuel? It may seem surprising that something may be regarded as waste depending on how it is used, but this reflects the approach of the European Court. This is because sometimes substances may be used in circumstances where it is clear they are being discarded. If you produce an excess of a substance that burns well, a decision to burn that substance may reflect a wish to simply get rid of it.

In other cases, it may be more difficult to determine the correct comparator and so industry and regulators should work together. Industry will try and demonstrate why the substance should be considered a marketable product. This will involve comparison with the alternatives used in the intended market – in terms of price, performance and environmental impact. Some fuels are not used in practice because of the damage they cause to industrial plant and it would not make sense to attempt to use those fuels as comparators. The substance must of course meet the regulatory controls that apply to the use of primary fuels – for instance the air quality legislation applying to the plant used.

A key test is whether the holder of waste discards or intends to discard the waste. This is to be inferred from all the circumstances – not just the fact that the recovered substance is being burned. The key question is whether the substance in question is commonly used as a fuel in that particular context. If so, that might suggest that the use as a fuel does not amount to evidence that the substance is being discarded. Otherwise, its use as a fuel would be evidence that the substance is being discarded and is waste.

CASE SUMMARIES

Case C-1/03 Van de Walle (Contaminated Soil)

2.67 On 7 September 2004 the European Court delivered its judgment on Case C-1/03 *Paul Van de Walle, Daniel Laurent, Thierry Mersch and Texaco Belgium SA*.

The Belgian court sought a preliminary ruling on the interpretation of Council Directive 75/442. The commercial lessee of a service station Texaco (T), a petroleum company, had engaged a manager Mersch under an 'operating agreement' which provided that the land, building, equipment and moveable property for the operation were made available to Mersch by Texaco. Mersch then operated the service station on his own behalf. As a result of defects in the service station's storage facilities an accidental leak of hydrocarbons occurred. The owner of the property adjacent to the service station complained that water contaminated with hydrocarbons was leaking into the cellar of its building. Texaco subsequently terminated their manager's contract for serious negligence and, whilst disclaiming liability, decontaminated the soil. The adjacent landowner paid for further decontamination as Texaco's actions were considered to have been insufficient. Criminal proceedings were brought against Texaco alleging a prohibited abandonment of waste and claiming damages. In the course of those proceedings, the national court asked the European Court of Justice to rule as to (1) whether hydrocarbons which had been spilled unintentionally were 'waste' within the meaning of Art1(a) of the Directive; (2) whether the soil thus contaminated was also to be considered as waste even when it had not been excavated; (3) whether Texaco was to be regarded as the producer or holder of such waste within the meaning of Art1(b) and Art1(c).

Held, giving a preliminary ruling, that (1) the mere fact of accidental spillage should not prevent a substance, which caused contamination from being considered waste within the meaning of Art1(a) of the Directive. T was considered to have discarded the hydrocarbons since they could not be reused without processing and so were unintentionally produced, Case C-418/97 *ARCO Chemie Nederland Ltd v Minister van Volkshuisvesting, Ruimtelijke Ordening en Milieubeheer* [2002] QB 646 and Case C-9/00 *Palin Granit Oy v Lounais-Suomen Ymparistokeskus* [2002] 1 WLR 2644 followed. (2) The contaminated soil was also waste since it could not be separated from the land contaminated by the hydrocarbons until decontamination operations had been carried out. The fact that the soil had not been excavated had no bearing on its classification as waste. (3) The hydrocarbons were in the possession of the manager of the service station since, at the time they became waste, he had had them in stock for use in the operation of the service station. Thus the manager was the holder of the hydrocarbons for the purposes of Art1(c). However, if it became apparent that the leakage was due to T disregarding its contractual obligations, T could be found to have 'produced waste' within the meaning of Art1(b) and would thus be regarded as the holder of that waste.

> 'Hydrocarbons which are unintentionally spilled and cause soil and groundwater contamination are waste within the meaning of Article 1(a) of Council Directive 75/442/EEC of 15 July 1975 on waste, as amended by Council Directive 91/156/EEC of 18 March 1991. The same is true for soil contaminated by hydrocarbons, even if it has not been excavated. In circumstances such as those in the main proceedings, the petroleum undertaking which supplied the service station can be considered to be the holder of that waste within the meaning of Article 1(c) of Directive 75/442 only if the leak from the service station's storage facilities which gave rise to the waste can be attributed to the conduct of that undertaking.'

Zanetti – Federal Republic of Germany

2.68

(a) 'The concept of waste, within the meaning of Article 1 of Directive 75/442/EEC and Article 1 of Directive 78/319/EEC1[A], is not to be understood as excluding substances and objects which are capable of economic re-utilization. The concept does not presume that the holder disposing of a substance or object intends to exclude all economic re- utilization of the substance or object by others.'

(b) 'National legislation which defines waste as excluding substances and objects which are capable of economic re-utilization is not compatible with Council Directives 75/442 and 78/319.'

A On toxic and dangerous gases

In the judgment delivered on 10 May 1995 the ECJ held that their finding on *Zanetti* was not affected by the amendments made to Directive 75/442/EEC by Council Directive 91/156/EEC (Case C-422/92 *Commission of the European Communities v Federal Republic of Germany*, paragraphs 22–23).

Joined Cases C-304/94, C-330/94 and C-2224/95 Tombesi – Savini

2.69 On 25 June 1997 the European Court delivered its judgment on Joined Cases C-304/94, C-330/94 and C-2224/95 *Criminal proceedings against Euro Tombesi and Others*. In doing so the European Court held that:

Tombesi and others were charged with transporting, discharging, disposing of, or incinerating urban and special waste produced by third parties without obtaining prior authorisation by the competent region. The criminal court made a reference to the ECJ concerning the interpretation of Council Directive 75/442 and Council Regulation 259/93.

The Court held that Council Directive 75/442 Art 3(1) required Member States to adopt measures which encouraged the prevention or reduction of waste production, together with waste recovery aimed at extracting secondary raw materials or using the waste as an energy source. Council Regulation 259/93 Art 2(a) set down a common definition of the concept of waste which was directly applicable, even to shipments within a Member State. National legislation which excluded substances and objects capable of economic re-utilisation from their definition of waste was incompatible with the relevant EC law. The system of supervision, which included the requirement for authorisation and registration of undertakings concerned with waste collection, disposal or recovery, was intended to include all objects or substances discarded by their owners, even where they had a commercial value and were collected on a commercial basis for the purpose of recycling, reclamation or re-use:

> 'The concept of "waste" in Article 1 of Council Directive 75/442/EEC of 15 July 1975 on waste, as amended by Council Directive 91/156/EEC of 18 March 1991, referred to in Article 1(3) of Council Directive 91/689/EEC of 12 December 1991 on hazardous waste and Article 2(a) of Council Directive Regulation (EEC) No 259/93 of 1 February 1993 on the supervision and control of shipments of waste within, into and out of the European Community, is not to be understood as excluding substances and objects which are capable of economic reutilization, even if the materials in question may be the subject of a transaction or quoted on public or private commercial lists. In particular, a deactivation process intended merely to render waste harmless, landfill tipping in hollows or embankments and waste incineration constitute disposal or recovery operations falling within the scope of the above mentioned Community rules. The fact that a substance is classified as a re-usable residue without its characteristics or purpose being defined is irrelevant in that regard. The same applies to the grinding of a waste substance.'

Case C-129/96 Wallonie

2.70 On 18 December 1997 the European Court delivered its judgment on Case C-129/96 *Inter-Environment Wallonie v Règion Wallone*. In doing so, the European Court held that in relation to the question on the definition of waste:

> 'A substance in not excluded from the definition of waste in Article 1(a) of council Directive 75442, as amended, by the mere fact that it directly or indirectly forms and integral part of an industrial process'.

The European Court of Justice was asked for a preliminary ruling on the circumstances in which a plan or programme within the meaning of Directive 2001/42 which had not been the subject of an environmental assessment as required by that Directive could provisionally be kept in force.

Directive 91/676 had the objective of reducing water pollution from agricultural use of nitrates, while Directive 2001/42 required the assessment of the effects of certain plans and programmes on the environment. When implementing the 1991 Directive, the respondent in the main proceedings, a regional government in Belgium, had adopted an order amending the law on the sustainable management of nitrogen in agriculture. The applicants sought annulment of the order, claiming it was a programme within the meaning of the 2001 Directive and so should have been the subject of an environmental assessment. Regional government maintained that the programme did not fall within the 2001 Directive. The Belgian court referred the matter to the ECJ, which ruled that an action programme adopted pursuant to Art 5(1) of the 1991 Directive was in principle a plan or programme covered by Art 3(2)(a) of the 2001 Directive. The Belgian court then held that the order was a plan or programme within Art 3(2)(a), and as it had not been subjected to an environmental assessment as required by the 2001 Directive and the ECJ had not limited the temporal effects of its ruling, it ought to be annulled. However, the Belgian court was concerned that annulling the order with retroactive effect would mean there was no measure transposing the 1991 Directive in Y's region and so Belgium would be failing to comply with its obligations under that Directive. It therefore asked the ECJ whether, in an action for annulment of a national measure constituting a plan or programme within the meaning of the 2001 Directive, which was adopted in breach of the obligation under that Directive to carry out a prior environmental assessment, but which implemented the 1991 Directive appropriately, some of the past effects of the measure could be maintained under national provisions until the date on which measures designed to remedy the irregularity came into force.

> 'A substance is not excluded from the definition of waste in Article 1(a) of Council Directive 75/442, as amended, by the mere fact that it directly or indirectly forms an integral part of an industrial production process'.

Case C-418/97 and C-419/97 ARCO Chemie

2.71 On 15 June 2000 the European Court delivered its judgment on Case C-418/97 and C-419/97 *ARCO Chemie Nederland Ltd etc*.

The Dutch court referred questions to the ECJ on the interpretation of the Waste Directive 75/422, as amended by the Waste Directive 91/156, and considered whether industrial by-products which were going to be used as fuel were 'waste' within the meaning of the Directive. Annex IIB to the Directive specified what was meant by the disposal and recovery of waste and listed relevant operations as they occurred 'in practice'. The listed operations included 'use principally as a fuel or other means to generate energy'. The ECJ was required to rule on:

(1) whether it might be inferred from the fact that the substances underwent an operation referred to in Annex IIB to the Directive that the substances had been discarded so as to enable them to be regarded as 'waste' for the purposes of the Directive;

(2) in determining whether a substance had been so discarded, whether it was necessary to take into consideration,

(a) the fact that those substances were commonly regarded as waste or the fact that the substances might be recovered in an environmentally responsible manner for use as a fuel without substantial treatment;
(b) whether the use as a fuel related to a main product or a by-product;
(c) the fact that the substance had already undergone, prior to burning, operations to render it suitable for use as a fuel, and if so, whether that operation might be regarded as a recovery of waste only if it was expressly mentioned in Annex IIB, or whether it might also be so regarded if it was analogous to an operation mentioned in that Annex.

The preliminary ruling said that:

(1) in the light of the aim of the Directive, the concept of waste could not be interpreted restrictively. Also the scope of the term 'waste' relied on the meaning attributed to the term 'discard'. In the light of Art 4 and Annexes IIA and IIB, the term 'discard' included the disposal or recovery of a substance or an object. It did not necessarily follow from the fact that certain methods of disposing or recovering waste were described in the Annexes that any substance treated by one of those methods was to be regarded as waste. Therefore, the inclusion of a process in Annex IIB did not give rise to an inference that a substance had been discarded.
(2) (a) the fact that use as a fuel was a common method of recovering waste and the fact that substances were commonly regarded as waste might be taken as evidence that the holder had discarded or intended to discard them within the meaning of Art 1(a). However, any judgment had to be made in the light of all the circumstances, with regard to the Directive's aim of protecting human health and the environment; (b) the fact that a substance used as a fuel was the residue of the manufacturing process of another substance, that no use for that substance other than disposal could be envisaged, that the composition of the substance was not suitable for the use made of it, or that special environmental precautions had to be taken when it was used might be regarded as evidence that the holder had discarded the substance or intended to discard it within the meaning of Art 1(a). Regard was given to all the circumstances and the aim of the Directive; and (c) the fact that a substance was the result of a recovery operation within the meaning of Annex IIB was only one of the factors to be taken into consideration for the purposes of determining whether the substance was still waste. Regard was to be had to all the circumstances and the aim of the Directive.

Case C-418/97

> '1. It may not be inferred from the mere fact that a substance such as LUWA-bottoms undergoes an operation listed in Annex IIB to Council Directive 75/442/EEC of 15 July 1975 on waste, as amended by Council Directive 91/156/EEC of 18 March 1991, that that substance has been discarded so as to enable it to be regarded as waste for the purposes of that Directive.

2. For the purpose of determining whether the use of a substance such as LUWA-bottoms as a fuel is to be regarded as constituting discarding, it is irrelevant that that substance may be recovered in an environmentally responsible manner for use as fuel without substantial treatment.
3. The fact that that use as fuel is a common method of recovering waste and the fact that that substance is commonly regarded as waste may be taken as evidence that the holder has discarded that substance or intends or is required to discard it within the meaning of Article 1(a) of Directive 75/442, as amended by Directive 91/156. However, whether it is in fact waste within the meaning of the Directive must be determined in the light of all the circumstances, regard being had to the aim of the Directive and the need to ensure that its effectiveness is not undermined.
4. The fact that a substance used as fuel is the residue of the manufacturing process of another substance, that no use for that substance other than disposal can be envisaged, that the composition of the substance is not suitable for the use made of it or that special environmental precautions must be taken when it is used may be regarded as evidence that the holder has discarded that substance or intends or is required to discard it within the meaning of Article 1(a) of that Directive. However, whether it is in fact waste within the meaning of the Directive must be determined in the light of all the circumstances, regard being had to the aim of the Directive and the need to ensure that its effectiveness is not undermined.'

Case C-419/97

'1. It may not be inferred from the mere fact that a substance such as wood chips undergoes an operation listed in Annex IIB to Directive 75/442, as amended by Directive 91/156, that that substance has been discarded so as to enable it to be regarded as waste for the purposes of the Directive.
2. The fact that a substance is the result of a recovery operation within the meaning of Annex IIB to that Directive is only one of the factors which must be taken into consideration for the purpose of determining whether that substance is still waste, and does not as such permit a definitive conclusion to be drawn in that regard. Whether it is waste must be determined in the light of all the circumstances, by comparison with the definition set out in Article 1(a) of Directive 75/442, as amended by Directive 91/156, that is to say the discarding of the substance in question or the intention or requirement to discard it, regard being had to the aim of the Directive and the need to ensure that its effectiveness is not undermined.
3. For the purpose of determining whether the use of a substance such as wood chips as a fuel is to be regarded as constituting discarding, it is irrelevant that that substance may be recovered in an environmentally responsible manner for use as fuel without substantial treatment.

4. The fact that that use as fuel is a common method of recovering waste and the fact that that substance is commonly regarded as waste may be taken as evidence that the holder has discarded that substance or intends or is required to discard it within the meaning of Article 1(a) of Directive 75/442, as amended by Directive 91/156. However, whether it is in fact waste within the meaning of that Directive must be determined in the light of all the circumstances, regard being had to the aim of the Directive and the need to ensure that its effectiveness is not undermined.'

Case C-9/00 Palin Granit Oy

2.72 On 18 April 2002 the European Court delivered its judgment on Case C-9/00 *Palin Granit Oy and Vehmassalon kansanwerveystyön kuntayhtymän hallitus*.

The Supreme Administrative Court of Finland, made a reference to the ECJ for a preliminary ruling under Art 234 EC during appeal proceedings which challenged the granting of an environmental licence to Palin Granit to operate a granite quarry. The issue was whether leftover stone resulting from quarrying was to be regarded as waste and this depended on the question as to which authority could grant Palin Granite the required licence. The national court asked if leftover stone resulting from stone quarrying was to be regarded as waste within the meaning of Council Directive 75/442 Art 1(a) taking account of:

(1) whether the leftover stone was stored on a site adjoining the place of quarrying to await subsequent use or was stored further away;
(2) the fact that the leftover stone was the same composition as the basic rock from which it was quarried, maintaining its composition regardless of how long or how it was stored;
(3) the fact that the leftover stone was harmless to human health and the environment;
(4) that it was intended to transfer the leftover stone away from the storage site for use, for example for landfill or breakwaters, without further processing.

The subsidiary question asked was if attention should be paid to how definite the plans were for such use and to how soon it would take place.

In a preliminary ruling it was held that the concept of waste could not be interpreted restrictively[76]. In the instant case, the leftover stone might have had a subsidiary use but was an extraction residue. It was therefore classified as waste within the meaning of the Directive. The uncertainty surrounding the proposed uses of the leftover stone and the impossibility of reusing it entirely

76 Case C-418/97 *ARCO Chemie Nederland Ltd v Minister van Volkshuisvesting, Ruimtelijke Ordening en Milieubeheer* [2002] QB 646.

supported that conclusion that its place of storage and the conditions under which it was kept were irrelevant. The composition of the leftover stone also had no impact on its definition as waste, nor did the fact that it posed no risk to public health or the environment.

> '1. The holder of leftover stone resulting from stone quarrying which is stored for an indefinite length of time to await possible use discards or intends to discard that leftover stone, which is accordingly to be classified as waste within the meaning of Council Directive 75/442/ EEC of 15 July 1975 on waste.
>
> 2. The place of storage of leftover stone, its composition and the fact, even if proven, that the stone does not pose any real risk to human health or the environment are not relevant criteria for determining whether the stone is to be regarded as waste.'

Case C-444/00 Mayer Parry Recycling Ltd

2.73 On 19 June 2003 the European Court delivered its judgment on Case C-444/00 *Mayer Parry Recycling Ltd.*

For the purposes of Council Directive 94/62 Art 3(7), 'recycling' excluded the reprocessing of metal packaging waste when it was transformed into a secondary raw material but included the reprocessing of such waste when it was used to produce ingots, sheets or coils of steel.

The Court of Justice of the European Communities considered a reference from the English High Court which sought a preliminary ruling as to the meaning of 'recycling' under the Waste Directive 75/442 and Council Directive 94/62 on packaging waste. Mayer processed packaging waste and ferrous scrap metal for resale to steelmakers and had applied to EA for official recognition as a reprocessor which status would have been economically advantageous to Mayer. EA had refused the application upon the grounds that Mayer's processes were not considered to be 'recycling' pursuant to the Directives. Mayer sought judicial review which proceedings were stayed pending the ruling of the ECJ.

Held, giving a preliminary ruling, that whilst Council Directive 75/442 set out the general scheme of the waste management regime, Council Directive 94/62 sought to regulate a specific waste category. Accordingly the former was relevant to all instances not expressly otherwise provided for by the latter. According to the Directives, recycling had the same objectives as recovery namely the ultimate conservation of natural resources. The recycling of metal packaging waste therefore had to produce a new product suitable for the production of metal packaging. Only some of M's processes would satisfy this requirement.

> '1. "Recycling" within the meaning of Article 3(7) of European Parliament and Council Directive 94/62/EC of 20 December 1994 on packaging and packaging waste is to be interpreted as not including the reprocessing of metal packaging waste when it is transformed into

a secondary raw material such as material meeting the specifications of Grade 3B, but as covering the reprocessing of such waste when it is used to produce ingots, sheets or coils of steel.

2. That interpretation would be no different if the concepts of "recycling" and "waste" referred to by Council Directive 75/442/ EEC of 15 July 1975 on waste were taken into account.'

Case C-114/01 AvestaPolarit Chrome Oy

2.74 On 11 September 2003 the European Court delivered its judgment on Case C-114/01 *AvestaPolarit Chrome Oy*.

The Finnish national court referred questions regarding the meaning of waste within Council Directive 75/442 Art 2(1)(b). AvestarPolarit carried out mining operations on a site and applied to the Finnish national authority for an environmental licence to cover its operations. The licensing authority granted the licence subject to certain conditions, some of which covered measures to be taken regarding the disposal of waste and waste products. The licence categorised leftover rock and ore dressing sand (sand used to clean or remove waste material from the ore) as waste. AvestaPolarit appealed against that decision and argued that the rock and sand should not be characterised as waste as they were not taken from the site but were stored on the mining site, that both products would have uses at some later stage when mining ceased on the site and that, in the case of rock, there had been no change to the composition of the material. The Finnish courts referred to the European Court of Justice two questions; (1) whether leftover rock and ore dressing sand was to be regarded as waste within the meaning of the Directive and, (2) whether controls made under 'other legislation' referred to in the Directive included controls made under national legislation or whether that referred only to Community legislation.

Held, that (1) where the leftover rock and ore dressing sand were discarded or were intended to be discarded they were to be classified as waste unless they were used lawfully for necessary filling in of the mine galleries with sufficient guarantees as to the identification and actual use of those substances for that purpose. The place of storage of materials, their composition and the fact that they did not pose any real risk to human health or the environment are not relevant criteria for determining whether such materials are to be regarded as waste. (2) The words 'already covered by other legislation' used in Art 2(1)(b) to exclude certain types of waste from the scope of the Directive can include national legislation so long as that legislation related to the management of waste within the meaning of the Directive and resulted in a level of environmental protection that was at least equivalent to that envisaged by the Directive.

'1. In a situation such as that at issue in the main proceedings, the holder of leftover rock and residual sand from ore-dressing operations from the operation of a mine discards or intends to discard those substances, which must consequently be classified as waste within the meaning of

Council Directive 75/442/EEC of 15 July 1975 on waste, as amended by Council Directive 91/156/EEC of 18 March 1991, unless he uses them lawfully for the necessary filling in of the galleries of that mine and provides sufficient guarantees as to the identification and actual use of the substances to be used for that purpose.

2. In so far as it does not constitute a measure of application of Directive 75/442, as amended by Directive 91/156, and in particular Article 11 of that Directive, national legislation must be regarded as other legislation within the meaning of Article 2(1)(b) of that Directive covering a category of waste mentioned in that provision, if it relates to the management of that waste as such within the meaning of Article 1(d) of Directive 75/442, and if it results in a level of protection of the environment at least equivalent to that aimed at by that Directive, whatever the date of its entry into force.'

Case C-235/02 Mario Antonio Saetti and Andrea Frediani (Petroleum Coke)

2.75 On 14 January 2004 the European Court delivered its judgment on Case C-235/02 *Mario Andonio Saetti and Andrea Frediani.*

The national court sought a preliminary ruling as to whether petroleum coke used as a fuel to generate electricity was 'waste' under Council Directive 75/442 Art 1(a). Criminal proceedings were commenced against Saetti, a director of the refinery, for failing to hold the requisite waste management permits in relation to certain activities carried on at the refinery. It was discovered during a technical survey carried out at the refinery on the instructions of the public prosecutor that petroleum coke, which was produced in the course of crude oil refining operations, was being used as a fuel for generating electricity.

Held, determining the preliminary issues, that petroleum coke did not constitute 'waste' for the purposes of Council Directive 75/442 as it was not a residue of production but rather was intentionally produced as a fuel during refinery operations, Case C-9/00 *Palin Granit Oy v Lounais-Suomen Ymparistokeskus* [2002] 1 WLR 2644 considered. The costs of producing petroleum coke were lower than the costs of other fuels that could be used for generating electricity as the coke was fully consumed during the generation process and all electricity surplus to the refinery's needs was sold. The use of petroleum coke for energy production purposes, a standard waste recovery method, was irrelevant, given that the purpose of the refinery was to produce different forms of fuel from crude oil. Further, the fact that the coke was considered to be waste by the refinery was not sufficient to justify a finding that it was waste for the purposes of the Directive.

> 'Petroleum coke which is produced intentionally or in the course of producing other petroleum fuels in an oil refinery and is certain to be used as fuel to meet the energy needs of the refinery and those of other industries does not constitute waste within the meaning of Council Directive 75/442/EEC of 15 July 1975 on waste, as amended by Council Directive 91/156/EEC of 18 March 1991.'

Case C-1/03 Van de Walle (Contaminated Soil)

2.76 On 7 September 2004 the European Court delivered its judgment on Case C-1/03 *Paul Van de Walle, Daniel Laurent, Thierry Mersch and Texaco Belgium SA*.

The Belgian court sought a preliminary ruling on the interpretation of Council Directive 75/442. The commercial lessee of a service station Texaco (T), a petroleum company, had engaged a manager Mersch under an 'operating agreement' which provided that the land, building, equipment and moveable property for the operation were made available to Mersch by Texaco. Mersch then operated the service station on his own behalf. As a result of defects in the service station's storage facilities an accidental leak of hydrocarbons occurred. The owner of the property adjacent to the service station complained that water contaminated with hydrocarbons was leaking into the cellar of its building. Texaco subsequently terminated their manager's contract for serious negligence and, whilst disclaiming liability, decontaminated the soil. The adjacent landowner paid for further decontamination as Texaco's actions were considered to have been insufficient. Criminal proceedings were brought against Texaco alleging a prohibited abandonment of waste and claiming damages. In the course of those proceedings, the national court asked the European Court of Justice to rule as to (1) whether hydrocarbons which had been spilled unintentionally were 'waste' within the meaning of Art 1(a) of the Directive; (2) whether the soil thus contaminated was also to be considered as waste even when it had not been excavated; (3) whether Texaco was to be regarded as the producer or holder of such waste within the meaning of Art 1(b) and Art 1(c).

Held, giving a preliminary ruling, that, (1) the mere fact of accidental spillage should not prevent a substance which caused contamination from being considered waste within the meaning of Art 1(a) of the Directive. T was considered to have discarded the hydrocarbons since they could not be reused without processing and so were unintentionally produced, Case C-418/97 *ARCO Chemie Nederland Ltd v Minister van Volkshuisvesting, Ruimtelijke Ordening en Milieubeheer* [2002] QB 646 and Case C-9/00 *Palin Granit Oy v Lounais-Suomen Ymparistokeskus* [2002] 1 WLR 2644 followed. (2) The contaminated soil was also waste since it could not be separated from the land contaminated by the hydrocarbons until decontamination operations had been carried out. The fact that the soil had not been excavated had no bearing on its classification as waste. (3) The hydrocarbons were in the possession of the manager of the service station since, at the time they became waste, he had had them in stock for use in the operation of the service station. Thus the manager was the holder of the hydrocarbons for the purposes of Art 1(c). However, if it became apparent that the leakage was due to T disregarding its contractual obligations, T could be found to have 'produced waste' within the meaning of Art 1(b) and would thus be regarded as the holder of that waste.

> 'Hydrocarbons which are unintentionally spilled and cause soil and groundwater contamination are waste within the meaning of Article 1(a) of Council Directive

75/442/EEC of 15 July 1975 on waste, as amended by Council Directive 91/156/EEC of 18 March 1991. The same is true for soil contaminated by hydrocarbons, even if it has not been excavated. In circumstances such as those in the main proceedings, the petroleum undertaking which supplied the service station can be considered to be the holder of that waste within the meaning of Article 1(c) of Directive 75/442 only if the leak from the service station's storage facilities which gave rise to the waste can be attributed to the conduct of that undertaking.'

Case C-457/02 Antonio Niselli

2.77 On 11 November 2004 the European Court delivered its judgment on Case C-457/02 *Antonio Niselli.*

'1. The definition of "waste" in the first subparagraph of Article 1(a) of Council Directive 75/442/EEC of 15 July 1975 on waste, as amended by Council Directive 91/156/EEC of 18 March 1991 and by Commission Decision 96/350/EC of 24 May 1996, cannot be construed as covering exclusively substances or objects intended for, or subjected to, the disposal or recovery operations mentioned in Annexes IIA and IIB to that Directive or in the equivalent lists, or to which their holder intends or is required to subject them.

2. The meaning of "waste" for the purposes of the first subparagraph of Article 1(a) of Directive 75/442, as amended by Directive 91/156 and by Decision 96/350, is not to be interpreted as excluding all production or consumption residues which can be or are reused in a cycle of production or consumption, either without prior treatment and without harm to the environment, or after undergoing prior treatment without, however, requiring a recovery operation within the meaning of Annex IIB to that Directive.'

Case C-416/02 European Commission v Kingdom of Spain

2.78 On 8 September 2005 the European Court delivered its judgment on Case C-416/02 *Commission of the European Communities v Kingdom of Spain.*

Livestock Effluent (Manure/slurry)

'89. As the United Kingdom Government correctly maintains in its statement of intervention, livestock effluent may, on the same terms, fall outside classification as waste, if it is used as soil fertiliser as part of a lawful practice of spreading on clearly identified parcels and if its storage is limited to the needs of those spreading operations.

90. Contrary to the Commission's submission, it is not appropriate to limit that analysis to livestock effluent uses as a fertiliser on land forming part of the same agricultural holding as that which generated the effluent. As the Court has already held, it is possible for a substance

not to be regarded as waste within the meaning of Directive 75/442 if it is certain to be used to meet the needs of economic operators other than that which produced it (see, to that effect, *Saetti and Frediani*, cited above, paragraph 47).'

Animal Carcasses

'91 On the other hand, the analysis which allows, in certain situations, a production residue to be regarded not as waste but as a by-product or a raw material reusable within the continuing process of production cannot apply to carcasses of animals being reared, where those animals died on the farm and were not slaughtered for human consumption.

92 Such carcasses cannot, as a general rule, be reused for the purposes of human consumption. They are regarded by Community legislation, in particular by Council Directive 90/667/EEC of 27 November 1990 laying down the veterinary rules for the disposal and processing of animal waste, for its placing on the market and for the prevention of pathogens in feedstuffs of animal or fish origin and amending Directive 90/425/EEC (OJ 1990 L363, p.51; Directive 90/667 was repealed, after the date fixed by the reasoned opinion, by Article 37 of Regulation (EC) No 1774/2002 of the European Parliament and of the Council of 3 October 2002 laying down health rules concerning animal by-products not intended for human consumption (OJ 2002 L273, p.1)), as "animal waste" and, furthermore, as waste within the category of "high-risk materials", which must be processed in factories approved by the Member States or disposed of by incineration or burial. Directive 90/667 provides that such matter may be used in feedstuffs for animals which do not enter the human food chain, but only by virtue of authorisations issued by the Member States and under the veterinary supervision of the competent authorities.

93 In no case may carcasses of animals which die on the farm in question therefore be used in conditions which would enable them not to be defined as waste within the meaning of Directive 75/442. The holder of those carcasses is certainly obliged to discard them, with the result that that matter must be regarded as waste.'

Case C-176/05 KVZ retec GmBH

2.79 On 1 March 2007 the European Court delivered its judgment on Case C-176/05 *KVZ retec GmBH v Republik Österrich*.

'Under Article 1(3)(a) of Council Regulation (EEC) No 259/93 of 1 February 1993 on the supervision and control of shipments of waste within, into and out of the European Community, as amended by Commission Regulation (EC) No 2557/2001 of 28 December 2001, the shipment of meat-and-bone meal classified as waste on account of a requirement or intention to discard it, which is destined for recovery only and listed in Annex II to that Regulation, is excluded from the scope of the provisions of the Regulation except as provided for in Article 1(3)(b) to (e), Article 11 and Article 17(1) to (3) thereof. However, it is for the national court to ensure that that shipment takes place in compliance with the requirements arising from the provisions of Regulation (EC) No 1774/2002 of the European Parliament and of the Council of 3 October 2002 laying down health rules concerning animal by-products not intended for human consumption, as amended by Commission Regulation (EC) No 808/2003 of 12 May 2003, amongst which those of Articles 7, 8 and 9 and of Annex II to the Regulation may prove to be relevant.'

Case C-252/05 Thames Water Utilities Limited

2.80 On 10 May 2007 the European Court delivered its judgment on Case C-252/05 *Thames Water Utilities Limited.*

'1. Waste water which escapes from a sewerage network maintained by a statutory sewerage undertaker pursuant to Council Directive 91/271/EEC of 21 May 1991 concerning urban waste water treatment and the legislation enacted to transpose that Directive constitutes waste within the meaning of Council Directive 75/442/EEC of 15 July 1975 on waste, as amended by Council Directive 91/156/EEC of 18 March 1991.

2. Directive 91/271 is not "other legislation" within the meaning of Article 2(1)(b) of Directive 75/442, as amended by Directive 91/156. It falls to the national court to ascertain whether, in accordance with the criteria set out in the present judgment, the national rules may be regarded as "other legislation" within the meaning of that provision. Such is the case if those national rules contain precise provisions organising the management of the waste in question and if they are such as to ensure a level of protection of the environment equivalent to that guaranteed by Directive 75/442, as amended by Directive 91/156, and, more particularly, by Articles 4, 8 and 15.

3. Directive 91/271 cannot be considered, as regards the management of waste water which escapes from sewerage network, to be special legislation (a lex specialis) vis-à-vis Directive 75/442, as amended by Directive 91/156, and cannot therefore be applied pursuant to Article 2(2) of Directive 75/442.'

Case C-188/07 Commune de Mesquer (Municipality of Mesquer) v Total France SA, Total International Ltd

2.81 On 24 June 2008 the European Court delivered its judgment on Case C-188/07 *Commune de Mesquer v Total France SA, Total International Ltd.*

'1. A substance such as that at issue in the main proceedings, namely heavy fuel oil sold as a combustible fuel, does not constitute waste within the meaning of Council Directive 75/442/EEC of 15 July 1975 on waste, as amended by Commission Decision 96/350/EC of 24 May 1996, where it is exploited or marketed on economically advantageous terms and is capable of actually being used as a fuel without requiring prior processing.

2. Hydrocarbons accidentally spilled at sea following a shipwreck, mixed with water and sediment and drifting along the coast of a Member State until being washed up on that coast, constitute waste within the meaning of Article 1(a) of Directive 75/442, as amended by Decision 96/350, where they are no longer capable of being exploited or marketed without prior processing.
3. For the purposes of applying Article 15 of Directive 75/442, as amended by Decision 96/350, to the accidental spillage of hydrocarbons at sea causing pollution of the coastline of a Member State:
   - The national court may regard the seller of those hydrocarbons and charterer of the ship carrying them as a producer of that waste within the meaning of Article 1(b) of Directive 75/442, as amended by Decision 96/350, and as a "previous holder" for the purposes of applying the first part of the second indent of Article 15 of that Directive, if that court, in the light of the elements which it alone is in a position to assess, reaches the conclusion that seller-charterer contributed to the risk that the pollution caused by the shipwreck would occur, in particular if he failed to take measures to prevent such an incident, such as measures concerning the choice of ship;
   - If it happens that the cost of disposing of the waste produced by an accidental spillage of hydrocarbons at sea is not borne by the International Oil Pollution Compensation Fund, or cannot be borne because the ceiling for compensation for that accident has been reached, and that, in accordance with the limitations and/or exemptions of liability laid down, the national law of a Member State, including the law derived from international agreements, prevents that cost from being borne by the ship-owner and/or the charterer, even though they are to be regarded as "holders" within the meaning of Article 1(c) of Directive 75/442, as amended by Decision 96/350, such a national law will then, in order to ensure that Article 15 of that Directive is correctly transposed, have to make provision for that cost to be borne by the producer of the product from which the waste thus spread came. In accordance with the "polluter pays" principle, however, such a producer cannot be liable to bear that cost unless he has contributed by his conduct to the risk that the pollution caused by the shipwreck will occur.'

Case C-283/07 European Commission v Italian Republic

2.82 On 22 December 2008 the European Court delivered its judgment on Case C-283/07 *Commission of the European Communities v Italian Republic.*

'(1) The Italian Republic, by adopting and maintaining in force provisions such as:

- Article 1(25) to (27) and (29)(a) of Law No 308 of 15 December 2004 delegating power to the Government to reform, coordinate and supplement legislation in environmental matters and direct implementation measures, and
- Article 1(29)(b) of Law No 308 of 15 December 2004 and Articles 183(1)(s) and 229(2) of Legislative Decree No 152 of 3 April 2006 laying down rules in environmental matters,

under which certain scrap intended for use in iron and steel and metallurgical activities and high-quality refuse-derived fuel (RDF-Q) respectively are excluded a priori from the scope of the Italian legislation on waste transposing Council Directive 75/442/EEC of 15 July 1975 on waste, as amended by Council Directive 91/156/EEC of 18 March 1991, the Italian Republic has failed to fulfil its obligations under Article 1(a) of that Directive.'

Castle Cement Ltd v Environment Agency 22 March 2001 [2001] EWHC Admin 224

2.83 Castle Cement sought a declaration that the Environment Agency (E) had exceeded its powers in issuing notices pursuant to the Hazardous Waste Incineration Direction 1998 so as to require reduced air emission limits and increased monitoring of air emissions in connection with the burning of 'Cemfuel' during the cement manufacturing process. Castle Cement contended that Cemfuel was not 'waste' for the purposes of Council Directive 74/442 as amended, but a raw fuel, and that the burning of it as an alternative to coal did not amount to the burning of 'hazardous waste' for the purposes of the 1998 Directive. The court held that Cemfuel remained waste within the meaning of the Waste Framework Directive and for the purposes of Directive 2000/76 on the incineration of waste, and had been rightly treated as such by Environment Agency.

Castle Cement, a cement manufacturer, sought a declaration that E had exceeded its powers in issuing notices pursuant to Council Directive 94/67. The notices required reduced air emission limits and increased monitoring of air emissions in connection with the use of 'Cemfuel', an alternative fuel manufactured from waste materials, which was burnt during the cement manufacturing process. Castle Cement contended that Cemfuel was not 'waste' for the purposes of Council Directive 75/442, as amended, but was a raw fuel, and that burning it as an alternative to coal did not amount to the burning of 'hazardous waste' for the purposes of Council Directive 75/442. E maintained that the burning of Cemfuel constituted part of the waste recovery process and that Cemfuel accordingly retained its character as waste until burnt.

Held, refusing the application, that Cemfuel remained 'waste' within the meaning of Council Directive 75/442. The fact that a 'complete' recovery process had been performed did not necessarily mean that any particular substance ceased to be waste. It was clear that the aim on the part of both the manufacturer of Cemfuel and Castle Cement was to 'discard' Cemfuel by utilising it as a fuel and having regard to its potentially toxic nature, it would undermine the effectiveness of the Directive if it were not categorised as waste, Case C-418/97 *ARCO Chemie Nederland Ltd v Minister van Volkshuisvesting, Ruimtelijke Ordening en Milieubeheer* [2002] QB 646 applied, and *Mayer Parry Recycling Ltd v Environment Agency* [1999] 1 CMLR 963 considered.

Scottish Power Generation Ltd v Scottish Environment Protection Agency 2005 SLT 98 P

2.84 The operators of a power station burning coal and waste derived fuel (P), petitioned for judicial review of a decision by the Scottish Environment Protection Agency to issue a notice of variation of the power station's authorisation, and sought declaration that waste derived fuel was not waste as defined by the Waste Council Directive 75/442 Art 1(a). The waste derived fuel was produced from the sewage emanating from the public sewerage network operated by a water company. P submitted that the notice was based on an error of law, was ultra vires and should be reduced as it proceeded on the basis that the power station fell within the ambit of the Waste Incineration (Scotland) Regulations 2003, reg 3(2) when it was not a 'waste incineration installation' since the activities carried on there did not include the co-incineration of waste. P argued that waste derived fuel was not 'waste' as defined in Art 1(a) of the Directive, nor was it waste as defined in the Pollution Prevention and Control (Scotland) Regulations 2000, Sch 1, Part I, s 5(1). P contended that the fuel was produced intentionally as a result of a complex industrial process and was designed for a particular purpose in response to a demand in the electricity generating market. Its production was subject to a specification and to quality control and rejection procedures, its use replaced fossil fuels which would otherwise be burnt and in line with EC environmental policy, that which was produced was certain to be used as fuel at the power station and the burning of it was strictly regulated.

Held, refusing to grant declaration that waste derived fuel was not waste as defined by Art 1(a) of the Directive and putting the case out by order, that (1) any material was capable of constituting 'waste' within the meaning of Art 1(a) of the Directive in the event that it was discarded, which in that context, included in particular the use of waste as a means of generating electricity, the recycling of waste and the reclamation from it of substances which were intended for re-use and it followed that waste might be of economic value which its holder might be said to 'discard' notwithstanding that it was put to some commercially valuable use. (2) There was no doubt that the sewage sludge received at the treatment centre was 'waste' within the meaning of the Directive and while P maintained that the operation carried out at the centre was a complete recovery operation, the waste derived fuel could not be used under the same conditions of environmental protection as coal, or without any greater danger of harm to human health or the environment, thus as the processing of the sludge could not be regarded as a complete recovery operation, its status as 'waste' was not altered within the meaning of the Directive.

Environment Agency v Thorn International UK Ltd [2008] EWHC 2595 (Admin)

2.85 The appellant agency appealed against a decision that refrigerators and computers held by the respondent company (T) were not 'waste' within the meaning of the Environmental Protection Act 1990, ss 33 and 75 and Directive 75/442.

T had bought the appliances, which consumers had exchanged for new appliances from retailers. T had selected the appliances with a view to repairing and reselling them. While awaiting repair, some appliances were left outside in T's yard. The agency said that those items were waste, requiring T to have a waste disposal licence.

The agency argued that an item was discarded by the consumer and became waste, at the moment the consumer made a contract with the retailer to exchange his old appliance for a new one; it did not cease to be waste by virtue of the process of selection by T. It only ceased to be waste once it had been repaired.

Held: Appeal dismissed

The instant case was not one where the items in question were undoubtedly waste from the time they were no longer wanted by the consumer. In Case C-418/97 *ARCO Chemie Nederland Ltd v Minister van Volkshuisvesting, Ruimtelijke Ordening en Milieubeheer* [2002] QB 646 and *OSS Group Ltd v Environment Agency* [2007] EWCA Civ 611, [2007] Bus LR 1732, the items had undoubtedly become waste. Although they could be converted into new items, they were no longer required for their original purpose. By contrast, the items in the instant case were to be repaired for their original use, *ARCO* and *OSS* considered. They had therefore not been discarded, Case C-129/96 *Inter-Environnement Wallonie ASBL v Region Wallonie* [1997] ECR I-7411 considered. The fact that an item which was undoubtedly waste remained waste until its character was changed by recycling did not establish a rule of law that any item requiring repair was waste until that repair was concluded.

OSS Group Ltd v Environment Agency [2007] EWCA Civ 611

2.86 The appellant company (O) appealed against a decision ([2006] EWHC 3023 (Admin)) that waste material prepared for use as a fuel only ceased to be 'waste' for the purposes of Directive 2006/12 when it was actually burnt.

O sold products that were derived from waste. They marketed their products on the basis that they had ceased to be waste as a result of a particular recovery process. O's product was 'clean fuel oil', which it sold competitively to customers on the basis that it was not waste and that therefore customers would not have to be subject to the waste regime. That regime had significant cost implications. The court below agreed with the respondent Environment Agency that O was subject to the waste regime contained in Directive 2006/12, Directive 2000/76 and Directive 75/439. Permission to appeal was granted on the limited issue of whether a lubricating oil, not originally used as fuel, that became waste could thereafter be burnt as anything other than waste.

The agency submitted that if the intended end use was combustion, the materials remained waste until that was completed, regardless of any prior treatment.

Appeal allowed.

The agency's view was too narrow and O's products could be burnt otherwise than as waste. A review of European case law did not demonstrate a logical coherence. A fundamental problem was the European Court of Justice's professed adherence to the definition in Directive 2006/12 Art1(a), even where it could be of no practical relevance. The subjective 'intention to discard' might be a useful guide to the status of material in the hands of the original producer but it was hard to apply to the status of material in the hands of someone who bought it for recycling or reprocessing, or put it to some other valuable use. In no ordinary sense was such a person discarding or getting rid of the material. His intention was precisely the opposite. The ECJ had understandably held that a material did not cease to be waste merely because it had passed into the possession of someone who intended to put it to a new use. That should not be because it still met the Art 1(a) definition in their hands but rather because, in accordance with the aims of the Directive, material that was originally waste needed to continue to be treated as such until acceptable recovery or disposal had been achieved. Lip-service had been paid by the court to the discarding test but case law revealed a series of objective indicators derived from the policy of the Directive.

Chapter 3

Controlled Waste

CONTROLLED WASTE

3.1 Controlled waste is waste that is subject to legislative control in either its handling or disposal. Controlled waste is a term which applies only to the UK, although other countries have a similar concept.

DEFINITIONS WITHIN CONTROLLED WASTE REGULATIONS

3.2 Schedule 1 of both the Controlled Waste (England and Wales) Regulations 2012[1] and the Controlled Waste (Northern Ireland) Regulations define what is considered to be household, industrial and commercial waste.

The Northern Ireland Regulations provide a list of waste under the different classifications, while the England and Wales legislation instead defines the places where waste may arise from and classifies the waste by the place of production. The legislation also looks at the nature of the waste producing activities and again classifies the waste according to these.

There are differences between the definitions – so they are explained in detail here. It is worth noting that as each part of the UK can set the definition separately and at different times, it is important to look at the latest amendments to be sure that you have the correct one – also to look closely at legislation in each part of the UK to ensure that you understand the differences.

ENGLAND AND WALES

3.3 The Controlled Waste (England and Wales) Regulations 2012 revoke and replace the Controlled Waste Regulations 1992[2] and apply to both England and Wales.

They classify waste as:

- Household;
- Industrial;
- Commercial.

They allow for local authorities who are waste collection authorities under the Environmental Protection Act 1990 to charge under s 45, for the act of the disposal of waste which arises from a wider range of non-domestic premises than the 1992 Regulations permitted.

They also clarify the definitions and classifications to bring them in line with recent legislation and also provide that certain litter and refuse is to be treated under Part 2 of the Environmental Protection Act 1990 in the same way as waste collected under s 45 of the Act.

Environmental Protection Act 1990

3.4 Part 2 of the Environmental Protection Act 1990 is the primary legislation relating to waste.

1 The Controlled Waste (England and Wales) Regulations 2012, SI 2012/811.
2 Controlled Waste Regulations 1992, SI 1992/588.

Section 75 provides for definitions. Subsection (2) defines waste and subsections (5), (6) and (7) classify certain types of waste as 'household waste', 'industrial waste' and 'commercial waste'.

Subsection (8) gives power to make Regulations providing that waste of a prescribed description is to be treated for the purposes of provisions of Part 2 as being or not being household, industrial or commercial waste.

Classification affects the duties and powers of local authorities under Part 2 of the 1990 Act in relation to waste. In particular, local authority duties and powers in s 45 in relation to collection of waste and charging for collection and subsequent disposal are defined by the classification.

The principal EU measure on waste is Directive 2008/98/EC and some provisions of Part 2 of the 1990 Act implement parts of that Directive.

These provisions use the term 'controlled waste', which is defined in s 75(4) of the 1990 Act to depend on the terms 'household waste', 'industrial waste' and 'commercial waste'. The Regulations ensure that the term 'controlled waste' has the same effect as the meaning of 'waste' in the Directive. The definition of waste is discussed in Chapter 2.

In s 45 of the1990[3] Act, the duties and powers of waste collection authorities in relation to the collection of waste are set out. Subsection (3) provides that no charge can be made for the collection of household waste except in cases prescribed in regulations (in which case a reasonable charge may be made).

Subsection (4) empowers a reasonable charge to be made for the collection and disposal of waste other than household waste. Regulation 4 of the 1992 Regulations prescribed certain types of waste in relation to which a collection charge could be made under s 45(3). Regulations 5 to 7A of the 1992 Regulations provided that certain descriptions of waste were to be treated as industrial and commercial waste for the purposes of Part 2 of the 1990 Act.

Waste Which is not to be Treated as Household Waste, Industrial Waste or Commercial Waste

3.5 Waste, which is not Directive waste, is not to be treated as household waste, industrial waste or commercial waste for the purposes of Part 2 of the Act.

The following waste (where it is Directive waste) is not to be treated as household waste, industrial waste or commercial waste for the purposes of Part 2 of the Act:

(a) sewage, sludge or septic tank sludge which is treated, kept or disposed of (otherwise than by means of mobile plant) within the curtilage of a sewage treatment works as an integral part of the operation of those works;

3 Section 45(1) and (2) of the Environmental Protection Act 1990.

(b) sludge, which is supplied or used in accordance with the Sludge (Use in Agriculture) Regulations 1989;
(c) septic tank sludge, which is used on agricultural land within the meaning of those Regulations.

Animal by-products[4] (where they are Directive waste), which are collected and transported[5], are not to be treated as household waste, industrial waste or commercial waste for the purposes of s 34 of the Act.

In this regulation:

(a) 'animal by-products' has the meaning given in Article 3(1) of the Animal By-Products Regulation;
(b) 'the Animal By-Products Regulation' means Regulation (EC) No 1069/2009 of the European Parliament and of the Council laying down health rules as regards animal by-products and derived products not intended for human consumption and repealing Regulation (EC) No 1774/2002 (Animal by-products Regulation).

Definitions Within Controlled Waste Regulations

Sites From Which Waste Could Arise

3.6 Schedule 1 goes into detail to describe the different types of sites where waste could arise:

- 'camp site' means land on which tents are pitched for the purpose of human habitation and land the use of which is incidental to that purpose;
- 'charity shop' means a hereditament used wholly or mainly for the sale of goods donated to a charity where the proceeds of sale (after deduction of expenses) are applied for the purposes of a charity;
- 'composite hereditament' has the same meaning as in section 64(9) of the Local Government Finance Act 1988(2) (inherited property);
- 'factory' has the meaning given in section 175 of the Factories Act 1961(3);
- 'general medical practitioner' means a person registered in the General Practitioner Register kept by the General Medical Council under section 34C of the Medical Act 1983(4);
- 'gypsies and travellers';
- in relation to England, has the same meaning as in regulation 2 of the Housing (Assessment of Accommodation Needs) (Meaning of Gypsies and Travellers) (England) Regulations 2006;
- in relation to Wales, has the same meaning as in regulation 2 of the Housing (Assessment of Accommodation Needs) (Meaning of Gypsies and Travellers) (Wales) Regulations 2007;
- 'not for profit body' means a body which, by virtue of its constitution or any enactment:

4 Controlled Waste (England and Wales) Regulations 2012, SI 2012/811, reg 3(3).
5 In accordance with Article 21(1)–(3) of Regulation (EC) No 1069/2009, the Animal By-Products Regulation.

  - is required (after payment of outgoings) to apply the whole of its income, and any capital which it expends, for charitable or public purposes; and
  - is prohibited from directly or indirectly distributing among its members any part of its assets (otherwise than for charitable or public purposes);
- 'place of worship' means a hereditament exempt from local non-domestic rating by virtue of paragraph 11 of Schedule 5 to the Local Government Finance Act 1988;
- 'scientific research association' has the meaning given in section 469 of the Corporation Tax Act 2010;
- 'vessel' includes a hovercraft within the meaning of section 4(1) of the Hovercraft Act 1968.

Sources of Household, Industrial and Commercial Waste

3.7 In Schedule 1, the sources of waste types are considered and the waste they produce is classified according to a table. See Table below on Classification by Place of Production. The waste is classified according to where it is produced. The second column describes the places where the waste is produced and the third column describes how the waste should be treated – but subject to any comments in the fourth column.

For example in No 4, the description of the waste is a vehicle or vessel used wholly for the purposes of living accommodation so the waste is to be treated as household waste. However, if the vehicle or vessel is used in the course of a business for the provision of self-catering accommodation then the waste is to be treated as commercial waste.

Table – Classification by Place of Production

No	*Description*	*Classification*	*Exceptions*
1	Private storage premises used wholly or mainly for the storage of articles of domestic use	Household waste	
2	Land belonging to or used wholly or mainly in connection with domestic property or a caravan where waste from that property or caravan is to be treated as household waste	Household waste	

No	*Description*	*Classification*	*Exceptions*
3	A private garage	Household waste	Where the garage has a floor area exceeding 25m^2 or is not used wholly or mainly for the accommodation of a private motor vehicle, the waste is to be treated as commercial waste
4	A vehicle or vessel used wholly for the purposes of living accommodation	Household waste	Where the vehicle or vessel is used in the course of a business for the provision of self-catering accommodation, the waste is to be treated as commercial waste
5	A place of worship	Household waste	
6	A residential hostel which provides accommodation only to persons with no other permanent address or who are unable to live at their permanent address	Household waste	
7	A penal institution	Household waste	
8	A charity shop selling donated goods originating from domestic property	Household waste	
9	A caravan or mobile home site for gypsies and travellers	Household waste	
10	Premises used wholly or mainly for public meetings	Household waste	
11	Domestic property used in the course of a business for the provision of self-catering accommodation	Commercial waste	

No	*Description*	*Classification*	*Exceptions*
12	A caravan—	Commercial waste	
	(a) used in the course of a business for the provision of self-catering accommodation, or (b) which is not allowed to be used for human habitation throughout the year by virtue of a licence or planning permission		
13	Premises occupied by a charity and wholly or mainly used for charitable purposes	Commercial waste	Where the waste is from a place of worship or from premises used wholly or mainly for public meetings, it is to be treated as household waste
14	A camp site or a tent pitched on land other than a camp site	Commercial waste	Where the waste is from domestic premises at a camp site, it is to be treated as household waste
15	A royal palace	Commercial waste	
16	Premises occupied by a club, society or any association of persons in which activities are conducted for the benefit of the members	Commercial waste	

No	*Description*	*Classification*	*Exceptions*
17	Premises occupied by— (a) a court; (b) a government department; (c) a local authority; (d) a person appointed by or under any enactment to discharge public functions; (e) a body incorporated by Royal Charter	Commercial waste	Waste classified as household waste or industrial waste because it is from a place— (a) otherwise described in this table (except for entry 27); or (b) described in section 75(5) or (6) of the Act (household waste or industrial waste)
18	A hotel	Commercial waste	
19	Any part of a composite hereditament used for the purposes of a trade or business	Commercial waste	
20	A market or fair	Commercial waste	
21	The practice of a general medical practitioner	Commercial waste	
22	A workshop or similar premises which is not a factory only because— (a) those working there are not employees; or (b) the work carried on there is not carried on by way of trade or for purposes of gain	Industrial waste	Where the principal activities at the premises are computer operations or the copying of documents by photographic or lithographic means the waste is to be treated as commercial waste
23	Waste from a laboratory	Industrial waste	
24	Waste from a scientific research association	Industrial waste	
25	Waste from premises used for the breeding, boarding or stabling of animals	Industrial waste	

No	*Description*	*Classification*	*Exceptions*
26	Waste imported into England or Wales	Industrial waste	
27	Directive waste from a place (including any vehicle, vessel or aircraft) not otherwise described in this table or in section 75(5) or (7) of the Act (household waste and commercial waste)	Industrial waste	

Nature of Waste Activities Producing Waste

3.8 This describes waste, which is to be treated as household waste, commercial waste or industrial waste because of its nature or the activity which produces it, notwithstanding the place where it is produced.

This is in the form of a table where the second column of the table describes the nature of the waste or the activity, which produces it. The waste must be treated for the purposes of Part 2 of the Act as the type of waste mentioned in the third column, subject to any exception in the fourth column.

For example, waste which is classified as waste oil, waste solvent or scrap metal is treated as industrial waste. However, if it is produced in the following places then it would be treated as household waste:

(a) a residential home;
(b) domestic property, a caravan or land belonging to or used in connection with such a place;
(c) a private garage which has a floor area of 25m^2 or less or is used wholly or mainly for the accommodation of a private motor vehicle;
(d) private storage premises used wholly or mainly for the storage of articles of domestic use;
(e) a vehicle or vessel used wholly for the purposes of living accommodation, except where it is used in the course of a business for the provision of self-catering accommodation.

Household waste is waste from:

- Collection of litter or refuse[6];
- Requirement to ensure that highways and roads are kept clean[7].

6 EPA 1990, s 89(1)(a),(c) or (f).
7 EPA 1990, s 89(2).

Commercial waste is waste from:

- Collection of litter and refuse[8];
- Street cleaning[9].

Industrial waste comes from all the other activities defined in the table. See Table on Classification by Waste Producing Activity.

Table – Classification by Nature of Waste or Activity Producing Waste

No	*Description*	*Classification*	*Exceptions*
1	Waste arising from the discharge of duties under section 89(1) (a), (c) or (f) of the Act (duties to collect litter and refuse)	Household waste	
2	Waste arising from the discharge by a local authority of its duty under section 89(2) of the Act (duty to ensure highways and roads are kept clean)	Household waste	
3	Waste arising from the discharge of duties under sections 89(1)(d), 92(9), 92C(3) and 93 of the Act (duties and powers in relation to the collection of litter and refuse)	Commercial waste	
4	Waste collected under section 22(3) of the Control of Pollution Act 1974 (street cleaning)	Commercial waste	
5	Waste arising from the discharge of duties under section 89(1) (b) and (e) of the Act (duties to collect litter and refuse)	Industrial waste	

8 EPA 1990, ss 89(1)(d), 92C(3) and 93.
9 Control of Pollution Act 1974.

No	*Description*	*Classification*	*Exceptions*
6	Waste arising from the discharge by the Secretary of State or the Welsh Ministers of the duty under section 89(2) of the Act (duty to ensure highways and roads are kept clean)	Industrial waste	
7	Hazardous waste arising from the following activities carried on at premises used for the purposes of a trade or business— (a) mixing or selling paints; (b) sign writing; (c) laundering or dry cleaning; (d) developing photographic film or making photographic prints; (e) selling petrol, diesel fuel, paraffin, kerosene, heating oil or similar substances; (f) selling pesticides, herbicides or fungicides	Industrial waste	
8	Waste oil, waste solvent or scrap metal	Industrial waste	Where the waste is produced at the following places it is to be treated as household waste— (a) a residential home; (b) domestic property, a caravan or land belonging to or used in connection with such a place;

No	*Description*	*Classification*	*Exceptions*
8	Waste oil, waste solvent or scrap metal – *contd*	*Industrial waste – contd*	(c) a private garage which has a floor area of 25m^2 or less or is used wholly or mainly for the accommodation of a private motor vehicle; (d) private storage premises used wholly or mainly for the storage of articles of domestic use; (e) a vehicle or vessel used wholly for the purposes of living accommodation, except where it is used in the course of a business for the provision of self-catering accommodation
9	Waste from construction or demolition works, including preparatory works	Industrial waste	The waste is to be treated as household waste for the purposes of section 34(2) and (2A) of the Act only (disapplication of section 34(1) and duty on the occupier of domestic property to transfer household waste only to an authorised person or for authorised transport purposes)
10	Septic tank sludge	Industrial waste	The waste is to be treated as household waste for the purposes of section 34(2) and (2A) of the Act only

No	*Description*	*Classification*	*Exceptions*
11	Sewage	Industrial waste	
12	Clinical waste and offensive waste	Industrial waste	Clinical waste and offensive waste produced at domestic property, a residential home, a caravan or a vehicle or vessel used wholly for the purposes of living accommodation is to be treated as household waste, except that where such a vehicle or vessel is used in the course of a business for the provision of self-catering accommodation, such waste is to be treated as commercial waste. Clinical waste and offensive waste collected under section 22(3) of the Control of Pollution Act 1974 or section 89, 92(9), 92C(3) or 93 of the Act is to be treated as household waste or commercial waste in accordance with entries 1 to 6

Charges for Collection and Disposal of Household Waste

3.9 This is defined under Schedule 1, para 4[10] of the Controlled Waste (England and Wales) Regulations 2012. It is important to note that paragraphs 5 to 8 only apply to England and not Wales.

Again the approach has been to create a Table, which in column 2 describes the household waste, column 3 lists whether the charges can be made for collection and column 4 where a further change can be made for disposal. There is a difference between England and Wales when looking at waste from premises forming part of a university, school or other educational establishment. In both England and Wales a charge can be made for the collection of this type of waste. In Wales a charge can also be made for the disposal of this type of waste – but in England a disposal charge can only be made for waste that satisfies paragraph 8.

10 Controlled Waste (England and Wales) Regulations 2012, SI 2012/811, Sch 1, para 4:

4. Household waste for which collection and disposal charges may be made

(1) The second column of the table in this paragraph describes waste which is household waste (by virtue of section 75(5) of the Act) or waste which is to be treated as household waste (by virtue of paragraph 2 or 3).

(2) The third column indicates which household waste is prescribed for the purposes of section 45(3) of the Act (cases in respect of which a charge for collection may be made).

(3) Subject to sub-paragraphs (5) and (8), the fourth column indicates which waste must be treated as commercial waste for the purposes of the following provisions of the Act—

(a) section 45(4) to the extent that it concerns liability to pay a reasonable charge for disposal of commercial waste and the recovery of such a charge; and

(b) section 52(9) (which entitles waste disposal authorities to reimbursement by waste collection authorities in relation to costs incurred in arranging the disposal of commercial and industrial waste).

(4) Sub-paragraphs (5) to (8) apply in England only.

(5) Nothing in sub-paragraph (3) has the effect of making a person liable to pay a charge under section 45(4) of the Act for the disposal of waste if the first and second conditions are satisfied.

(6) The first condition is that no charge would have been payable under section 45(4) of the Act for the disposal of that waste had it been collected under section 45(1) of the Act before the coming into force of these Regulations.

(7) The second condition is that the waste is from a hereditament as regards which the person is subject to a non-domestic rate under section 43(1) of the Local Government Finance Act 1988(19), for which the chargeable amount payable is calculated in accordance with section 43(4A)(a) of that Act(20).

(8) In relation to entry 15 of the table (waste from premises forming part of a university, school or other educational establishment), nothing in sub-paragraph (4) has the effect of making a person liable to pay a charge under section 45(4) of the Act for the disposal of waste from premises forming part of—

(a) a school maintained by a local authority (within the meaning of section 142(1) of the School Standards and Framework Act 1998(21)),

(b) an institution within the further education sector (within the meaning of section 91 of the Further and Higher Education Act 1992(22)), or

(c) an Academy (within the meaning of section 1(10) of the Academies Act 2010(23)),

if, immediately before the coming into force of these Regulations, the collection of waste from those premises was being undertaken pursuant to arrangements made under section 45(1) of the Act.

(8) In relation to entry 15 of the table (waste from premises forming part of a university, school or other educational establishment), nothing in sub-paragraph (4) has the effect of making a person liable to pay a charge under section 45(4) of the Act for the disposal of waste from premises forming part of:

(a) a school maintained by a local authority (within the meaning of section 142(1) of the School Standards and Framework Act 1998(21)),
(b) an institution within the further education sector (within the meaning of section 91 of the Further and Higher Education Act 1992(22)), or
(c) an Academy (within the meaning of section 1(10) of the Academies Act 2010(23)).

No	*Description of household waste*	*Collection charge*	*Disposal charge*
1	Any article of waste which exceeds 25kg in weight	Yes	No, subject to entries 11 to 17
2	Any article of waste which does not fit or cannot be fitted into— (a) a receptacle for household waste provided in accordance with section 46 of the Act; or (b) where no such receptacle is provided, a cylindrical container 750mm in diameter and 1m in length	Yes	No, subject to entries 11 to 17
3	Garden waste	Yes	No, subject to entries 11 to 17
4	Dead domestic pets	Yes	No, subject to entries 11 to 17
5	Waste oil or grease	Yes	No, subject to entries 11 to 17
6	Asbestos	Yes	No, subject to entries 11 to 17
7	Waste which may not be put into a receptacle provided under section 46 of the Act because of a notice served under that section	Yes	No, subject to entries 11 to 17
8	Waste from premises used wholly or mainly for public meetings	Yes	No
9	Clinical waste and offensive waste produced at a domestic property, a caravan or a vehicle or vessel used wholly for the purposes of living accommodation	Yes	No

No	*Description of household waste*	*Collection charge*	*Disposal charge*
10	Waste from a residential hostel which provides accommodation only to persons with no other permanent address or who are unable to live at their permanent address	Yes	No
11	Waste from a charity shop selling donated goods originating from domestic property	Yes	Yes, but only to the extent that the waste originated from non-domestic property
12	Waste from premises occupied by— (a) a community interest company (being a company which is registered as such with the registrar of companies), or (b) a charity or other not for profit body, which collects goods for re-use or waste to prepare for re-use from domestic property	Yes	Yes, but only to the extent that the waste originated from non-domestic property
13	Litter and refuse collected under section 89(1)(f) of the Act	Yes	Yes
14	Waste from a residential home or land belonging to or wholly or mainly used in connection with a residential home	Yes	Yes
15	Waste from premises forming part of a university, school or other educational establishment	Yes	In Wales: yes
			In England: yes, subject to paragraph 4(8)
16	Waste from premises forming part of a hospital or nursing home except for waste from a residential hostel forming such part which provides accommodation only to persons with no other permanent address	Yes	Yes
17	Waste from a penal institution	Yes	Yes

SCOTLAND[11]

3.10 Scotland uses the definition of waste contained in the Controlled Waste Regulations 1992.

Waste Not to be Treated as Household Waste[12]

3.11 Waste of the following descriptions shall not be treated as household waste for the purposes of section 33(2) (treatment, keeping or disposal of household waste within the curtilage of a dwelling): That is the area of land attached to a house and forming one enclosure with it.

(a) any mineral or synthetic oil or grease;
(b) asbestos; and
(c) clinical waste.

(2) Scrap metal shall not be treated as household waste for the purposes of section 34 at any time before 1st April 1993.

See **Chapter 11** – Scrap Metal Dealers.

Waste Not to be Treated as Industrial or Commercial Waste[13]

3.12 Some types of waste are not to be treated as industrial waste or commercial waste for the purposes of Part II of the Act—

(a) sewage, sludge or septic tank sludge which is treated, kept or disposed of (otherwise than by means of mobile plant) within the curtilage of a sewage treatment works as an integral part of the operation of those works
(b) sludge, which is supplied or used in accordance with the 1989 Regulations
(c) septic tank sludge, which is used in accordance with the 1989 Regulations

(2) Scrap metal shall not be treated as industrial waste or commercial waste for the purposes of section 34 at anytime before 1st April 1993.

Types of Household Waste for Which a Charge for Collection May be Made

3.13 Under Schedule 2, there is a list of waste that even though this is household waste then a charge can be levied for its removal.

11 Controlled Waste Regulations 1992, SI 1992/588.
12 SI 1992/588, Sch 3.
13 SI 1992/588, Sch 7.

1. Any article of waste which exceeds 25 kilograms in weight
2. Any article of waste which does not fit, or cannot be fitted into—
    a. receptacle for household waste provided in accordance with section 46; or
    b. where no such receptacle is provided, a cylindrical container 750 millimetres in diameter and 1 metre in length.
3. Garden waste.
4. Clinical waste from a domestic property, a caravan or from a moored vessel used wholly for the purposes of living accommodation.
5. Waste from a residential hostel, a residential home or from premises forming part of a university, school or other educational establishment or forming part of a hospital or nursing home.
6. Waste from domestic property or a caravan used in the course of a business for the provision of self-catering holiday accommodation.
7. Dead domestic pets.
8. Any substances or articles which, by virtue of a notice served by a collection authority under section 46, the occupier of the premises may not put into a receptacle for household waste provided in accordance with that section.
9. Litter and refuse collected under section 89(1)(f).
10. Waste from—
    a. in England and Wales, domestic property forming part of a composite hereditament;
    b. in Scotland, the residential part of part residential subjects.
11. Any mineral or synthetic oil or grease.
12. Asbestos.
13. Waste from a caravan which in accordance with any licence or planning permission regulating the use of the caravan site on which the caravan is stationed is not allowed to be used for human habitation throughout the year.
14. Waste from a camp site, other than from any domestic property on that site.
15. Waste from premises occupied by a charity and wholly or mainly used for charitable purposes, unless it is waste falling within paragraph 1 of Schedule 1.
16. Waste from a prison or other penal institution.
17. Waste from a hall or other premises used wholly or mainly for public meetings.
18. Waste from a royal palace.

Waste to be Treated as Industrial Waste

3.14 Schedule 3 contains a list of waste which is to be treated as industrial waste.

Subject to paragraph (2) and regulation 7[14], waste of the descriptions set out in Schedule 3 shall be treated as industrial waste for the purposes of Part II of the Act.

1. Waste from premises used for maintaining vehicles, vessels or aircraft, not being waste from a private garage to which paragraph 4 of Schedule 1 applies.

2. Waste from a laboratory.

3.(1) Waste from a workshop or similar premises not being a factory within the meaning of section 175 of the Factories Act 1961(**1**) because the people working there are not employees or because the work there is not carried on by way of trade or for purposes of gain.

(2) In this paragraph, "workshop" does not include premises at which the principal activities are computer operations or the copying of documents by photographic or lithographic means.

4. Waste from premises occupied by a scientific research association approved by the Secretary of State under section 508 of the Income and Corporation Taxes Act 1988.

5. Waste from dredging operations.

6. Waste arising from tunnelling or from any other excavation.

7. Sewage not falling within a description in regulation 7 which—

(a) is treated, kept or disposed of in or on land, other than by means of a privy, cesspool or septic tank;
(b) is treated, kept or disposed of by means of mobile plant; or
(c) has been removed from a privy or cesspool.

8. Clinical waste other than—

(a) clinical waste from a domestic property, caravan, residential home or from a moored vessel used wholly for the purposes of living accommodation;
(b) waste collected under section 22(3) of the Control of Pollution Act 1974; or
(c) waste collected under sections 89, 92(9) or 93.

9. Waste arising from any aircraft, vehicle or vessel, which is not occupied for domestic purposes.

10. Waste which has previously formed part of any aircraft, vehicle or vessel and which is not household waste.

11. Waste removed from land on which it has previously been deposited and any soil with which such waste has been in contact, other than:

(a) waste collected under section 22(3) of the Control of Pollution Act 1974; or
(b) waste collected under sections 89, 92(9) or 93.

14 7. Waste not to be treated as either industrial or commercial waste.

12. Leachate from a deposit of waste.

13. Poisonous or noxious waste arising from any of the following processes undertaken on premises used for the purposes of a trade or business:

(a) mixing or selling paints;
(b) sign writing;
(c) laundering or dry cleaning;
(d) developing photographic film or making photographic prints;
(e) selling petrol, diesel fuel, paraffin, kerosene, heating oil or similar substances; or
(f) selling pesticides, herbicides or fungicides.

14. Waste from premises used for the purposes of breeding, boarding, stabling or exhibiting animals.

15. (1) Waste oil, waste solvent or (subject to regulation 7(2)) scrap metal, other than:

(a) waste from a domestic property, caravan or residential home;
(b) waste falling within paragraphs 3 to 6 of Schedule 1.

(2) In this paragraph:

"waste oil" means mineral or synthetic oil which is contaminated, spoiled or otherwise unfit for its original purpose; and

"waste solvent" means solvent, which is contaminated, spoiled or otherwise unfit for its original purpose.

16. Waste arising from the discharge by the Secretary of State of his duty under section 89(2).

17. Waste imported into Great Britain.

18.(1) Tank washings or garbage landed in Great Britain.

(2) In this paragraph—

"tank washings" has the same meaning as in regulation 2(4) of the Control of Pollution (Landed Ships' Waste) Regulations 1987(5) ; and

"garbage" has the same meaning as in regulation 1(2) of the Merchant Shipping (Reception Facilities for Garbage) Regulations 1988(6).

(3) Waste of the following descriptions shall be treated as industrial waste for the purposes of Part II of the Act (except section 34(2))—

(a) waste arising from works of construction or demolition, including waste arising from work preparatory thereto;
(b) septic tank sludge not falling within regulation 7(1)(c).

Waste to be Treated as Commercial Waste

3.15 Schedule 4 lists waste that is to be treated as commercial waste.

1. Waste from an office or showroom.
2. Waste from a hotel within the meaning of:

   (a) in England and Wales, section 1(3) of the Hotel Proprietors Act 1956; and
   (b) in Scotland, section 139(1) of the Licensing (Scotland) Act 1976.

3. Waste from any part of a composite hereditament, or, in Scotland, of part residential subjects, which is used for the purposes of a trade or business.
4. Waste from a private garage, which either has a floor area exceeding 25 square metres or is not used wholly, or mainly for the accommodation of a private motor vehicle.
5. Waste from premises occupied by a club, society or any association of persons (whether incorporated or not) in which activities are conducted for the benefit of the members.
6. Waste from premises (not being premises from which waste is by virtue of the Act or of any other provision of these Regulations to be treated as household waste or industrial waste) occupied by—

   (a) a court;
   (b) a government department;
   (c) a local authority;
   (d) a body corporate or an individual appointed by or under any enactment to discharge any public functions; or
   (e) a body incorporated by a Royal Charter.

7. Waste from a tent pitched on land other than a campsite.
8. Waste from a market or fair.
9. Waste collected under section 22(3) of the Control of Pollution Act 1974.

Waste not to be Treated as Industrial or Commercial Waste

3.16 SI 1992/558, reg 7

(1) Waste of the following descriptions shall not be treated as industrial waste or commercial waste for the purposes of Part II of the Act—

(a) sewage, sludge or septic tank sludge which is treated, kept or disposed of (otherwise than by means of mobile plant) within the curtilage of a sewage treatment works as an integral part of the operation of those works;
(b) sludge, which is supplied or used in accordance with the 1989 Regulations;
(c) septic tank sludge, which is used in accordance with the 1989 Regulations.

(2) Scrap metal shall not be treated as industrial waste or commercial waste for the purposes of section 34 at anytime before 1st April 1993.

NORTHERN IRELAND

3.17 In Northern Ireland controlled waste was defined by the Controlled Waste Regulations (Northern Ireland) 2002, NISR 2002/248 (the 2002 Regulations) as amended in 2003. In 2013 the Controlled Waste and Duty of Care Regulations (Northern Ireland) 2013, NISR 2013/255 (the 2013 Regulations) were introduced. This was to provide clarity in the classification of 'household', 'commercial' and 'industrial' wastes, identify for which of these wastes a charge may be levied for collection and disposal and to put in place more stringent controls under the Duty of Care System as a deterrent to waste crime.

Waste of the following descriptions shall not be treated as household waste for the purposes of Article 4(2) of the Order (treatment, keeping or disposal of household waste within the curtilage of a dwelling)–

(a) any mineral or synthetic oil or grease;
(b) asbestos; and
(c) clinical waste.

Northern Ireland – Charges for Collection and Disposal of Household Waste

3.18 Article 20(3) of the Waste and Contaminated Land (Northern Ireland) Order 1997, NIOC 1997/2778 (the 1997 Order) establishes that district councils shall not charge for the collection of household waste except in prescribed cases. The purpose of the 2002 Regulations was to provide clarity as to which types of waste are treated as household waste, commercial waste, industrial waste or any other type of controlled waste for the purposes of Part II of the 1997 Order and to lay down which of those wastes the councils may recover a charge for. This was set out in Schedule 2 to the 2002 Regulations.

The 2002 Regulations were out of date and failed to recognise developments in the management of waste and no longer provided clarity on the classification of wastes, therefore creating a 'loophole' in relation to charging for the collection and disposal of certain wastes, which was against the 'polluter pays principle'. They also created a distortion in the waste management market. Accordingly, the 2013 Regulations revoke and replace the 2002 Regulations.

The 2013 Regulations also amend the Controlled Waste (Duty of Care) Regulations (Northern Ireland) 2002, NISR 2002/271 (the DoC Regulations) to provide that better information is contained in Waste Transfer Notes (WTNs) to assist in the tracking of certain wastes and to strengthen controls around the movement of waste, which may assist in tackling environmental crime.

Waste Which is not to be Treated as Household Waste, Industrial Waste or Commercial Waste

3.19 Regulation 4 describes waste which is not to be treated as household waste.

Waste, which is not Directive waste, is not to be treated as household waste, commercial waste or industrial waste for the purposes of Part 2 of the 1997 Order.

The following waste (where it is Directive waste) is not to be treated as household waste, industrial waste or commercial waste for the purposes of Part 2 of the 1997 Order:

(a) sewage sludge or septic tank sludge which is treated, kept or disposed of (otherwise than by means of mobile plant) within the curtilage of a sewage treatment works as an integral part of the operation of those works;
(b) sludge which is supplied or used in accordance with the Sludge (Use in Agriculture) Regulations (Northern Ireland) 1990 ('the 1990 Regulations');
(c) septic tank sludge, which is used on agricultural land within the meaning of the 1990 Regulations.

(3) Animal by-products[15] [16](where they are Directive waste) which are collected and transported are not to be treated as household waste, industrial waste or commercial waste for the purposes of Article 5 of the 1997 Order.

Sites from which Waste could Arise

3.20 Under the interpretation of the Schedule, SI 2013/255, Schedule.

- 'agricultural waste' means waste from premises used for agriculture within the meaning of the Agriculture Act (Northern Ireland) 1949.
- 'camp site' means land on which tents are pitched for the purpose of human habitation and land the use of which is incidental to that purpose.
- 'charity shop' means a premises used wholly or mainly for the sale of goods donated to a charity where the proceeds of sale (after deduction of expenses) are applied for the purposes of a charity.
- 'clinical waste' means waste from a healthcare activity (from medical, nursing, dental, veterinary, pharmaceutical or similar practice, investigation, treatment, care, teaching or research) that:

  (a) contains viable micro-organisms or their toxins, which are known or reliably believed to cause disease in humans or other living organisms;

15 In accordance with Art 21(1)–(3) of the Animal By-Products Regulation (EC) No 1069/2009.
16 The 'Animal By-Products Regulation' means Regulation (EC) No 1069/2009 of the European Parliament and of the Council laying down health rules as regards animal by-products and derived products not intended for human consumption and repealing Regulation (EC) No 1774/2002 (Animal By-Products Regulation).

(b) contains or is contaminated with drugs or medicine that contains a biologically active pharmaceutical agent; or
(c) is a sharp, swab or dressing, or a body fluid or other biological material (including human and animal tissue) containing or contaminated with a dangerous substance within the meaning of Council Directive 67/548/EEC on the approximation of laws, regulations and administrative provisions relating to the classification, packaging and labelling of dangerous substances, and waste of a similar nature from a non-healthcare activity.

- 'construction' includes improvement, repair or alteration.
- 'controlled waste' means household, industrial and commercial waste or any such waste.
- 'Directive waste' has the same meaning as in regulation 1(3) of the Waste Management Licensing Regulations (Northern Ireland) 2003.
- 'European Waste Catalogue' means the list of waste belonging to the categories listed in the List of Wastes Regulations (Northern Ireland) 2005.
- 'factory' has the meaning given in Section 175 of the Factories Act (Northern Ireland) 1965.
- 'general medical practitioner' means a person registered in the General Practitioner Register kept by the General Medical Council under Section 34 of the Medical Act 1983.
- 'hazardous waste' has the same meaning as given in regulation 6 of the Hazardous Waste Regulations (Northern Ireland) 2005 (waste which displays one or more of the hazardous properties listed in Annex III of the Waste Framework Directive).
- 'Irish Traveller Community' means the community of people as described in Article 5 of the Race Relations (Northern Ireland) 1997 Order.
- 'mixed hereditament' means a hereditament which is used partly for the purposes of a dwelling house as defined in Schedule 5 to the Rates (Northern Ireland) Order 1977and partly for other purposes.
- 'not for profit body' means a body which, by virtue of its constitution or any enactment:

  (a) is required (after payment of outgoings) to apply the whole of its income, and any capital which it expends, for charitable or public purposes; and
  (b) is prohibited from directly or indirectly distributing among its members any part of its assets (otherwise than for charitable or public purposes).

- 'offensive waste' means waste that—

  (a) is not clinical waste;
  (b) contains body fluids, secretions or excretions; and
  (c) falls within code 18 01 04, 18 02 03 or 20 01 99 in Schedule 1 to the List of Wastes (Northern Ireland) Regulations 2005.

- 'scientific research association' has the meaning given in Section 469 of the Corporation Tax Act 2010.

- 'scrap metal' means any waste metal, and any broken, worn out, defaced or partly manufactured articles made wholly or partly of metal and any metallic wastes, and also includes old, broken, worn out or defaced tooltips or dies made of any of the materials commonly known as hard metal or of cemented or sintered metallic carbides.
- 'septic tank sludge' means residual sludge from septic tanks and other similar installations for the treatment of sewage.
- 'sludge' means residual sludge from sewage plants treating domestic or urban waste waters and from other sewage plants treating waste waters of a composition similar to domestic and urban waste waters.
- 'The Waste Directive' means Directive 2008/98/EC of the European Parliament and of the Council, on waste.
- 'transferor' and 'transferee' means respectively, in relation to a transfer of controlled waste by a person who is subject to the duty imposed by Article 5 of the 1997 Order, the person who in compliance with that Article transfers a written description of the waste and the person who receives that description.
- 'vessel' includes a hovercraft within the meaning of Section 4(1) of the Hovercraft Act 1968.
- 'waste oils' means any mineral or synthetic lubrication or industrial oils which have become unfit for the use for which they were originally intended, such as used combustion engine oils and gearbox oils, lubricating oils, oils for turbines and hydraulic oils.
- 'waste solvent' means solvent, which is contaminated, spoiled or otherwise unfit for its original purpose.

Sources of Household, Industrial and Commercial Waste

3.21 The Table describes the places where waste is produced in column 2 and then classifies the type of waste it is in column 3. The classification is also subject to exceptions contained in column 4 which can change the classification. See Table below on Classification by Place of Production.

For example: the place producing the waste is a private garage and this has waste classified as household. However if the garage has a floor area of greater than 25m^2 or is not used wholly for mainly for the accommodation of a private motor vehicle, the waste is to be treated as commercial waste.

Table – Classification by Place of Production (SI 2013/255, Sch 2)

No	*Description*	*Classification*	*Exceptions*
1	Private storage premises used wholly or mainly for the storage of articles of domestic use	Household waste	

No	*Description*	*Classification*	*Exceptions*
2	Land belonging to or used wholly or mainly in connection with domestic property or a caravan where waste from that property or caravan is to be treated as household waste.	Household waste	
3	A private garage.	Household waste	Where the garage has a floor area exceeding 25^{m2} or is not used wholly or mainly for the accommodation of a private motor vehicle, the waste is to be treated as commercial waste
4	A vehicle or vessel used wholly for the purposes of living accommodation	Household waste	Where the vehicle or vessel is used in the course of a business for the provision of self-catering accommodation, the waste is to be treated as commercial waste
5	A place of worship	Household waste	
6	A residential hostel which provides accommodation only to persons with no other permanent address or who are unable to live at their permanent address	Household waste	
7	A 'Bed and Breakfast'	Household waste	
8	Prisons, Borstals or other such penal institution	Household waste	
9	A charity shop selling donated goods originating from domestic property	Household waste	

No	*Description*	*Classification*	*Exceptions*
10	A caravan or mobile home site for members of the Irish Traveller Community	Household waste	
11	Premises used wholly or mainly for public meetings	Household waste	
12	University, school or other educational establishment	Household waste	
13	Domestic property used in the course of a business for the provision of self-catering accommodation	Commercial waste	
14	A caravan— (a) used in the course of a business for the provision of self-catering accommodation; or (b) which is not allowed to be used for human habitation throughout the year by virtue of a licence or planning permission	Commercial waste	
15	Premises occupied by a charity and wholly or mainly used for charitable purposes	Commercial waste	Where the waste is from a place of worship or from premises used wholly or mainly for public meetings, it is to be treated as household waste
16	A camp site or a tent pitched on land other than a camp site	Commercial waste	Where the waste is from domestic premises at a camp site, it is to be treated as household waste

No	*Description*	*Classification*	*Exceptions*
17	Premises occupied by a club, society or any association of persons in which activities are conducted for the benefit of the members	Commercial waste	
18	Premises occupied by— (a) a court; (b) a government department; (c) a district council; (d) a person appointed by or under any enactment to discharge public functions; (e) a body incorporated by Royal Charter; and (f) A public body or Trust.	Commercial waste	Waste classified as household waste or industrial waste because it is from a place— (a) otherwise described in this table (except for entry 27); or (b) described in Article 2(2) of the 1997 Order (household waste or industrial waste)
19	A hotel	Commercial waste	
20	Any part of a 'mixed hereditament' used for the purposes of a trade or business	Commercial waste	
21	A market or fair	Commercial waste	
22	The practice of a general medical practitioner	Commercial waste	
23	An office or showroom	Commercial waste	
24	A workshop or similar premises which is not a factory only because— (a) those working there are not employees; or (b) the work carried on there is not carried on by way of trade or for purposes of gain	Industrial waste	Where the principal activities at the premises are computer operations or the copying of documents by photographic or lithographic means the waste is to be treated as commercial waste

No	*Description*	*Classification*	*Exceptions*
25	A laboratory	Industrial waste	
26	A scientific research association	Industrial waste	
27	Premises used for the breeding, boarding or stabling of animals	Industrial waste	
28	Waste imported into Northern Ireland	Industrial waste	
29	Directive waste from a place (including any vehicle, vessel or aircraft) not otherwise described in this table or in Article 2(2) of the 1997 Order (household waste and commercial waste)	Industrial waste	

Nature of Waste and Activities Producing Waste

3.22 This describes how the waste is to be classified depending on the nature of the activity which produces the waste. The second column of the table lists the ways in which the waste can be produced and the third column classifies the waste – but this is subject to the exception, which might be listed, in the fourth column. See Table below on Classification by Nature of Waste or Activity Producing Waste.

If the waste produced is clinical waste and offensive waste – this is to be treated as industrial waste except for:

> 'Clinical waste and offensive waste produced at domestic property, a residential home, a caravan or a vehicle or vessel used wholly for the purposes of living accommodation is to be treated as household waste, except that where such a vehicle or vessel is used in the course of a business for the provision of self-catering accommodation, such waste is to be treated as commercial waste. Clinical waste and offensive waste collected under Article 7(1) and (2) of the 1994 Order is to be treated as household waste or commercial waste in accordance with entries 1 to 6.' (SI 2013/255 Sch 2)

Table – Classification by Nature of Waste or Activity Producing Waste

No	*Description*	*Classification*	*Exceptions*
1	Waste arising from the discharge of duties under Article 7(1)(a) or (e) of the 1994 Order (duty to collect litter)	Household waste	
2	Waste arising from the discharge by a district council of its duty under Article 7(2) of the 1994 Order (duty to ensure roads are kept clean)	Household waste	
3	Waste arising from the discharge of duties under Article 7(1)(c), 12(10), 12C(3) and 13 of the 1994 Order (duty and powers in relation to the collection of litter)	Commercial waste	
4	Waste arising from the discharge of duties under Article 7(1)(b) and (d) of the 1994 Order (duties to collect litter)	Industrial waste	
5	Waste arising from the discharge by a district council or the Department for Regional Development of the duty under Article 7(2) of the 1994 Order (duty to ensure roads are kept clean)	Industrial waste	

No	*Description*	*Classification*	*Exceptions*
6	Hazardous waste arising from the following activities carried on at premises used for the purposes of a trade or business— (a) mixing or selling paints; (b) sign writing; (c) laundering or dry cleaning; (d) developing photographic film or making photographic prints; (e) selling petrol, diesel fuel, paraffin, kerosene, heating oil or similar substances; or (f) selling pesticides, herbicides or fungicides	Industrial waste	
7	Waste oil, waste solvent or scrap metal	Industrial waste	Where the waste is produced at the following places it is to be treated as household waste— (a) a residential home; (b) domestic property, a caravan or land belonging to or used in connection with such a place; (c) a private garage which has a floor area of 25^{m2} or less or is used wholly or mainly for the accommodation of a private motor vehicle;

No	*Description*	*Classification*	*Exceptions*
7	Waste oil, waste solvent or scrap metal – *contd*	Industrial waste – *contd*	(d) private storage premises used wholly or mainly for the storage of articles of domestic use; or (e) a vehicle or vessel used wholly for the purposes of living accommodation, except where it is used in the course of a business for the provision of self-catering accommodation
8	Waste from construction or demolition works, including preparatory works	Industrial waste	The waste is to be treated as household waste for the purposes of Articles 5(2) and (2A) of the 1997 Order only (dis-application of Article 5(1) and duty on the occupier of domestic property to transfer household waste only to an authorised person or for authorised transport purposes)
9	Septic tank sludge	Industrial waste	The waste is to be treated as household waste for the purposes of Articles 5(2) and (2A) of the 1997 Order
10	Sewage	Industrial waste	

No	*Description*	*Classification*	*Exceptions*
11	Clinical waste and offensive waste	Industrial waste	Clinical waste and offensive waste produced at domestic property, a residential home, a caravan or a vehicle or vessel used wholly for the purposes of living accommodation is to be treated as household waste, except that where such a vehicle or vessel is used in the course of a business for the provision of self-catering accommodation, such waste is to be treated as commercial waste. Clinical waste and offensive waste collected under Article 7(1) and (2) of the 1994 Order is to be treated as household waste or commercial waste in accordance with entries 1 to 6
12	Agricultural Waste	Industrial waste	Applies only where this waste is Directive waste

Household Waste for which Collection and Disposal Charges may be Made

3.23 Waste is described in the Table – the nature of the waste in Column 2, Column 3 then indicates if a collection charge can be made and Column 4 indicates which waste must be treated as commercial waste for the provisions of the 1997 order[17] and whether a disposal charge can be made. See Table below on the Nature of the Waste

17 Article 20(4) to the extent that it concerns liability to pay a reasonable charge for disposal of commercial waste and the recovery of such a charge.

For example: waste that comes from a charity shop selling donated goods originating from domestic property attracts a collection charge but only to the extent that the waste originated from non-domestic property.

Table on the Nature of the Waste

No	*Description of household waste*	*Collection charge*	*Disposal charge*
1	Any article of waste which exceeds 25kg in weight	Yes	No, subject to entries 11 to 17
2	Any article of waste which does not fit or cannot be fitted into— (a) a receptacle for household waste provided in accordance with Article 21 of the 1997 Order; or (b) where no such receptacle is provided, a cylindrical container 750mm in diameter and 1m in length	Yes	No, subject to entries 11 to 17
3	Garden waste	Yes	No, subject to entries 11 to 17
4	Dead domestic pets	Yes	No, subject to entries 11 to 17
5	Waste oil or grease	Yes	No, subject to entries 11 to 17
6	Asbestos	Yes	No, subject to entries 11 to 17
7	Waste which may not be put into a receptacle provided under Article 21 of the 1997 Order because of a notice served under that Article	Yes	No, subject to entries 11 to 17
8	Waste from premises used wholly or mainly for public meetings	Yes	No
9	Clinical waste and offensive waste produced at a domestic property, a caravan or a vehicle or vessel used wholly for the purposes of living accommodation	Yes	No
10	Waste from a residential hostel which provides accommodation only to persons with no other permanent address or who are unable to live at their permanent address	Yes	No

No	*Description of household waste*	*Collection charge*	*Disposal charge*
11	Waste from a Bed and Breakfast	Yes	No
12	Waste from a charity shop selling donated goods originating from domestic property	Yes	Yes, but only to the extent that the waste originated from non-domestic property
13	Waste from premises occupied by: (a) a community interest company (being a company which is registered as such with the registrar of companies); or (b) a charity or other not for profit body, which collects goods for re-use or waste to prepare for re-use from domestic property	Yes	Yes, but only to the extent that the waste originated from non-domestic property
14	Litter collected under Article 7(1)(e) of the 1994 Order	Yes	Yes
15	Waste from a residential home or land belonging to or wholly or mainly used in connection with a residential home	Yes	Yes
16	Waste from premises forming part of a university, school or other educational establishment	Yes	Yes
17	Waste from premises forming part of a hospital or nursing home (except for waste from a residential hostel forming such part which provides accommodation only to persons with no other permanent address)	Yes	Yes
18	Waste from Prisons, Borstals or other such penal institution	Yes	Yes

Chapter 4

Duty of Care

DUTY OF CARE

4.1 Anyone who produces, imports, keeps, stores, transports, treats or disposes of waste must take all reasonable steps to ensure that waste is managed properly. This duty of care is imposed under s 34 of the Environmental Protection Act 1990. It also applies to anyone who acts as a broker and has control of waste. A breach of the duty of care could lead to a penalty of up to £5,000 if convicted in the magistrates court or an unlimited fine if convicted in the crown court. There is also a duty of care relating to household waste.

How Long Does the Duty of Care Last?

4.2 The duty of care lasts from the point that the waste is produced until it is given to a licensed waste business to deal with. The waste producer is still responsible for checking that the waste is being dealt with correctly and any mishandling of waste should be reported.

Business waste includes any waste that comes from:

— any commercial activity – including any you run from your home;
— construction;
— demolition;
— industry;
— agriculture.

CLASSIFICATIONS OF WASTE

4.3 Waste should be classified as hazardous and non-hazardous. Premises which store waste, must be registered (see **Chapter 8**). If an organisation stores, treats, transports or disposes of waste they must have a permit to do so. Waste must be stored safely and securely – there are extra responsibilities for dealing with hazardous waste (or special waste in Scotland).

As part of a Smarter Environmental Regulation Review (SERR), Defra has removed the need to register premises from 1 April 2016. This only applies to businesses in England. As a consequence of this, the consignment notes have changed. Waste which is to be transported from England to Scotland, Wales or Northern Ireland will need to use the new consignment code.

In England and Wales, the Waste (England and Wales) Regulations 2011, SI 2011/988 (the 2011 Regulations) repealed the Environmental Protection (Duty of Care) Regulations 1991, SI 1991/2839 and apply the duty of care requirements brought in by the Environmental Protection Act 1990.

Part 5[1] of the 2011Regulations details what duties a waste producer has and what is required.

Prior to these Regulations, most implementation of the predecessor Directives was dealt with by the following legislation:

— The Environmental Protection Act 1990, particularly s 34 and s 44A requiring a national waste strategy and the Environmental Protection (Duty of Care) Regulations 1991, SI 1991/2839 as amended), dealing with the 'duty of care' on those responsible for waste;

1 The Waste (England and Wales) Regulations 2011, SI 2011/988, Part 5 Duties in Relation to Waste Management and Improved Use of Waste as a Resource.

— The Control of Pollution (Amendment) Act 1989 and the Controlled Waste (Registration of Carriers and Seizure of Vehicles) Regulations 1991, SI 1991/1624 as amended), dealing with the registration of waste carriers;
— The registration of waste carriers and with the registration of dealers and brokers of waste;
— The Environmental Permitting (England and Wales) Regulations 2010, SI 2010/675 as amended), dealing with permitting of waste disposal and recovery operations;
— The Hazardous Waste (England and Wales) Regulations 2005, SI 2005/894 as amended) and the Hazardous Waste (Wales) Regulations 2005, SI 2005/1806 as amended) (W138), dealing with hazardous waste; and
— Town and Country Planning Act 1990 and the Planning and Compulsory Purchase Act 2004 and subordinate legislation made under those Acts, dealing with waste plans.

The implementation of the revised Waste Framework Directive continues to rely on most of this legislation, amended where necessary by the draft instrument. However, these Regulations also reduce the fragmentation of waste legislation to some extent and streamlines and replaces some waste regulation, in particular the subordinate legislation relating to the registration of waste carriers and brokers and to the 'duty of care'.

Duty in Relation to the Waste Hierarchy

4.4 An establishment or undertaking which imports, produces, collects, transports, recovers or disposes of waste, or which as a dealer or broker has control of waste must, on the transfer of waste, take all such measures available to it as are reasonable in the circumstances to apply the following waste hierarchy as a priority order:

(a) prevention;
(b) preparing for re-use;
(c) recycling;
(d) other recovery (for example energy recovery);
(e) disposal.

However, an establishment or undertaking may depart from the hierarchy, so as to achieve the best overall environmental outcome and where this is justified by life-cycle thinking on the overall impacts of the generation and management of the waste.

When considering the overall impacts of departing from the hierarchy, the following considerations must be taken into account:

(a) the general environmental protection principles of precaution and sustainability;
(b) technical feasibility and economic viability;
(c) protection of resources;
(d) the overall environmental, human health, economic and social impacts.

Duties in Relation to Collection of Waste

4.5 An establishment or undertaking which collects waste paper, metal, plastic or glass must, from 1 January 2015, take all such measures to ensure separate collection of that waste as are available to the establishment or undertaking in that capacity and are:

(a) technically, environmentally and economically practicable; and
(b) appropriate to meet the necessary quality standards for the relevant recycling sectors.

For the avoidance of doubt, co-mingled collection (being the collection together with each other but separately from other waste of waste streams intended for recycling with a view to subsequent separation by type and nature) is a form of separate collection.

Every waste collection authority must, when making arrangements for the collection of waste paper, metal, plastic or glass, ensure that those arrangements are by way of separate collection.

Duty in Relation to Collected Waste

4.6 An establishment or undertaking which collects, transports or receives waste paper, metal, plastic or glass must, from 1 January 2015, take all such measures available to it in that capacity as are reasonable in the circumstances to ensure that where that waste has been separately collected it is not mixed with other waste or other material with different properties.

This duty applies only where keeping waste separate, facilitates or improves recovery.

Guidance

4.7 The appropriate authority may give guidance on the discharge of the duties in regs 12 to 14 of the 2011 Regulations[2].

An establishment or undertaking discharging any of the duties in regs 12 to 14 of the 2011 Regulations must, in doing so, have regard to any such guidance.

DUTIES ON PLANNING AUTHORITIES

4.8 The Waste (England and Wales) Regulations 2011 provide for the Policy Planning Statement No 10 to take account of the revised Directive's transposition.

2 SI 2011/988, reg 12 – Duty in relation to waste hierarchy.

A planning authority must have regard to the provisions of the Waste Framework Directive when exercising its planning function that relate to waste management and have regard to Article 13, the first paragraph of Article 16(1) ignoring the words 'in co-oporation with other member states where this is necessary and advisable' and 'taking into account best available techniques'. Also Article 16(2) and (3).

Exercise of Planning Functions

4.9 A planning authority must have regard to the following provisions of the Waste Framework Directive when exercising its planning functions to the extent that those functions relate to waste management:

(a) Article 13;
(b) the first paragraph of Article 16(1), ignoring the words 'in cooperation with other Member States where this is necessary or advisable' and 'taking into account best available techniques';
(c) Article 16(2) and (3).

Inspections

4.10 This regulation applies where a planning authority has planning functions in relation to establishments or undertakings carrying on disposal or recovery of waste.

The planning authority must ensure that appropriate periodic inspections of those establishments or undertakings are made.

Further Duties in Relation to Planning Permission

4.11 Within the Waste (England and Wales) Regulations 2011, a planning authority must also not grant planning permission or development consent for a landfill unless it has taken into account the requirements of paragraph 1.1 Annex to Council Directive 1999/31/EC on Landfill Waste[3] and paragraph 5 of the Annex in respect of nuisances and hazards arising from traffic beyond the site of the landfill.

3 1. Location
1.1. The location of a landfill must take into consideration requirements relating to:
(a) the distances from the boundary of the site to residential and recreation areas, waterways, water bodies and other agricultural or urban sites;
(b) the existence of groundwater, coastal water or nature protection zones in the area;
(c) the geological and hydrogeological conditions in the area;
(d) the risk of flooding, subsidence, landslides or avalanches on the site;
(e) the protection of the nature or cultural patrimony in the area.

A mineral planning authority must not grant planning permission for a mining waste facility to which Article 7 of Directive 2006/21/EC of the European Parliament and of the Council on the management of waste from extractive industries applies, unless it is satisfied that the operator of that facility will meet the requirements of Article 11(2)(a) of that Directive; and the management of waste at that facility will not conflict directly or otherwise interfere with the implementation of the plans referred to in Article 7(3)(b) of that Directive.

> 'landfill' has the meaning given in Article 2(g) of Directive 1999/31/EC, but does not include any operation excluded from the scope of that Directive by Article 3(2);
>
> 'mining waste facility' means a 'waste facility' as defined in Article 3(15) of Directive 2006/21/EC, but does not include those facilities mentioned in Article 24(2) or in the first paragraph of Article 24(4).'

PART 7 DEPOSITS AT SEA

4.12 Where the Act mentions off-shore licensing functions it applies to:

— functions under Part 2 of the Food and Environment Protection Act 1985;
— functions under Part 4 of the Marine and Coastal Access Act 2009(2).

Inspections

This regulation applies where an appropriate authority has offshore licensing functions in relation to establishments or undertakings carrying on disposal or recovery of waste.

The appropriate authority must ensure that appropriate periodic inspections of those establishments or undertakings are made.

SCOTLAND

4.13 Scotland has a Duty of Care Guidance Document[4].

Section 34 of the Environmental Protection Act 1990 (as amended) lays out a number of duties with respect to the management of waste. Waste must be managed correctly by storing it properly, only transferring it to the appropriate persons and ensuring that when it is transferred it is sufficiently well described to enable its safe recovery or disposal without harming the environment.

The Waste (Scotland) Regulations 2012, SSI 2012/148 amended s 34 to implement a number of actions in the Scottish Government's Zero Waste Plan. Under these amendments, holders of waste, including producers, have a duty

4 Duty of Care – The Scottish Government 2012 available at: http://www.gov.scot/resource/0040/00404095.pdf

to take reasonable steps to increase the quantity and quality of recyclable materials.

This 'Duty of Care: A Code of Practice' (the Code) explains these duties which apply to anyone who produces, keeps, imports or manages controlled waste in Scotland. The Code is made under s 34(7) of the Environmental Protection Act 1990 (as amended) and replaces all previous versions applicable in Scotland. This Code applies in Scotland only.

Under s 34(10) of the 1990 Act, this Code is admissible as evidence in court and the court shall take it into account in determining any questions to which it appears to be relevant. The intention is that the Code will assist the courts, when hearing cases under s 34 of the 1990 Act, in determining whether persons subject to the duty took reasonable measures to comply with it.

The Environmental Protection (Duty of Care) (Scotland) Regulations 2014[5] are made under s 34 of the Environmental Protection Act 1990, which imposes a range of duties on persons who import, produce, carry, keep, treat or dispose of controlled waste for the purposes of that Act. The Regulations provide for persons subject to duties of care to make, retain and produce documents and also make provision for a waste transfer note to accompany waste that is being transferred.

They change the existing paper transfer notes requirements of the 1991 Duty of Care Regulations and revoke them. These Regulations allow the use of electronic duty of care transfer notes and electronic signatures. This change has facilitated and supported the use of the electronic Duty of Care (eDoC) which was launched for use on a voluntary basis by UK authorities (including SEPA) early in 2014. It had prevsiously run from January 2011 until December 2014 and was co-financed by EU's LIFE+ Environment Programme[6].

The Regulations also introduce a requirement for SIC codes to be recorded on transfer notes. This is consistent with transfer note requirements in other parts of the UK and has been a requirement in England and Wales since 2011.

The United Kingdom Standard Industrial Classification of Economic Activities (SIC) is used to classify business establishments and other standard units by the type of economic activity in which they are engaged. It provides a framework for the collection, tabulation, presentation and analysis of data and its use promotes uniformity. In addition, it can be used for administrative purposes and by non-government bodies as a convenient way of classifying industrial activities into a common structure.

SIC information is therefore valuable to help make strategic planning decisions, for example on the need for waste infrastructure. It is also required for providing data to the European Commission.

5 SSI 2012/4.
6 Laymans Report of the Edoc project, LIFE09 ENV/UK/000023.

Section 34(5) of the 1990 Environmental Protection Act enables the Scottish Ministers to make regulations imposing requirements on a person subject to a duty under s 34(1) of that Act, in respect of the making and retention of documents and the furnishing of documents or copies of documents.

Within the Environmental Protection (Duty of Care) (Scotland) Regulations 2014 there are the following regulations.

Regulation 3 provides that persons transferring and receiving waste must, at the same time as the written description of waste is transferred, complete and sign a transfer note in respect of the waste.

Regulation 4 provides an exception to the requirement for a transfer note where the waste transferred is special waste.

Regulation 5 provides for persons transferring and receiving waste to keep the written description and the transfer note for at least two years.

Regulation 6 provides for persons who are required under regulation 4 to keep a document to produce the document (or a copy) to SEPA or to a waste collection authority under the 1990 Act on request.

Regulation 7 revokes the Environmental Protection (Duty of Care) Regulations 1991.

Section 34(6) of the 1990 Act provides for it to be an offence to fail to comply with a requirement in these Regulations.

Interpretation (Regulation 2)

4.14

2 In these Regulations:

"the Act" means the Environmental Protection Act 1990;
"transfer note" has the same meaning as in regulation 3(1);
"transferor" and "transferee" mean, in relation to a transfer of controlled waste by a person who is subject to the duty in section 34(1) of the Act(1), the person who transfers the waste and the person who receives the waste; and
"written description" means the written description for the purposes of section 34(1)(c)(ii) of the Act.

In these Regulations a reference to a document:

being in writing includes an electronic communication within the meaning of section 15 (general interpretation) of the Electronic Communications Act 2000; and
being signed means, where the document is in electronic form, an electronic signature (and "electronic signature" means data in electronic form which are attached to or logically associated with the document and which serve as a method of authentication).

Transfer Note (Regulation 3)

4.15

3 (1) The transferor and the transferee must ensure that a document as described in paragraphs (3) and (4) is completed in writing and signed by each of them in respect of the waste being transferred ("a transfer note").

(2) A transfer note must be prepared at the same time as the written description is transferred in accordance with section 34(1)(c) of the Act.

(3) A transfer note must:

(a) give the name and address (including the postcode) of the transferor and the transferee

(b) give the date and place (including the postcode) of the transfer

(c) state whether the transferor is the producer of the waste

(d) state whether the transferor is the importer of the waste

(e) describe the type, composition and quantity of the waste being transferred (including, where the waste is in a container, the type of container)

(f) identify the waste being transferred by reference to the appropriate six-digit code in the European Waste Catalogue; and

(g) identify the activity carried out by the transferor in respect of the waste being transferred by reference to the SIC code for that activity.

(4) A transfer note must where the transferor or transferee is a person described in column 1 of the Table—

(a) state that the transferor or transferee is such a person; and

(b) include the additional information where specified in column 2 of the Table.

Category of person	Additional information
A waste collection authority.	
A person who holds a— (a) waste management licence under section 35 of the Act[7]; or (b) disposal licence under section 5 of the Control of Pollution Act 1974.	If the waste being transferred is to be kept, treated or disposed of by that person, the relevant licence number and the name of the licensing authority.
A person who holds a permit under the Pollution Prevention and Control (Scotland) Regulations 2012[8] which authorises the carrying out of a specified waste management activity within the meaning of those Regulations.	

Category of person	Additional information
A person to whom section 33(1)(a) and (b) of the Act[9] does not apply by virtue of regulations under subsection (3) of that section.	
A person registered as a carrier of controlled waste under— (a) section 2 of the Control of Pollution (Amendment) Act 1989; or (b) paragraph 12 of Part I of Schedule 4 to the Waste Management Licensing (Scotland) Regulations 2011[10].	The name of the waste regulation authority with whom that person is registered and the registration number.

(5) In this regulation:

"European Waste Catalogue" means the list set out in Commission Decision 2000/532/EC establishing a list of wastes, as that instrument may be amended from time to time; and

"SIC code" means the code for an activity in the UK Standard Industrial Classification of Economic Activities 2007 (SIC 2007)(7), as amended from time to time.

For Special Waste see **Chapter 4**.

Special waste: exception to requirement to prepare a transfer note (Regulation 4)

4.16

4 Regulation 3 does not apply where—

(a) the waste transferred is special waste within the meaning of the Special Waste Regulations 1996(1); and

(b) a consignment note (and if appropriate schedule) is completed and dealt with in accordance with those Regulations.

7 As amended by the Environment Act 1995, Sch 2, para 66(2) and SSI 2000/323 and SSI 2011/226.
8 SSI 2012/360.
9 As amended by SSI 2009/247.
10 SSI 2011/228.

Duty to keep a written description of waste or a transfer note (Regulation 5)

4.17

5 The transferor and the transferee must keep the written description and the transfer note (or copies of them) for a period of two years from the date of transfer of the waste.

Duty to produce a written description of waste or a transfer note (Regulation 6)

4.18

6 (1) A competent authority may request any person required to keep a written description, or a transfer note under regulation 5, to produce the description or note (or a copy) to that authority.

(2) The person must produce the description or note (or the copy) as soon as is reasonably practicable, and in any case:
  (a) within 7 days from the date of the request; or
  (b) such longer period as is permitted by that authority.

(3) The description or note (or copy) must be:
  (a) given to an officer of that authority; or
  (b) sent to the address specified by that authority when making the request.

(4) In this regulation, "competent authority" means:
  (a) the Scottish Environment Protection Agency; or
  (b) a waste collection authority for the purposes of Part 2 of the Act.

NORTHERN IRELAND

4.19 There are several pieces of legislation covering duty of care in Northern Ireland.

Controlled Waste and Duty of Care (Amendment) Regulations (Northern Ireland) 2014, NISR 2014/117

4.20 The Controlled Waste and Duty of Care (Amendment) Regulations (Northern Ireland) 2014, SR 2014/117 amends the Controlled Waste Duty of Care Regulations (Northern Ireland) 2013, SR 2013/255 in that they provide clarification in relation to the classification of a number of items contained in the tables to the schedule, including the sources of household, industrial and commercial waste and the nature of waste and activities producing waste.

Amendments to the Controlled Waste and Duty of Care Regulations (Northern Ireland) 2013, SI 2013/255

4.21

2 The Controlled Waste and Duty of Care Regulations (Northern Ireland) 2013(1) shall be amended as follows.

(2) In the Schedule:

(a) in paragraph 1 (Interpretation of the Schedule) insert in the appropriate alphabetical order:

""bed and breakfast" means an establishment meeting the criteria set out in Schedule 3 to the Categories of Tourist Establishment (Statutory Criteria) Regulations (Northern Ireland) 1992(2);

"guest House" means an establishment meeting the criteria set out in Schedule 2 to the Categories of Tourist Establishment (Statutory Criteria) Regulations (Northern Ireland) 1992;""

(b) in paragraph 2 (Sources of household, industrial and commercial waste) in Table 1 at item 7 for "A 'Bed and Breakfast'" substitute "A 'Bed and Breakfast' or a 'Guest House'";

(c) in paragraph 2 (Sources of household, industrial and commercial waste) in Table 1 at item 22 for "The practice of a general medical practitioner" substitute "The practice premises of a general medical practitioner";

(d) in paragraph 3 (Nature of waste and activities producing waste)

(i) in Table 2 at item 2 after "(duty to ensure roads are kept clean)" insert "where it relates to household waste as classified in Table 1"; and

(ii) in item 5 after "(duty to ensure roads are kept clean)" insert "where it relates to industrial waste as classified in Table 1".

Controlled Waste and Duty of Care Regulations (Northern Ireland) 2013, SR 2013/255

4.22 These Regulations revoke and replace the Controlled Waste Regulations (NI) 2002, SR 2002/271 (the 2002 Regulations). The Regulations clarify classification of wastes as 'household', 'commercial', or 'industrial', identify which of these wastes incur a charge for collection and disposal and introduce more stringent controls under the waste Duty of Care system.

Article 20(3) of the Waste and Contaminated Land (NI) Order 1997, SI 1997/2778 (NI 19) establishes that district councils shall not charge for the collection of household waste except in prescribed cases. The purpose of 2002 Regulations was, therefore, to provide clarity in relation to which types of waste are treated as household waste, commercial waste, industrial waste or any other type of controlled waste for the purposes of Part II of the 1997 Order and to lay down which of those wastes the councils may recover a charge for. This was set out in Schedule 2 to the 2002 Regulations.

> 3.2. However, the 2002 Regulations were out of date, failed to recognise developments in the management of waste and no longer provided clarity on the classification of wastes, thus creating a 'loophole' in relation to charging for the collection and disposal of certain wastes, which was in contravention of the 'polluter pays principle' and created a distortion in the waste management market. Accordingly, the Controlled Waste and Duty of Care Regulations (Northern Ireland) 2013 (the 2013 Regulations) revoke and replace the 2002 Regulations.
>
> 3.3. The 2013 Regulations also amend the Controlled Waste (Duty of Care) Regulations (NI) 2002 (the DoC Regulations) to provide for better information to be contained in Waste Transfer Notes (WTNs) to assist in the tracking of certain wastes and to strengthen controls around the movement of waste, which may assist in tackling environmental crime.

The Controlled Waste (Amendment) Regulations (Northern Ireland) 2003, SR 2003/404 amend SR 2002/248 to include animal by-products.

> Amendment of the principal Regulations
>
> (1) Regulations 3, 4, 6 and 8 of the principal Regulations as originally enacted shall be renumbered as Regulations 3(1), 4(1), 6(1) and 8(1).
> (2) For paragraphs (3) and (4) of regulation 8 (waste not to be treated as industrial or commercial waste) of the Controlled Waste Regulations (Northern Ireland) 2002 there shall be substituted –
> (3) Animal by-products which are collected and transported in accordance with Article 6 of the Animal By-Products Order (Northern Ireland) 2002 shall not be treated as industrial waste or commercial waste for the purpose of Article 5 (duty of care, etc. as respects waste).
> (4) In this regulation "animal by-products" has the same meaning as in Article 2 of the Animal By-Products Order (Northern Ireland) 2002;".

The 2002 Regulations create a duty of care for controlled waste that requires all producers, carriers and managers of waste to keep records and use waste transfer notes.

Regulation 2(1) and (2) requires the transferor and the transferee to complete and sign a transfer note at the same time as the written description of the waste is transferred. The transfer note must identify the waste in question and state its quantity, how it is stored, the time and place of transfer, the name and address of the transferor and the transferee, whether the transferor is the producer or importer of the waste, which (if any) authorised transport purpose applies, in which category of person the transferor and the transferee are and certain additional information. Regulation 2(3) disapplies regulation 2(1) and (2) where special waste (within the meaning of the Special Waste Regulations (Northern Ireland) 1998) is transferred and a consignment note and (where appropriate) the schedule are completed and dealt with in accordance with those Regulations.

Regulation 3 requires the transferor and the transferee to keep the written description of the waste and the transfer note or copies of them for two years from the transfer.

Regulation 4 imposes a duty on a person who is under a duty to keep any document by virtue of regulation 3, to furnish a copy of that document to the Department if he is required by notice to do so by the Department. The document must be furnished to the Department within the period specified in the notice (not being less than 7 days).

Regulation 5 amends the Controlled Waste Regulations (Northern Ireland) 2002 to bring certain wastes within the scope of the duty of care.

Citation, commencement and interpretation (Regulation 1)

1 (1) These Regulations may be cited as the Controlled Waste (Duty of Care) Regulations (Northern Ireland) 2002 and shall come into operation on 1st October 2002.

(2) In these Regulations:

"the 1997 Order" means the Waste and Contaminated Land (Northern Ireland) Order 1997;

"European Waste Catalogue" means the list of waste belonging to the categories listed in Commission Decision 2000/532/EC, as amended by Commission Decision 2001/118/EC and Commission Decision 2001/119/EC;

"scrap metal" means any old metal, and any broken, worn out, defaced or partly manufactured articles made wholly or partly of metal and any metallic wastes, and also includes old, broken, worn out or defaced tooltips or dies made of any of the materials commonly known as hard metal or of cemented or sintered metallic carbides;

"transferor" and "transferee" mean respectively, in relation to a transfer of controlled waste by a person who is subject to the duty imposed by Article 5(1) of the 1997 Order, the person who in compliance with that Article transfers a written description of the waste and the person who receives that description.

Amendment of the Controlled Waste Regulations (Northern Ireland) 2002 (Regulation 5)

5 The Controlled Waste Regulations (Northern Ireland) 2002(1) shall be amended as follows:

(a) at the end of regulation 3 there shall be added:

"(2) Waste of the following descriptions shall be treated as household waste for the purposes only of Article 5(2) (household waste produced on domestic property) –

(a) waste arising from works of construction or demolition, including waste arising from work preparatory thereto; and

(b) septic tank sludge.";

(b) at the end of each of regulations 4 and 8 there shall be added –

"(2) Scrap metal shall not be treated as household or, as the case may be, industrial or commercial waste, for the purposes of Article 5 at any time before the day appointed under Article 1(2) of the Order for the repeal of Articles 5 to 12 of the 1978 Order in relation to controlled waste which is scrap metal.";

(c) in regulation 6

(i) for "Subject to regulation 8" there shall be substituted "Subject to paragraph (2) and regulation 8"; and

(ii) at the end there shall be added –

"(2) Waste described in paragraphs 7 and 8 of Schedule 3 shall not be treated as industrial waste for the purposes of Article 5(2) of the Order.";

(d) at the end of regulation 8 there shall be added –

"(3) Animal by-products which are collected and transported in accordance with Schedule 2 to the Animal By-Products Regulations (Northern Ireland) 1993 shall not be treated as industrial waste or commercial waste for the purpose of Article 5 (duty of care etc. as respects waste).

(4) In this regulation "animal by-products" has the same meaning as in regulation 2(1) of the Animal By-Products Regulations (Northern Ireland) 1993."; and

(e) in Schedule 3, paragraph 17(1) for "Waste oil or waste solvent" there shall be substituted "Waste oil, waste solvent or (subject to regulation 8(2)) scrap metal".

Chapter 5

Transfrontier Shipment of Waste

INTRODUCTION

5.1 Economic growth and globalisation have led to a worldwide increase in waste being transported across borders, whether by road, rail, air or sea. In order to control waste shipments, which can create risks for human health and the environment, certain procedures and requirements have been introduced in international and European Union law.

SAFE WASTE SHIPMENTS WITHIN THE EU AND WITH NON-EU COUNTRIES

5.2 The European Union (EU) has a system to supervise and control shipments of waste within its borders and with the countries of the European Free Trade Association (EFTA), the Organisation for Economic Co-operation and Development (OECD) and non-EU countries that have signed the Basel Convention[3].

The Regulation on shipment of waste[4] lays down rules for controlling waste shipments in order to improve environmental protection. It also incorporates the provisions of the Basel Convention and the revision of the OECD's 2001 decision on the control of transboundary movements of wastes destined for recovery operations (ie, where a waste is processed to recover a usable product or converted into a fuel) in EU law.

1 Inserted by the Environmental Civil Sanctions (Miscellaneous Amendments) (England) Regulations) 2010, SI 2010/1159.

2 SI 2014/861, reg 15.

3 The Basel Convention on the Control of Transboundary Movements of Hazardous Wastes and their Disposal Available at: http://www.basel.int/TheConvention/Overview/tabid/1271/Default.aspx.

4 Regulation (EC) No 1013/2006 of the European Parliament and of the Council of 14 June 2006 on shipments of waste.

The law applies to shipments of waste[5]:

- between EU countries within the EU or transiting via non-EU countries;
- imported into the EU from non-EU countries;
- exported from the EU to non-EU countries;
- in transit through the EU, on the way from or to non-EU countries.

It covers almost all types of waste[6], with the exception of radioactive waste, waste generated on board ships, shipments subject to the approval requirements of the animal by-product regulation, certain shipments of waste from the Antarctic, imports into the EU of certain waste generated by armed forces or relief organisations in situations of crisis, etc.

CONTROL PROCEDURES

5.3 There are two control procedures for the shipment of waste:

- the general information requirements of Article 18[7], which is normally applicable to shipments for recovery of wastes listed in Annex III ('green' listed wastes – non-hazardous, such as paper or plastics) or IIIA; and
- the procedure of prior written notification and consent for other types of shipments of wastes, including:
  - shipments of wastes listed in Annex IV ('amber' listed wastes containing both hazardous and non-hazardous parts) or in Part 2 of Annex V (European list of wastes, eg, wastes from mining, quarrying and physical and chemical treatment of minerals); and
  - shipments for disposal of wastes listed in Annex III ('green' listed wastes);
  - other provisions.

All parties involved must ensure that waste is managed in an environmentally sound manner, respecting EU and international rules throughout the shipment process and when it is recovered or disposed of.

The notification procedure requires the prior written consent of the competent authorities of the countries concerned by the shipment (country of dispatch, country of transit and country of destination) to be given within 30 days and the notifier has a duty to take back waste shipments that are found to be illegal or cannot be provided as intended (including the recovery or disposal of waste).

5 EC 1013/2006, Article 1(2).
6 Defined under Article 1(1)(a) of Directive 2006/12/EC and Article 1(14) of Council Directive 91/689/EEC of 12 December 1991 on Hazardous Waste.
7 EC 1013/2006, Article 18, waste to be accompanied by certain information.

Trade Bans

5.4 Exports[8] to non-EU countries of waste for disposal are prohibited, except to EFTA countries that are party to the Basel Convention. Exports for recovery of hazardous waste (ie, that poses a risk to human health and the environment) are prohibited, except those directed to countries to which the OECD decision applies.

Imports from non-EU countries of waste for disposal or recovery are prohibited, except for imports:

- from countries to which the OECD decision applies;
- non-EU countries that are party to the Basel Convention;
- countries that have concluded a bilateral agreement with the EU or EU countries; or
- other areas during situations of crisis.

Inspection Systems

5.5 The law was amended in 2014 to strengthen EU countries' inspection systems. It sets out minimum inspection requirements with a focus on problematic waste streams (such as hazardous waste and waste sent illegally for dumping or sub-standard treatment). EU countries will have to prepare inspection plans by 2017.

When it became apparent that hazardous waste from developed countries was being exported and dumped in developing countries the European Community introduced the first a directive on control procedures for the shipment of hazardous waste in 1984. The Basel Convention of 1989 established worldwide notification requirements for the movement of hazardous waste and required those involved to minimise the generation of such waste and to ensure its environmentally sound management.

Regulations governing transfrontier shipment of waste (TFS) were finalised by two further pieces of European legislation and are now implemented in the UK by:

- Transfrontier Shipment of Waste Regulations 2007, which detail the UK procedures, offences, penalties and relevant enforcement authorities.
- UK Plan on Shipments of Waste which sets out Government policy in relation to imports and exports for disposal.

The Revision of the Waste Shipments Regulation (WSR)

5.6 Both the Basel Convention and OECD Decision have been substantially revised since the WSR came into force in 1994. The existing WSR is therefore

8 EC 1013/2006, Articles 39 and 40.

unaligned with some of the requirements of the Basel Convention and the relevant OECD Decision. Consequently, the WSR needed to be revised to ensure that these changes were implemented within the Community. This process offered the opportunity to improve the structure and clarity of the existing WSR and the degree of environmental protection afforded.

The European Commission published their proposal for a revised Waste Shipments Regulation in 2003. In October 2005 an agreement was reached between the Council of the European Union and the European Parliament on a new Waste Shipments Regulation. Regulation (EC) No 1013/2006 of the European Parliament and of the Council of 14 June 2006 on shipments of waste was published in June 2006 and applied from 12 July 2007.

The Transfrontier Shipment of Waste Regulations 1994, SI 1994/1137 (TFS Regulations) were revised by the 2007 Regulations (SI 2007/1711) to be consistent with the revised WSR. The revised TFS Regulations came into force at the same time that the WSR applied, on 12 July 2007. Whilst the broad policy direction is already set by the WSR, there is some flexibility in relation to how it is supported in national legislation. In the light of recent developments regarding illegal and potentially illegal exports of waste, the review of the TFS Regulations has considered what can be done to strengthen enforcement of the revised WSR.

The revision of the WSR and the subsequent changes to the TFS Regulations will affect enterprises who are involved in the shipment of waste or who have shipped waste consigned to them. To a lesser extent, the changes will also have an impact upon anyone who has any involvement with producing, holding, transporting, or arranging such transport of waste that is destined for, or that has been subject to, a transboundary movement. The effects of the proposed changes to the TFS Regulations will be variable, depending on the types of wastes.

Key Changes to the System Under the Revised WSR

5.7 The review of the WSR had a number of objectives, the principal ones being:

- to implement the OECD Decision C(2001)107 of 14 June 2001 in Community legislation;
- to address the problems encountered in the application, administration and enforcement of the 1993 Regulation and establish greater legal clarity;
- to pursue global harmonisation in the area of trans-boundary shipments of waste;
- to enhance the structure of the articles of the Regulation.

In order to achieve this, various sections and aspects of the WSR have been amended. There have been changes to its structure, changes and clarifications as regards definitions and clarification of its scope, changes to the procedures applicable to shipments of waste depending on where it is coming from or going to and finally, changes in other provisions of the Regulation.

Rationale for Government Intervention

5.8 The TFS Regulations must give full effect to the provisions of the WSR in the UK and thereby provide the controls to prevent the shipment of waste which is uncontrolled, or ineffectively controlled and which could result in damage to human health and/or the environment. There is a risk of legal uncertainty and confusion if the provisions of the revised WSR are not supported in national legislation and implemented in a timely manner.

The provisions of the WSR ensure the competent authorities have prior notification of any proposed import or export of hazardous waste and a tracking system for shipments to ensure such wastes are treated in an environmentally sound manner. By giving full effect to the provisions of the WSR, the TFS Regulations ensure the competent authorities have a legal basis to oppose hazardous waste imports or exports and requires them to object to shipments notified to them that are not in accordance with the Plan. Given the potential toxicity of hazardous waste, the possibility that its identity and responsibility for its management can be more readily lost when shipped over longer distances between countries and different operators and the length of time that some forms can remain toxic within the environment, it is reasonable to conclude that a substantial risk would be posed to human health and the environment without this legislation. The recent example of the release of toxic waste in the Ivory Coast leading to the deaths of a number of people, and the hospitalisation of thousands underlines the risks involved in the movement and management of waste.

The rules governing waste imports and exports that are subject to notification procedures are designed to ensure that any potentially adverse environmental consequences of their treatment and management are considered before they are moved and that a system is in place to track their movement. This system also provides protection of the environment and accountability for its proper management by, for example, requiring that waste is sent back to its origin if it is found to have been moved illegally. However, there has been rising concern that compliance with the rules is being disregarded, whether deliberately or by accident.

Tackling the illegal shipment of waste is one of the many priorities in ensuring environmentally sound waste management. However, there is a growing body of evidence that suggests some operators are sending waste illegally to other countries. Companies and individuals involved in the import and export of waste, particularly 'Green List' waste, often conduct their activities outside the framework of control systems provided by the WSR and hence risk causing such shipments to be made illegally. The results of the IMPEL 'Seaport' project demonstrated the problem of illegal waste export across Europe. In addition to this, inspections by UK competent authorities at a number of UK ports have shown many of the waste shipment containers to be non-compliant.

The legislation governing this area makes it difficult to identify and enforce against illegal activity because of the relative ease with which unscrupulous exporters and importers can circumvent the requirements. Consequently, the

UK competent authorities currently have no means to obtain information about movements of 'Green List' waste, even those that are legitimately made. The revision of the WSR and necessary changes to the TFS Regulations offers an excellent opportunity to tackle this issue.

Non-compliance needs to be addressed as effectively and efficiently as possible to ensure that the UK meets its international obligations and to avoid legitimate business from being undermined. It has become clear that some traders are unaware of the relevant controls or are deliberately trying to avoid them. Whilst there is a need to raise awareness generally of the controls that apply, it is also clear that action needs to be taken against specific actors who are deliberately mis-classifying waste to avoid controls. There is no doubt that illegal waste is a problem.

European Union Network for the Implementation and Enforcement of Environmental Law also damage businesses that work to ensure that their waste is dealt with legally and in an environmentally sound manner.

In light of the publication of the revised WSR, the necessary review of domestic legislation has offered Defra the opportunity to work with regulators to see where we can strengthen enforcement activities. Consequently, there have been changes to the structure of the revised TFS Regulations to support the work that is being undertaken by the UK competent authorities, strengthen enforcement of the TFS regime and establish greater legal clarity.

Issues and Options

5.9 There is limited flexibility in terms of how the WSR may be transposed into national legislation. However, several Articles do present Member States with certain options in terms of how they can be implemented. In this context, there are four key issues to address in the TFS Regulations:

I. Reporting of information that accompanies shipments of 'green list' waste for recovery
II. Financial guarantees and equivalent insurance
III. Powers of HM Revenue and Customs to detain waste
IV. Charges for notified shipments

Under the revised WSR, movements of green waste are included within the scope of the Regulation and are now subject to certain procedural requirements, set out in Article 18 of the revised WSR. This places an obligation on anyone shipping 'green list' waste (recyclables) for recovery to complete certain information (as set out in Annex VII of the revised WSR) and requires that this information shall accompany the waste. This is to assist the tracking of shipments of such waste. The information required in Annex VII includes: the composition, quantity and destination of the waste, and a declaration that a contract with the consignee has been entered into.

Article 18(3) also allows Member States to require that the information contained in the Annex VII document be provided for inspection, planning and statistical purposes in accordance with national legislation. Should the UK wish to take advantage of this provision, it is necessary to include the relevant obligations and requirements in the TFS Regulations.

Old System for Notifications/Current Notifications

5.10 Under the existing WSR where notifiers have a shipment(s) comprised of different types of wastes they can include all the relevant waste codes on the notification form.

Notifiers can also submit 'general' notifications to cover several shipments if in the case of each shipment: the type of waste is the same; the waste is shipped to the same consignee and the same facility and the route of each shipment is as indicated on the notification.

Competent authorities may grant 'pre-consented' status to specific recovery facilities under their jurisdiction. Such consents must be notified to the Commission. There are currently nine pre-consented facilities in the UK.

Three key changes in the WSR will affect the pattern of notifications received under the new control system:

- Only one code can be used per notification when the general notification procedure is used (Article 6(4));
- Competent authorities can grant up to a maximum of three years (compared to one under the existing system) to pre-consented recovery facilities (Article 14(2));
- Waste destined for interim recovery or disposal operations will require separate, additional, notification to any subsequent interim and/or final recovery or disposal operations (Article 15).

Taking into account the changes to general notification, pre-consented notification and interim disposal or recovery operations, it is very difficult to predict how the current spread of notifications will change. The impact of distinguishing between interim and final operations, as set out in Article 15 of the WSR, is particularly difficult to model. However a 5% increase in notifications received in England and Wales, per annum, is a possibility.

New Structure for Charges

5.11 Due to the types of notifications most likely to be affected, continuing with the current charge structure could result in a significant increase in income. However, as a result of the planned efficiency changes, the Environment Agency predicts that they will be able to handle the increase in notifications without an increase in the total income. TFS income in 2005/06 was £1.7m.

SEPA does not currently fully recover regulatory, monitoring and reporting costs for this scheme. However, subject to approval of the revised charges SEPA will reduce the shortfall in cost recovery and be able to handle the predicted increase in notifications.

The proposed new charges are structured to more closely reflect the differences in regulatory effort required by the different types of notification. The proposed charges differ according to where the shipment is going to, or coming from and whether it is destined for recovery or disposal. It is proposed to replace the charges per notification and per shipment with charge per notification based on a banding scheme of number of shipment per notification. A more proportionate approach to transit notifications means that these can be processed free of charge.

THE TRANSFRONTIER SHIPMENT OF WASTE REGULATIONS 2007, SI 2007/1711 (AS AMENDED)

5.12 These Regulations ensure full implementation and enforcement of the revised Waste Shipments Regulation 2006 (EC 1013/2006), which was adopted in the European Union in June 2006 and which makes provision for the supervision and control of shipments of waste within, into and out of the European Union.

Background

5.13 The Regulations are made under s 2(2) of the European Communities Act 1972 and are made for the purpose of implementing and enforcing the provisions of Regulation (EC) No 1013/2006 on shipments of waste.

The Regulations replace current transposition through controls imposed by:

- Transfrontier Shipment of Waste Regulations 1994, SI 1994/1137 amended by Transfrontier Shipment of Waste (Amendment) Regulations 2005, SI 2005/187.
- Transfrontier Shipment of Waste (Fee) Regulations (Northern Ireland) 2005, SI (NI) 2005/90.

Extent and Application (Regulation 2)

5.14 Regulation 2 has been revoked by the Transfrontier Shipment of Waste (Amendment) Regulations 2014, SI 2014/861.

This applies to all of the United Kingdom including the territorial sea adjacent to the United Kingdom, including the area on the landward side of the baselines submerged at mean high water springs, the seabed and subsoil situated within the areas designated by Order in Council under section 1(7) of the Continental Shelf Act 1964 and the waters superjacent to the seabed and the seabed and its subsoil situated within the area[9].

The UNEP Basel Convention[10] on transboundary movements of waste and their disposal sets out the global control framework for the import and export of wastes. The related OECD Decision[11] also applies to the control of transboundary movements of waste destined for recovery operations between OECD countries. The provisions of the Convention and the OECD decision have been implemented in the European Union through Council Regulation (EEC) No 259/93 on the supervision and control of shipments of waste within, into and out of the European Community (the Waste Shipments Regulation). The Waste Shipments Regulation is supplemented in the UK by the Transfrontier Shipment of Waste Regulations 1994, SI 1994/1137 and the UK Management Plan for Exports and Imports of Waste.

Following changes to the Basel Convention and the OECD decision, the Waste Shipments Regulation was revised to ensure that the changes were implemented within the Community. The revised Regulation[12] will apply in Member States from 12 July 2007. The provisions of the Transfrontier Shipment of Waste Regulations 1994 are replaced by the Transfrontier Shipment of Waste Regulations 2007 to take into account the changes in the EC's Waste Shipments Regulation.

These Regulations were amended in 2014[13] by the Transfrontier Shipment of Waste (Amendment) Regulations 2014, SI 2014/861 and by The Environmental Civil Sanctions (Miscellaneous Amendments) (England) Regulations) 2010, SI 2010/1159. These changes have been included in this chapter and all provisions appear, as amended.

Definitions (Regulation 4)

5.15 The amended definitions are:

Community Regulations means Regulation (EC) No 1013/2006 of the European Parliament and of the Council on shipments of waste,

The 'competent authority' means a competent authority within the meaning of regulation 6[14] or 7[15].

9 Order in Council under s 84(4) of the Energy Act 2004.

10 Basel Convention on the transboundary movements of hazardous wastes and their disposal – Protocol on liability and compensation for damage resulting from transboundary movements of hazardous wastes and their disposal UNEP August 2014 UNEP/BRS/2014/3

11 Decision of the Council concerning the Control of Transboundary Movements of Wastes Destined from Recovery Operations C(92)39/FINAL.

12 Regulation (EC) No 1013/2006 of the European Parliament and of the Council.

13 The Transfrontier Shipment of Waste (Amendment) Regulations 2014, SI 2014/861.

14 SI 2014/86 includes Natural Resources Body for Wales.

15 SI 2014/861 definition competent body for transit.

Competent Authorities (Regulation 6)

5.16 The Transfrontier Shipment of Waste Regulations 2007 set out offences and penalties and designate the competent authorities for enforcing the Waste Shipments Regulations in the UK[16].

(a) Environment Agency for England;
(b) Natural Resources Body for Wales[17], for Wales and the Welsh area;
(c) Scottish Environment Protection Agency (SEPA) in Scotland;
(d) Department of the Environment in Northern Ireland.

Competent Authority for Transit (Regulation 7)

5.17

(1) The Environment Agency is the competent authority for the transit[18] for the United Kingdom for the purpose of Community Regulations.
(2) The Natural Resources Body for Wales and the Scottish Environment Protection Agency and the Department of the Environment Northern Ireland must supply the Environment Agency with any information that the Environment Agency may require in connection with its functions as the competent authority in transit.

The Regulations also amend the Environment Act 1995 to allow the Environment Agency and the Scottish Environment Protection Agency to prescribe fees for the carrying out of their functions under the Community Regulation.

Notices (Regulation 9)

5.18 Any notice given under these regulations must be in writing and can be subject to conditions, which may be amended, suspended or revoked in writing at any time.

Waste Management Plan (Regulation 11)

5.19 The Secretary of State must prepare a waste management plan in accordance with Article 7 of Directive 2006/12/EC of the European Parliament and of the Council on waste containing his policies on the bringing into, or dispatch from, the United Kingdom of waste for disposal.

16 SI 2007/1711, Part 1, reg 6.
17 SI 2014/861, reg 6(b).
18 SI 2014/861, reg 7.

Consultation

Public Participation Procedures (Regulation 12)

5.20

(1) As soon as reasonably practicable after preparing proposals for a waste management plan, the Secretary of State must:

  (a) take such steps as he considers appropriate to bring the proposals to the attention of the persons who in his opinion are affected or likely to be affected by, or have an interest in, the plan (in this Part, referred to as "public consultees");
  (b) inform public consultees of the address from which a copy of the proposals may be obtained;
  (c) invite public consultees to express their opinions on the proposals, specifying the address to which, and the period within which, opinions must be sent;
  (d) the period referred to must be sufficient to ensure that public consultees are given an effective opportunity to express their opinions on the proposals.

(2) The Secretary of State must keep a copy of the proposals for inspection by the public at all reasonable times free of charge.
(3) The Secretary of State may make a reasonable charge for any copies.

Procedures Following Public Participation (Regulation 13)

5.21

(1) Before decisions on a waste management plans are made, the Secretary of State must take account of any opinions expressed by public consultees in accordance with regulation 12.
(2) As soon as reasonably practicable after making decisions on a waste management plan, the Secretary of State must—

  (a) inform public consultees of the matters referred to in paragraph xx;
  (b) take such steps as he considers appropriate to bring those matters to the attention of the public;
  (c) if he has adopted the waste management plan, make a copy of it available for inspection at all reasonable times free of charge.

(3) The matters are—

  (a) the decisions made by the Secretary of State on the waste management plan;
  (b) the reasons and considerations upon which those decisions are based; and
  (c) information about the public participation procedure.

Provision of Plan to Competent Authorities (Regulation 14)

5.22 The Secretary of State must send a copy of the waste management plan to—

(a) Environment Agency for England;
(aa) Natural Resources Body for Wales[19];
(b) Scottish Environment Protection Agency; and
(c) Department of the Environment in Northern Ireland.

Objection to Shipments of Waste (Regulation 15)

5.23 To prevent shipments of waste that are not in accordance with the waste management plan—

(a) a competent authority for the destination must object to any shipment notified in accordance with the Community Regulation which the plan indicates should not be brought into the United Kingdom; and
(b) a competent authority for dispatch must object to any shipment notified in accordance with the Community Regulation which that plan indicates should not be dispatched from the United Kingdom.

Waste Management Licensing Regulations 1994 (Regulation 16)

5.24 Paragraph 4(1)(b) of Schedule 4 to the Waste Management Licensing Regulations 1994 and paragraph 4(1)(b) of Part 1 of Schedule 3 to the Waste Management Licensing Regulations (Northern Ireland) 2003 have effect as if the references in those paragraphs to any plan made under the plan-making provisions include references to a waste management plan made under this Part.

General Requirement for Shipment of Waste

Protection of the Environment (Regulation 17)

5.25 A person commits an offence if they fail to comply with Article 49(1)[20] (the management of shipments of waste in an environmentally sound manner and without endangering human health).

19 SI 2014/861, reg 9. This now includes within reg 14(aa) the Natural Resources Body for Wales.
20 EC 1013/2009, Article 49(1). The producer, the notifier and other undertakings involved in a shipment of waste and/or its recovery or disposal shall take the necessary steps to ensure that any waste they ship is managed without endangering human health and in an environmentally sound manner throughout the period of shipment and during its recovery and disposal. In particular, when the shipment takes place in the EU, the requirements of Article 4 of Directive 2006/12/EC and other EU legislation on waste shall be respected.

Compliance with Notification (Regulation 18)

5.26 A person commits an offence if they transport waste other than in accordance with—

(a) the notification document or movement document, in the case of notifiable waste; or
(b) the Annex VII document, in the case of waste subject to the procedural requirements of Article 18(1)[21].

Shipment of Waste Within the Community

Requirements for Shipments of Notifiable Waste (Regulation 19)

5.27

(1) This applies to waste which is specified in Article 3(1)(a)[22] or (b)[23] that originates in the Community and is either brought into the United Kingdom or destined for another Member State. Regardless of whether or not the waste passes through a third country.
(2) A person who transports such waste will be committing an offence if they do so without—
    (a) notifying the competent authority of dispatch in accordance with Article 4[24];
    (b) having a contract with a declaration made in accordance with Article 5[25];

21 EC 1013/2009, Article 18(1). Waste as referred to in Article 3(2) and (4) that is intended to be shipped shall be subject to the following procedural requirements:
(a) In order to assist the tracking of shipments of such waste, the person under the jurisdiction of the country of dispatch who arranges the shipment shall ensure that the waste is accompanied by the document contained in Annex VII.
(b) The document contained in Annex VII shall be signed by the person who arranges the shipment before the shipment takes place and shall be signed by the recovery facility or the laboratory and the consignee when the waste in question is received

22 (a) if destined for disposal operations: all wastes.

23 (b) if destined for recovery operations:
(i) wastes listed in Annex IV, which include, inter alia, wastes listed in Annexes II and VIII to the Basel Convention,
(ii) wastes listed in Annex IVA,
(iii) wastes not classified under one single entry in either Annex III, IIIB, IV or IVA,
(iv) mixtures of wastes not classified under one single entry in either Annex III, IIIB, IV or IVA unless listed in Annex IIIA.

24 EC 1013/2006, Article 4. Where the notifier intends to ship waste as referred to in Article 3(1) (a) or (b), he/she shall submit a prior written notification to and through the competent authority of dispatch and, if submitting a general notification, comply with Article 13.

25 EC 1013/2006, Article 5(1). All shipments of waste for which notification is required shall be subject to the requirement of the conclusion of a contract between the notifier and the consignee for the recovery or disposal of the notified waste.

(c) a financial guarantee or equivalent insurance being in place and approved by the competent authority in accordance with Article 6[26];

(d) a consent issued by the competent authority of dispatch and destination in accordance with Article 9[27] and a consent issued by the competent authority of transit, unless there is tacit consent to the transit in accordance with Article 9(1)[28];

(e) the movement document having been—

(i) completed in accordance with Article 16[29], first paragraph and Article 16(a)[30]; or

(ii) sent to the competent authorities concerned and the consignee in accordance with Article 16(b)[31];

(f) the waste being accompanied by the movement document and notification document in accordance with Article 16(c)[32]; or

(g) the competent authorities and consignee having been notified of any change in the details or conditions of the shipment in accordance with Article 17[33] and, in such a case, a new notification having been submitted in accordance with that Article.

(3) An offence is committed if:

(a) a condition imposed under Article 10[34] is breached;

26 EC 1013/2006, Article 6, Financial Guarantee.

27 EC 1013/2006, Article 9. Consents by the competent authorities of destination, dispatch and transit and time periods for transport, recovery or disposal.

28 EC 1013/2006, Article 9(1). The competent authorities of destination, dispatch and transit shall have 30 days following the date of transmission of the acknowledgement by the competent authority of destination in accordance with Article 8 in which to take one of the following duly reasoned decisions in writing as regards the notified shipment:
(a) consent without conditions;
(b) consent with conditions in accordance with Article 10; or
(c) objections in accordance with Articles 11 and 12.
Tacit consent by the competent authority of transit may be assumed if no objection is lodged within the said 30-day time limit.

29 EC 1013/2006, Article 16. Requirements following consent to a shipment.

30 EC 1013/2006, Article 16(a). Completion of the movement document by the notifier: once the notifier has received consent from the competent authorities of dispatch, destination and transit or, in relation to the competent authority of transit, can assume tacit consent, he/she shall insert the actual date of shipment and otherwise complete the movement document to the extent possible.

31 EC 1013/2006, Article 16(b). Prior information regarding actual start of shipment: the notifier shall send signed copies of the then completed movement document, as described in point (a), to the competent authorities concerned and to the consignee at least three working days before the shipment starts.

32 EC 1013/2006, Article 16(c), Documents to accompany each transport: the notifier shall retain a copy of the movement document. The movement document and copies of the notification document containing the written consents and the conditions of the competent authorities concerned shall accompany each transport. The movement document shall be retained by the facility which receives the waste.

33 EC 1013/2006, Article 17, Changes in the shipment after consent.

34 EC 1013/2006, Article 10, Conditions of shipment.

(b) In the case of a shipment covered by a general notification, they do so without the competent authorities concerned having been notified of a change in route in accordance with Article 13(2)[35];
(c) in the case of a shipment of waste destined for an interim recovery or disposal operation, he does so without the notification document having been completed in accordance with Article 15(a)[36]; or
(d) they fail to comply with Article 19[37].

Requirements of Shipment of Waste Specified in Article 3(2) and (4) (Regulation 20)

5.28

(1) This regulation applies to waste specified in Article 3(2)[38] or (4)[39] that originates in the Community and is either brought into the United Kingdom or is destined for another Member State, whether or not that waste passes through a third country.
(2) An offence is committed by the person transporting waste if:

35 EC 1013/2006, Article 13(2), If, owing to unforeseen circumstances, the same route cannot be followed, the notifier shall inform the competent authorities concerned as soon as possible and, if possible, before the shipment starts if the need for modification is already known.
36 EC 1013/2006, Article 15(a).
37 EC 1013/2006, Article 19, **Prohibition on mixing waste during shipment**
From the start of the shipment to the receipt in a recovery or disposal facility, waste, as specified on the notification document or as referred to in Article 18, shall not be mixed with other waste.
38 EC 1013/2006, Article 3(2), Shipments of the following wastes destined for recovery shall be subject to the general information requirements laid down in Article 18, if the amount of waste shipped exceeds 20 kg:
(a) waste listed in Annex III or IIIB;
(b) mixtures, not classified under one single entry in Annex III, of two or more wastes listed in Annex III, provided that the composition of these mixtures does not impair their environmentally sound recovery and provided that such mixtures are listed in Annex IIIA, in accordance with Article 58.
39 EC 1013/2006, Article 3(4), Shipments of waste explicitly destined for laboratory analysis to assess either its physical or chemical characteristics or to determine its suitability for recovery or disposal operations shall not be subject to the procedure of prior written notification and consent as described in paragraph 1. Instead, the procedural requirements of Article 18 shall apply. The amount of such waste exempted when explicitly destined for laboratory analysis shall be determined by the minimum quantity reasonably needed to adequately perform the analysis in each particular case, and shall not exceed 25 kg.

(a) they do so without the waste being accompanied by a completed Annex VII document in accordance with Article 18(1)(a)[40], that is signed in accordance with Article 18(1)(b)[41];
(b) in the case of waste specified in Article 3(2)[42], they do so without a contract as referred to in Article 18(2)[43] having been entered into or
(c) they fail to comply with Article 19[44].

Export of Waste to Third Countries

Prohibition of Export of Waste for Disposal Other than EFFTA Countries (Regulation 21)

5.29 An offence is committed if a person transports waste destined for disposal in a third country in breach of Article 34 (prohibition on export except, in certain circumstances, to EFTA[45] countries which are parties to the Basel Convention).

Requirements for Export of Waste for Disposal to EFTA Countries (Regulation 22)

5.30

(1) This regulation applies to waste destined for disposal in any EFTA country, party to the Basel Convention.
(2) It is an offence if a person who transports such waste does so—
    (a) without the provisions of Article 35(1)[46] having been complied with (procedural requirements for export to EFTA countries) or

40 EC 1013/2006, Article 18(1)(a), In order to assist the tracking of shipments of such waste, the person under the jurisdiction of the country of dispatch who arranges the shipment shall ensure that the waste is accompanied by the document contained in Annex VII.
41 EC 1013/2006, Article 18(1)(b), The document contained in Annex VII shall be signed by the person who arranges the shipment before the shipment takes place and shall be signed by the recovery facility or the laboratory and the consignee when the waste in question is received.
42 EC 1013/2006, Article 3(2), Shipments of the following wastes destined for recovery shall be subject to the general information requirements laid down in Article 18, if the amount of waste shipped exceeds 20 kg:
(a) waste listed in Annex III or IIIB;
(b) mixtures, not classified under one single entry in Annex III, of two or more wastes listed in Annex III, provided that the composition of these mixtures does not impair their environmentally sound recovery and provided that such mixtures are listed in Annex IIIA, in accordance with Article 58.
43 EC 1013/2006, Article 18(2), Waste to be accompanied by certain information.
44 EC 1013/2006, Article 19, Prohibition on mixing waste during shipment.
45 EFTA – European Free Trade Association.
46 EC 1013/2006, Article 35(1), Where waste is exported from the Community and destined for disposal in EFTA countries Parties to the Basel Convention, the provisions of Title II shall apply mutatis mutandis, with the adaptations and additions listed in paragraphs 2 and 3.

(b) in breach of Article 35(5)[47] (requirement for consignment only to facilities operating or authorised to operate under the applicable national law of the country of destination).

Prohibition on Export of Certain Waste for Recovery to Non-OECD Countries (Regulation 23)

5.31 An offence is committed if a person in breach of Article 36(1)[48], transports waste specified in that Article that is destined for recovery in a country to which the OECD Decision does not apply.

Requirements for Export of Waste for Recovery to OECD Decision Countries (Regulation 24)

5.32

(1) This applies to waste specified in Article 38(1)[49] destined for recovery in any country to which the OECD Decision applies (with or without transit through any country to which that Decision applies).

(2) An offence is committed if a person transports such waste:

(a) without the provisions of Article 38(1) having been complied with (procedural requirements for the export of waste listed in Annexes III, IIIA, IIIB, IV and IVA); or

(b) in breach of Article 38(6)[50] (requirement for consignment only to facilities operating or authorised to operate under the applicable national law of the country of destination).

Miscellaneous Prohibitions on Export of Waste (Regulation 25)

5.33 An offence is committed if a person:

47 EC 1013/2006, Article 35(5), Where waste is exported, it shall be destined for disposal operations within a facility which, under applicable national law, is operating or is authorised to operate in the country of destination.

48 EC 1013/2006, Article 36(1), Exports Prohibition.

49 EC 1013/2006, Article 38(1), Where waste listed in Annexes III, IIIA, IIIB, IV and IVA, waste not classified or mixtures of wastes not classified under one single entry in either Annex III, IV or IVA are exported from the Community and destined for recovery in countries to which the OECD Decision applies, with or without transit through countries to which the OECD Decision applies, the provisions of Title II shall apply mutatis mutandis, with the adaptations and additions listed in paragraphs 2, 3 and 5.

50 EC 1013/2006, Article 38(6), Where waste is exported, it shall be destined for recovery operations within a facility which, under applicable national law, is operating or is authorised to operate in the country of destination.

(a) transports waste to the Antarctic;
(b) transports waste destined for disposal in an overseas country or territory in breach of Article 40(1)[51];
(c) transports waste destined for recovery in an overseas country or territory in breach of Article 40(2)(prohibition on the export of certain waste); or
(d) transports waste specified in Article 40(3)[52] destined for recovery in an overseas country or territory without the provisions of that Article having been complied with (procedural requirements for exports to overseas countries or territories).

Import of Waste from Third Countries

Prohibition on the Import of Waste for Disposal (Regulation 26)

5.34 An offence is committed if, in breach of Article 41(1)[53], he transports waste destined for disposal that has come from a third country or area other than those specified in that Article.

Procedure for Import of Waste for Disposal from a Country Party to the Basel Convention (Regulation 27)

5.35

(1) This applies to waste destined for disposal that has come from a country party to the Basel Convention.
(2) An offence is committed if a person does so without the provisions of Article 42(1)[54] having been complied with (procedural requirements for the import of waste destined for disposal from countries parties to the Basel Convention).

51 EC 1013/2006, Article 40(1), Exports from the Community of waste destined for disposal in overseas countries or territories shall be prohibited.

52 EC 1013/2006, Article 40(3), As regards exports of waste destined for recovery in overseas countries or territories not covered by the prohibition set out in paragraph 2, the provisions of Title II shall apply mutatis mutandis.(*when comparing two or more cases or situations*).

53 EC 1013/2006, Article 41(1), Imports prohibited except from a country Party to the Basel Convention or with an agreement in place or from other areas during situations of crisis or war.

54 EC 1013/2006, Article 42(1), Where waste is imported into the Community and destined for disposal from countries Parties to the Basel Convention, the provisions of Title II shall apply mutatis mutandis, with the adaptations and additions listed in paragraphs 2 and 3.

Prohibition on Import of Waste for Recovery (Regulation 28)

5.36 An offence is committed if a person in breach of Article 43(1)[55], transports waste destined for recovery that has come from a third country or area other than specified in that Article.

Requirements for Import of Waste for Recovery from or through an OECD Decision Country (Regulation 29)

5.37

(1) This applies to waste that is destined for recovery that has come from or through a country to which the OECD Decision applies.

(2) An offence is committed if a person who transports such waste does so without the provisions of Article 44(1)[56] having been complied with (procedural requirements for imports of waste destined for recovery from, or through, an OECD Decision country).

Requirements for Import of Waste for Recovery from a Non-OECD Decision Country Party to the Basel Convention (Regulation 30)

5.38

(1) This applies to waste destined for recovery that has come

  (a) from a country to which the OECD Decision does not apply; or

  (b) through a country to which the OECD Decision does not apply and that is party to the Basel Convention.

(2) An offence is committed if the person who transports such waste does so without the provisions of Article 45[57] having been complied with (procedural requirements for imports from a non-OECD Decision country party to the Basel Convention).

55 EC 1013/2006, Article 43(1), Imports prohibited except from an OECD Decision country or a country Party to the Basel Convention or with an agreement in place or from other areas during situations of crisis or war.

56 EC 1013/2006, Article 44(1), Where waste destined for recovery is imported into the Community from countries and through countries to which the OECD Decision applies, the provisions of Title II shall apply mutatis mutandis, with the adaptations and additions listed in paragraphs 2 and 3.

57 EC 1013/2006, Article 45, Procedural requirements for imports from a non-OECD Decision country Party to the Basel Convention or from other areas during situations of crisis or war.

Requirements for Import of Waste from Overseas Countries or Territories (Regulation 31)

5.39 A person who transports waste that has come from any overseas country or territory commits an offence if they do so without the provisions of Article 46(1)[58] having been complied with (procedural requirements for the import of waste from overseas countries or territories).

Transit of Waste

Requirements for Transit of Waste for Disposal (Regulation 32)

5.40

(1) This applies to waste destined for disposal that

(a) originates from, and is destined for, a third country; and
(b) is transported through the United Kingdom.

(2) An offence is committed when a person who transports such waste does so without the provisions of Article 47[59] having been complied with (procedural requirements for shipments of waste destined for disposal from and to a third country).

Requirements for Transit of Waste for Recovery: Non-OECD Decision Countries (Regulation 33)

5.41

(1) This applies to waste destined for recovery that—

(a) originates from, and is destined for, a country to which the OECD Decision does not apply; and
(b) is transported through the United Kingdom.

(2) An offence is committed if a person who transports such waste does so without the provisions of Article 48(1)[60] having been complied with (procedural requirements for shipments of waste destined for recovery from and to a country to which the OECD Decision does not apply).

58 EC 1013/266, Article 46(1), Where waste is imported into the Community from overseas countries or territories, Title II shall apply mutatis mutandis.
59 EC 1013/2006, Article 47, Transit through the Community of waste destined for disposal.
60 EC 1013/2006, Article 48(1), Where waste destined for recovery is shipped through Member States from and to a country to which the OECD Decision does not apply, Article 47 shall apply mutatis mutandis.

Requirements for Transit of Waste for Recovery: OECD Decision Countries (Regulation 34)

5.42

(1) This applies to waste destined for recovery that—

(a) originates from, and is destined for, a country to which the OECD Decision applies; and
(b) is transported through the United Kingdom.

(2) An offence is committed if a person who transports such waste does so without the provisions of Article 48(2)[61] having been complied with (procedural requirements for shipments of waste destined for recovery from and to a country to which the OECD Decision applies).

Requirements for Transit of Waste Between OECD Decision Countries and Non-OECD Decision Countries (Regulation 35)

5.43

(1) This applies to waste destined for recovery that—

(a) originates from a country to which the OECD Decision does not apply and is destined for a country to which the OECD Decision does apply; or
(b) originates from a country to which the OECD Decision applies and is destined for a country to which the OECD Decision does not apply

and is transported through the United Kingdom.

(2) An offence is committed if a person who transports such waste does so without the provisions of Article 48(3)[62] having been complied with (procedural requirements for shipments of waste destined for recovery between OECD Decision and non-OECD Decision countries).

61 EC 1013/2006, Article 48(2), Where waste destined for recovery is shipped through Member States from and to a country to which the OECD Decision applies, Article 44 shall apply mutatis mutandis, with the adaptations and additions listed below:
(a) the first and last competent authority of transit in the Community shall, where appropriate, send a stamped copy of the decisions to consent to the shipment or, if they have provided tacit consent, a copy of the acknowledgement in accordance with Article 42(3)(a) to the customs offices of entry into and exit from the Community respectively; and
(b) as soon as the waste has left the Community, the customs office of exit from the Community shall send a stamped copy of the movement document to the competent authority(ies) of transit in the Community, stating that the waste has left the Community.

62 EC 1013/2006, Article 48(3), Where waste destined for recovery is shipped through Member States from a country to which the OECD Decision does not apply to a country to which the OECD Decision applies or vice versa, paragraph 1 shall apply as regards the country to which the OECD Decision does not apply and paragraph 2 shall apply as regards the country to which the OECD Decision applies.

Additional Duties

Illegal Shipments of Waste Brought to a Facility (Regulation 36)

5.44 If the operator of a facility knows or has reasonable grounds to suspect that waste brought to that facility is an illegal shipment of waste, they must notify the competent authority immediately and comply with their instructions. Failure to do so is an offence.

Duties on Operators of Facilities in Respect of Notifiable Waste (Regulation 37)

5.45

(1) This applies to the operator of a facility that receives notifiable waste.
(2) An offence is committed if they fail to—

  (a) ensure that the recovery or disposal of such waste is completed within the time period in Article 9(7)[63];
  (b) comply with any condition of a consent imposed on him by the competent authority of destination in accordance with Article 10(1)[64];
  (c) comply with any condition, imposed by the competent authority of destination in accordance with Article 10(5)[65], requiring him to keep records specified in that Article and send them to that competent authority in accordance with that Article;
  (d) retain the movement document in accordance with Article 16(c)[66];

63 EC 1013/2006, Article 9(7), The recovery or disposal of waste in relation to a planned shipment shall be completed no later than one calendar year from the receipt of the waste by the facility, unless a shorter period is indicated by the competent authorities concerned.

64 EC 1013/2006, Article 10(1), The competent authorities of dispatch, destination and transit may, within 30 days following the date of transmission of the acknowledgement of the competent authority of destination in accordance with Article 8, lay down conditions in connection with their consent to a notified shipment. Such conditions may be based on one or more of the reasons specified in either Article 11 or Article 12.

65 EC 1013/2006, Article 10(5), The competent authority of destination may also, within the 30-day time limit referred to in paragraph 1, lay down a condition that the facility which receives the waste shall keep a regular record of inputs, outputs and/or balances for wastes and the related recovery or disposal operations as contained in the notification, and for the period of validity of the notification. Such records shall be signed by a person legally responsible for the facility and be sent to the competent authority of destination within one month of completion of the notified recovery or disposal operation.

66 EC 1013/2006, Article 16(c), Documents to accompany each transport: the notifier shall retain a copy of the movement document. The movement document and copies of the notification document containing the written consents and the conditions of the competent authorities concerned shall accompany each transport. The movement document shall be retained by the facility which receives the waste.

(e) provide, in accordance with Article 16(d)[67], confirmation that waste has been received;
(f) certify, in accordance with Article 16(e)[68], that non-interim recovery or disposal has been completed;
(g) keep any document sent to or by the competent authorities in relation to a notified shipment in accordance with Article 20(1)[69]; or
(h) inform the competent authority of destination in accordance with Article 22(1)[70] if a shipment of waste is rejected.

Interim Recovery or Disposal Operations (Regulation 38)

5.46

(1) This applies to the operator of a facility that carries out interim recovery or disposal operations.
(2) They commit an offence when they fail to:

67 EC 1013/2006, Article 16(d), Written confirmation of receipt of the waste by the facility: within three days of receipt of the waste, the facility shall provide confirmation in writing that the waste has been received.
This confirmation shall be contained in, or annexed to, the movement document.
The facility shall send signed copies of the movement document containing this confirmation to the notifier and to the competent authorities concerned.

68 EC 1013/2006, Article 16(e), Certificate for non-interim recovery or disposal by the facility: as soon as possible, but no later than 30 days after completion of the non-interim recovery or disposal operation, and no later than one calendar year, or a shorter period in accordance with Article 9(7), following receipt of the waste, the facility carrying out the operation shall, under its responsibility, certify that the non-interim recovery or disposal has been completed.
This certificate shall be contained in, or annexed to, the movement document.
The facility shall send signed copies of the movement document containing this certificate to the notifier and to the competent authorities concerned.

69 EC 1013/2006, Article 20(1), All documents sent to or by the competent authorities in relation to a notified shipment shall be kept in the Community for at least three years from the date when the shipment starts, by the competent authorities, the notifier, the consignee and the facility which receives the waste.

70 EC 1013/2006, Article 22(1), Where any of the competent authorities concerned becomes aware that a shipment of waste, including its recovery or disposal, cannot be completed as intended in accordance with the terms of the notification and movement documents and/or contract referred to in the second subparagraph, point 4 of Article 4 and in Article 5, it shall immediately inform the competent authority of dispatch. Where a recovery or disposal facility rejects a shipment received, it shall immediately inform the competent authority of destination.

(a) provide confirmation to the notifier or competent authorities concerned of the receipt of waste for interim recovery or disposal operations in accordance with Article 15(c)[71]; or
(b) certify, in accordance with Article 15(d)[72], that interim recovery or disposal has been completed.

(3) An offence is committed if, in the case of waste sent to another facility for subsequent interim or non-interim recovery or disposal operation, they fail to:

(a) obtain from that facility, in accordance with Article 15(e), first paragraph, a certificate that such operation has been completed; or
(b) transmit that certificate to the notifier and competent authorities concerned in accordance with Article 15(e)[73], second paragraph.

(4) They commit an offence if they deliver waste for subsequent interim or non-interim recovery or disposal operations in accordance with Article 15(f)[74], without complying with the notification requirements of that Article.

71 EC 1013/2066, Article 15(c), Within three days of the receipt of the waste by the facility which carries out this interim recovery or disposal operation, that facility shall provide confirmation in writing that the waste has been received.
This confirmation shall be supplied on, or annexed to, the movement document. The said facility shall send signed copies of the movement document containing this confirmation to the notifier and to the competent authorities concerned.

72 EC 1013/2006, Article15(d), As soon as possible, but no later than 30 days after completion of the interim recovery or disposal operation, and no later than one calendar year, or a shorter period in accordance with Article 9(7), following the receipt of the waste, the facility carrying out this operation shall, under its responsibility, certify that the interim recovery or disposal has been completed.
This certificate shall be contained in, or annexed to, the movement document.
The said facility shall send signed copies of the movement document containing this certificate to the notifier and to the competent authorities concerned.

73 EC 1013/2006, Article 15(e), When a recovery or disposal facility which carries out an interim recovery or disposal operation delivers the waste for any subsequent interim or non-interim recovery or disposal operation to a facility located in the country of destination, it shall obtain as soon as possible but no later than one calendar year following delivery of the waste, or a shorter period in accordance with Article 9(7), a certificate from that facility that the subsequent non-interim recovery or disposal operation has been completed.
The said facility that carries out an interim recovery or disposal operation shall promptly transmit the relevant certificate(s) to the notifier and the competent authorities concerned, identifying the shipment(s) to which the certificate(s) pertain.

74 EC 1013/2006, Article 15(e), When a delivery as described in subparagraph (e) is made to a facility respectively located:
(i) in the initial country of dispatch or in another Member State, a new notification shall be required in accordance with the provisions of this Title, or
(ii) in a third country, a new notification shall be required in accordance with the provisions of this Regulation, with the addition that the provisions concerning the competent authorities concerned shall also apply to the initial competent authority of the initial country of dispatch.

Duties on Operators of Facilities in Respect of Waste Subject to General Information Requirements (Regulation 39)

5.47

(1) This applies to the operator of a recovery facility that receives waste subject to the procedural requirements of Article 18(1)[75].
(2) An offence if committed if they fail to—

   (a) sign the Annex VII document, in accordance with Article 18(1)(b); or
   (b) keep the information given pursuant to Article 18(1), in accordance with Article 20(2)[76].

Duties on Consignees in Respect of Notifiable Waste (Regulation 40)

5.48

(1) This applies to a consignee of notifiable waste.
(2) An offence is committed if he fails to keep any document sent to or by the competent authorities in relation to a notified shipment in accordance with Article 20(1)[77].

Duties on Consignees in Respect of Waste Subject to General Information Requirements (Regulation 41)

5.49

(1) This applies to a consignee of waste subject to the procedural requirements of Article 18(1).
(2) An offence is committed if they fail to—

   (a) sign the Annex VII document, in accordance with Article 18(1)(b);

75 EC 1013/2006, Article 18(1), Waste as referred to in Article 3(2) and (4) that is intended to be shipped shall be subject to the following procedural requirements:
   (a) In order to assist the tracking of shipments of such waste, the person under the jurisdiction of the country of dispatch who arranges the shipment shall ensure that the waste is accompanied by the document contained in Annex VII.
   (b) The document contained in Annex VII shall be signed by the person who arranges the shipment before the shipment takes place and shall be signed by the recovery facility or the laboratory and the consignee when the waste in question is received.

76 EC 1013/2006, Article 20(2), Information given pursuant to Article 18(1) shall be kept in the Community for at least three years from the date when the shipment starts, by the person who arranges for the shipment, the consignee and the facility which receives the waste.

77 EC 1013/2006, Article 20(2), Information given pursuant to Article 18(1) shall be kept in the Community for at least three years from the date when the shipment starts, by the person who arranges for the shipment, the consignee and the facility which receives the waste.

(b) provide a copy of the contract referred to in Article 18(2), second paragraph, to the competent authority upon request, in accordance with that Article; or
(c) keep information given pursuant to Article 18(1), in accordance with Article 20(2).

Duties on Laboratories (Regulation 42)

5.50

(1) This applies to the operator of a laboratory that receives waste subject to the procedural requirements of Article 18(1).
(2) An offence is committed if they fail to—

(a) sign the Annex VII document, in accordance with Article 18(1)(b)[78]; or
(b) keep the Annex VII document for three years from the date the shipment started.

Notifier's Duty to Keep Documents (Regulation 43)

5.51 A notifier commits an offence if they fail to keep—

(a) a copy of the movement document in accordance with Article 16(c); or
(b) any document sent to or by the competent authorities in relation to a notified shipment in accordance with Article 20(1).

Duties on Persons Who Arrange the Shipment of Waste (Regulation 44)

5.52

(1) This regulation applies to a person who arranges the shipment of waste subject to the procedural requirements of Article 18(1).
(2) They commit an offence if they fail to—

(a) provide the competent authority with a copy of the contract referred to in Article 18(2) upon request, in accordance with the second paragraph of that Article; or
(b) keep the information given pursuant to Article 18(1), in accordance with Article 20(2).

78 EC 1013/2006, Article 18(1)(b), The document contained in Annex VII shall be signed by the person who arranges the shipment before the shipment takes place and shall be signed by the recovery facility or the laboratory and the consignee when the waste in question is received.

Annex VII Document (Regulation 45)

5.53

(1) This applies in Scotland and Northern Ireland only.
(2) A person who arranges the shipment of waste subject to the procedural requirements of Article 18(1) must ensure that a copy of the Annex VII document that he has signed in accordance with Article 18(1)(b) is received by the competent authority before the shipment takes place and failure to do so is an offence.

Miscellaneous

Fees (Regulation 46)

5.54 Schedule 3 sets out the fees that will apply in Northern Ireland. Provision is also made for the competent authorities to recover the costs incurred in taking back illegal shipments and shipments that cannot be completed as intended.

There is also provision for enforcement to be by the competent authorities and Schedule 5 sets out powers that may be used by any person authorised by a competent authority. These include the power to serve enforcement and prohibition notices and powers to seize wastes. There are also powers to competent authorities to enable them to carry out their take-back functions under the Community Regulations as well as powers to officers of HM Revenue and Customs to detain waste for up to five days (as opposed to the current three), where requested to do so by a competent authority.

Costs of Take-Back etc (Regulation 47)

5.55 A competent authority may charge fees to meet any costs arising from

(a) the return of waste from a shipment that cannot be completed (including costs of transport, recovery, disposal and storage) in accordance with Article 23; and
(b) the take-back, recovery or disposal of waste from an illegal shipment (including costs of transport and storage) in accordance with Article 25.

Approval of a Financial Guarantee (Regulation 48)

5.56 Schedule 4 (application for approval of a financial guarantee) has effect.

This Schedule sets out the process for the approval of a financial guarantee.

Supply of the Financial Guarantee (Regulation 49)

5.57 The notifier must supply

(a) the financial guarantee or equivalent insurance; or
(b) evidence of that guarantee or insurance or a declaration certifying its existence,

as part of the notification document at the time of notification or, with the consent of the competent authority, at any other time before the waste is shipped.

Enforcement[79] (Regulation 50)

5.58 **The original Regulation 50 has been substituted by the following to add clarity as to how the different bodies should; work together and share information.**

The Community Regulation and these Regulations are enforced by

(a) the Environment Agency for England and the English area
(b) the Natural Resources Body for Wales for Wales and the Welsh area
(c) the Scottish Environment Protection Agency; Scotland and the Scottish area
(d) the Department of the Environment in Northern Ireland for Northern Ireland and the Northern Irish area

There is nothing in the regulations that authorises a competent authority to bring proceedings in Scotland for an offence.

Authorised Persons (50A)

A competent authority may authorise persons in writing for the purposes of enforcing the Community Regulation and these Regulations.

An authorisation under either of the following is an authorisation unless the body which issued the authorisation specifies to the contrary

(a) section 108 of the Environment Act 1995;
(b) Article 72 of the Waste and Contaminated Land (Northern Ireland) Order 1997.

Assistance by the Secretary of State in relation to offshore installations (50B)

The Secretary of State may assist a competent authority in its enforcement of the Community Regulation and these Regulations on or in relation to offshore installations, in particular by inspecting such installations and providing a competent authority with information to facilitate the effective enforcement of the Community Regulation and these Regulations.

79 SI 2014/861 added reg 49A, Power to share information.

The Secretary of State may authorise persons in writing for the purposes of exercising the function as described above.

An authorisation under either of the following is an authorisation for the purposes of the above, unless the Secretary of State specifies to the contrary—

(a) regulation 16 of the Offshore Chemicals Regulations 2002;
(b) regulation 12 of the Offshore Petroleum Activities (Oil Pollution Prevention and Control) Regulations 2005.

Enforcement Powers (Regulation 51)

5.59 Schedule 5 (enforcement powers) has effect.

Compliance with Notices (Regulation 52)

5.60

(1) An offence is committed if a person on whom a notice is served under these Regulations fails to comply with the provisions of that notice.
(2) Compliance with the notice is at the expense of the person on whom it is served and, if it is not complied with, the competent authority may arrange for it to be complied with at the expense of that person.
(3) Enforcement of the notice by a competent authority may be by a notice or an injunction or, in Scotland, by way of an order under section 45(b) of the Court of Session Act 1988.

Obstruction (Regulation 53)

5.61 An offence is committed if a person

(a) intentionally obstructs any person acting in the execution of these Regulations;
(b) gives any information which he knows to be false or misleading to any person acting in the execution of these Regulations; or
(c) fails, without reasonable excuse—
    (i) to give any assistance or information that any person acting in the execution of these Regulations may require him to give; or
    (ii) to produce any record that any person acting in the execution of these Regulations may require him to produce,

    for the performance of that person's functions under these Regulations.

Falsely Obtaining Consent etc (Regulation 54)

5.62 An offence is committed if a person, for the purpose of obtaining consent to a shipment or approval of a financial guarantee or equivalent insurance,

(a) makes any statement or declaration that he knows to be false or misleading; or
(b) endeavours to obtain such consent or approval by deception.

Offences by Bodies Corporate (Regulation 55)

5.63

(1) If an offence under these Regulations committed by a body corporate is shown—

(a) to have been committed with the consent or connivance of an officer; or
(b) to be attributable to any neglect on his part,

then the officer as well as the body corporate is guilty of the offence and is liable to be proceeded against and punished accordingly.

(2) If the affairs of a body corporate are managed by its members, paragraph (1) applies in relation to the acts and defaults of a member in connection with his functions of management as if he were a director of the body.
(3) "Officer", in relation to a body corporate, means a director, member of the committee of management, chief executive, manager, secretary or other similar officer of the body, or a person purporting to act in any such capacity.

Offences by Scottish Partnerships (Regulation 56)

5.64 Where a Scottish partnership commits an offence under these Regulations, and that offence is proved to have been committed with the consent or connivance of, or to have been attributable to any neglect on the part of—

(a) any of the partners; or
(b) any person who was purporting to act in such capacity,

they, as well as the partnership, is guilty of the offence and is liable to be proceeded against and punished accordingly.

Acts of Third Parties (Regulation 57)

5.65 Where the commission by any person of an offence is due to the act or default of some other person, that other person is guilty of the offence and a person may be charged with and convicted of an offence by virtue of this regulation whether or not proceedings are taken against the first-mentioned person.

Penalties (Regulation 58)

5.66 A person guilty of an offence under these Regulations is liable—

(a) on summary conviction, to a fine not exceeding the statutory maximum or to imprisonment not exceeding three months or to both;
(b) on conviction on indictment, to a fine or to imprisonment for a term not exceeding two years or to both.

Fixed Penalties (Regulation 59)

5.67

(1) If an authorised person has reason to believe that a person has committed an offence under these Regulations, except an offence under regulation 49A(7)[80] he may give that person a notice (a "fixed penalty notice") offering him the opportunity of discharging any liability to conviction for that offence by payment of a fixed penalty.
(2) Where a person is given a fixed penalty notice in respect of an offence—
    (a) no proceedings may be instituted for that offence before the expiry of 28 days following the date of the notice; and
    (b) they may not be convicted of that offence if he pays the fixed penalty before the expiry of that period.
(3) A fixed penalty notice must:
    (a) be in the form set out in Schedule 6[81];
    (b) give such particulars of the circumstances alleged to constitute the offence as are necessary for giving reasonable information of the offence; and
    (c) state:
        (i) the period during which, by virtue of paragraph (2), proceedings will not be taken for the offence;
        (ii) the amount of the fixed penalty; and
        (iii) the name of the person to whom and the address at which the fixed penalty may be paid.
(4) Payment of the fixed penalty may be made by pre-paying and posting to that person at that address a letter containing the amount of penalty.
(5) Where a letter is sent in accordance with paragraph (4), payment is deemed to be made at the time at which that letter would be delivered in the ordinary course of post.

80 SI 2014/861 added *except an offence under regulation 49A(7)*.
81 SI 2007/1711, Sch 6 Form of Fixed Penalty Notices.

(6)[82] The amount of fixed penalty payable in pursuance of a notice under this regulation is £300, which is to be paid in the case of a notice served by an authorised person from—

(a) the Environment Agency or the Scottish Environment Protection Agency, to the Secretary of State;
(b) the Natural Resources Body for Wales, to the Welsh Ministers;
(c) the Department of the Environment in Northern Ireland, to that Department.

(7) In any proceedings, a certificate which purports to be signed by or on behalf of the chief finance officer in England and Scotland, in Wales the principle accounting officer[83], or in Northern Ireland an authorised person; and states that payment of a fixed penalty was or was not received by a date specified in the certificate, is evidence of the facts stated.

Civil Sanctions, England[84] (Regulation 59A)

5.68 The Environment Agency may impose a variable monetary penalty, or accept an enforcement undertaking, in relation to an offence specified in the Table of civil sanctions England as if it were an offence under a provision specified in relation to that sanction in Schedule 5 to the Environmental Civil Sanctions (England) Order 2010.

This regulation applies only in England and the marine area adjacent to English area[85].

Civil Sanctions, Wales[86] (Regulation 59B)

5.69 The Natural Resources Body for Wales may impose a variable monetary penalty, or accept an enforcement undertaking, in relation to an offence specified in the following Table of civil sanctions, as indicated in that Table, as if it were an offence under a provision specified in relation to that sanction in Schedule 5 to the Environmental Civil Sanctions (Wales) Order 2010.

Review (Regulation 59C)

5.70

(1) The Secretary of State must from time to time—

82 SI 2014/861, reg 13.
83 SI 2014/861, reg 13.
84 Inserted by the Environmental Civil Sanctions (Miscellaneous Amendments) (England) Regulations) 2010, SI 2010/1159.
85 SI 2014/861, reg 14.
86 SI 2014/861, reg 15.

(a) carry out a review of these Regulations;
(b) set out the conclusions of the review in a report; and
(c) publish the report.

(2) The report must in particular—

(a) set out the objectives intended to be achieved by these Regulations;
(b) assess the extent to which those objectives are achieved; and
(c) assess whether those objectives remain appropriate and, if so, the extent to which they could be achieved in a less burdensome way.

(3) The first report under this regulation must be published before the end of the period of five years beginning with 1st May 2014.
(4) Reports under this regulation are afterwards to be published at intervals not exceeding five years.

Revocations (Regulation 60)

5.71

(1) The following are revoked—

(a) Transfrontier Shipment of Waste Regulations 1994;
(b) Transfrontier Shipment of Waste (Amendment) Regulations 2005; and
(c) Transfrontier Shipment of Waste (Fee) Regulations (Northern Ireland) 2005.

(2) Notwithstanding the revocation of the Transfrontier Shipment of Waste Regulations 1994, shipments to which Article 62 (transition rules) applies are subject to the provisions of those Regulations.

CASE LAW

5.72 During a series of prosecutions by the Environment Agency new case law has emerged about how to interpret the UK regulations in particular reg 23 of the revised TFS Regulations 2007[87], which makes it an offence if, in breach of Article 36 of the EU WSR, a person transports waste that is destined for recovery in a country to which the OECD Decision does not apply.

87 SI 2007/1711, reg 23, Prohibition of certain waste for recovery to non-OECD Decision countries.

R v V and Others [2011] EWCA Crim 2342[88]

5.73 Charges were made against companies under the Transfrontier Shipment of Waste Regulations 2007, reg 23[89] for the illegal transport of waste.At a preparatory hearing, the judge ruled that reg 23 applied to all those who were involved in transporting waste for export, from the point of origin where the waste was collected and stored for onward transmission to another country, through to the point where the waste was delivered to that country.

The defendants' argument was that an export did not occur until the waste actually left the EU or alternatively, at the earliest when the waste started on the final journey.

The appeal court dismissed the defendants' argument.

Under Article 36[90] there are two key concepts. The waste is destined for recovery in a foreign country and that the 'export' meant that the waste would leave the EU. Therefore waste could be destined for recovery in a non-OECD Decision country long before it reached the point of leaving the EU.

5.74 The appellants were charged with various counts of transporting waste for recovery in China.

The waste had allegedly been treated at a municipal recycling facility and transported in containers to Southampton, ready for export to China. The containers were inspected and there was evidence that they contained waste paper and cardboard but also certain items of household waste. In reg 23 a person committed an offence if, in breach of Regulation 1013/2006, Article 36(1), they transported waste specified in Article 36 that was destined for recovery in certain countries including China. Article 36(1)(b) of the 2006 Regulation prohibited exports of waste, including household or category Y46[91] waste, in such circumstances. Article 36(1)(f) also prohibits the export of wastes, the import of which had been prohibited by the country of destination. China had prohibited the import of Y46 household waste.

88 *KV and others v R* [2011] EWCA Crim 2342.

89 SI 2007/1711, reg 23, Prohibition on export of certain waste for recovery to non-OECD Decision Countries.

23. A person commits an offence if, in breach of Article 36(1), he transports waste specified in that Article that is destined for recovery in a country to which the OECD Decision does not apply.

90 SI 2007/1711, reg 36, Illegal shipments of waste brought to a facility.

If the operator of a facility knows or has reasonable grounds to suspect that waste brought to that facility is an illegal shipment of waste, he must notify the competent authority immediately and comply with their instructions and failure to do so is an offence.

91 EC 1013/2006, List A Basel Convention Y46 Waste collected form households.

Under the 2006 WSR, paper, paperboard and paper product wastes, which were categorised as B3020[92] waste, were excluded from the export prohibition unless they were contaminated by other materials to an extent that either sufficiently increased the risks associated with the wastes to render it appropriate for submission to the prior written notification and consent procedure when taking into account certain specified hazardous characteristics, or prevented the recovery of the waste in an environmentally sound manner. Y46 household waste could, by appropriate sorting, become B3020 waste.

The Environment Agency had stated that there was no '0 per cent' contamination requirement for recovered paper for export, and that it was not going to specify a percentage in any guidance it produced. X's defence was that the waste, having been treated, was no longer Y46 household waste, but B3020 waste.

However, the appellants argued that the waste they were transporting was 'paper... and paper product wastes', which is not subject to the prohibition on export for recovery unless it is contaminated by other materials to such an extent that it either:

- increases the risks associated with the waste to render it appropriate for submission to the written procedure; or
- prevents the recovery of the waste in an environmentally sound manner.

An inspection of the transportation containers found that they contained not only waste paper and cardboard, but items of household waste, plastics, metal cans, tetrapak cartons and rotting meat. The appellants argued that the absence of any guidance from the EA or the setting of a standard to enable a judgment to be made by an operator as to whether its activities are likely to be treated as criminal or not, meant that the proceedings against them amounted to an abuse of process.

It was argued by the prosecution that it was a question of fact for the jury whether a particular consignment is household waste and the burden is on the prosecution to establish that it is still household waste. There is clearly difficulty in specifying by weight or percentage what degree of contamination, if any, can be permitted.

92 EC 1013/2006, B3 Wastes containing principally organic constituents, which may contain metals and inorganic materials
B3020 Paper, paperboard and paper product wastes
The following materials, provided they are not mixed with hazardous wastes:
Waste and scrap of paper or paperboard of:
— unbleached paper or paperboard or of corrugated paper or paperboard;
— other paper or paperboard, made mainly of bleached chemical pulp, not coloured in the mass;
— paper or paperboard made mainly of mechanical pulp (for example, newspapers, journals and similar printed matter);
— other, including but not limited to;
1. laminated paperboard;
2. unsorted scrap.

The Court of Appeal dismissed the appeal, stating that a very high standard is required of operators in this field, but made no comment on the need for operators and the EA to create some guidance.

R v Ezeemo and others [2012] EWCA Crim 2064[93]

5.75 In this case there were seven appellants who were convicted of 21 charges of transporting hazardous waste for recovery in Nigeria. The offence was transporting hazardous waste to a country that was not a member of the OECD, contrary to the Transfrontier Shipment of Waste Regulations 2007, reg 23 which is one of strict liability. The case was that the defendants had transported to Nigeria (a non-OECD country), cathode ray tube televisions and fridge/freezers containing ozone depleting substances. The defendants had recovered the appliances from waste disposal sites and the items could not be economically repaired in the United Kingdom. They could however, be repaired and re-used at a market in Nigeria. The defendants claimed that their process of selection and testing before shipment meant that by the time the appliances were consigned to Nigeria, they were no longer waste[94].

The defendants argued that:

- 'waste' in the Regulation should be read, in light of the OECD decision in 2001, with the definition of the Basel Convention on waste, and that the appliances did not fulfill that definition;
- the test that the jury applied as to whether they intended to discard was subjective and the judge directed that the defendants' intention was irrelevant;
- the prosecution could not prove that the process to which the appliances were to be subjected to in Nigeria was 'recovery';
- the judge had wrongly ruled that the offence was one of strict liability.

They were granted leave to appeal – but the appeal was dismissed.

(1) The 2007 Regulations were specific as to the meanings of terms also used in the 2006 Regulation, and the OECD decision had dealt only with the movement of waste within and between OECD countries. Further, the ECJ had adopted an autonomous meaning of 'waste' which it had applied since 1991, under which re-use did not exclude material from the definition of waste[95]. Had the EU legislature intended to abandon that definition, it would have said so in Article 2 of the Regulation.

93 *R v Ezeemo (Godwin Chukwnaenya), R v Reliance Export Ltd, R v BJ Electronics, R v Ezeemo (Chika), R v Benson (Joseph), R v Orient Export Ltd, R v Ezechukwu (Nnamdi Chinedu)* [2012] EWCA Crim 2064.

94 As defined by Directive 2006/12, Article 1(1)(a).

95 *ARCO Chemie Nederland Ltd v Minister van Volkshuisvesting, Ruimtelijke Ordening en Milieubeheer* (C-418/97) [2002] QB 646.

(2) There was a consistent line of authority holding that once a substance or object had been discarded, and had therefore become waste, whether it had changed its status was a question of objective fact[96]. The jury had to be sure that the substance or object was one which the holder discarded or intended to discard which could cause confusion. The trial judge had to adapt his directions so as to eliminate the apparent contradictions between the words used and their interpretation. What mattered was whether the holder of the objects took, or intended to take, some action which changed their status; whether the holder discarded or intended to discard the objects was judged by what he did with them, not by his subjective belief as to whether he was discarding them. There was little doubt that the appliances' original owners had discarded them; the jury had had to decide whether the process the defendants claimed to have subjected the appliances to had had the effect of rendering them non-waste.

(3) 'Recovery' meant any of the operations in Directive 2006/12 Annex II B. Article 3 showed that the Directive recognised re-use as a possible form of recovery of waste. The objective of Article 36 of the Regulation was to protect non-OECD countries from dangers to health and the environment by prohibiting the export of hazardous waste to those countries for recovery. The majority of the appliances had not been suitable for re-use without undergoing repair or regeneration. The re-use had fallen within Annex II B, generously and properly construed.

(4) No state of mind by the defendant was required by the words used in reg 23 unless that was to be presumed. English law presumed that an offence required mens rea unless by necessary implication it was excluded. The scheme of the 2007 Regulations was to create a series of offences comprising breaches of the requirements of the 2006 Regulation. The offences were created for failure to take the procedural steps required by the Regulation or taking a step prohibited by it. That was the language of strict liability. There were exceptions, but there the draftsman explicitly defined the required mens rea. Further, reg 57 implied that a person could commit an offence, without fault on his own part, due to the actions of another person. That was consistent with at least some of the offences being of strict liability. The offence created by reg 23 was one of strict liability[97].

Waste

In this case the appellants had argued that the visual inspection and selection of items for shipment and by carrying out portable appliance testing meant that items were no longer waste. Although the items were waste when they were discarded by owners at the waste amenity site, they, as intermediate holders of the waste had no intention of discarding the electrical items collected. The Court of Appeal examined in some detail previous authorities that held that, once an object has been discarded (and therefore become waste), the question

96 *OSS Group Ltd v Environment Agency* [2007] EWCA Civ 611, [2007] Bus LR 1732.
97 *R v V* [2011] EWCA Crim 2342, [2012] Eu LR 302.

of whether it has changed its status is one of objective fact. The fact of which the jury had to be sure was that the object was one of which the holder 'discarded' or 'intended to discard' and this was to be judged by what the holder did with it and not by the holder's subjective belief.

Recovery

The appellants argued that the prosecution could not establish what occurred in Nigeria was recovery rather than mere re-use. The key legislation for the definition of recovery is the list of operations in Annex IIB of the WFD, which includes recycling and reclamation of metals, metal compounds and other inorganic materials. The Court of Appeal agreed with the prosecution's contention that the list in the WFD should be construed generously so as to give Article 36 of the EU Regulation its intended effect. It cited Article 3 of the WFD, which requires Member States to take appropriate measures to encourage 'the recovery of waste by means of recycling, re-use or reclamation, or any other process with a view to extracting secondary raw materials...', thereby recognising re-use of waste as a possible form of 'recovery'. It also cited Article 5, which requires each Member State to take measures to ensure that its waste is recovered with minimal danger to health and the environment and to prohibit the abandonment, dumping or uncontrolled disposal of waste. It was concluded by the Court of Appeal that Article 36 is to protect non-OECD countries from the same dangers by prohibiting the export of hazardous waste to those countries for recovery.

Strict liability

The appellants said that it was not the intention of Parliament to create an offence, which, without fault, could result in criminal liability for all those who contributed to the transport of an unlawful shipment. The Court of Appeal said that all the offences created under the UK Regulations used the language of close regulation to ensure the safe management and shipment of waste. The offences were created for a failure to take the steps required by the EU Regulation or the taking of a step that was prohibited by the EU Regulation. This was the language of strict liability. If a 'guilty mind' was required for the offence, in the absence of a specified lesser culpability, this would have to be knowledge of the material facts and this would substantially deprive the legislation of its intended effect, which was to deter those in the business of handling such material for export from taking risks with the environment.

Chapter 6

Environmental Permitting and Waste Management Licences

INTRODUCTION

6.1 The Environmental Permitting Regulations were introduced in the UK to minimise the impact of activities that have the potential to damage the environment.

ENVIRONMENTAL PERMITTING

6.2 Some facilities could harm the environment or human health[1] unless they are controlled. The environmental permitting regime requires that operators obtain permits for some facilities, to register others as exempt and provides for on-going supervision by regulators.

The aim of the regime is to:

- protect the environment so that statutory and Government policy environmental targets and outcomes are achieved;
- deliver permitting and compliance with permits and certain environmental targets effectively and efficiently in a way that provides increased clarity and minimises the administrative burden on both the regulator and the operators;
- encourage regulators to promote best practice in the operation of facilities;
- continue to fully implement European legislation.

ENGLAND AND WALES

6.3 The Environmental Permitting (England and Wales) Regulations 2010, SI 2010/675 (EPR) integrated existing permitting regimes. They covered facilities previously regulated under the Pollution Prevention and Control (England and Wales) Regulations 2000, SI 2000/1973 and Waste Management Licensing and Exemptions Schemes[2] (as superseded by the Environmental Permitting (England and Wales) Regulations 2007, SI 2007/3538), some parts of the Water Resources Act 1991, the Radioactive Substances Act 1993 and the Groundwater (England and Wales) Regulations 2009, SI 2009/2902.

1 'Protect the environment' also includes human health.
2 Environmental Protection Act 1990, Part 2 and the Waste Management Licensing Regulations 1994, SI 1994/1056.

In England and Wales, only one type of permit is required which can be obtained from either the Local Authority or Environment Agency depending on the activity carried out.

Permits for activities listed in SI 2010/675, Sch 1, Part A(1) and Part A(2) are issued by the Environment Agency. Permits for activities listed in Sch 1, Part B are issued by the Local Authority

In order to receive a permit, an organisation must comply with certain obligations including:

- Using all appropriate pollution-prevention measures;
- Preventing all large-scale pollution;
- Preventing, recycling or disposing of waste in the least polluting way possible;
- Using energy efficiently;
- Ensuring accident prevention and damage limitation;
- Returning sites to their original state when the activity is over.

This extends to England and Wales and also covers the adjacent sea as far as the seaward boundary of the territorial sea.

The Legal Framework

6.4 The Regulations set out the following:

- the facilities that need environmental permits or need to be registered as exempt;
- the process for registering exempt facilities;
- how to apply for and determine permit applications;
- requirements that environmental permits contain conditions to protect the environment as required by Directives and, where applicable, national policy;
- how environmental permits can be changed and ultimately be surrendered;
- a simplified permitting system called standard rules;
- compliance obligations backed up by enforcement powers and offences provisions for public participation in the permitting process;
- the powers and functions of regulators, the Secretary of State and the Welsh Ministers;
- a simple transition to the new regime; and
- provisions for appeals against permitting decisions.

The principal offences under the Regulations are, operating a regulated facility without a permit, causing or knowingly permitting a water discharge activity or groundwater activity without a permit, and failing to comply with a permit or enforcement related notice.

The Natural Resources Body for Wales (Establishment) Order 2012, SI 2012/1903 established a new statutory body, the Natural Resources Body for Wales, for the management of Wales' natural resources. For Environmental

Permitting from 1 April 2013, the Natural Resources Body for Wales will exercise the Welsh devolved functions of the Environment Agency.

European Directives

6.5 European Directives (Directives) and international agreements have a variety of requirements, some of which can be delivered through a permitting and compliance system and some of which are delivered in other ways. The majority of environmental quality and specific permitting standards and other related requirements for environmental and human health protection come from Directives.

The Regulations place duties on regulators to exercise their permit-related functions and deliver the requirements of the relevant Directives and, in some cases, national policy. This means that where a permit is granted the regulator will ensure that permit conditions achieve the objectives and outcomes of any of the Directives or national policy, which apply. The Regulations also give regulators powers in relation to their permit-related functions.

The Schedules to the Regulations generally identify particular requirements, which must be delivered through the permitting system. In some cases requirements to be delivered through the permitting system are located in other legislation[3]. Each Directive or policy area covered by the regime has a specific Schedule:

- Industrial Emissions Directive Chapter II (Integrated Pollution Prevention and Control). Schedules 7 and 7A – Part A Installations;
- Part B Installations and Part B Mobile Plant. Schedule 8 – Part B Installations and Part B Mobile Plant;
- Waste Framework Directive Schedule 9 – Waste Operations;
- Landfill Directive Schedule 10 – Landfill;
- End-of Life Vehicles Directive Schedule 11 – Waste Motor Vehicles;
- Waste Electrical and Electronic Equipment Directive Schedule 12 – Waste Electrical and Electronic Equipment;
- Industrial Emissions Directive Chapter IV Schedules 13 and 13A – Waste Incineration;
- Industrial Emissions Directive Chapter V;
- Schedule 14 – Solvent Emission Activities;
- Industrial Emissions Directive Chapter III Schedules 15 and 15A – Large Combustion Plants;
- Asbestos Directive Schedule 16 – Asbestos;
- Industrial Emissions Directive Chapter VI Schedules 17 and 17A – Titanium Dioxide;
- Petrol Vapour Recovery Directive Schedule 18 – Petrol Vapour Recovery;
- Batteries Directive;
- Schedule 19 – Waste Batteries and Accumulators;

3 An example of this is the Water Environment (Water Framework Directive) Regulations 2003, SI 2003/3242.

- Mining Waste Directive;
- Schedule 20 – Mining Waste Operations;
- Water Discharge Activities;
- Schedule 21 – Water Discharge Activities;
- Groundwater Activities;
- Schedule 22 – Groundwater Activities;
- Radioactive Substances Activities;
- Schedule 23 – Radioactive Substances Activities.

Where a facility comes under more than one Schedule then each set of Schedule requirements must be met. For example, in England, most waste incinerators (in Wales all incinerators) must meet the requirements of Chapters II and IV of the Industrial Emissions Directive and of the revised Waste Framework Directive. Schedules 7/7A, 13/13A and 21 contain the provisions that relate to most incinerators. This is because there is an overlap between the European Directives and national policy.

Facilities That Require an Environmental Permit

6.6 The Regulations specify which facilities require an environmental permit and also allow that some facilities can be exempt[4] from those requirements.

Exemptions provide for a lighter touch of the regulations for small scale, low risk waste operations and aim to encourage waste recycling and recovery. The facilities that require a permit are described as 'regulated facilities'. There are different kinds of regulated facility and each is known as a 'class' of regulated facility.

The Nine Classes of Regulated Facility

6.7 These are defined under the Regulations as[5]:

(a) an installation[6];
(b) mobile plant[7];
(c) a waste operation[8];
(d) a mining waste operation[9];
(e) a radioactive substances activity[10];
(f) a water discharge activity[11];

4 EPR 2010, SI 2010/675, Pt 1, reg 4.
5 EPR, Pt 1, reg 8.
6 EPR, Sch 1, Pt 1, para 1.
7 Part A mobile plant; Part B mobile plant; waste mobile plant.
8 Waste operations – recovery or disposal of waste.
9 EPR, Sch 20, para 2(1).
10 EPR, Sch 23, Pt 2, para 5.
11 'Exempt waste operation' has the meaning given in EPR, reg 5.

(g) a groundwater activity[12];
(h) small waste incineration plant[13];
(i) solvent emission activity.

Some regulated facilities may be 'carried on as part of the operation of a regulated facility of another class'. These are:

(a) waste operations;
(b) mining waste operations;
(c) water discharge activities;
(d) groundwater activities;
(e) small waste incineration plant and solvent emission activities[14].

Some provisions of the Regulations apply to these classes of regulated facility differently. They may be carried on as part of the operation of another regulated facility. For example, any part of a permit which authorises a stand-alone water discharge activity may be surrendered by notification, whereas any part of a permit which authorises a water discharge activity that is carried on as part of the operation of another regulated facility (such as an installation), can only be surrendered by application.

A summary of the requirements for different activities, installations and operations can be found in the Defra Guidance[15].

This will be the case where a regulated facility is carried on as part of the operation of another regulated facility, but may also occur in other circumstances. In such cases there are arrangements in the Regulations to allow all such facilities to be regulated by the same regulator and to allow, in many cases, for a single permit.

Activities are listed in the Environmental Permitting (England and Wales) Regulations 2010 and are split into three categories A1, A2 and Part B.

Part A1 activities are administered by the Environment Agency for England and by WAG for Wales

- Part A(1) installations;
- waste mobile plant;
- waste operations, including those carried on at a Part B installation or by Part B mobile plant (unless the waste activity is a Part B activity);
- mining waste operations, including any carried on at a Part B installation;
- radioactive substances activities;
- water discharge activities, including those carried on at a Part B installation;
- groundwater activities, including those carried on at a Part B installation.

12 'Exempt groundwater activity' has the meaning given in EPR, reg 5.
13 The Environmental Permitting (England and Wales) (Amendment) Regulations 2013, SI 2013/390, reg 8(4) added (h) small waste incineration plant, (i) a solvent emission activity.
14 SI 2013/390, reg 8(4) added (h) small waste incineration plant, (i) a solvent emission activity.
15 Environmental Permitting Guidance Defra March 2013 Annex 2.

Part A2 activities are administered by the local authority and are:

- Part A(2) installations including any waste operations, water discharge activities or groundwater activities carried on as part of the installation or mobile plant;
- Part B installations and Part B mobile plant (except as set out above)
- small waste incineration plants;
- solvent emission activities.

Part B activities are administered by the local authority but only regulate emissions to the air. In general Part A1 and A2 processes tend to be of a more complex nature.

Defra and the Welsh Government jointly provide guidance on local authority air pollution control. Guidance on Part A(1) and Part A(2) installations can be found in the Guidance on Part A installations.

The Different Classes of Regulated Facility and Their Related Exempt Facilities

Installations

6.8 Schedule 1, Pt 2 to the EPR[16] provides a list of specified activities. Exclusions from the list and other rules for interpretation are contained in Sch 1, Pt 1.

An installation is made up of any stationary technical unit where one or more activities listed in Sch 1 and any directly associated activities are carried on (Sch 1, Pt 1, para 1).

Waste Operations

6.9 A waste operation is defined in EPR, reg 2 as recovery or disposal of waste. Any recovery or disposal of waste is a waste operation.

Further guidance on the meaning of waste operation is provided in the revised Waste Framework Directive guidance. See **Chapter 2** – Definition of Waste. Some waste operations not carried on at an installation, are excluded from the regime because there is already appropriate environmental regulation under other regimes.

These excluded operations are those that are carried on under a FEPA permit[17] or the disposal or recovery of sludges which are not to be treated as industrial

16 Reproduces in the main ,Sch 1 of the Pollution Prevention and Control (England and Wales) Regulations 2000, SI 2000/1973.
17 Food and Environmental Protection Act 1985.

or commercial waste under the Controlled Waste Regulations are excluded from the requirement for a permit[18]. See **Chapter 3**.

Installations and Waste Operations – Exempt Waste Operations

6.10 Waste operations not carried on at an installation are capable of being exempt waste operations[19]. An exempt waste operation does not require an environmental permit but will need to comply with the rules laid down for each type of exempt activity and must be registered with the relevant registration authority.

No other type of waste operation carried on at an installation, for example as a directly associated activity, is capable of being an exempt waste operation.

Mobile Plant

6.11 Mobile plant is defined in EPR, reg 2 as Part B mobile plant or waste mobile plant. The 2010 Regulations require that to be mobile, plant must be designed to move or be moved. This movement can be by road, rail or water, eg by canal. There is Environment Agency guidance on factors to consider when assessing whether the plant it regulates is mobile, such as its environmental impact, degree of mobility and the intention of the operator.

Plant that carries out a Part A activity will be regarded as stationary. A stationary technical unit forms the basis of an installation and must be, by definition, stationary.

European Commission guidance explains the definition of installation and considers the meaning of 'stationary'[20]. The Commission guidance considers the question of whether plant that is designed to be moved periodically but which in practice operates from the same location for some time, should be considered to be 'stationary'. For example, how long is the plant expected to be or does remain stationary; the nature of the activities and their environmental impact; and the degree of physical installation involved in moving and establishing the plant.

The Commission guidance also concludes that whilst the term 'stationary' means that the installation as a whole should be stationary[21], it may still include plant or equipment, which is mobile.

18 Regulation 3(2) of the Controlled Waste (England and Wales) Regulations 2012, SI 2012/811.
19 EPR, reg 5.
20 Guidance on the Interpretation of 'Installation' and 'Operator' for the Purposes of the IPPC Directive http://ec.europa.eu/environment/archives/air/stationary/ippc/pdf/installation_guidance.pdf.
21 Paragraph 3 of the Guidance on the Interpretation of 'Installation' and 'Operator' for the Purposes of the IPPC Directive.

Mobile Plant – Waste Mobile Plant

6.12 Waste mobile plant is defined in EPR, reg 2 as mobile plant which is used to carry on a waste operation and is designed to move or be moved whether on roads or on other land and is not an installation, Part A mobile plant or Part B mobile plant. Waste operations are described above at para **6.9**. Any waste operation, other than those which are part of an installation, can theoretically be carried on by mobile plant.

Mining Waste Operations

6.13 A mining waste operation is defined as[22] the management of extractive waste, whether or not involving a mining waste facility, but does not include activities in Article 2(2)(c) of the Mining Waste Directive[23].

Radioactive Substances Activities

6.14 A radioactive substances activity[24] is one involving the keeping and use of radioactive material (including mobile radioactive apparatus) or the accumulation and disposal of radioactive waste and which is not excluded or exempted from regulation as a radioactive substances activity. For nuclear site licensees, the keeping and use of radioactive material and accumulation of radioactive waste are regulated by the Office for Nuclear Regulation (ONR).

Water Discharge Activities

6.15 A water discharge activity is[25], in summary:

- the discharge or entry to inland freshwaters, coastal waters or relevant territorial waters of any: poisonous, noxious or polluting matter, waste matter or trade effluent or sewage effluent;
- the discharge from land through a pipe into the sea outside the seaward limits of relevant territorial waters of any trade effluent or sewage effluent;
- the removal from any part of the bottom, channel or bed of any inland freshwaters of a deposit accumulated by reason of any dam, weir or sluice holding back the waters, by causing it to be carried away in suspension in the waters, unless the activity is carried on in the exercise of a power

22 EPR 2010, SI 2010/675, Sch 10, para 2(1).
23 Directive 2004/21/EC ([2004] OJ L 158, 30.4.2004) p 7-49 on the management of waste from extractive industries.
24 EPR, Sch 23, Pt 2, para 5.
25 EPR, Sch 21, para 3.

conferred by or under any enactment relating to land drainage, flood prevention or navigation;
- the cutting or uprooting of a substantial amount of vegetation in any inland freshwaters or so near to any such waters that it falls into them and failure to take reasonable steps to remove the vegetation from these waters.

A discharge or an activity that might lead to a discharge is not a 'water discharge activity' – if the discharge is made, or authorised to be made, by or under any prescribed statutory provision; or if the discharge is of trade effluent or sewage effluent from a vessel.

In determining whether a discharge or an activity is a water discharge activity, no account must be taken of any radioactivity possessed by any substance or article or by any part of any premises.

A water discharge activity does not include a discharge of trade effluent or sewage effluent from a vessel, certain types of discharge unless a notice has been served, or certain discharges made under prescribed statutory provisions.

Certain types of water discharge activity are capable of being exempt water discharge activities.

Sometimes a water discharge activity forms part of the operation of a regulated facility of another class (eg, an installation). When it does not, it is referred to in this guidance as a 'stand-alone water discharge activity'. Some provisions of the Regulations apply only to stand-alone water discharge activities.

Groundwater Activities

6.16 A groundwater activity[26] is, in summary:

- the discharge of a pollutant that results in the direct input of that pollutant to groundwater;
- the discharge of a pollutant in circumstances that might lead to an indirect input of that pollutant to groundwater;
- any other discharge that might lead to the direct or indirect input of a pollutant to groundwater;
- an activity in respect of which a notice under paragraph 10[27] has been served and has taken effect;
- an activity that might lead to a discharge mentioned in paragraph (a), (b) or where that activity is carried on as part of the operation of a regulated facility of another class;
- a discharge or an activity that might lead to a discharge is not a 'groundwater activity' if the discharge is made, or authorised to be made, by or under any prescribed statutory provision; or of trade effluent or sewage effluent from a vessel;

26 EPR 2010, SI 2010/675, Sch 22, para 3.
27 Notice requiring an environmental permit.

- the regulator may determine that a discharge, or an activity that might lead to a discharge, is not a groundwater activity if the input of the pollutant is the consequence of an accident or exceptional circumstances of natural cause that could not reasonably have been foreseen, avoided or mitigated;
- is or would be of a quantity and concentration so small as to obviate any present or future danger of deterioration in the quality of the receiving groundwater; or is or would be incapable, for technical reasons, of being prevented or limited without using—
  - measures that would increase risks to human health or to the quality of the environment as a whole, or
  - disproportionately costly measures to remove quantities of pollutants from, or otherwise control their percolation in, contaminated ground or subsoil.

Certain types of groundwater activity are capable of being exempt groundwater activities.

Sometimes a groundwater activity forms part of the operation of a regulated facility of another class (eg, an installation). When it does not, it is referred to in this guidance as a 'stand-alone groundwater activity'. Some provisions of the Regulations apply only to stand-alone groundwater activities.

Small Waste Incineration Plant

6.17 A small waste incineration plant is one whose capacity is below the relevant threshold in Section 5.1 of Part 2 of Sch 1 to the EPR. In England it will be subject only to the requirements of Schs 13/13A unless it is exempt from those requirements because it burns only the wastes which are specified in paragraph (a) of Part B of Section 5.1. In Wales, small waste incinerators are, like their larger counterparts, subject to Chapters II and IV of the Industrial Emissions Directive and of the revised Waste Framework Directive (Schs 8, 13/13A and 21 of the EPR).

Solvent Emission Activity

6.18 A solvent emission activity is one, which is listed in Annex VII of the Industrial Emissions Directive and hence subject to Sch 14 of the EPR. In Wales, solvent emission activities are subject to Sch 8 of the EPR.

A Single Permit

6.19 A single environmental permit can cover more than one regulated facility[28], where:

28 EPR 2010, SI 2010/675, reg 17.

- the regulator is the same for each facility;
- the operator is the same for each facility; and
- all the facilities are on the same site (the exceptions to this are set out below).

Where the regulator and operator are the same, a single environmental permit can be granted to an operator for more than one mobile plant. Mobile plant does not have to be operating on the same site in order to be included in a single permit.

Where the regulator and operator are the same, a single environmental permit can be granted to an operator for more than one regulated facility to which standard rules apply. Standard facilities do not have to be on the same site in order to be included in a single permit.

A permit, which authorises the keeping or use of mobile radioactive apparatus, may authorise the carrying on of that activity on more than one site.

Regulated facilities have to be operated on the same site in order to be covered by the same permit (with the exceptions of mobile plant, standard facilities and radioactive substances activities involving the keeping or use of mobile radioactive apparatus). The regulator should consider the following factors in determining whether the facilities are operated on the same site:

- Proximity: There should however be no simple 'cut off' distance since some industrial complexes cover very large areas but still can be regarded as one site for permitting purposes;
- Coherence of a site: Some regulated facilities will be operated within a single fenced area or may share security or emergency systems;
- Management systems: The extent to which the regulated facilities share a common management system is a relevant consideration.

When More Than One Permit is Required

6.20 A single environmental permit cannot cover regulated facilities with different regulators. A separate permit is required covering the facilities for which each regulator is responsible.

A single permit cannot cover mobile plant combined with any other class of regulated facility. Mobile plants are not associated with a particular geographical site. The differences in the requirements for mobile plant mean that a single permit cannot cover mobile plant and other classes of regulated facility.

Environmental Permit Application

The Operator

6.21 Only the person who has control over the operation of a regulated facility may obtain or hold an environmental permit. This person is the 'operator'[29] (EPR, reg 7).

An operator will have to obtain one or more environmental permits for each regulated facility he operates. For example, to understand the relationship between 'operator', 'regulated facility' and 'installation':

- a 'regulated facility' is a facility, which falls within one of the classes listed in reg 8;
- one of those classes is an 'installation';
- an 'installation' can include one or more other regulated facilities (eg, a waste operation' or 'water discharge activity') but will only require one permit unless the next point applies;
- if different parts of a single installation are operated by different operators, each part of the installation with a separate operator constitutes a separate regulated facility;
- if the number of operators operating different parts of the installation changes over time, the number of regulated facilities will therefore also change;
- where there are different operators for different parts of one installation, each will be responsible for complying with their permit conditions. In such cases, regulators should ensure that there is no ambiguity over which operator has responsibility for which part of the installation.

Pre-Application Discussions

6.22 Pre-application discussions between operators and regulators can be helpful in improving the quality of the formal application and are encouraged. In order for such discussions to make the best use of time, the operator is expected to have read the relevant published guidance. The regulator should not be expected to provide advice that might prejudice its determination of an application.

29 'Operator' is defined in reg 7 as the person who has control over the operation of a regulated facility.
If a regulated facility has not been put into operation, the person who will have control over it when it is in operation is the operator.
If a regulated facility authorised by an environmental permit has ceased to be in operation, the person who holds the permit is the operator.
Legal obligations may be imposed on an operator during the pre- and post-operational phases. The operator must demonstrably have the authority and ability to ensure the environmental permit is complied with.

Operators and regulators may use the discussions to clarify whether a permit is likely to be needed. The regulator may also give operators general advice on how to prepare their applications, focus on the key issues, and tell them what additional guidance is available. Other parties may be invited to join these discussions – for example, a public consultee (see **Chapter 10** Regulators) can reassess competence at any time and if not satisfied can revoke the permit. In particular, if the operator of a relevant waste operation fails to comply with an approved competence scheme, the regulator may revoke the permit.

Consultation and Public Participation

6.23 Participation of other parties might be subject to national security restrictions or limited because of commercial confidentiality issues (See **Chapter 11** on Public Registers).

Using Existing Data

6.24 Operators may draw upon or attach other sources of information in their applications such as extracts from:

- Environmental Impact Assessments;
- documents relating to an installation's regulation under the Control of Major Accident Hazards (COMAH) Regulations 2015, SI 2015/483
- externally certified environmental management systems;
- site reports prepared for planning purposes;
- reports to meet Quarries Regulations 1999, SI 1999/2024.

It should be clear which parts of any attachments are relevant to the environmental permit applications and should demonstrate how they relate to the relevant requirements.

Timing of Applications

6.25 Where proposals involve substantial expenditure, whether on construction work, equipment, software, procedures or training, operators should normally make an application when they have drawn up full designs but before the work commences (whether on a new regulated facility or when making changes to an existing one). Where facilities are not particularly complex or novel, the operator should usually be able to submit an application at the design stage containing all information the regulator needs. If, in the course of construction or commissioning, the operator wants to make any changes, which mean that the permit conditions have to be varied, the operator may apply for this in the normal way.

There is nothing in the Regulations to stop an operator from beginning construction before a permit has been issued. However, regulators may not agree with the design and infrastructure put in place. Therefore, to avoid any expensive delays and re-work, it is in the operator's interest to submit applications at the design stages. Any investment or construction work that an operator carries out before they have an environmental permit will be at their own risk and will in no way affect the regulator's decision.

Novel Applications

6.26 If an operator is planning an innovative process for which the regulator has not produced relevant guidance, the operator should, in consultation with the regulator, assemble details of the process, including the potential environmental impact, before making an application. When determining the application the regulator must consider the predicted environmental outcome rather than focusing on the novel nature of the process.

For some novel and complex installations, with long lead times and multiple design and construction phases, the regulator and the operator may agree to a staged application procedure.

Planning and Environmental Permit Applications

6.27 If a regulated facility also needs planning permission, it is recommended that the operator should make both applications in parallel whenever possible. This will allow the environmental regulator to start its formal consideration early on, thus allowing it to have a more informed input to the planning process.

Applicants for complex proposals are likely to find pre-application discussions particularly beneficial. Wherever possible in such cases, operators should engage in pre-application discussions with the regulator before submitting an application for an environmental permit. This can potentially avoid significant costs and delay in the course of the permitting process by identifying any issues of fundamental concern at an early stage and ensuring these are addressed at the design stage if possible.

For certain waste operations and certain mining waste facilities, where planning permission is required it must be in force before an environmental permit can be granted (EPR, Sche 9, para 3).

Specific Procedures for Different Types of Applications

Variation Applications

6.28 Once an operator has an environmental permit, changes in the operation of the regulated facility may require the operator to apply to vary the permit.

The operator must apply to the regulator to vary the permit conditions when proposing a change that would mean that a permit condition could no longer be complied with. Other aspects of the environmental permit may also require a variation application – for example, to change the name of the operator on the permit (though not when the operator's legal identity changes, for example a change to its unique identifier at Companies House; this would require a transfer application).

A variation application may include an increase to the extent of the site over which the regulated facility operates, as covered by the permit. Where this occurs, issues such as the protection of the land must be addressed.

However, a variation cannot reduce the extent of the site covered by the environmental permit if the permit requires consideration of the condition of the land. Since this is not required for Part B activities (regulated for emissions to air only), or for 'stand alone' water discharge or groundwater activities, this restriction does not apply to them (EPR, reg 20(2) and (3)). Where the restriction does apply, an operator wishing to reduce the extent of the site of the regulated facility must apply for partial surrender (see para **6.13** on Surrender Applications and Notifications).

For applications to vary an environmental permit, public participation is required in two situations (EPR, Sch 5, Pt 1, para 5). First where there is a substantial change to the operation of an installation (see guidance for Part A installations) and mining waste facilities. Secondly, the regulator may also require consultation in cases (whether or not relating to an installation) that do not involve substantial changes. In these cases, the regulator will notify the operator of its decision and the consultation will proceed as if there were a substantial change

Transfer Applications and Notifications

6.29 The Regulations allow for permit transfers (EPR, reg 21). Whenever the operator's legal identity changes, for example a change to its unique identifier at Companies House; or if the permit holder ceases to be the operator, this requires a transfer application or notification.

A permit can be transferred:

- completely, or
- partially, so that the regulated facility becomes two regulated facilities with the original operator retaining control of some of the original facility, which then becomes a separate regulated facility and another operator taking over the operation of the transferred part of the original facility, which then becomes a second separate regulated facility.

Transfer by notification is only possible in the case of any part of a permit that relates to a 'stand alone' water discharge or groundwater activity. In these cases joint notification by the operator and proposed transferee must be made on a form provided by the regulator, include information specified on that form and specify an effective date for the transfer to take place. Where more than one person is the current permit holder only one such person need notify a transfer, together with the proposed operator.

In all other cases where an operator wants to transfer all, or part, of a permit to someone else, they must make a joint application with the proposed new operator. For a partial transfer, where the original operator retains part of the permit, the application must include a plan identifying which parts of the site and which regulated facility (or facilities) the operator proposes to transfer.

Where the regulator grants an application to transfer the whole permit, it will issue a new permit to the new operator.

For partial permit transfers, the regulator will issue a new permit to the new operator. This will cover the parts of the operation that have been transferred. At the same time, the regulator will issue a new permit to the original operator, reflecting the extent of the transfer.

Regulators should vary permit conditions, where necessary, as a result of a transfer. For example, further conditions may become necessary to reflect the shared operation. This will ensure that there is co-operation between the operators.

Surrender Applications and Notifications

6.30 There are two separate methods for surrender. The operators of some regulated facilities may simply notify the regulator but others must make an application to the regulator (EPR, regs 24 and 25).

Surrender of the environmental permit by notifying the regulator is restricted to the operators of Part B installations (except to the extent they relate to a waste operation), stand-alone water discharge activities and groundwater activities, and mobile plant.

The pollution control measures, which the regulator is required to apply to Part B installations, water discharge activities and groundwater activities, do not include measures to address pollution of the land. There is therefore no requirement to consider the condition of the land prior to surrendering the permit.

For mobile plant the position is similar in that there is no geographical site associated with the environmental permit. There cannot therefore be a consideration of the condition of the land before a surrender takes place. It should be noted that, where relevant, the permit conditions for mobile plant should be in place to ensure the protection of the land on which they operate.

Enforcement

6.31 Regulation 36 of the EPR allows the regulator to serve an 'enforcement notice' if it believes an operator has contravened, is contravening, or is likely to contravene any permit conditions. Enforcement notices will specify the steps required to remedy the problem and the timescale in which they must be taken. Enforcement notices may include steps to remedy the effects of any harm and to bring a regulated facility back into compliance.

If the operation of a regulated facility involves a risk of serious pollution, the regulator may serve a 'suspension notice' under reg 37(2) of the EPR. This applies whether or not the operator has breached a permit condition. A suspension notice may also be served by a local authority regulator for non-payment of a charge. Guidance on this is contained in the Manual.

The suspension notice must describe the nature of the risk of pollution and the actions necessary to remove that risk. The notice must specify the deadline for taking actions. When the regulator serves a suspension notice, the permit ceases to authorise the operation of the entire facility or specified activities depending upon what is included in the notice.

A suspension notice should allow activities to continue unless their cessation is necessary to address the risk of pollution. While the suspension notice is in force, additional steps may need to be taken in relation to any activities that are allowed to continue. Where this is the case the suspension notice must set out these additional steps.

When the operator has taken the remedial steps required by the notice, the regulator must withdraw the notice.

See **Chapter 10** on Powers of the Agencies.

Charging

6.32 Two separate but related sets of charging schemes apply to regulated facilities:

- for facilities where the Environment Agency is the regulator, under s 41 of the Environment Act 1995 and approved by the Secretary of State, and
- for facilities where the local authority is the regulator, set by the Secretary of State and the Welsh Ministers under EPR, reg 65.

Within this overall arrangement, different charges are payable at different regulatory stages. They will also vary across sectors. Further details are available from regulators or from Defra and the Welsh Government.

Applications will normally incur charges. Where an operator must pay a charge when submitting an application, the regulator must receive this before the application can be considered duly made. If the regulator judges that an application accompanied by payment is not duly made, it would normally return the charge to the applicant.

Regulators must allow for different charges for different categories of variation application. This is to allow an approach that can reflect the:

- amount of effort that the regulator has to put into determining the application;
- potential environmental impact or risk; and
- necessary degree of public participation.

Subsistence

6.33 Operators must pay subsistence charges to support the regulator's ongoing costs for such things as checking and monitoring data or compliance assessment. If an operator fails to pay a subsistence charge, the regulator may revoke the permit.

Appeal

Appeal Procedures

6.34 A person may appeal (EPR, reg 31[30]) when:

- the regulator has refused their application for the grant of a permit[31];
- the regulator has refused their application for a variation of a permit[32];
- the regulator has refused their application to transfer or surrender the permit[33];
- they disagree with the conditions imposed by the regulator in their permit[34];
- an application is deemed to have been refused[35];
- the regulator has deemed an application to be withdrawn[36];
- the regulator has decided not to authorise the closure procedure mentioned in Article 13 of the Landfill Directive or Article 12 of the Mining Waste Directive;
- the regulator has served a revocation, enforcement, suspension or prohibition notice, or a landfill or mining waste facility closure notice on them[37];
- the regulator has determined that certain information in relation to their application or permit must be included on the public register [38].

Time limits for making appeals vary according to the basis of the appeal[39]. The Secretary of State or the Welsh Ministers have the power to extend some of the limits, but would only do so in the most compelling circumstances.

30 EPR 2010, SI 2010/675, Pt 2, Ch 5, reg 31, Appeals to the appropriate authority.
31 EPR, Pt 2, Ch 5, reg 31(2)(a).
32 EPR, Pt 2, Ch 5, reg 31(4)(b).
33 EPR, Pt 2, Ch 5, reg 31(3).
34 EPR, Pt 2, Ch 5, reg 31(2)(b), (c).
35 EPR, Sch 5, Pt 1, para 15(1).
36 EPR, Pt 2, Ch 5, reg 31(2)(d).
37 EPR, Pt 2, Ch 5, reg 31(2)(f).
38 EPR, Pt 5, reg 53, Appeals in relation to confidentiality.
39 EPR, Sch 6, para 3, Time for making an appeal.

An appeal must be made:

- before a revocation notice takes effect;
- within two months from the date of a suspension, enforcement, mining waste facility closure or landfill closure notice or a regulator initiated variation;
- no later than 15 working days after receiving notice that an application is deemed to be withdrawn[40] or receiving notice that certain information must be included on the public register [41], or unless the refusal was made[42] in relation to an existing Category A mining waste facility.
- in all other cases within six months of the date of the decision or deemed decision.

Appeals must be made to the Secretary of State or the Welsh Ministers.

The Secretary of State and the Welsh Ministers may appoint the Planning Inspectorate (PINS) to determine appeals. The Secretary of State or the Welsh Ministers can however recover any appeal. This would be an exceptional step and likely to be taken only if the appeal meets particular criteria.

The Planning Inspectorate has produced separate guidance on appeals for environmental permitting[43]. A failure to follow the procedure set out in that guidance may lead to adverse costs implications.

If a person appeals against a revocation notice, the revocation does not take effect until the appeal has been determined or withdrawn[44] except in certain circumstances in relation to 'stand alone' water discharge activities.

If the appeal is against a permit condition, variation, enforcement, suspension, prohibition, landfill closure or mining waste facility closure notice, then the notice or condition must be complied with until the appointed person or the Secretary of State or the Welsh Ministers has determined the appeal, except in certain circumstances in relation to 'stand alone' water discharge activities.

Who Should be Informed That an Appeal Has Been Made?

6.35 Following receipt of the notice of appeal (other than against a decision on commercial confidentiality), the regulator should contact anyone it thinks may be affected by or has a particular interest in the matter (EPR, Sch 6,

40 EPR 2010, SI 2010/675, Sch 5, Pt 1, para 4(2).

41 EPR Pt 5, reg 53.

42 EPR Sch 20, para14(2).

43 Environmental Permitting (England and Wales) Regulations 2010 Environmental Permits – The Appeal Guidance Procedure. The planning Inspectorate March 2013 available at: https://www.gov.uk/government/uploads/system/ uploads/attachment_data/file/471594/environmental_permitting_guidance_notes.pdf.

44 EPR, Pt 5, reg 31(9).

para 4). The regulator should ensure that the relevant people are informed as to the nature of the appeal and are made aware that representations can be made in writing to the appointed person.

Secretary of State or the Welsh Ministers

6.36 The regulator should tell the appointed person or the Secretary of State or the Welsh Ministers who it has notified of the appeal and when[45].

Conducting an Appeal

6.37 An appeal may be conducted by written representations, or through a hearing or inquiry under the control of the Secretary of State or the Welsh Ministers' appointee.

All parties have the right to request a hearing and one will also be held if the Secretary of State or the Welsh Ministers or their appointee decides one is necessary. Each appeal procedure will be conducted in the spirit of the rules and Regulations for planning appeals.

Withdrawing an Appeal

6.38 An appellant may withdraw an appeal at any time by giving notice in writing to the Secretary of State's or the Welsh Ministers' appointee, copied to the regulator. The Secretary of State's or the Welsh Ministers' appointee should tell anyone with an interest in the appeal that it has been withdrawn.

The regulator should inform all the people it notified of the appeal[46].

Costs

6.39 The appellant and regulator will normally be expected to pay their own expenses during an appeal. However where a hearing or inquiry is held, either the appellant or the regulator can apply for costs.

Applications for costs will only be allowed if the party claiming them can show that the other side behaved unreasonably and put them to unnecessary expense.

45 EPR, Sch 6, para 4 (3).
46 EPR 2010, SI 2010/675, Sch 6, para 4(4).

Appeal Decisions

6.40 After the exchange of written representations has been completed or the hearing or inquiry held, the appointed person will either make a decision or report their conclusions and recommendations to the Secretary of State or the Welsh Ministers for determination.

The appointed person or the Secretary of State or the Welsh Ministers may dismiss the appeal or may uphold the appeal in total or in part. They may quash or vary any notice. They may also direct the regulator on what permit conditions to impose.

If the appeal is dismissed, the original decision by the regulator continues in force. Where the original decision has been ineffective during the appeal (ie, for a revocation notice), the regulator's decision becomes effective from the end of the day on which the appeal is dismissed or withdrawn.

If the appeal is upheld, either in total or in part, the regulator has a duty to give effect to that decision. Consequently, where the regulator grants a permit or issues a variation notice following an appeal, such a notice should include reference to the fact that the decision is giving effect to a determination by the appointed person or the Secretary of State or the Welsh Ministers.

The regulator should take into account relevant appeal decisions when reviewing and developing the conditions to be attached to permits.

Appeal decisions may be challenged by judicial review on a point of law.

SCOTLAND

6.41 In May 2012 a Consultation on Proposals for an integrated Framework of Environmental Regulation[47] was created as a discussion point for change within the Environmental processes in Scotland. The proposals outlined in the consultation aim to deliver a simpler legislative framework and provide a more consistent range of enforcement tools so that swift, proportionate and effective action can be taken against those who would damage the environment for criminal profit or through wilful negligence.

SEPA[48] is the primary regulator for environmental regulation and enforcement regimes covering water, waste, radioactive substances and pollution, prevention and control. These are referred to as the four main regimes. The others include regulating areas such as the Emissions Trading Scheme, contaminated land, chemicals, packaging and producer responsibility and movement of waste and radioactive waste.

47 Consultation on Proposals for an integrated Framework of Environmental Regulation The Scottish Government ISBN 978-1-78045-808-3 available at: http://www.gov.scot/Resource/0039/00392549.pdf.

48 SEPA – Scottish Environment Protection Agency.

The outcome which is required of a new legislative framework is one that is:

- a single, proportionate and risk-based permissioning structure;
- a single consistent regulatory procedure;
- a flexible approach to permissioning; and
- a flexible and proportionate approach to enforcement.

The current set of environmental permissions varies in nature and content so the proposed framework would seek to:

- comply with European and national legislative requirements;
- provide effective controls to manage impacts on and limit damage to the environment;
- be proportionate to the environmental risk arising from uncontrolled activity;
- be core to SEPA's regulatory approach;
- contain conditions which are fair, reasonable and achievable (where conditions are necessary);
- be a clearer and simpler system for permissioning;
- be outcome oriented;
- be capable of being administered efficiently; and
- take account of the degree of uniformity of the activity.

Current Regulatory Framework

6.42 The four main regimes fall under separate regulatory regimes and are set out in 33 pieces of primary legislation at the national level.

Water

6.43 This falls under the Water Environment and Water Services (Scotland) Act 2003 and the Water Environment (Controlled Activities) (Scotland) Regulations 2011, SI 2011/209.

This requires a licence, registration and has general binding rules.

Pollution Prevention and Control (PPC)

6.44 This comes under the Pollution Prevention and Control (Scotland) Regulations 2000, SSI 2000/323 and requires a permit and has standard rules.

Waste

6.45 This comes under the Environmental Protection Act 1990 and the Waste Management Licensing (Scotland) Regulations 2011, SSI 2011/228 and requires a licence form on which there are some exemptions.

Radioactive Substances

6.46 These come under the Radioactive Substances Act 1993 and the Radioactive Substances Exemption (Scotland) Order 2011, SSI 2011/147 and require authorisation/registration and there are some exemptions.

The new regulatory framework will look at having a single permissioning structure with a single set of procedures.

This would be a permit, registration and general binding rules. The general binding rules are the simplest level of permission for the simplest low risk activities. The activity would need to be carried out in accordance with rules/conditions but the operator would not be required to notify SEPA. This is similar to Water Environment (Controlled Activities) (Scotland) Regulations 2011, SSI 2011/209 activities, which are covered by the WFD[49] and the IPPC Directive[50].

The next level of permission would be one of registration against a specific set of activity rules and conditions. Registrations would involve simple site specific screening which, if passed, would allow the activity to be registered against standard conditions. Registrations would normally be subject to ongoing active regulation but SEPA may still carry out targeted inspections or campaigns for these activities.

A permit would be a written document, which authorises an environmental activity to be carried out subject to a number of conditions. Operators would need to apply to SEPA for a permit; providing such information as SEPA reasonably requires by relevant directives (for example, the IPPC Directive requires that an application include a description of the installation and its activities, the raw and auxiliary materials, other substances and the energy used in or generated by the installation, the sources of emissions from the installation, the conditions of the site of the installation etc). There are also a number of conditions under the relevant directives, which must be included in a permit. The application would be subject to the full requirements of the regulatory regime including advertising and consultation (where required), variation and review.

Permits would range from a basic, generic permit for simpler, more standard and lower risk activities that still require a permit, through to permits for complex/high risk activities which would require site specific assessment and some site specific conditions.

Standard conditions and rules are an approach, which has been taken with the PPC (Scotland) Regulations 2000[51], which allow the Scottish ministers, or SEPA to develop and consult on a set of standard rules, which apply to installations or mobile plant as described in the rules.

49 Waste Framework Directive 2008/98/EC.
50 Integrated Pollution and Control 2008/1/EC.
51 Pollution Prevention and Control (Scotland) Regulations 2000, SSI 2000/323.

The single regulatory procedure would also provide an enforcement notice, which deals with compliance and restoration (including the removal of waste, a suspension/stop notice and a revocation notice.

The Regulatory Reform (Scotland) Act 2014

6.47 The Regulatory Reform (Scotland) Act 2014 has the primary purpose of further improving the way in which regulation is developed and applied. It is also there to support SEPAs transformation programme and improve the way environmental regulations are applied across Scotland. This will introduce:

- A new integrated framework for environmental regulation, to help simplify procedures for SEPA and businesses.
- New enforcement tools to allow SEPA and criminal courts to better tackle environmental crime and be tougher with those who disregard their environmental obligations and measures to extend protection against intimidation or physical assaults to SEPA officers.
- A new statutory purpose for SEPA.

Environmental Regulation

6.48 Part 3[52] of the Act deals with Environmental Regulation. It allows Scottish Ministers to make provision for or in connection with protecting and improving the environment, which includes regulating environmental activities and implementing EU obligations relating to protecting and improving the environment. It also allows SEPA to impose fixed and variable monetary penalties and to accept and enforce undertakings from a person whom SEPA reasonably suspects has committed a 'relevant offence'.

Timeline

6.49 Introduced to the Scottish Parliament on 27 March 2013, the Regulatory Reform (Scotland) Bill was approved on 16 January 2014 and received Royal Assent on 19 February 2014. It was first announced by the First Minister in the Programme for Government in September 2012 (as the Better Regulation Bill), taking forward the 2011 commitment to improve further the way regulation is developed and applied in practice across Scotland.

Until the regulations change then this describes the current process in Scotland.

The Pollution Prevention and Control (Scotland) Regulations 2012, SSI 2012/360 exercise the powers in s 2 of and Sch 1 to the Pollution Prevention and Control Act 1999 and s 2(2) of the European Communities Act 1972 and all other powers enabling them to do so.

52 Regulatory Reform (Scotland) Act 2014, Pt 3.

The Industrial Emissions Directive takes seven existing Directives:

- those concerning integrated pollution prevention and control ('IPPC') (2008/1/EC);
- large combustion plants (2001/80/EC);
- waste incineration (2000/76/EC);
- solvent emissions (1999/13/EC);
- three concerning waste from the titanium dioxide industry[53].

These 'component Directives' are currently transposed in Scotland through the Pollution Prevention and Control (Scotland) Regulations 2000, SSI 2000/323 – usually referred to as 'the PPC Regulations'.

Much of the material in the recast Directive remains substantively unchanged from the component Directives. The PPC Regulations provide the most appropriate vehicle through which to transpose the Industrial Emissions Directive.

The PPC Regulations have already been amended by no fewer than 25 other pieces of legislation, up to and including the Waste (Scotland) Regulations 2012, SSI 2012/148. These have ranged from significant amendments dealing with large combustion plants, waste incineration, solvent emissions, and petrol vapour recovery, to more minor amendments updating cross-references and the like. The result is a very complex patchwork of legislation, making it difficult for all users to determine what the current requirements are.

Transposition of the Directive inevitably requires a large number of further amendments to the PPC Regulations, compounding the difficulties of use. The PPC Regulations, incorporate the changes required to transpose the Directive. While this makes it a little more difficult to see precisely what has changed, the end result is a single clear set of Regulations.

Water

6.50 This is under the Water Environment and Water Services (Scotland) Act 2003 and the Water Environment (Controlled Activities) (Scotland) Regulations 2011, SSI 2011/209.

In Scotland, if you discharge anything other than clean, uncontaminated water you must have:

- an authorisation under the Controlled Activity Regulations (CAR);
- an integrated pollution control authorisation or pollution prevention and control permit.

These authorisations and permits are issued by the Scottish Environment Protection Agency (SEPA).

53 Directives 78/176/EEC, 82/883/EEC and 92/112/EEC.

CAR has three levels of authorisation:

- general binding rules;
- registration;
- water use licences.

Pollution Prevention and Control (PPC)

6.51 The Pollution Prevention and Control (Scotland) Regulations 2012, SSI 2012/360 (PPC Regulations 2012) came into force on 7 January 2013 to implement the requirements of the Industrial Emissions Directive[54] (IED), as well as consolidating the Pollution Prevention and Control (Scotland) Regulations 2000, SSI 2000/323 (PPC 2000). These revisions remove the ambiguities and inconsistencies.

The PPC Regulations 2012, apply an integrated environmental approach to the regulation of certain industrial activities. This means that emissions to air, water (including discharges to sewer) and land, plus a range of other environmental effects, must be considered together.

Operators of installations that fall under the PPC Regulations 2012 must have a permit in order to operate. Permit conditions are set with the aim of achieving a high level of protection for the environment as a whole. The use of Best Available Techniques (BAT) to prevent and minimise emissions, aims to balance the costs to the operator against the benefits to the environment as much as possible.

Permits are considered based on the nature of the activity and whether it is defined as a Part A activity or a Part B activity.

Part A activities[55] are defined in Sch 1 of the PPC Regulations 2012.

Operators of installations where the highest risk industrial activities are undertaken are required to have a permit under Part A of the Regulations. The regulation of PPC Part B activities focuses on control of emissions to air.

Waste

6.52 This comes under the Environmental Protection Act 1990 and the Waste Management Licensing (Scotland) Regulations 2011, SSI 2011/228 and requires a licence from which there are some exemptions.

Radioactive Substances

6.53 These come under the Radioactive Substances Act 1993 and the Radioactive Substances Exemption (Scotland) Order 2011, SSI 2011/147

54 Industrial Emissions Directive 2010/75/EU.

55 Pollution Prevention and Control (Scotland) Regulations 2000, SSI 2000/323, Sch 1.

and require authorisation/registration and there are some exemptions. These regulations regulate the keeping of radioactive materials and the accumulation and disposal of radioactive waste. Issues of contaminated land and historic uses of radioactive substances come under the Radioactive Contaminated Land (Scotland) Regulations 2007, SSI 2007/179 as amended.

Radioactive Water Advisors

6.54 Radioactive waste advisers are specialists in radioactive waste disposal and environmental radiation protection. There is a requirement under European law for people who keep or use radioactive material, or who accumulate or dispose of radioactive waste, to appoint advisers, known as 'radiation protection experts' to advise them on radiological protection.

The Radioactive Waste Adviser scheme was developed jointly by SEPA, the Environment Agency and the Northern Ireland Environment Agency with extensive stakeholder engagement.

Anyone who has a permit under the Environmental Permitting (England and Wales) Regulations 2010, SI 2010/675 or an authorisation under the Radioactive Substances Act 1993 to accumulate or dispose of radioactive waste needs to appoint a Radioactive Waste Adviser.

A RWA is a specialist in radioactive waste disposal and environmental radiation protection who has demonstrated competence in the RWA syllabus.

An individual RWA is someone who has demonstrated competence in all parts of the RWA syllabus, whereas a Corporate RWA is a group of individuals who can collectively demonstrate competence in the Radioactive Waste Adviser syllabus for giving advice to a nuclear permit holder.

The permit holder is responsible for assessing the suitability of RWAs. The environment agencies may check the permit holder's procedures for determining and assessing suitability during routine inspections, as for any other permit condition.

NORTHERN IRELAND

6.55 Authorisation in the form of permits, licences, consents, registrations, notifications or exemptions, are required in order to carry out certain types of business activity. These come with conditions, which must be complied with and aim to control the impact of activities on the environment.

Most applications are made to the Northern Ireland Environment Agency (NIEA), which also acts as the main environmental regulator in Northern Ireland. District councils and water companies are in charge of some types of authorisation.

The relevant legislation is the Pollution Prevention and Control (Northern Ireland) Regulations 2003, NISR 2003/46 and the Waste Management Licensing (Northern Ireland) Regulations 2003, NISR 2003/493.

Authorisations are required to:

- carry out certain listed activities at an installation or mobile plant, such as industrial activities, or carry out waste disposal or recovery activities, such as metals recycling or waste transfer stations;
- discharge anything other than clean water into surface waters or groundwater;
- discharge trade effluent into the public foul sewer;
- take water from surface waters or groundwater, or obstruct them in any way;
- transport waste, or arrange for someone else to do so (Waste Broker and Dealer registration);
- produce or move hazardous waste;
- keep, use or dispose of radioactive substances;
- produce greenhouse gas emissions that are covered by the European Union emissions trading system or the CRC Energy Efficiency Scheme.

Pollution Prevention and Control Permits and Waste Management Licences

6.56 Pollution, Prevention and Control (PPC) permits and waste management licences regulate business activities that could have an impact on the environment or human health.

PPC Permits

6.57 A PPC permit may be required from the Northern Ireland Environment Agency (NIEA) or the district council because of the type and size of the business activities. For example, a PPC permit may be required for:

- energy activities, such as combustion and refining;
- metals production and processing;
- mineral activities, such as production of cement, lime, glass and ceramics;
- production of chemicals, including fertilisers and pharmaceuticals;
- waste activities, such as operating a landfill site or incinerator;
- activities involving the use of solvents, such as coating, printing and dry cleaning.

Conditions of the permit must be complied with.

There are strict controls on discharging substances and materials to land, surface waters or groundwater. Surface waters include rivers, lakes, loughs, lochs, streams, reservoirs and canals. Groundwater is all water located below the water table.

Discharging to Water

6.58 In Northern Ireland, anything which is discharged other than clean, uncontaminated water must have:

- discharge consent for discharges to surface water;
- a ground water authorisation for discharges to groundwater.

These consents and authorisations are issued by the Northern Ireland Environment Agency (NIEA). Conditions of the discharge consent or groundwater authorisation must be complied with.

Trade Effluent Consents and Agreements

6.59 A trade effluent consent or a trade effluent agreement with the water and sewerage company must be obtained before any discharge of trade effluent into a public foul sewer is made. This is to avoid causing serious damage to the sewage system and creating a risk of danger to public health.

Trade effluent is any liquid waste resulting from non-domestic or industrial activity. It includes:

- wastewater contaminated with oils, chemicals and solvents;
- liquid process wastes;
- detergents;
- condensate water from compressed air installation;
- cooling and wash water;
- biodegradable liquids;
- contaminated mine or quarry water.

If the effluent is hazardous, you must dispose of it as hazardous/special waste. See **Chapter 8**.

Water Abstraction and Impoundment Authorisations

6.60 If you take or store surface water or groundwater from any source, you are abstracting or impounding water.

If you abstract or impound water, you may need authorisation from the Northern Ireland Environment Agency (NIEA). The type of authorisation required will depend on how much water is taken. Authorisation may be needed even if the water is moved and then temporarily returned to the source.

Surface water and groundwater sources include:

- rivers;
- streams;
- loughs and lochs;
- reservoirs;
- estuaries;
- coastal waters;
- wells;

- springs;
- boreholes.

If water is taken from the mains supply then an authorisation or licence is not required.

Abstraction and Impoundment in Northern Ireland

6.61 If more than 20 cubic metres (m^3) of water is abstracted a day from surface waters or groundwater, then an abstraction licence from the NIEA will need to be obtained.

If $20m^3$ or less of water is abstracted a day then you must:

- be able to demonstrate the volume of water you abstract;
- minimise water leaks;
- prevent any contamination or pollution.

Abstraction of amounts between $10m^3$ and $20m^3$ of water a day requires the NIEA to be notified.

If water is impounded on a watercourse, for example to create a reservoir, then an impoundment licence from the NIEA may be required.

The NIEA should be notified as early as possible if a licence is being transferred. The licence will remain the responsibility of the current licence holder until it is transferred.

Radioactive Substances Certificates

6.62 You must have a certificate of registration from the Northern Ireland Environment Agency (NIEA) if you keep or use radioactive materials.

The following need to be measured:

- open or unsealed sources on your premises;
- closed or sealed sources on your premises;
- mobile apparatus incorporating a sealed radioactive source;
- mobile apparatus (open sources) for use in environmental studies;
- radioactive packages stored in transit.

A certificate of authorisation from NIEA must be obtained if you are accumulating radioactive waste on your premises or disposing of radioactive waste from your premises, unless an exemption order applies. Disposal of radioactive waste includes disposal of radioactive gases, vapours, liquids and solids to the air, land, sea, surface waters (such as rivers and lakes) and sewers. You are also disposing of radioactive waste if you transfer it to other premises.

Radioactive materials must not be kept or used, or disposed of or accumulated, before there is a certificate of registration or authorisation. Any condition in the certificate must be complied with.

Waste Management Licences and Exemptions

6.63 A waste management licence from NIEA may be required for example:

- to store other people's waste;
- to treat waste, carry out recycling or use waste mobile plants;
- to carry out final disposal of waste.

It is unlikely that a licence will be needed if the only waste that is stored is waste that you produce and an authorised waste carrier removes it from your site regularly. You must check that anyone who handles your waste has the correct permit, licence or exemption.

An activity may qualify for an exemption from waste management licensing – but whether a licence is needed will depend on:

- how long the waste is stored;
- the types and quantities of wastes handled;
- the activities carried out on site.

An activity that is exempt may still need to be registered with the NIEA. Controls will still need to be complied with to prevent pollution and harm to human health.

Chapter 7

Exclusions and Exemptions from Waste Licensing

INTRODUCTION

7.1 Since devolution – the regulations which govern licensing and exemptions are different in England and Wales from Northern Ireland and Scotland.

The Regulations covering exemptions and exclusion in England and Wales are the Environmental Permitting (England and Wales) Regulations 2010, SI 2010/675 as amended.

For Scotland the regulations are the Waste Management Licensing (Scotland) Regulations 2011, SSI 2011/228 as amended and the Pollution Prevention and Control (Scotland) Regulations 2000, SSI 2000/323. For Northern Ireland the regulations are, the Waste Management Licensing Regulations (Northern Ireland) 2003, NISR 2003/493 as amended and the Pollution Prevention and Control (Northern Ireland) Regulations 2003, NISR 2003/46.

These Regulations list the type of waste and the exemptions, which are relevant to the different areas that the Regulations apply to.

These exemptions and exclusions and the relevant sections of the Regulations are discussed by region below.

Quality Protocols

7.2 The Waste Protocols Project is a joint venture between the Environment Agency and WRAP initiative in collaboration with industry. It is funded by the Department for the Environment, Food and Rural Affairs (Defra), the Welsh Assesmbly Government (WAG) and the Northern Ireland Environment Agency (NIEA) and is a business resource efficiency activity.

The Quality Protocols (QPs) form part of voluntary end of waste frameworks for specific wastes and end uses based on the relevant end of waste law. The Quality Protocols are there to explain when a waste derived material can be regarded as a non-waste product and is no longer subject to waste controls.

Where there is no quality protocol of a material then the end of waste test guidance need to be applied. Quality protocols (QPs): rules for QPs is a guidance document produced by the Environment Agency[1].

Waste Management Control Rules

7.3 Waste controls must be applied until the material is no longer a waste and the correct authorisation for transport, storage, handling and treatment of waste must be in place and complied with, otherwise an offence is committed.

There is no legal duty to apply with a QP – but if they are not complied with then any waste derived material must be treated as waste,

Non-Waste Product Rules

7.4 When a non-waste product is produced then there is an exemption from, for example, the REACH (Registration, Evaluation, Authorisation and Restriction of Chemicals) Regulations.

1 Quality protocols (QPs): rules for all QPs https://www.gov.uk/government/publications/quality-protocols-qps-rules-for-all-qps/quality-protocols-qps-rules-for-all-qps.

Imports and Exports

7.5 Even though a QP product may cease to be waste in England, Wales and Northern Ireland, other countries may have different rules. This can be ascertained by looking at the Waste Shipment Regulation (EC/1013/2066) ([2006] OJ L 190, 12.7.2006, p 1–98) and with the country's competent authority.

European Union Equivalent Standards

7.6 Countries within the EU may recycle waste derived material using their own codes for production and processing provided that:

- A relevant standard or code of practice developed by that country;
- An international standard or technical regulation adopted by that country.

These must provide product performance levels and protection of human health and the environment – equal to that of the UK QPs.

Updating the QPs

7.7 QPs may be reviewed by the environmental regulators when there is:

- Pollution incident;
- Change in the market;
- Change to legislation and law;
- Change to agreed European Standard;
- Change to the understanding of the chemical composition or physical properties of the material.

A QP can also be withdrawn if it is being misapplied and/or misused. There are a number of QP documents and each one indicates whether it applies to England, Northern Ireland and/or Wales.

Quality Protocol	England	Wales	Northern Ireland
Non-packaging plastics: Quality Protocol	Yes	Yes	No
Recycled gypsum from waste plasterboard: Quality Protocol	Yes	Yes	Yes
Biodiesel: Quality Protocol	Yes	Yes	Yes
Aggregate from waste steel slag: Quality Protocol	Yes	Yes	No
Quality Protocol: Flat glass	Yes	Yes	No

Quality Protocol	England	Wales	Northern Ireland
Quality Protocol: Type-derived rubber materials	Yes	Yes	No
Quality Protocol: Anaerobic digestate	Yes	Yes	Yes
Quality Protocol: Processed Fuel Oil	Yes	Yes	Yes
Position statement: Processed Fuel Oil (PFO)			
Quality Protocol: Biomethane from waste	Yes	Yes	No
Quality Protocol: Aggregates from inert waste	Yes	Yes	Yes
Quality Protocol: Poultry Litter Ash (PLA)	Yes	Yes	Yes
Quality Protocol: Compost	Yes	Yes	Yes
Quality Protocol: Pulverised Fuel Ash (PFA) and Furnace Bottom Ash (FBA)	Yes	Yes	Yes

ENGLAND AND WALES

Introduction

7.8 In England and Wales, the Regulations which govern the exemptions and exclusions of permits are the Environmental Permitting (England and Wales) Regulations 2007, SI 2007/3538. These were first introduced in 2007 but were replaced by the Environmental Permitting (England and Wales) Regulations 2010, SI 2010/675. Since their introduction there have been a number of amendments – and in particular two amendments, which relate only to England and not to Wales.

There are some rules to follow if an exemption is to be obtained:

- The waste exemption may need to be registered;
- A waste operation can only be registered as exempt if it meets all the conditions of an exemption;
- The waste must not harm the environment or human health.

If these rules are not followed then the waste operation is not exempt and the Environment Agency (EA) can deregister the exemption.

The EA can also take enforcement action or prosecute if the conditions of an exemption are not met, or if harm has occured to people or the environment. If it is not possible to meet the conditions of an exemption then an environmental permit must be applied for.

Registering for a waste exemption is free except for a T11, which is a waste electrical and electronic equipment (WEEE) exemption, which costs £840.

Each waste exemption registration lasts for three years so an exemption will need to be re-registered if the waste operation is still to be carried out after three years.

When Registration is not Required

7.9 A registration is not required where there is an exemption. These are called the Non Waste Framework Directive (NWFD) exemptions. These exemptions cover the temporary storage of waste, but the limits and conditions set out must be followed.

NWFD 2 – Temporary Storage at a Place of Production

7.10 This exemption allows the temporary storage of any waste at the place of production before it is collected.

Although this exemption does not have to be registered, the terms and conditions of the exemption must still be complied with.

Types of activity that can be carried out:

- Waste can only be stored in a secure place at the place of production;
- Some treatments can be carried out on the waste to help with storage and collection.

These treatments are only to make the waste produced easier to store and collect for recovery or disposal somewhere else.

The treatments must not result in a change in the characteristics of the waste. Treatment must be purely to help with transport or collection of different wastes. Examples of such treatments include:

- compacting paper and cardboard to increase the amount of waste that can be stored within a container;
- shredding confidential papers for security purposes;
- separating recyclables such as paper, card, plastic and glass from mixed wastes into separate containers.

Types of activity that cannot be carried out:

Waste cannot be stored at a site if it was produced somewhere else although any waste can be temporarily stored at the place of production before it is collected. The waste must be stored in a secure place and not be stored for longer than 12 months.

NWFD 3 – Temporary Storage of Waste at a Place Controlled by a Producer

7.11 This exemption allows the temporary storage of waste that has been produced when working away from the normal business premises.

It applies to organisations such as a medical or veterinary practice or a builder who carries out work at other premises and generates waste. Then the waste is brought back to the normal business premises and collected together.

Types of activity that can be carried out:

- the storage of waste that has been produced working elsewhere, in a secure place that the producer has control over;
- carrying out some treatments on the waste to help with storage and collection, such as separating wastes into different types.

These treatments are only to make waste that has been produced easier to store and collect for recovery or disposal somewhere else.

The treatments must not result in a change in the characteristics of the waste. Treatment must be purely to help with transport or collection of different wastes.

Examples of such treatments include:

- compacting paper and cardboard to increase the amount of waste that can be stored within a container;
- shredding confidential papers for security purposes;
- separating recyclables such as paper, card, plastic and glass from mixed wastes into separate containers.

Types of activity that cannot be carried out:

- provision of a waste management service to others for these operations;
- mixing different types of waste.

Types of waste that can be stored:

Any waste that the producer of the waste has control over that does not contain or consist of:

- unbonded asbestos;
- any substance that has a flash point of less than 21°C.

Quantity of waste that can be stored:

At any one time storage of up to:

- 50 cubic metres of solid waste;
- 1,000 litres of liquid waste.

Main conditions

If more than one type of waste is stored, the different types must not be mixed and the waste must not be stored for longer than three months.

NWFD 4 – Temporary Storage at a Collection Point

7.12 This exemption allows the temporary storage of waste at a collection point before recovering or disposing of the waste elsewhere.

For example:

- a pharmacy which takes back and stores out of date medicines returned by patients or operates a needle exchange scheme;
- a shop which provides a container for customers to deposit waste batteries;
- a shop which provides a take back arrangement for Waste Electrical and Electronic Equipment (WEEE).

Types of activity that can be carried out:

- only collect and store waste in a secure container at a collection point;
- carry out some treatments on the waste to help with storage and collection.

These treatments are only to make waste that has been produced, easier to store and collect for recovery or disposal somewhere else.

The treatments must not result in a change in the characteristics of the waste. Treatment must be purely to help with transport or collection of different wastes. Examples of such treatments include:

- compacting paper and cardboard if it increases the amount of waste that can be stored within a container;
- shredding confidential papers for security purposes;
- separating recyclables such as paper, card, plastic and glass from mixed wastes into separate containers.

Types of activity that cannot be carried out:

- collection of waste as the main business activity;
- receive payment for collecting the waste;
- store mixed waste.

Types of waste that can be stored:

Any waste that doesn't contain or consist of:

- asbestos;
- any substance that has a flash point of less than 21°C.

Quantity of waste that can be stored

At any one time storage of up to:

- 30 cubic metres of Waste Electrical and Electronic Equipment (WEEE);
- 50 cubic metres of non-hazardous waste that isn't WEEE that will be recovered elsewhere;
- 5 cubic metres of waste, other than those mentioned.

Main conditions

If more than one type of waste is stored then the different types must not be mixed and waste can only be stored temporarily; generally no longer than three months.

Definitions

'Collection' – gathering waste, including preliminary sorting and storage of waste to transport it to a waste treatment facility.

'Collection point' – a place used for collecting waste where the organisation collecting the waste doesn't receive payment for collecting the waste or collects waste as its main business activity.

'Impermeable surface' – a surface or pavement constructed and maintained to a standard sufficient to prevent the transmission of liquids beyond the surface.

'Place of production' – where the waste was originally produced.

'Sealed drainage' – a drainage system with an impermeable surface which makes sure that:

- no liquid will run off the surface apart from via the system;
- except where they may be lawfully discharged, all liquids entering the system are collected in a sealed sump.

'Secondary containment' – a bund or any other system for preventing waste that has leaked from its main container from escaping from the place where it is stored or treated.

'Secure' – a container, lagoon or other place is secure in relation to waste kept in it if:

- all reasonable precautions are taken to make sure the waste can't escape;
- members of the public can't gain access to the waste.

Exemption Categories

7.13 Under the Regulations there are four exemption categories each of which is divided into specific categories, which reference specific wastes.

Using Waste

7.14

U1 use of waste in construction;
U2 use of baled end-of-life tyres in construction;
U3 construction of entertainment or educational installations;
U4 burning of waste as a fuel in a small appliance;
U5 using biodiesel produced from waste as fuel;

U6 using sludge to re-seed a waste water treatment plant;
U7 using effluent to clean a highway gravel bed;
U8 using waste for a specified purpose;
U9 using waste to manufacture finished goods;
U10 spreading waste to benefit agricultural land;
U11 spreading waste to benefit non-agricultural land;
U12 using mulch;
U13 spreading plant matter to provide benefits;
U14 incorporating ash into soil;
U15 pig and poultry ash.
U16 using depolluted end-of-life vehicles for parts.

Treating Waste

7.15

T1 cleaning, washing, spraying or coating relevant waste;
T2 recovering textiles;
T4 preparatory treatments (baling, sorting, shredding etc);
T5 screening and blending waste;
T6 treating waste wood and waste plant matter by chipping, shredding, cutting or pulverising;
T8 mechanically treating end-of-life tyres;
T9 recovering scrap metal;
T10 sorting mixed waste;
T11 repairing or refurbishing waste electrical and electronic equipment (WEEE);
T12 manually treating waste;
T13 treating waste food;
T14 crushing and emptying vehicle waste oil filters;
T15 treating waste aerosol cans;
T16 treating waste toner and ink cartridges;
T17 crushing waste fluorescent tubes;
T18 using flocculants to remove water from waste;
T19 physical and chemical treatment of waste edible oil and fat to produce biodiesel;
T20 treating waste at a water treatment works;
T21 recover waste at a waste water treatment works;
T23 aerobic composting and associated prior treatment;
T24 anaerobic digestion at premises used for agriculture and burning resulting biogas;
T25 anaerobic digestion at premises not used for agriculture and burning resulting biogas;
T26 treatment of kitchen waste in a wormery;
T27 treatment of sheep dip using organophosphate-degrading enzyme;
T28 sort and denature controlled drugs for disposal;

T29 treatment of non-hazardous pesticide washings by carbon filtration for disposal;
T30 recover silver;
T31 recover monopropylene glycol from aircraft antifreeze fluids;
T32 treatment of waste in a biobed or biofilter;
T33 recover central heating oil by filtration.

Disposing of Waste

7.16

D1 depositing waste from dredging inland waters;
D2 depositing waste from a railway sanitary convenience;
D3 depositing waste from a portable sanitary convenience;
D4 depositing agricultural waste consisting of plant tissue under a Plant Health Notice;
D5 depositing waste samples for testing or analysis;
D6 disposal by incineration;
D7 burning waste in the open;
D8 burning waste at a port under a Plant Health Notice.

Storing Waste

7.17

S1 storing waste in secure containers;
S2 storing waste in a secure place;
S3 storing sludge.

The Environmental Permitting (England and Wales) Regulations 2007, SI 2007/3538

7.18 In England and Wales the main Regulations are the Environmental Permitting Regulations 2007, SI 2007/3538 (as amended) (the EP Regulations 2007). The amendments where they affect exemptions and exclusion are listed.

- Exempt facilities (reg 4);
- Interpretation: exempt facilities (reg 5);
- Regulations 12 and 8 read together;
- Requirement for an environmental permit (reg 12);
- Interpretation: regulated facility and class of regulated facility (reg 8);
- Schedule 2: Exempt facilities: general;
- Schedule 3: Exempt facilities: descriptions and conditions.

The Environmental Permitting (England and Wales) (Amendment) Regulations 2009, SI 2009/1799

7.19 Regulation 4 of SI 2009/1799 amends reg 8 (Interpretation: regulated facility) of the EP Regulations 2007 by adding mining waste operations, whether or not involving a mining waste facility, to the list of regulated facilities, and amending the provision, which specifies when an exempt waste operation is not a regulated facility.

SI 2009/1799, reg 5 amends reg 12 (Requirement for an environmental permit) of the EP Regulations 2007 to make compliance with the duty to hold a permit for a regulated facility subject to the transitional provisions applicable to mining waste operations. The transitional provisions are inserted into the EP Regulations 2007 by reg 17.

Regulation 19 of SI 2009/1799 amends paras 5 and 29 of Sch 3 (descriptions: exempt waste operations and other operations to which s 33(1)(a) of the 1990 Act does not apply) to the EP Regulations 2007 in order to correct certain omissions relating to the application of the Waste Incineration Directive (2000/76/EC).

The Environmental Permitting (England and Wales) (Amendment) (No 2) Regulations 2009, SI 2009/3381

7.20 These Regulations revise the definition of an exempt waste operation (reg 4). They also insert a new definition into reg 2 of the EP Regulations 2007 (reg 3).

They provide for the revocation of an existing environmental permit (in whole or in part) where, as a result of the amendments made by these Regulations, the permit covers an exempt waste operation (reg 5).

The substituted Sch 2 of the EP Regulations 2007 defines who is an exemption registration authority – in most cases, this is the Environment Agency but in some cases, local authorities are exemption registration authorities (Sch 2, para 2). Schedule 2, para 3(1) sets out the requirements that must be met by a waste operation in order for it to be an exempt waste operation. Schedule 2, para 4 sets out the procedure for registering exempt waste operations and Sch 2, para 5 contains provisions relating to restrictions on the registration of multiple waste operations at the same place. Schedule 2, para 6 contains provisions for fees in relation to the registration of a WEEE operation (repair or refurbishment of WEEE). Schedule 2, para 7 contains provisions relating to the period of validity of registrations and the renewal of registrations. Schedule 2, para 8 contains provisions relating to changes to information appearing on the register (referred to as relevant particulars). Schedule 2, para 9 sets out the requirements for the exemption registration authority as regards establishing and maintaining a register of exempt waste operations. Schedule 2, para 10 sets out the circumstances under which such an authority must remove an entry from the register. Schedule 2, para 11 makes provision for information affecting national security to be excluded from the register. Schedule 2, para 12 sets out record keeping requirements and para 13 sets out inspection requirements.

The substituted Sch 3 sets out revised descriptions and conditions for exempt waste operations (SI 2009/3381, reg 11). The provisions of Sch 3 in the 2007 Regulations that relate to other waste operations to which s 33(1)(a) of the 1990 Act does not apply are revised and set out in new Sch 3A to the EP Regulations 2007.

The Environmental Permitting (England and Wales) Regulations 2010, SI 2010/675

7.21 These Regulations provide a consolidated system of environmental permitting in England and Wales. They replace the Environmental Permitting (England and Wales) Regulations 2007, SI 2007/3538 (which provided a system for permitting waste operations, mining waste operations, mobile plant and installations), the system of consenting of water discharges in Chapter 2 of Pt 3 of the Water Resources Act 1991, the groundwater permitting system in the Groundwater (England and Wales) Regulations 2009, SI 2009/2902 and the system of radioactive substances regulation in the Radioactive Substances Act 1993.

These Regulations transpose provisions of 18 Directives which impose obligations required to be delivered through permits or capable of being delivered through permits.

Regulation 8 defines the term 'regulated facility' and reg 12 requires every regulated facility to be operated under the authority of an environmental permit.

The combined classes of 'regulated facility' include (unless they are exempt or excluded) every installation, mobile plant, waste operation, mining waste operation, radioactive substances activity, water discharge activity and groundwater activity, whether or not carried on as part of the operation of another regulated facility. Those terms are defined by reg 2(1) and in Schs 1 (Activities, installations and mobile plant), 9 (Waste operations) and 20 (Mining waste operations) to 23 (Radioactive substances activities).

The term 'exempt facility' is defined in reg 5. Schedule 2 contains procedure in relation to exempt facilities, including registration requirements.

The Environmental Permitting (England and Wales) (Amendment) Regulations 2010, SI 2010/676

7.22 These Regulations correct an error in the Environmental Permitting (England and Wales) Regulations 2010, SI 2010/675 relating to the storage limit for waste oils in the table in para 1(2) of Section 2 of Ch 5 of Pt 1 of Sch 3 to those Regulations.

The Environmental Permitting (England and Wales) (Amendment) (No 2) Regulations 2010, SI 2010/2172

7.23 These Regulations amend the Environmental Permitting (England and Wales) Regulations 2010, SI 2010/675 (the EP Regulations). They make certain amendments to the EP Regulations 2010 relating to the activities listed in Sch 1 that require an environmental permit. Regulation 2(1)–(4) amends Pt 2 of Sch 1 to the EP Regulations 2010 by making amendments to the regulation of certain activities involving the melting of non-ferrous metals, the manufacture of powder coating materials which do not use lead chromate or triglycidyl isocyanurate and adding the drying of green crops to the list of activities excluded from regulation.

The Environmental Permitting (England and Wales) (Amendment) Regulations 2011, SI 2011/2043

7.24 Regulations 4 (Exempt facilities), 7 (Interpretation: operate a regulated facility and operator), 8 (Interpretation: regulated facility and class of regulated facility) and 13 (Grant of an environmental permit) amend the EP Regulations in consequence of the insertion of the new Sch 23 by these Regulations.

Regulation 5 makes an amendment to extend the provisions in the EP Regulations 2010 relating to groundwater activities to allow a tracer test involving a radioactive substance to fall within the exemption.

Regulation 12 inserts a new regulated activity into Sch 1 (activities, installations and mobile plant) of the EP Regulations relating to the capture of carbon dioxide.

The Environmental Permitting (England and Wales) (Amendment) (No 2) Regulations 2011, SI 2011/2933

7.25 These Regulations amend the Environmental Permitting (England and Wales) Regulations 2010, SI 2010/675 (EP Regulations 2010) to implement Directive 2009/126/EC of the European Parliament and of the Council on Stage II petrol vapour recovery during refuelling of motor vehicles at service stations ([2009] OJ No L285, 31.10.2009, p 36). Regulation 2(2) amends Sch 1, Pt 2 of the EP Regulations 2010 to alter the motor vehicle refuelling activities to which environmental permitting requirements are applied. Regulation 2(4) substitutes a new Sch 18 which includes a new Pt 2 requiring regulators to observe the requirements of Directive 2009/126/EC in relation to permits for motor vehicle refuelling activities.

The Environmental Permitting (England and Wales) (Amendment) Regulations 2012, SI 2012/630

7.26 These Regulations amend the Environmental Permitting (England and Wales) Regulations 2010, SI 2010/675 (the EP Regulations):

(a) to clarify the position relating to single site permits for certain radioactive substances activities (reg 4);
(b) to provide for the vesting of an environmental permit in the personal representatives of a deceased operator (reg 9);
(c) to reduce regulatory requirements for those who operate certain anaerobic digestion installations or burn in appliances waste-derived fuel that has ceased to be waste (reg 12).

The Environmental Permitting (England and Wales) (Amendment) Regulations 2013, SI 2013/390

7.27 Regulations 4 to 9 amend various definitions in accordance with the Industrial Emissions Directive. Regulation 10 (Giving notices, notifications and directions, and the submission of forms) inserts a new provision into reg 12 of the EP Regulations whereby the requirement for an environmental permit will not apply to certain activities until 7 July 2015. Regulation 15 (Conditions in relation to certain land) makes provision for incidents and accidents in accordance with Articles 7 and 8 of the Industrial Emissions Directive.

The Environmental Permitting (England and Wales) (Amendment) Regulations 2014, SI 2014/255

7.28 To amend provision relating to the variation by the regulator of the terms of an environmental permit when it is partially transferred from one operator to another (and to require the regulator to grant a new permit to the transferee, varied in consequence of the transfer), and to provide that where a suspension notice applies to a permit it continues to apply when the permit is transferred to another operator (regs 4 and 5);

To clarify the position relating to the exercise of functions in relation to waste mobile plant (reg 8), and a consequential amendment is made to the Natural Resources Body for Wales (Functions) Order 2013, SI 2013/755 (reg 21).

The Environmental Permitting (England and Wales) (Amendment) (England) Regulations 2014, SI 2014/2852

7.29 These Regulations amend the Environmental Permitting (England and Wales) Regulations 2010, SI 2010/675 (the EP Regulations) to remove the registration requirements for certain exempt facilities in England.

These Regulations make provision—

(a) to remove the requirement to register small discharges of sewage effluent (which are discharges from a sewage treatment plant into surface water and discharges from a septic tank or sewage treatment plant into groundwater), in order to be exempt from the requirement for an environmental permit;
(b) for exemption authorities to replace exemption registration authorities in carrying out certain functions in relation to small discharges of sewage effluent;
(c) for revisions to the conditions that must be satisfied in order for small discharges of sewage effluent to be exempt from the need for a permit, including removing the requirements to keep records of maintenance and to notify the exemption registration authority if an exempt facility ceases to be in operation;
(d) to clarify the position relating to the relevant person who is responsible for satisfying the conditions referred to in (c);
(e) to clarify which provisions of the principal Regulations apply to England and which apply to Wales.

These regulations only apply to England.

The Environmental Permitting (England and Wales) (Amendment) (England) Regulations 2015, SI 2015/324

7.30 These Regulations amend the Environmental Permitting (England and Wales) Regulations 2010, SI 2010/675 in relation to England. They provide for the Environment Agency to accept an enforcement undertaking in respect of an offence under reg 38(1), (2), (4)(a), (5)(a) or (6) of those Regulations, committed in relation to a regulated or exempt facility in England. An enforcement undertaking is an undertaking from a person whom the Environment Agency suspects of having committed an offence to take certain action within a specified period (for example action to secure that the offence does not continue or recur). A person from whom an enforcement undertaking is accepted may not be convicted of the offence to which it relates unless that person fails to comply with the undertaking.

These changes apply to England only.

The Environmental Permitting (England and Wales) (Amendment) Regulations 2015, SI 2015/918

7.31 These Regulations amend the Environmental Permitting (England and Wales) Regulations 2010, SI 2010/675 (the EP Regulations), to transpose Article 14(5)–(9) of Directive 2012/27/EU on energy efficiency ([2012] OJ No L 315, 14.11.2012, p 1). Article 14(5)–(9) of the Directive requires Member States to ensure that a cost-benefit analysis is carried out to assess the installation of cogeneration or the utilisation of waste heat from energy producing installations, with the aim to increase energy efficiency.

Regulation 4 amends Pt 2 of Sch 1 to the EP Regulations to provide for the aggregation of the net rated thermal input of appliances operated by the same operator on the same site. Where the aggregated net rated thermal input exceeds 20 megawatts, this means they are brought within the scope of the definition of 'Part B installations' for the purposes of applying the cost-benefit analysis requirements introduced by the new Sch 8A.

Regulation 6 inserts new Sch 8A, which provides for a cost benefit analysis to be included in applications for environmental permits and for the variation of environmental permits for specified installations.

The cost-benefit analysis must assess the cost and benefits of installations producing electricity, operating as cogeneration installations. In the case of other installations, the cost-benefit analysis must assess the cost and benefits of utilising waste heat for the purposes of district and cooling networks.

Regulations 8 and 9 contain transitional provisions. Regulation 8 provides that a Part B installation that is a regulated facility under the EP Regulations due to the application of the aggregation provisions in reg 4 and is in operation before the coming into force of these regulations, is to be deemed an exempt facility until it is substantially refurbished.

Regulation 9 provides that Sch 8A does not apply to applications for an environmental permit or for a variation of an environmental permit received by the regulator prior to the coming into force of these regulations.

The Environmental Permitting (England and Wales) (Amendment) (No 2) Regulations 2015, SI 2015/934

7.32 These Regulations correct an error in the Environmental Permitting (England and Wales) Regulations 2010, SI 2010/675 relating to the application of the exemption to installations in para 11 of Sch 8A to those Regulations, which was inserted by the Environmental Permitting (England and Wales) (Amendment) Regulations 2015, SI 2015/918.

The Environmental Permitting (England and Wales) (Amendment) (No 3) Regulations 2015, SI 2015/1756

7.33 These Regulations amend the Environmental Permitting (England and Wales) Regulations 2010, SI 2010/675 (the EP Regulations).

Regulation 3 amends reg 37 of the EP Regulations to allow the regulator (the Environment Agency in relation to England, the Natural Resources Body for Wales in relation to Wales and local authorities in both countries) to serve a suspension notice where it considers that there has been a contravention of an environmental permit condition and such contravention involves a risk of pollution. It also allows the regulator to require the operator to put up a sign to make clear to the public that no further waste of a specified description may be accepted at that facility.

Regulation 4 amends reg 42 of the EP Regulations to make clear that the regulator may make an application to the High Court whether or not it has taken other steps to secure compliance with an enforcement notice or other specified type of notice.

Regulation 5 amends reg 57 of the EP Regulations to allow the regulator to arrange for steps to be taken to remove the risk of serious pollution, which arises as a result of the operation of a regulated facility or an exempt facility.

The Environmental Permitting (England and Wales) (Amendment) Regulations 2016, SI 2016/149

7.34 These Regulations implement Commission Directive 2014/99/EU ([2014] OJ No L 304, 23.10.2014, p 89), which amends Directive 2009/126/EC ([2009] OJ No L 285, 31.10.2009, p 36) as regards the test methods to certify petrol vapour recovery systems for use in service stations and the test methods to verify the operation of such systems.

Regulation 3 amends the Environmental Permitting Regulations (England and Wales) 2010, SI 2010/675 to substitute the reference to Directive 2009/126/EC in Section 1.2 of Part 2 of Schedule 1, so as to take account of the amendment of that Directive by Commission Directive 2014/99/EU. Regulation 4 amends those Regulations to substitute the definition of 'PVR II' in para 2 of Pt 2 of Sch 18 so that the updated definition refers to that Directive as amended by Commission Directive 2014/99/EU.

The Environmental Permitting (England and Wales) (Amendment) (No 2) Regulations 2016, SI 2016/475

7.35 These Regulations amend the Environmental Permitting (England and Wales) Regulations 2010, SI 2010/675 (the EP Regulations) in order to extend the requirement for an environmental permit to flood risk activities.

Regulation 3 amends the EP Regulations so as to provide for relevant definitions. In particular the term 'flood risk activity' is defined by reference to a new Sch 23ZA to the EP Regulations (inserted by Sch 2 to these Regulations).

Regulation 4 amends reg 5 of the EP Regulations (exempt facilities) to provide for exempt flood risk activities. Regulations 5–7 amend definitions in the EP Regulations in order to bring flood risk activities within the class of operations that require an environmental permit.

Regulation 8 substitutes a new reg 15 in the EP Regulations in connection with provision for permit applications to be granted subject to conditions relating to ongoing maintenance or access requirements by the regulator. Regulations 9 and 10 extend to flood risk activities current flexibilities in the EP Regulations concerning the grant by the regulator of a permit covering various activities by a single operator.

Regulations 11–13 extend the current provisions in the EP Regulations on variation, transfer and surrender of environmental permits so that they provide for flood risk activities. Regulation 14 makes provision for appeals in relation to permits authorising flood risk activities. Regulation 15 adds flood risk activities to the list of specific provisions applying to environmental permits, set out in reg 35 of the EP Regulations.

Regulations 16–20 extend the current provisions in the EP Regulations on enforcement and offences relating to permits so that they provide for flood risk activities. Regulation 21 gives the regulator power to arrange for steps to be taken to remove a risk of serious flooding, detrimental impact on drainage or harm to the environment, corresponding to the current power under the EPRs for the regulator to prevent or remedy pollution.

Regulation 22 requires the Environment Agency and the Natural Resources Body for Wales to consult each other before exercising a function relating to a flood risk activity, which may affect a flood or coastal erosion risk in Wales or England respectively. It also requires the appropriate authority to have regard to the interests of fisheries, including sea fisheries, when exercising a power under the EP Regulations that relates to a flood risk activity.

Regulations 23 and 24 amend Schs 2 and 3 respectively to make provision in respect of exempt facilities for flood risk activities. Regulation 25 amends Sch 5 to the EP Regulations in order to make provision in respect of the grant, variation and surrender of environmental permits that include flood risk activities (including provision as to fees for the grant of such permits). Regulation 26 adds provision in respect of flood risk activity enforcement notices to Sch 6 to the EP Regulations (appeals to the appropriate authority). Regulation 28 amends Sch 23A to the EP Regulations in order to exclude the option of accepting an enforcement undertaking in relation to flood risk activities. Regulation 30 repeals certain provisions of the Water Resources Act 1991.

Regulations 31–34 make transitional and saving provision in respect of existing flood defence consents, outstanding applications for flood defence consents, existing notices and arbitration matters.

Schedule 1 to these Regulations adds a new Pt 4 to Sch 3 to the EP Regulations (descriptions and conditions for exempt flood risk activities). Schedule 2 makes specific provision for flood risk activities included in environmental permits. This includes definitions, enforcement notices and excluded flood risk activities. Schedule 3 makes consequential amendments to primary and subordinate legislation.

Exempt Waste – By Category

Using

U1: Use of Waste in Construction

7.36 This exemption allows the use of suitable waste rather than virgin raw material or material which has ceased to be waste – for example by complying with a Quality Protocol. See para 7.2 – Quality Protocols.

'Construction" means building or engineering work – including repairing, altering, maintaining or improving existing work and preparatory or landscaping work. Land reclamation is only allowed when it is an integral part of the construction activity.

Types of activity that can be carried out:

- using crushed bricks, concrete, rocks and aggregate to create a noise bund around a new development and then using soil to landscape the area;
- using road planings and rubble to build a track, path or bridleway;
- using woodchip to construct a track, path or bridleway;
- bringing in soil from somewhere else to use in landscaping at housing developments.

Not included:

- treatment of waste to make it suitable to use. See related exemptions T5, T6 and T7 for treatment options;
- dispose of waste by using waste that is unsuitable or by using more waste than is needed. The amount needed will need to be justified;
- use waste for land reclamation, for example by infilling a hollow;
- register this exemption more than once at the same place for three years after registration;
- de-register and then re-register this exemption at the same place within a three-year period;
- store the waste for longer than 12 months before it is used.

Types of waste that can be used:

The types of waste that can be used under this exemption have been grouped together into five tables. These show the maximum quantities that can be used and conditions for using the specific types of waste.

The waste codes are those listed in the List of Wastes (LoW) Regulations. The waste needs to fit within the waste code and the description in the table.

Up to a total of 5,000 tonnes of the waste from this list for any type of construction:

Waste Code	Types of waste
010102	Waste from mineral non-metalliferous excavation
010408	Waste gravel and crushed rock not containing hazardous substances
010409	Waste sand and clays
020202	Shellfish shells from which the soft tissue or flesh has been removed only
101208	Waste ceramics, bricks, tiles and construction products (after thermal processing)
101314	Waste concrete and concrete sludge
170101	Concrete
170102	Bricks
170103	Tiles and ceramics
170107	Mixtures of concrete, bricks, tiles and ceramics not containing hazardous substances
170508	Track ballast not containing hazardous substances
191205	Glass
191209	Minerals (for example, sand and stones)
191212	Aggregates only

Within the 5,000 tonnes total, the waste described below can only be used for drainage work carried out under the Land Drainage Act 1991, the Water Resources Act 1991 or the Environment Act 1995. This is work that can be carried out only by drainage authorities such as Inland Drainage Boards, local authorities or the Environment Agency:

Waste code	Types of waste
170506	Dredging spoil not containing hazardous substances

Up to a total of 1,000 tonnes of the waste can be used in this list for construction:

Waste code	Type of waste
020399, 020401	Soil from cleaning and washing fruit and vegetables only
170504	Soil and stones not containing hazardous substances
170506	Dredging spoil not containing hazardous substances
191302	Solid waste from soil remediation not containing hazardous substances
200202	Soil and stones

Within the 1,000 tonnes total, you caan only use the waste to build tracks, paths, bridleways or car parks. The waste must be processed into chips before it is used.

Waste code	Type of waste
170302	Bituminous mixtures not containing coal tar
020103	Plant tissue waste
030101, 030301	Untreated waste bark, cork and wood only
030105	Untreated wood, including sawdust, shavings and cuttings from untreated wood only
170201	Untreated wood only
191207	Untreated wood not containing hazardous substances
200138	Untreated wood not containing hazardous substances

Up to a total of 50,000 tonnes of the waste in this list can be used for building roads only. The road should be constructed to a specific engineering standard and have a sealed surface in order to qualify for this larger limit:

Waste code	Type of waste
170302	Bituminous mixtures not containing coal tar
170504	Road sub base only

A combination of waste from each table can be used providing it does not exceed the limits for each table.

As well as registering this exemption, there may also be the need to comply with other legislation. This could include:

- planning permission;
- flood defence consent;
- if building is proposed in the floodplain or near an ordinary watercourse, that contact should be made with the lead local flood authority (county council, unitary authority or Internal Drainage Board) to discuss whether any consent other than planning permission is needed;
- this exemption can be carried out along a linear place such as a highway.

The meaning of 'place' is contained in guidance[2].

U1 Related Exemptions

These are related exemptions for the use or treatment of waste:

2 Environment Agency Position statement The meaning of 'place' under the new waste exemption system https://www.gov.uk/government/uploads/system/uploads/attachment_data/file/294249/Business_Guidance_on_meaning_of_place.pdf.

U8: using waste for a specified purpose.

T5: screening and blending of waste.

T6: treatment of waste wood and plant tissue by chipping.

T7: treating waste bricks, tiles and concrete by crushing, grinding or reducing size. (This needs to be registered with a local authority).

Using More Waste Than Your Exemption Allows

If there is a need to use more waste than is allowed under this exemption, then an application for a permit will need to be made. Standard rules permits are available for using waste in construction not covered by this exemption[3].

Register a U1 Exemption

The exemption must be registered with the Environment Agency if they meet the requirements.

U2: Use of Baled End-of-Life Tyres in Construction

7.37 This exemption allows the use of a small number of end-of-life tyre bales in construction.

Types of activity that can be carried out could include use of tyre bales to provide a free draining layer behind a retaining wall.

Not included:

- use other waste in construction under this exemption;
- use of tyre bales that are unsuitable for the project (meaning they do not comply with PAS 108 standard 'specification for the production of tyre bales for use in construction'[4]);
- storage or use of more than 50 tonnes of tyre bales during construction;
- use tyre bales which are not suitable, or more bales than are needed: this would be disposal and is not covered by an exemption;
- use the tyre bales more than once;
- storage of the tyre bales for more than three months before they are used.

Type of waste that can be used:

The waste code is listed in the List of Wastes (LoW) Regulations. It is important to make sure the waste fits within this waste code and the description.

Waste code	Type of waste
160103	Baled end-of-life tyres only

3 Standard Rules: environmental permitting https://www.gov.uk/government/collections/standard-rules-environmental-permitting.

4 PAS108:2007 Specification for the production of tyre bales used in construction BSI ISBN 978 0 580 50349 8.

Amount of waste that can be used:

- store or use up to 50 tonnes of baled tyres in construction.

Key conditions:

- the bales must comply with standard PAS 108;
- the waste cannot be used more than once;
- the waste cannot be stored for longer than three months before use.

U2 Related Exemptions

If there is a requirement to use other waste in construction, then consideration can be given to exemption U1: using waste in construction.

If the requirement is to bale tyres or treat them in another way it could be registered as T8: mechanical treatment of end-of-life tyres.

Register a U2 Exemption

This exemption must be registered with the Environment Agency if the requirements are met.

If there is a requirement to use more than 50 tonnes of baled tyres on a civil engineering project, there will normally be a need to apply for an environmental permit from the Environment Agency. However, there is a current Modern Regulation Position Statement, MWRP RPS 085 which supports areas if they wish to agree a local enforcement position for use of PAS 108 baled tyres in civil engineering or landfill infrastructure schemes.

U3: Construction of Entertainment or Educational Installations

7.38 This exemption allows groups such as schools, colleges and theatres to use waste such as off-cuts of wood creatively.

Types of activity that can be carried out include:

- creating a sculpture from metal during an art class;
- building a theatrical set using recycled wood and bricks.

Not included:

- using this exemption to construct buildings or similar constructions.

Types of waste that can be used:

The waste codes are those listed in the List of Wastes (LoW) Regulations. It is important to make sure the waste fits within this waste code and the description.

Waste code	Type of waste
170102	Bricks
170103	Tiles and ceramics

Waste code	Type of waste
170201, 200138	Wood
170203 and 200139	Plastic
170401 to 170407, 170411, 200140	Metals, including their alloys
200101	Paper and cardboard
200111	Textiles

Amount of waste that can be used:

Up to 20 tonnes of waste can be stored or used at any one time.

Conditions:

- this exemption can only be used for entertainment, education or similar projects;
- the waste used must be suitable for the project;
- no more waste than is needed should be used.

When there is no need for the remaining waste, it should be taken to an appropriately authorised facility to be reused, recovered or disposed of.

U4: Burning of Waste as a Fuel in a Small Appliance

7.39 This exemption allows the use of waste as fuel to produce heat or power.

The types of activity that can be carried out include:

- a plant nursery burning dead plants and untreated wood waste in a boiler to produce heat for its greenhouses;
- a business burning wooden pallets that have not been treated with timber preservatives in a boiler that also provides hot water for a building.

Not included:

- burn waste other than those listed;
- burn treated wood waste such as painted or stained pallets, door frames;
- use this exemption to burn waste for disposal – for example, waste should not be burned for heating purposes when the external temperature is comfortably warm;
- use this exemption to burn waste in the open – a brazier is not an appliance, it provides a physical containment for what would otherwise be an open bonfire.

Waste code	Type of waste
020103, 200201	Plant tissue waste only
020107, 170201	Untreated wood only
020304	Vegetable waste unsuitable for consumption or processing
030101	Waste bark and cork
030105	Untreated sawdust and wood shavings other than those containing hazardous substances
030301	Fibre rejects (fibrous vegetable waste from virgin pulp preparation or paper pulp production) only
150103	Untreated wooden packaging only

Amount of waste:

- burn up to 50kg of waste as fuel in 1 hour;
- store up to 10 tonnes of waste at any one time before using it.

Conditions:

- the appliance (for example, boiler) must have a net rated thermal input of less than 0.4MW;
- if there is more than one appliance, the total net rated thermal input must be less than 0.4MW (eg, four 0.1 MW appliances);
- the waste must be stored in a secure place;
- the waste used must be suitable for burning as fuel;
- no more waste than is needed should be used.

U4 Related Exemptions

If waste is not used as a fuel then some types of waste can be disposed of by burning in an incinerator under exemption D6: disposal by incineration (see para 7.90).

This disposal exemption applies only at the place where waste is produced.

Disposing of wood by burning should be a last resort and if suitable, wood should be recycled or recovered (including burning as a fuel) before disposal.

It may be possible to spread the ashes from burning under this exemption on agricultural land (U10) or non-agricultural land (U11).

The exemption will need to be registered with the Environment Agency.

An environmental permit from either the local authority or the Environment Agency will be required if:

- the appliance or group of appliances have a net rated thermal input of more than 0.4MW;
- the burn will be for more than 50kg of waste over any 1 hour period;

- burning any other type of waste not listed under this exemption.

U5: Using Biodiesel Produced from Waste as Fuel

7.40 This exemption allows the storage and use biodiesel produced from waste as fuel in portable generators and motor vehicles.

Types of activity that can be carried out include:

- running a delivery van or tractor on biodiesel produced from waste;
- running a portable generator to provide electricity for a power washer.

Not included:

- use this exemption to use any other type of waste as fuel.

Waste code	Type of waste
190210	Biodiesel produced from waste only

Quantity of waste:

- store up to 5,000 litres at any one time at a place owned or occupied by the owner or user of the biodiesel;
- store or use up to 1,000 litres of biodiesel in a portable generator or in a motor vehicle at any one time.

Conditions:

- the biodiesel must be stored with secondary containment;
- using a portable generator, it must have a net rated thermal input of less than 0.4MW;
- using more than one portable generator, the total net rated thermal input of all the portable generators must be less than 0.4MW (for example, four 0.1MW appliances);
- no more waste than is needed should be used.

If a private individual is running a car on biodiesel made from waste, they do not need to register this exemption, but they must comply with its requirements.

Secondary containment is defined in the Control of Pollution (Oil Storage) (England) Regulations 2001[5].

Quality biodiesel produced and supplied under the Quality Protocol (QP) is not waste. An exemption will not need to be registered.

U5 Related Exemptions

Biodiesel can be made from waste edible oils and fats under exemption T19: physical and chemical treatment of waste edible oil and fat to produce biodiesel.

5 SI 2001/2954.

U6: Using Sludge to Re-Seed a Waste Water Treatment Plant

7.41 This exemption allows the import of sludge from another waste water treatment plant to re-seed the biological process at a waste water treatment plant.

The activity that can be carried out involves using sludge from one waste water treatment plant to re-seed the biological process (for example, filter bed) at another waste water treatment plant.

There is no exemption to store, treat or use septic tank or cesspool sludge under U6.

The type of waste that can be used is in the table below.

Waste code	Type of waste
190805	Sludge from treating urban waste water
190812	Sludge from biologically treating industrial waste water not containing hazardous substances

Amount of waste that can be used:

- store or use up to 1,000 cubic metres at any one time at the receiving waste water treatment plant.

Conditions:

No more waste than is needed or should be used or this will be considered as waste disposal and will not be exempt.

Waste should be stored in a secure place before using it so the public cannot have access to the waste and the effluent cannot enter controlled waters.

U7: Using Effluent to Clean a Highway Gravel Bed

7.42 This exemption allows the use of effluent from water and waste treatment plants to clean highway gravel beds. This reduces the amount of tap water used where high quality water is not needed.

The activity that can be carried out involves cleaning sealed gravel beds that capture highway run-off using effluent from water and wastewater treatment plants.

Not included:

- use of effluent that does not comply with the discharge consent issued to the water treatment works or waste water treatment works that it came from;
- clean highway gravel beds that do not have sealed drainage;
- clean gravel beds that are not part of a highway drainage system.

Type of waste that can used is in the table below.

Waste code	Type of waste
190899	Effluent only

Amount of waste can be up to 10 cubic metres of effluent in any 24-hour period.

Conditions:

- the gravel bed must have sealed drainage so that effluent can only enter groundwater or surface water through a lawful discharge or collect in a sealed sump;
- all effluent used or produced during the cleaning operation must be collected and remove it to be treated at an appropriately authorised site;
- any effluent used must comply with the discharge consent issued to the water treatment works or waste water treatment works that it came from;
- no more waste than is needed should be used.

U8: Using Waste for a Specified Purpose

7.43 This exemption allows the use of waste that does not need treating, to reduce the use of virgin or non-waste materials.

Types of activity that can be carried out:

- paper pulp or sawdust as animal bedding;
- waste paint in community projects;
- compost produced for growing mushrooms;
- waste wood or stones to construct walls or structures;
- waste water treatment sludge or solutions for treating water.

Not included:

- use waste that needs treating before it can be used;
- treat waste under this exemption;
- mix hazardous waste with other hazardous or non-hazardous waste.

Types and quantities of waste are listed in the table below.

Waste code	Type of waste	Purpose	Quantity
030105, 191207	Untreated wood shavings, woodchip and sawdust, and oversized compost only	Horse manèges	1,000 tonnes

Waste code	Type of waste	Purpose	Quantity
191204	Shredded or granulated rubber and end-of-life tyres only	Horse manèges	1,000 tonnes
191201	Shredded paper and cardboard	Animal bedding	100 tonnes
030305, 030310, 030311	Paper fibre, de-inked paper pulp and de-inked paper sludge from paper manufacturing only	Animal bedding	100 tonnes
030105, 191207	Untreated wood shavings, woodchip and sawdust, and oversized compost only	Animal bedding	100 tonnes
160103	End-of-life tyres	Weight on cover sheeting on farms or as crash barriers	40 tonnes
150102	Geotextile bags (flexible intermediate bulk containers) only	Reinforcement in construction	100 bags
020202	Shellfish shells (flesh removed)	Ornamental purposes	50 tonnes
191205	Crushed glass only	Ornamental purposes	50 tonnes
200127 *, 200128, 080111 *, 080112	Paints (excluding industrial paint, wood preservatives, aerosol and spray paints, inks, adhesives and resins)	Paint	1,000 litres
190599	Compost produced for growing mushrooms only	Growing mushrooms	1,000 tonnes

Waste code	Type of waste	Purpose	Quantity
170102, 170904	Stones and bricks in their existing state only	Construction of buildings, fencing, barriers, containment or similar above ground construction	100 tonnes
170201, 191207, 200138	Non-hazardous wood, including telegraph poles and railway sleepers. Lock gates including balance beams	Construction of buildings, fencing, barriers, containment or similar above ground construction	100 tonnes
110105 *, 110107 *	Ferric chloride and aluminium hydroxide only	Treating municipal waste water effluent	50 cubic metres
110105 *, 110107 *	Ferric chloride and aluminium hydroxide only	Treating drinking water	50 cubic metres
190902, 190903, 190906	Sludge/solution from treating water only	Treating municipal waste water effluent	50 cubic metres
100201, 100202, 170504, 191209	Blast furnace slag and stones only	Filter media at waste water treatment works	50,000 tonnes
170204 *, 191206 *, 200137 *	Hazardous wooden telegraph poles and railway sleepers, lock gates including balance beams	Construction of buildings, fencing, barriers, containment or similar above ground construction	100 tonnes

Hazardous Waste

Conditions:

- no more waste than is needed should be used or this will be considered as waste disposal and will not be exempt;
- the waste must not be treated or need to be treated before being used.

In relation to using aluminium hydroxide, ferric chloride and sludges/solutions from treating water:

- the waste must be stored within secondary containment and in a location with sealed drainage;
- no more than 10 tonnes of ferric chloride or aluminium hydroxide may be used per day within the limit of 50 cubic metres;
- in relation to waste tyres, within the 40 tonnes that can be used for the specified purpose, only store piles up to 10 tonnes, each with a gap as a fire break between each pile.

U8 Related Exemptions

- S2: storage of waste in a secure place.
- T4: preparatory treatments – shredding paper and cardboard.
- T6: shredding waste wood.
- T12: manually treating waste telegraph poles or lock gates.

An exemption will need to be registered with the Environment Agency.

U9: Using Waste to Manufacture Finished Goods

7.44 This exemption allows the use of waste in place of raw materials to manufacture a finished product.

Types of activity include using:

- boiler ash to manufacture breeze blocks;
- waste textiles to manufacture carpet underlay.

The exemption cannot be used if the operation involves any activity described in Part A(1) or Part A(2) of Pt 2 of Sch 1 of the Environmental Permitting (England and Wales) Regulations 2010, SI 2010/675 (as amended). These are industrial scale processes which need an installations permit.

For the purposes of this exemption, 'finished goods' means goods that are ready for use by an end consumer without any further processing.

A manufacturing process should result in a product that is significantly different from the raw materials that it was made from.

Wood chip is not considered to be finished goods.

Types and quantities of waste are listed in the table below.

Waste code	Type of waste	Quantity (tonnes) at any one time
100101, 100102	Ash only	500
101208	Ceramics	100
101112, 150107, 191205, 200102	Glass	5,000
100105	Gypsum only	500
200199	Lion faeces only	5

Waste code	Type of waste	Quantity (tonnes) at any one time
120101, 120103, 150104, 160117, 160118, 191001, 191002, 191202, 191203, 200140	Metals	500
150101, 191201, 200101	Paper and cardboard	15,000
070213, 120105, 150102, 191204, 200139	Plastics	500
191204	Rubber only	30
040221, 040222, 150109, 191208, 200110, 200111	Textiles	1,000
030105, 030301, 150103, 191207, 200138	Wood, bark, cork, sawdust, shavings, cuttings, particle board	100

- the waste must be stored in a secure location at the place where the manufacturing is carried out;
- ash, ceramics, glass, gypsum, metals must be stored at a secure location with sealed drainage;
- lion faeces, textiles and wood, bark etc. must be stored indoors or in a secure container;
- paper and cardboard must be stored in bales, in a container or indoors;
- for paper and cardboard, up to 1,000 tonnes can be stored outdoors in an enclosure designed and maintained to prevent escape of litter;
- heat treatment is not permitted for plastics and rubber;
- within the 30 ton limit, rubber can be stored in piles no bigger than 10 tonnes, with a gap as a fire break between each pile;
- no more waste than is needed should be used or this will be considered as waste disposal and will not be exempt.

U9 Related Exemptions

T1: cleaning, washing, spraying or coating relevant waste;

T2: recovering textiles;

T4: preparatory treatments (baling, sorting, shredding);

To use waste without treatment for a specified purpose that an exemption for:

U8: using waste for a specified purpose

could be used.

For the use of more waste than is allowed or a type of waste which is not allowed, then an environmental permit will be needed.

U10: Spreading Waste to Benefit Agricultural Land

7.45 This exemption allows the spreading of specific waste on agricultural land to replace manufactured fertilisers or virgin materials such as agricultural lime used to improve or maintain soil.

Type of activity that can be carried out is the spreading of waste on agricultural land to improve or maintain the physical, chemical and biological properties of the soil to grow crops (for example, chalk can be added to the soil as a liming agent).

Not included:

- spread of waste that does not benefit the land;
- spread of waste on non-agricultural land;
- spread of waste that is not listed even if they would provide benefits;
- disposal of waste under this exemption, eg applying waste that will not benefit the land, or applying more than is needed to provide benefit.

Types and quantities of waste that can be used are in the table below.

Waste code	Type of waste	Quantity (tonnes per hectare per year)	Storage (tonnes)	Conditions
010102, 010408, 170504	Chalk only	50	200	A
020101	Sludge from cleaning fruit and vegetables on farm only	50	200	A
020199, 020399	Untreated water from cleaning fruit and vegetables on farm only	100	200	A, F
020305	Effluent from treating water used to clean fruit and vegetables on farm only	100	200	A, F
020401, 020399	Soil from cleaning fruit and vegetables only	50	200	A

Waste code	Type of waste	Quantity (tonnes per hectare per year)	Storage (tonnes)	Conditions
100101	Ash from wood chip boilers burning plant tissue and untreated wood only. Registered and operating under a U4 exemption	1	10	A
170506	Dredging spoil, not containing any dangerous substances, from creating or maintaining habitats, ditches or ponds in parks, gardens, fields and forests only	150	1,250	C
020199	Used compost from growing mushrooms only	50	500	A
190599	Compost produced only from the limited list of waste and in the proportion stated in the T23 exemption or worm compost from T26 exemption only	50	500	A
190604	Digestate produced only from the limited range of waste and conditions allowed under T24 and T25 exemptions only	50	200	A

Waste code	Type of waste	Quantity (tonnes per hectare per year)	Storage (tonnes)	Conditions
190812	Biobed or biofilter material produced from treatment registered and operated under T32 exemption only	50	200	A, D
020199	Milk from agricultural premises only	50 cubic metres of diluted milk per hectare in any 24-hour period	200	A, B, E

Conditions

Condition	Additional Requirement
A	The place where waste that is stored or land which is to be spread must be at least 10 metres from a watercourse and 50 metres from a spring, well or borehole.
B	Before spreading, the waste must be diluted with at least an equal amount of water or slurry, and waste must only be spread on the land once in any four week period.
C	The waste must be spread next to the place it was dredged from.
D	The waste must be stored at least 12 months before it is spread.
E	The waste must not be stored longer than 24 hours before it is spread.
F	The waste must be spread at the place where it is produced.

Conditions:

Waste must not be spread if the land is:

- waterlogged, frozen or covered with snow;
- has been frozen for 12 hours or more in the 24 hours before the time to start spreading.

Biobed or biofilter material must be stored for 12 months before it is spread.

Waste milk:

- must not be stored for more than 24 hours before it is spread;

- must be diluted with an equal or greater amount of water or slurry before it is spread;
- can only be spread on the same area once in any four-week period.

Waste other than milk or biobed material must not be stored for longer than 12 months before being spread.

Dredging spoil from maintaining ditches can only be spread on land next to the ditch from which it is dredged.

Sludges, wash waters or effluents from washing fruit or vegetables on a farm can only be spread on land at the same farm.

Records of the amount, nature and origin of all waste spread on the land must be kept for two years. These records must be made available to the Environment Agency if they ask to see them. If there is already a requirement to keep records under the Nitrate Pollution Prevention Regulations 2015[6] or the Nitrate Pollution Prevention (Wales) Regulations 2013, SI 2013/2506 and SI 2008/3243 (the NVZ Rules), then records do not need to be kept.

U10 Related Exemptions

U11: spreading waste on non-agricultural land to confer benefit.

U12: use of mulch.

U13: spreading plant matter to confer benefit.

T23 : aerobic composting and associated prior treatment.

T24: anaerobic digestion at premises used for agriculture and burning resulting biogas.

T25: anaerobic digestion at premises not used for agriculture and burning resulting biogas.

T26: treating kitchen waste in a wormery.

U11: Spreading Waste to Benefit Non-Agricultural Land

7.46 This exemption allows the spreading of different waste on non-agricultural land to replace manufactured fertilisers or virgin materials such as agricultural lime used to improve or maintain soil.

Types of activity that can be carried out:

- spreading compost produced under T23: aerobic composting and associated prior treatment to condition the soil in a communal garden;
- spreading small amounts of ash produced by wood chip boilers under U4: burning waste as a fuel in a small appliance.

Waste cannot be spread:

- to benefit agricultural land under this exemption – see U10 in related exemptions.

6 SI 2015/66, Pt 7.

- that is not listed in the table below;
- on land unless it provides a benefit.

Types and quantities of waste are listed in the table below.

Waste	Type of waste	Quantity (tonnes per hectare per year)	Storage (tonnes)	Conditions
010102, 010408, 170504	Chalk only	50	200	A
020202	Shellfish shells from which the soft tissue or flesh has been removed only	50	200	A
020399, 020401	Soil from cleaning fruit and vegetables only	50	200	A
100101	Ash from wood chip boilers registered and operating under U4 exemption only	1	10	A
170506	Dredging spoil not containing dangerous substances from creating or maintaining habitats, ditches or ponds in parks, gardens, fields and forests	150	1,250	B
020199	Used compost from growing mushrooms only	50	500	A
190599	Compost produced only from the limited list of waste and in the proportions stated in the T23 exemption or worm compost from the T26 exemption	50	500	A

Waste	Type of waste	Quantity (tonnes per hectare per year)	Storage (tonnes)	Conditions
190604	Digestate produced only from the limited range of waste and conditions allowed under T24 and T25 exemptions only	50	200	A
190812	Biobed or biofilter material produced from a treatment registered and operated under T32 exemption only	50	200	A, C
200108	Coffee grounds only	50	200	A

Conditions

Condition	Additional requirement
A	The place where waste is stored or land on which waste is to be spread must be at least 10 metres from a watercourse and 50 metres from a spring, well or borehole.
B	The waste must be spread next to the place it was dredged from.
C	The waste is stored at least 12 months before spreading.

Conditions:

Waste must be spread to benefit the land.

No more waste should be used than is needed or this will be considered as waste disposal and will not be exempt.

Waste must be stored in a secure location before it is used.

Waste must not be spread on land that is:

- waterlogged, frozen or covered with snow;
- has been frozen for 12 hours or more in the 24 hours before the start of spreading.

Biobed or biofilter material must be stored for 12 months before it is spread. Other types of waste must be stored for less than 12 months before they are spread.

Dredging spoil from maintaining ditches can only be spread next to the ditch from which it is dredged.

Records must be kept of the amount, nature and origin of all waste spread on the land for two years. And these records must be made available to the Environment Agency if they ask to see them. If records are already required to be kept under the Nitrate Pollution Prevention Regulations 2008, SI 2008/2349 or the Nitrate Pollution Prevention (Wales) Regulations 2008, SI 2008/3143 (the Nitrate Vulnerable Zones Rules), the others records are not required to be kept.

U11 Related Exemptions

U10: spreading of waste on agricultural land to confer benefit.

T23 : aerobic composting and associated prior treatment.

T24: anaerobic digestion at premises used for agriculture and burning resulting biogas.

T25: anaerobic digestion at premises not used for agriculture and burning resulting biogas.

T26: treating kitchen waste in a wormery.

U12: Using Mulch

7.47 This exemption allows landscapers and farmers or growers to spread mulch as a protective covering onto land around trees, bushes or plants.

Type of activity that can be carried out:

Wood chips which are made from waste as a mulch (eg, for the following) to:

- conserve soil moisture;
- moderate soil temperature;
- control growth of weeds;
- provide organic matter as the mulch breaks down.

Activities that cannot be carried out:

- use treated wood waste (apart from shredding);
- use any other waste apart from those listed below.

Types of waste are listed in the table below.

Waste code	Type of waste
020103, 191207, 200201	Untreated wood and plant matter

Conditions:

Waste must be stored in a secure location before it is used.

Up to 100 tonnes of mulch can be stored and used in any month.

Only as much waste as is needed should be used to achieve the intended benefit. If more mulch than appears reasonable is used, the Environment Agency may consider this as disposal of waste.

This exemption can be carried out along a linear place such as a highway.

The total amount used at one place in a specified period must not be exceeded.

U12 Related Exemptions

U10: spreading of waste on agricultural land to confer benefit.

U11: spreading waste on non-agricultural land to confer benefit.

T6: treatment of waste wood and waste plant matter by chipping, shredding, cutting or pulverising.

U13: Spreading Plant Matter to Provide Benefits

7.48 This exemption allows the spreading of cut plants at the place of production for weed suppression or to provide nutrients to the soil.

Types of activity that can be carried out include:

- strimming plants along the central reservation of a motorway and leaving the plants where they are cut to provide nutrients to the soil;
- strimming plants along the banks of a river or along a railway embankment and leaving the plants where they are cut to stop weeds growing.

Not included are:

- bringing in plant matter from other places to spread;
- moving plant matter from where it is grown to spread it somewhere else unless it is treated first.

Types of waste are listed in the table below.

Waste code	Type of waste
020103, 020107, 020304, 200201	Plant tissue only

Conditions

Plant matter can only be spread under this exemption if, at the time of spreading:

- the land has not been frozen for 12 hours or more in the preceding 24 hours;

- the land is not waterlogged, frozen or covered with snow.

The maximum amount that can spread over any 12 month period cannot exceed 50 tonnes per hectare.

Only the minimum amount needed to provide benefits should be spread:

- except for plant tissue spread on the bank of a waterway, waste must not be spread within 50 metres of a watercourse, spring, well or borehole;
- only as much waste should be used as is needed to achieve the intended benefit. If more material is applied than appears reasonable, the Environment Agency may consider this as disposal of waste.

U13 Related Exemptions

If the spreading of waste plant matter other than where it was grown is required, then it could be composted under:

- T23: anaerobic composting

and then spread under either:

- U10: spreading waste to benefit agricultural land.

Or exemption:

- U11: spreading waste to benefit non-agricultural land.

Or it could be shredded first under:

- T6: treating waste wood and waste plant matter by chipping, shredding, cutting or pulverising.

And then used under:

U12: using mulch.

U14: Incorporating Ash into Soil

7.49 This exemption allows the mixing of ash back into the soil to return some of the nutrients from the burnt crops and vegetation back to the soil.

Types of activity:

Ploughing or mixing into the soil ash produced from burning:

- cereal straw and cereal stubble;
- waste such as untreated wood and vegetation under exemption D7: burning waste in the open.

Not included:

- bringing ash onto a farm from other places and ploughing it into the soil;
- ploughing or mixing ash from burning other types of waste into the soil.

Type of waste are listed in the table below.

Waste code	Type of waste
020103, 020107, 200201	Ash from burning plant tissue only

Up to 10 tonnes of ash per hectare can be spread.

Only as much waste should be used as is needed to achieve the intended benefit. If more material is applied than appears reasonable, the Environment Agency may consider this as disposal of waste.

Conditions

Ash from burning cereal straw or stubble into the soil must be ploughed within 24 hours of it being burnt, unless the wind would cause excessive dust. If it is too windy, then the ash must be incorporated as soon as conditions allow.

The types of wood that can be burnt under D7: burning waste in the open are:

- waste bark and wood;
- plant tissue;
- sawdust or shavings from untreated wood.

Under U14 only ash from burning plant tissue can be incorporated, not wood from joinery.

If crop residues are being burnt, compliance with the Crop Residue (Burning) Regulations 1993, SI 1993/1366 is required.

U15: Pig and Poultry Ash

7.50 This exemption allows the mixing of ash from burning pig or poultry carcasses with slurry and/or manure and the spreading of it on farmland to provide the soil with nutrients.

Activities that can be carried out:

- mix manure or slurry with ash from the incineration of pig and poultry carcasses;
- spread the manure/slurry ash mix on the farm to benefit the soil;
- store the ash before mixing it with the manure or slurry, and store the manure/slurry ash mix before spreading it.

Activities that cannot be carried out:

- a person cannot take ash from neighbouring farms and mix it with manure and slurry on their farm;
- mix any other ashes with manure and slurry and spread the mix on land;
- plough ash straight into the soil.

Type of waste are listed in the table below.

Waste code	Type of waste
190112	Ash from the incineration of pig and poultry carcasses only

Amount of waste:

The maximum amount of ash that can be spread under this exemption is 150kg per hectare in a 12 month period. Only the amount that the soil needs should be spread. If more than is reasonable is applied then the Environment Agency may consider this as disposal of waste.

Conditions

The ash cannot be stored or mixed within 10 metres of a watercourse or 50 metres of a spring, well or borehole.

The ash must be mixed with an equal or larger amount of manure or slurry before spreading it.

The ash must be stored in a secure place before it is spread.

When the ash is spread, the land must not have been frozen for 12 hours or more in the preceding 24 hours.

The land must not be frozen, waterlogged or covered by snow.

The pig and poultry carcasses must be incinerated in compliance with the Animal By-Product Regulations.

This exemption can only take place on the same farm where the pigs or poultry died and were then incinerated.

U16: Using De-polluted End-of-Life Vehicles for Parts

7.51 This exemption allows the parts from de-polluted end-of-life vehicles to be re-used in other vehicles.

Types of activity is that of a small business that wants to restore a classic car using parts from de-polluted end-of-life vehicles (ELVs).

Not included:

- use waste other than those listed;
- deal with ELVs or parts that have not been depolluted at an authorised treatment facility – this means a site operating under an environmental permit for vehicle depollution and dismantling.

Types of waste are listed in the table below.

Waste code	Type of waste
160106	End-of-life vehicles containing no liquids or other hazardous parts
160122	Non-hazardous parts from end-of-life vehicles only

Amount of waste:

- up to 2 de-polluted ELVs at any one time;
- up to 5 cubic metres of non-hazardous parts from depolluted ELVs at any one time.

Conditions

- the waste must be stored in a secure place;
- the waste must be stored on an impermeable surface;
- the operation must be for the purpose of re-using that waste.

The waste used must be suitable for purpose.

Only as much waste should be used as is needed to achieve the intended benefit, if more material is applied than appears reasonable, the Environment Agency may consider this as disposal of waste.

U16 Related Exemptions – Using depolluted end-of-life vehicles for parts

T9: recovering scrap metal.

Registering an exemption

7.52 To register an exemption the following definitions must be used.

Definitions'Authorised treatment facility' – a site operating under an environmental permit for vehicle depollution and dismantling.

'Compost' – a solid particulate material resulting from composting, which has been sanitised and stabilised and which benefits the soil when added to it.

'De-polluted end-of-life vehicle' – a waste motor vehicle from which all hazardous substances or components, including fuels, lubricating and hydraulic oils, coolants, brake fluids, batteries, air conditioning fluids and air bags have been removed.

'Discharge consent' – a consent issued under Chapter II of Pt 3 of the Water Resources Act 1991 or the Environmental Permitting (England and Wales) Regulations 2010, SI 2010/675.

'Digestate' – output from the treatment of organic and biodegradable waste.

'End-of-life vehicle' – waste motor vehicle.

'Finished goods' – goods that are ready for an end consumer to use without any more processing.

'Impermeable surface' – a surface or pavement constructed and maintained to a standard sufficient to prevent the transmission of liquids beyond the surface.

'Net rated thermal input' – the rate at which fuel can be burned at the maximum continuous rating of the appliance multiplied by the net calorific value of the fuel, expressed as megawatts thermal.

'Place of production' – place where waste is originally produced.

'Plant tissue waste' – includes all materials of plant origin such as trees, shrubbery, and branches with or without foliage, wood and vegetation.

"Sealed drainage' – a drainage system with an impermeable surface that makes sure that:

- no liquid will run off the surface apart from via the system;
- except where they may be lawfully discharged, all liquids entering the system are collected in a sealed sump. In this case, any authorised outlet should be closed temporarily to allow the effluent to be collected and removed for treatment.

'Secondary containment' – means a bund or any other system for preventing waste that has leaked from its main container from escaping from the place where it is stored or treated.

'Secure' – a container, lagoon or other place is secure if:

- all reasonable precautions are taken to make sure the waste cannot escape;
- members of the public are unable to gain access to the waste.

'Untreated wood' – wood that has not been treated with oils, tar oil preservatives, waterborne preservatives, organic based preservatives, boron and organo-metallic based preservatives, boron and halogenated flame retardants and surface treatments.

Treatment of Waste

7.53 A waste exemption is a waste operation that is exempt from needing an environmental permit. Each exemption has specific limits and conditions that need to be operated within.

If a requirement is to operate under a waste exemption, that exemption needs to be registered with the Environment Agency – if not then a permit will be required.

Each registration lasts three years. The registration will then expire unless a re-registration of the exemptions is made.

Registering an exemption does not remove the need to apply for other permits or permissions. For example, planning permission may be required for the waste operation, or a water discharge permit.

There are three waste treatment exemptions that need to be registered with other authorities and not the Environment Agency. These are:

- T3 – treatment of waste metals and metal alloys by heating for the purposes of removing grease;
- T7 – treatment of waste bricks, tiles and concrete by crushing, grinding or reducing in size;

- T22 – treatment of animal by-product waste at a collection centre.

T3 and T7 are registered with the local authority where the principal place of business is carried out. If this is outside England and Wales, it should be with the local authority where the operation will be carried out.

T22 should be registered with the authority responsible for granting an authorisation under the Animal By-Products Regulations.

A full description of T3, T7 and T22 can be found in the Environmental Permitting (England and Wales) Regulations 2010, SI 2010/675.

Treating Waste Exemptions

T1: Cleaning, Washing, Spraying or Coating Relevant Waste

7.54 This exemption allows the treatment of waste packaging so that it can be re-used in its original form. Suitable waste can also be cleaned to allow it to be recycled.

Types of activity include:

- washing waste plastic drums so that they can be re-used or sold for re-use;
- cleaning waste plastic or glass to allow it to be recycled.

Not included:

- the use of this exemption if the operation falls within Sch 1, Pt 2, section 6.4, Pt B of the Environmental Permitting (England and Wales) Regulations 2010,SI 2010/675 (this is regulated by the local authority);
- treatment of hazardous waste;
- shredding or baling any of the types of waste under this exemption (see related exemption, T4 at para **7.66**);
- cleaning the waste before disposing of it to landfill or incinerating it;
- treat waste which is contaminated by more than 1% of its original volume.

Types of waste are listed in the table below.

Waste code	Type of waste
150105	Composite packaging
150107, 160120, 170202, 191205, 200102	Glass
150101	Paper and cardboard packaging
020104, 070213, 150102, 160119, 170203, 200139	Plastic, including plastic packaging from agriculture
150109	Textile packaging

Amount of waste:

- treat up to 300 tonnes of waste over any seven day period;
- store up to 300 tonnes of waste at any one time;
- store waste for up to three months before it is treated.

Where containers have stored a dangerous substance:

- one tonne of waste is the maximum amount that can be stored over any seven day period;
- the waste must not be contaminated by more than 1% of its original volume before it is treated.

Waste must be:

- stored and treated in a location with sealed drainage;
- with the correct consent before discharge of dirty water from washing waste to surface water, groundwater or sewer.

For consent to discharge to a sewer the local water company will need to contacted.

Related exemptions

To bale or shred suitable waste after cleaning it needs to be registered as an exemption T4 Preparatory treatments, such as baling, sorting and shredding.

T2: Recovering Textiles

7.55 This exemption allows the cleaning of waste clothes and textiles to recover them so they can be reused or recycled.

Types of activity include:

- laundering or cleaning waste clothes to recover them so they can be reused;
- laundering or cleaning waste textiles so they can be used to make carpet underlay under another exempt waste operation.

Not included:

- shredding or baling any of the types of waste under this exemption;
- clean textiles that are not going to be recovered or re-used;
- treat hazardous waste.

Types of waste are listed in the table below.

Waste code	Type of waste
191208	Textiles
200110	Clothes
200111	Textiles

Amount of waste:

Storage or treatment up to 20,000 tonnes of waste at any one time.

Conditions

The waste must be stored and treated in a location with sealed drainage.

Discharge is not allowed of dirty water from laundering to surface water, groundwater or sewers without the correct consent.

To discharge dirty water to sewer consent is required from the local water company.

Before baling waste textiles, consideration should be given to:

T4 – preparatory treatments (baling, sorting, shredding etc).

If the requirement is to make finished goods from waste textiles consideration should be given to:

U9 – use of waste to manufacture of finished goods.

An exemption that meets the requirements must be registered.

Definitions

'Discharge consent' – a consent issued under Pt 3, Ch IIof the Water Resources Act 1991 or the Environmental Permitting (England and Wales) Regulations 2010, SI 2010/675.

'Sealed drainage' – a drainage system with an impermeable surface that ensures:

- no liquid will run off the surface apart from via the system;
- all liquids entering the system are collected in a sealed sump except where they may be lawfully discharged.

T4: Preparatory Treatments (Baling, Sorting, Shredding etc)

7.56 This exemption allows treatment of waste to reduce its volume to help transport it to another site for re-use or recycling or make handling easier.

Types of activity include:

- baling loose paper and cardboard before transporting it to another site for recycling;
- baling and shredding aluminium cans;
- sorting different types of plastic bottles.

Not included:

- the use of heat to densify the waste;
- the combining of the storage limits of this exemption and S2 (Storage of Waste in a Secure Place) to increase the overall storage capacity (this exemption has storage limits for each type of waste so that there is no need to register two separate exemptions);

- mixing types of waste for treatment or storage;
- the treatment of hazardous waste;
- the baling of waste before it is sent to landfill or incineration.

Types of waste are listed in the table below.

Waste code	Type of waste
150104, 200140	Cans and foils only
070213, 150102, 150105	Food and drink cartons only
150107, 160120, 170202, 191205, 200102	Glass
030308, 030307, 150101, 191201, 200101	Paper and cardboard (excluding food and drink cartons)
020104, 070213, 120105, 150102, 160119, 170203, 191204, 200139	Plastic
040222, 150109, 191208, 200110, 200111	Textiles and clothes

Amount of waste that can be treated or stored:

Where the treatment involves pulverising waste up to five tonnes can be treated in any seven day period and the pulverising must take place indoors.

The waste can be stored for up to 12 months.

Cans and foils

- treat up to 100 tonnes outside in any seven-day period;
- treat up to 500 tonnes indoors in any seven-day period;
- store up to 500 tonnes at any one time.

Food and drink cartons and plastics

- treat up to 100 tonnes outside in any seven-day period;
- treat up to 3,000 tonnes indoors in any seven-day period;
- store up to 500 tonnes at any one time.

Glass

- treat up to 5,000 tonnes in any seven-day period;
- store up to 5,000 tonnes at any one time.

Textiles

- treat up to 1,000 tonnes outside in any seven-day period;
- treat up to 3,000 tonnes indoors in any seven-day period;
- store up to 1,000 tonnes at any one time.

Paper and cardboard

- treat up to 500 tonnes outside in any seven-day period;
- treat up to 3,000 tonnes indoors in any seven-day period;
- store up to 15,000 tonnes at any one time (up to 1,000 tonnes outside in an enclosure designed and maintained to prevent litter escaping).

The waste can only be stored:

- in a baled form;
- in a container or indoors.

Conditions

Different types of waste have to arrive separately and be kept apart throughout the treatment.

The waste must be stored and treated in a secure place.

An exemption does not have to be registered if only baling waste at the place where it is produced. This is accepted as ancillary treatment before the collection of waste.

Related exemptions

If types of waste are to be stored but not treated they could be registered:

S2 – storage of waste in a secure place

There is another exemption, which allows small-scale sorting of mixed recyclable wastes – T10 – sorting mixed wastes

Register a T4 exemption

Registration needs to be made to the Environment Agency.

Related permits

If the treatment is for more than the amount of waste allowed under this exemption, then a permit must be applied for.

T5: Screening and Blending Waste

7.57 This exemption allows the temporary treatment of waste on a small scale to produce aggregate or soil at a place such as a construction or demolition site.

Types of activity include:

- screening soil on a demolition site to remove wood and rubble before sending the soil to a construction site to be re-used;

- blending soil and compost that has been produced under an exemption on a construction site to produce better soil for landscaping on that site;
- crushing waste (except bricks, tiles and concrete) before screening or blending;
- grading waste concrete after it has been crushed to produce a certain type of aggregate.

Not included:

- import waste, treat it and then export it somewhere else (this applies even if the treated aggregate would meet the Quality Protocol standard and will no longer be considered waste);
- treat waste where the main purpose is to dispose of it to landfill or incinerate it;
- crush waste tiles, bricks or concrete (see related exemptions below);
- treat hazardous waste.

Types of waste are listed in the table below.

Waste code	Type of waste
010408	Waste gravel and crushed rocks not containing hazardous substances
010409	Waste sand and clay
020202	Shellfish shells from which the soft tissue or flesh have been removed only
030101	Untreated waste bark and cork only
030301	Untreated waste bark and wood
100101	Bottom ash, slag and boiler dust (excluding boiler dust containing oil fly ash and dust)
100115	Bottom ash, slag and boiler dust from co-incineration not containing hazardous substances
170101	Concrete
170102	Bricks
170103	Tiles and ceramics
170107	Mixtures of concrete, bricks, tiles and ceramics not containing hazardous substances
170201	Untreated wood only
170302	Bituminous mixtures not containing hazardous substances
170504	Soil and stones not containing hazardous substances
170506	Dredging spoil not containing hazardous substances
170508	Track ballast not containing hazardous substances

190599	Compost produced only by aerobic composting under the T23 exemption or standard rules permit SR2011 no.1 specifically, or by treating kitchen waste in a wormery under T26 exemption
191205	Glass
191209	Aggregates only
191212	Gypsum recovered from construction materials only
191302	Solid waste from soil remediation not containing hazardous substances
191304	Sludge from remediation not containing hazardous substances
200202	Soil and stones

The amount of waste:

- up to 50,000 tonnes of bituminous mixtures for making roadstone;
- up to 5,000 tonnes of other waste under this exemption;
- over a three-year period from the date of registering the exemption.

Waste cannot be stored for more than 12 months.

Conditions

Waste can only be treated at the place where it is to be used or where the waste was produced.

Once the waste has been treated there are two options:

- if the treated waste meets the requirements of a Waste Quality Protocol, it will no longer be considered waste;
- use the treated waste under an exemption or permit when it has not met the requirements of a Waste Quality Protocol.

Related exemptions

- T7 – treating waste bricks, tiles and concrete by crushing, grinding or reducing size (This will need to be registered with the local authority);
- U1 – use of waste in construction.

T6: Treating Waste Wood and Waste Plant Matter by Chipping, Shredding, Cutting or Pulverising

7.58 This exemption allows the chipping, shredding, cutting or pulverising of waste wood and waste plant matter to make it easier to store and transport, or to convert it into a suitable form to use.

The waste treated by these methods must be suitable for its intended use, which can include feedstock for producing products such as panel board, mulch, surfacing tracks (paths and bridleways) or fuel.

Types of activity include:

- chipping untreated waste wood to use as fuel in a boiler to provide heat;
- cutting waste wood into a suitable size to manufacture chipboard;
- shredding waste plant matter before moving it to an authorised facility for composting.

Not included:

- sort out waste wood or waste plant tissue from other types of waste as this exemption only allows wood and plant tissue waste;
- treat waste where the main purpose is to dispose of it to landfill or incinerate it;
- compost, burn or manufacture products from waste wood (see related exemptions below);
- treat wood that is hazardous waste.

There is no need to register a T6 for tree and forestry maintenance work where the wood is cut, chipped or shredded and is left on the ground or transported off-site for use elsewhere.

Types of waste are listed in the table below.

Waste code	Type of waste
020103, 200201	Plant tissue waste
030101, 030301, 170201	Wood
150103	Wooden packaging only

Wood cannot be treated that has been sorted from mixed sources at a waste treatment facility.

Amount of waste that can be treated or stored is up to 500 tonnes of waste over any seven-day period.

Waste can be stored for up to three months after treatment.

If waste has been registered for one of the following exemptions, and it is intended to chip or shred the waste before it is treated at the same place, then there is no need for registration:

T23 – aerobic composting and associated prior treatment;

T24 – anaerobic digestion at premises used for agriculture and burning of resultant biogas;

T25 – anaerobic digestion at premises not used for agriculture and burning biogas produced;

- when chipping treated or coated wood this must not be used for construction, burning as fuel, mulch, animal bedding or as feedstock for composting. The only suitable use for treated or coated wood is the U9 exemption for manufacturing finished goods;

- the T6 exemption has been modified to allow processing of non hazardous 'treated' wood but only from the specific sources (listed above). Wood that has been sorted from mixed sources at a waste treatment facility can be processed.

Related exemptions

Chipped or shredded untreated waste wood can be used under a number of exemptions:

- U1 – using waste in construction;
- U4 – burning waste as fuel in a small appliance;
- U8 – using waste for a specified purpose;
- U9 – using waste to manufacture finished goods;
- U12 – use of mulch;
- U13 – spreading plant matter to control the growth of weeds or provide nutrients or organic matters to soils.

The exemption will need to be registered with the Environment Agency.

Related permits

If it is required to treat more than the amount of waste allowed under this exemption, then an environmental permit will need to be applied for.

Definitions

'Untreated wood' – wood that has not been treated with oils, tar oil preservatives, waterborne preservatives, organic based preservatives, boron and organo-metallic based preservatives, boron and halogenated flame retardants and surface treatments.

T8: Mechanically Treating End-of-Life Tyres

7.59 This exemption allows the treatment of small amounts of waste end-of-life tyres by baling, shredding, peeling, shaving or granulating so they can be recovered.

Types of activity include:

- baling end-of-life tyres to use in construction;
- granulating end-of-life tyres to use in horse manége;
- re-treading end-of-life tyres so they can be reused as tyres;
- granulating end-of-life tyres again where the granulate is too big or needs further processing to comply with PAS 107.

Not included:

- the treatment of any other types of waste other than end-of-life tyres or shredded/granulated end-of-life tyres;
- the treatment of hazardous waste;

- the sorting of tyres under this exemption (we have adopted a Low Risk position LRW 414 when sorting tyres as an associated prior treatment when T8 is registered).

Types of waste are listed in the table below.

Waste code	Type of waste
160103	End-of-life tyres
191204	Shredded or granulated end-of-life tyres only

Amount of waste that can be stored or treated is 60 tonnes of truck tyres or 40 tonnes of any other tyres over any seven-day period.

Conditions

End-of-life tyres can be treated under this exemption, by:

- granulating;
- baling;
- peeling;
- shaving;
- shredding;
- re-treading.

End-of-life tyres can be treated by shredding or granulating providing the granulation is carried out indoors.

Tyres can be cleaned and separated from the rims before further treatment.

Within the seven-day limit the storage of the tyres should be in piles no bigger than 10 tonnes with a gap as a firebreak between each pile.

There are two relevant Publicly Available Specifications (PAS) for end-of-life tyres:

- PAS 108 standard – specification for production of tyre bales for use in construction;
- PAS 107 standard – specification for the manufacture and storage of size reduced tyre materials.

There is a Waste Quality Protocol for the use of waste derived rubber materials produced under PAS 107 Waste Quality Protocols.

If waste materials are recovered and then sold as products (recycled materials) rather than waste, it must comply with the REACH (Registration, Evaluation, Authorisation and Restriction of Chemicals) Regulations.

Related exemptions

End-of-life tyres that have been baled to PAS 108 standard can use U2 – Using baled end-of-life tyres in construction.

Granulated end-of-life tyres in horse manége can use U8 – Using waste for a specified purpose.

Where it is appropriate, treated end-of-life tyres can use U9 – Using waste to manufacture finished goods.

T9: Recovering Scrap Metal

7.60 This exemption allows the treatment of scrap metal by sorting, grading, shearing by manual feed, baling, crushing or cutting it with hand-held equipment to make it easier to handle and to help recover it.

Types of activity include:

- crushing and baling aluminium cans;
- crushing ferrous (containing iron) metal such as steel;
- shearing copper pipes by manual feed to make storage easier.

Not included:

- the receiving, storage or treatment of any end-of-life vehicles or components of end-of-life vehicles such as engine blocks, batteries or catalytic converters;
- the treatment of hazardous waste;
- the burning of any waste.

Types of waste are listed in the table below.

Waste code	Type of waste
020110	Waste metal
150104	Metallic packaging
160117, 191202	Ferrous metal
160118, 191203	Non ferrous metal
170401	Copper, bronze, brass
170402	Aluminium
170403	Lead
170404	Zinc
170405	Iron and steel
170406	Tin
170407	Mixed metals
170411	Cables not containing oil, coal tar or other dangerous substances
200140	Metals

Amount of waste:

- store or treat up to 1,000 tonnes at any one time;
- store or treat up to 50 tonnes of cables at any one time;
- store waste for up to 24 months;
- store any pile or stack of waste up to a maximum height of five metres.

Conditions

Treatments that can be carried out:

- sorting;
- grading;
- shearing by manual feed;
- baling;
- crushing;
- cutting it with hand-held equipment.

The treatment must be carried out at a location with sealed drainage.

Records must be kept of the quantity, nature, origin and where relevant, the destination and treatment method of all waste recovered and they must be retained for two years. During this time, they must be made available to the Environment Agency if they ask for them.

Related exemptions

If there is a requirement to use up to two de-polluted end-of-life vehicles as parts for vehicles, they should be registered as:

U16 – using de-polluted end-of-life vehicles for vehicle parts.

If the requirement is to store metal at a dock before it is exported or after it is imported to be recovered elsewhere, this should be registered as S2 – storing waste in a secure place.

Metal can be used under the following exemptions, where it is appropriate:

- U3 – using waste in the construction of entertainment or educational installations etc;
- U9 – using waste to manufacture finished goods.

The exemption will need to be registered with the Environment Agency.

Related permits

If the requirement is to treat more than the amount of waste allowed under this exemption or treat end-of-life vehicles then an environmental permit will need to be applied for.

T10: Sorting Mixed Waste

7.61 This exemption allows small organisations such as charities, to sort out separate recyclable waste so that it can be recovered.

Types of activity include:

- a charity collects mixed aluminium cans and paper to raise money and needs to separate them;
- small amounts of mixed recyclable waste are collected and sorted.

Not included:

- bale or shred the waste (see related exemptions);
- accept waste that is mixed with a type of waste that is not permitted;
- accept waste that is contaminated with biodegradable waste.

Types of waste are listed in the table below.

Waste code	Type of waste
150101, 200101	Paper and cardboard
150102, 200139	Plastics
150104, 200140	Metals
150105	Composite packaging
150106	Mixed packaging
150107, 200102	Glass
150109, 200110, 200111	Textiles and clothing only

Amount of waste:

- store up to 10 tonnes at any one time;
- treat up to 10 tonnes within any seven-day period.

Conditions:

- sort one type of the waste from another;
- the waste must be stored in a secure place.

Related exemptions

Registration of the following exemptions for further treating or using the waste once it has been sorted:

- U3 – using waste in the construction of entertainment or educational installations etc;
- U9 – manufacturing finished goods;
- T1 – cleaning, washing, spraying or coating relevant waste;
- T2 – recovering textiles;
- T4 – preparatory treatments (baling, sorting, shredding etc).

This exemption needs to be registered with the Environment Agency.

Related permits

If there is a requirement to treat more than the amount of waste allowed under this exemption or treat mixtures of waste not listed above, then an environment permit needs to be applied for.

T11: Repairing or Refurbishing Waste Electrical and Electronic Equipment (WEEE)

7.62 This exemption allows the repair, refurbishment or dismantling of various types of waste electrical and electronic equipment so that the whole WEEE item or any dismantled parts can be re-used for their original purpose or dismantled parts can be recovered.

Types of activity include:

- a business collects waste computers from householders and businesses that no longer want them;
- a reclamation group receives waste electrical goods collected by the local authority from businesses and households.

Both sort out the waste equipment into those that can be repaired and those that can't. Once sorted they repair some items and dismantle the rest to be re-used as parts.

Not included:

- degas ozone depleting substances;
- treat WEEE to be disposed of to landfill or incinerated;
- mix hazardous waste with other hazardous or non-hazardous waste;
- accept or treat fluorescent light bulbs or tubes.

Types of waste are listed in the table below.

Waste code	Type of waste
160211*	WEEE containing chlorofluorocarbons (CFCs), hydrochlorofluorocarbons (HCFCs) and hydrofluorocarbons (HFCs)
160213*	WEEE containing hazardous substances or components other than polychlorinated biphenyls, CFC, HCFC or HFC, or free asbestos. For example, a TV monitor containing a cathode ray tube
160214	WEEE not containing hazardous substances or components
160216	Non hazardous components removed from discarded equipment
200123*	WEEE containing CFCs
200135*	WEEE containing hazardous components other than fluorescent tubes and other mercury containing waste or CFCs
200136	WEEE not containing hazardous substances or components

Amount of waste:

- store or treat up to 1,000 tonnes over any 12-month period.

Key conditions:

- the best available treatment, recovery and recycling techniques (BATRRT) when treating the waste must be used – see BATRRT guidance for more information[7];
- the place where WEEE is stored or treated should have an impermeable surface with facilities for collecting spillages and where appropriate, decanters and cleanser-degreasers;
- the area where WEEE is stored should have a weatherproof covering – a covered container or roofed building;
- whole and dismantled WEEE components should be stored appropriately to prevent them being damaged that could stop them being reused or pose a risk to the environment;
- hazardous waste, such as batteries, should be stored in suitable containers;
- the purpose must be to ensure that the WEEE is put back to use as whole equipment if possible. If not possible, equipment can be taken apart so that the components can be used again. If that is not possible, equipment can be taken apart so the materials can be recycled.

Related exemptions

S2 – storing waste in a secure place.

T12: Manually Treating Waste

7.63 This exemption allows the sorting, repair or refurbishment of waste to re-use it, or to sort and dismantle it so it can be recovered.

Types of activity include:

- a charity collecting discarded furniture, bicycles and garden tools to repair or refurbish them so they can be used again;
- a company collecting mattresses to dismantle them to recover the various parts.

Not included:

- the treatment of the waste in any other way than by sorting, repairing, refurbishing or dismantling;
- the use of this exemption if the waste is being treated to dispose of it to landfill or to incinerate it;
- store or sort through general mixed waste under this exemption.

7 Guidance on Best Available Treatment Recovery and Recycling Techniques (BATRRT) and Treatment of Waste Electrical and Electronic Equipment (WEEE) Defra, Nov 2006.

Types and quantities of waste that can be treated:

It is important to make sure that the waste fits within the relevant waste code and description. In all examples allowed in T12, the types of waste are more restricted than the LoW code. This is indicated by the word 'only' in the types of waste description.

Waste for sorting, repairing or refurbishing

Waste code	Type of waste	Sorting, repairing or refurbishing (tonnes) at any one time	Storage (years)
200199	Bicycles only	100	2
200110	Clothing, fabrics	100	2
200111	Carpets only	100	2
200199	Footwear only	100	2
200307	Furniture only	100	2
200138, 200139, 200140	Garden tools only	100	2
170904, 170102, 170201, 200138	Stone, bricks, wood only	500	2
170201, 170202, 170203, 200102, 200138, 200139, 200140	Windows, doors only	100	2
150103	Wooden pallets only	100	2

Waste for sorting and dismantling

Waste code	Type of waste	Sorting and dismantling (tonnes) at any one time	Storage duration
200138, 200139, 200140	Coat hangers only	100	12 months
200140	Domestic pots and pans only	100	2 years
200138, 200139, 200140	Lock gates only	100	2 years
200307	Mattresses only	5	12 months

Waste code	Type of waste	Sorting and dismantling (tonnes) at any one time	Storage duration
200137*, 200138, 200140	Telegraph poles only	100	12 months
170201, 170202, 170203, 200102, 200138, 200139, 200140	Windows, doors only	10	12 months
150103	Wooden pallets only	100	12 months

Condition

Mattresses must be treated and stored indoors.

Related exemptions

Some types of waste can be used under:

- U8 – using waste for a specified purpose.

Register a T12 exemption

This exemption needs to be registered with the Environment Agency.

Related permits

If the requirement is to treat more than the amount of waste allowed under this exemption or any other types of waste not listed here, then an application for an environmental permit will need to be made.

T13: Treating Waste Food

7.64 This exemption allows the recovery of waste food by decanting or unwrapping it and recovering the packaging.

Types of activity include:

- unwrapping waste food from a supermarket that is past its sell-by date so it can be composted or put through a digestion process;
- bulking up waste food or packaging to reduce transportation costs.

Not included:

- the storage or treatment of waste covered by the Animal By-Products Regulations that are going to be fed to animals;
- the treatment of the waste food further, for example by composting under this exemption (see related exemptions).

Types of waste are listed in the table below.

Waste code	Type of waste
020304, 020501, 020601, 020704	Materials unsuitable for consumption
200199	Non liquid foods unsuitable for consumption or processing

Amount of waste that can be stored or treated is up to 30 tonnes of waste at any one time.

Conditions

The following treatments can be applied to the waste, including the packaging:

- decanting;
- unwrapping;
- bulking up;
- sorting;
- no waste must be stored for longer than seven days.

The treatment and storage must take place indoors, and in a secure place.

Measures should be taken to reduce odours by making sure that waste food is treated and moved on to be reused or recovered as soon as possible.

Related exemptions

T23 – aerobic composting and associated prior treatment.

T25 – anaerobic digestion at premises not used for agriculture and burning the resultant biogas.

The exemption will need to be registered with the Environment Agency.

T14: Crushing and Emptying Vehicle Waste Oil Filters

7.65 This exemption allows the recovery of oil from oil filters, then crushing them so they can be transported to be recovered.

Types of activity include:

- at a garage an oil filter can be changed during a vehicle service. The oil is then drained from the old filter and crushed to make it easier to store before it is collected to be recovered.

Not included:

- crush any other part of a vehicle under this exemption;
- mix hazardous waste with other hazardous or non-hazardous waste.

Type of waste is listed in the table below.

Waste code	Type of waste
160107*	Oil filters

Amount of waste that can be stored is up to one tonne of crushed and emptied waste oil filters at any one time.

Conditions

The equipment used to treat the waste is:

- designed for crushing oil filters;
- designed and maintained to make sure that oil doesn't escape during treatment.

The treatment of the waste oil filters takes place as soon as possible after the oil filter is removed from the vehicle.

The treatment of the waste oil filters must take place where the filters are removed.

Waste oil and waste oil filters should be stored in a container with secondary containment to prevent it escaping into the environment.

This exemption must be registered with the Environment Agency:

Related permits

If parts of a vehicle are to be treated as end-of-life vehicle, then an environmental permit needs to be applied for.

T15: Treating Waste Aerosol Cans

7.66 This exemption allows the treatment of aerosol cans by puncturing or crushing them using specialist treatment equipment so that the metal can be recovered.

Types of activity include:

- running a fleet of vehicles. Empty de-icer cans are collected and taken back to the depot and they could be treated before taking them off-site to be recovered;
- use aerosol cans treatment equipment (see definitions) to crush a single type of aerosol can to allow the metal can to be recycled.

Not included:

- collecting aerosol cans from different places and taking them back to a single place for treatment.
- the mixing of hazardous waste with other hazardous or non-hazardous waste.

The types of waste are listed in the table below.

Waste code	Type of waste
160504*	Aerosol cans containing residues of or contaminated by hazardous substances only
160505	Aerosol cans only

*may be hazardous waste

The amount of waste that can be treated or stored is up to 3,000 cans in any 12-month period.

Key Conditions

- treatment and storage can only be carried out at the place where aerosols are produced;
- waste must be stored in a secure location in vented containers before being treated;
- treatment must be carried out in a well-ventilated area;
- equipment used to treat the aerosols must be designed for the purpose.

If the activity is regulated under the Solvents Emissions Directive then this exemption cannot be registered. It should be treated as a directly associated activity under the permit.

T16: Treating Waste Toner and Ink Cartridges

7.67 This exemption allows the treatment of waste toner or ink cartridges by sorting, cleaning, dismantling or refilling them.

The activity includes the cleaning and refilling of waste toner cartridges so they can be reused.

What cannot be done:

- the treating of waste toner or ink cartridges that are hazardous waste;
- the treating of waste toner or ink cartridges to dispose of them to landfill or incinerate them.

The types of waste are listed in the table below.

Waste code	Type of waste
080313	Waste ink not containing hazardous substances
080318	Waste printing toner not containing hazardous substances
150102	Plastic packaging
160216	Cartridges not containing hazardous substances taken from discarded electrical equipment
200139	Plastics

The amount of waste that can be stored and treated is up to 150 tonnes of cartridges at any one time.

Conditions

The treatment must be carried out indoors.

The waste must be treated in a location with sealed drainage.

Any waste toner or ink produced from the treatment activity should be stored to prevent it being released into the environment.

Related exemptions

S2 – storage of waste in a secure place.

T17 Crushing Waste Fluorescent Tubes

7.68 This exemption allows the treatment of waste fluorescent tubes before collection for recovery.

Type of activity is for someone running a large office block with lots of fluorescent tubes used for lighting. Once a year they hire in a crusher specially designed to crush the tubes and capture any mercury emissions to make it easier to collect and transport the materials for recovery elsewhere.

Not included:

- bring in fluorescent tubes to site to crush them;
- mix hazardous waste with other hazardous or non-hazardous waste.

Type of waste is listed in the table below.

Waste code	Type of waste
200121*	Fluorescent tubes only

*hazardous waste

Amount of waste

- the crushing of up to three tonnes of waste in a 24-hour period;
- the mercury concentration in any resultant emissions must not exceed 25 microgrammes per cubic metre.

Conditions

Any glass or mercury released after crushing must be stored in a secure container under weatherproof covering.

The equipment used to treat the waste must be designed for that purpose and must be able to meet the above mercury concentration emission threshold.

The crushing can only be carried out at the place where the tubes are produced.

The exemption would require registration with the Environment Agency.

For the treatment of florescent tubes at any other location an environment permit will need to be applied for.

T18: Using Flocculants to Remove Water from Waste

7.69 This exemption allows the treatment of waste by using flocculants to remove water from the waste so that clay or water-based paints can be recovered.

Types of activity include:

- the removing of water from clay effluent produced from manufacturing ceramics to recover the clay and/or water;
- the recovery of water-based paints from wash water.

Not included:

- the use of this exemption if it is intended to dispose of the waste, for example by discharging it to a sewer;
- the treatment of hazardous waste.

Types of waste are listed in the table below.

Waste code	Type of waste
010409	Clay effluent from the manufacture of ceramics
080120	Water-based paint wash water

Quantity of waste that can be stored or treated is up to 30,000 litres at any one time.

Conditions

The waste must be stored in a container with secondary containment.

Dirty water must not be discharged to controlled waters or sewers without the correct consent.

The exemption must be registered with the Environment Agency.

T19: Physical and Chemical Treatment of Waste Edible Oil and Fat to Produce Biodiesel

7.70 This exemption allows the small-scale physical and chemical treatment of waste edible oils and fat to produce fuel.

Type of activity is the collecting of chip frying oil from chip shops and treating it to make biodiesel.

Not included:

- the acceptance of any other waste oils;
- the treatment of waste where the main purpose is to dispose of it to landfill or incineration.

Type of waste is listed in the table below.

Waste code	Type of waste
200125	Edible oil and fat

Quantity of waste:

- physically treat or store up to 5,000 litres at any one time;
- chemically treat up to 250 litres at any one time;
- store waste for up to three months.

The waste must be stored in a container with secondary containment. **Also:**

- there is a quality protocol for biodiesel;
- unless the biodiesel is produced and supplied under the quality protocol for biodiesel, it is still waste;
- if more than 2,500 litres of biodiesel is produced in a year, HM Revenue and Customs allows the accounting of excise duty at a lower rate.

Related exemptions

U5 – using waste-derived biodiesel as fuel.

T20: Treating Waste at a Water Treatment Works

7.71 This exemption allows the treatment of certain waste at a water treatment works to reduce the volume for transport or to make it easier to handle before the waste is recovered.

Types of activity that can be carried out include:

- storing and treating sludge from water clarification by removing water.

Not included:

- the use of this exemption to treat waste at locations other than at a water treatment works;
- the treatment of hazardous waste.

Types of waste are listed in the table below.

Waste code	Type of waste
190902	Sludge from water clarification
190903	Sludge from decarbonation
190906	Solutions and sludge from regeneration of ion exchangers
190999	Waste water and bore hole flushings only

Quantity of waste that can be treated:

- treat up to 10,000 cubic metres of waste over any 12 month period.

Conditions

The waste must be treated and stored in a secure location with sealed drainage.

Discharge of dirty water to controlled waters or sewers requires the correct consent.

Related exemptions

U8 – using waste for a specified purpose.

T21 – recovering waste at a wastewater treatment works.

T21: Recover Waste at a Waste Water Treatment Works

7.72 This exemption allows the recovery of waste, such as sludge from a septic tank or cesspool, which needs further treatment at a waste water treatment works.

Type of activity includes:

- septic tank sludge stored and treated at a sewage treatment works.

Not covered:

- the use of this exemption to treat waste at locations other than at a water treatment works;
- the treatment of hazardous waste.

Types of waste are listed in the table below.

Waste code	Type of waste
190801	Screenings
190802	Sewage grit (waste from de-sanding) only
190805	Sludge from treating urban waste water
190899	Centrate liquor only
190902	Sludge from water clarification
190903	Sludge from decarbonation
190906	Solutions and sludge from regeneration of ion exchangers
200304	Septic tank sludge
200306	Waste from sewage cleaning
200399	Cesspool waste and other sewage sludge only

Quantity of waste that can be treated is up to 100,000 cubic metres of waste over any 12-month period.

Conditions

The waste must be stored and treated in a secure location with sealed drainage.

Related exemptions

T20 – treating waste at a water treatment works.

This exemption must be registered with the Environment Agency.

T23: Aerobic Composting and Associated Prior Treatment

7.73 This exemption allows the composting of small volumes of vegetation, cardboard and food waste to produce compost that can be spread to benefit the land by adding nutrients or improving the structure of the soil. Waste can also be treated before it is composted, by chipping or similar activities.

Types of activity include:

- school composting kitchen and garden waste in its grounds;
- allotment association composting their old plants and trimmings;
- community composting group bringing locally produced vegetable peelings and garden waste to a central point for composting before it is used back in local gardens.

Not included:

- purposely treating the waste in the absence of oxygen;
- treating waste not listed under this exemption;
- treating waste that is an animal by-product unless there is the appropriate authorisation.

Types of waste that can be treated under this exemption are listed into two groups. This shows the maximum quantities that can be used and the conditions for using the specific types of waste.

The waste codes are listed in Table 1 below.

Table 1

Waste code	Type of waste	Quantity limit (tonnes)
170506	Plant tissue waste from inland waters only	see 'Quantity of waste that can be treated'
020103, 200201	Plant tissue waste	see 'Quantity of waste that can be treated'
020106	Horse manure and farmyard manure only	20
020107	Biodegradable waste from forestry only	see 'Quantity of waste that can be treated'
020199	Fully biodegradable animal bedding	see 'Quantity of waste that can be treated'
200101	Paper and cardboard	10
200201	Biodegradable waste plant matter only	see 'Quantity of waste that can be treated'

The waste in Table 1 can be stored for up to one month before treatment.

Table 2

Waste code	Type of waste
020202	Animal tissue waste
020501, 020601	Materials unsuitable for consumption or processing
200108	Biodegradable kitchen and canteen waste
200302	Biodegradable waste from markets only

Waste in Table 2:

- treated up to 10 tonnes in total of this waste at any time;
- stored for up to seven days before treatment.

After treatment:

- compost can be stored for up to 12 months, but this is included in the total quantity that can be stored or treated.

Quantity of waste that can be treated:

There are two separate limits for the amount of waste that can be stored or treated at a place at any one time.

If the waste is composted and used in the same place as it is produced then up to 80 tonnes of waste can be stored or treated at any one time. This could apply to a large estate such as a National Trust property, a plant nursery or an allotment association.

Up to 60 tonnes of waste can be stored or treated at any one time if:

- the waste is brought from other places to the place where it will be composted;
- the compost is used somewhere other than where it was composted.

This could apply to an allotment association or a community composting group.

Conditions

The treatment must produce a stable sanitised material (see compost definition) that can be spread onto land to add nutrients or improve the soil structure.

Associated prior treatment in this exemption means screening, chipping, shredding, cutting, pulverising or sorting waste for aerobic composting.

Composting can be done in open windows, in heaps that are turned regularly or in small closed vessels (known as In-Vessel Composting (IVC)).

Composting is an active process and must maintain the right conditions to prevent the treatment becoming anaerobic as this would lead to odours and poor compost being produced[8].

Animal by-products Regulations – animal by-products can include 020202 animal tissue waste or catering waste from kitchens and restaurants (200108 biodegradable kitchen and canteen waste).

If the intention is to compost animal by-products they may need approval from the Animal Health Veterinary Laboratory Agency (AHVLA). This approval is likely to be required if the intention is to supply compost to other users or use the compost at another place.

Farmed animals (including pet sheep, cattle, pigs or poultry) need to be excluded from the area where the composting takes place and where the compost will be used.

Related exemptions

T24 – anaerobic digestion at premises used for agriculture and burning of resulting biogas.

T25 – anaerobic digestion at premises not used for agriculture and burning of resulting biogas.

U10 – spreading waste to benefit agricultural land.

U11 – spreading waste on non-agricultural land.

T24: Anaerobic Digestion at Premises Used for Agriculture and Burning Resulting Biogas

7.74 This exemption allows farmers to anaerobically digest manure, slurry and vegetation on their farms to produce digestate that can be used as a fertiliser or soil conditioner.

The anaerobic digestion (AD) process also produces biogas, which can be burnt to generate energy, either to use on the farm or to export to the National Grid.

Types of activity include:

- anaerobic digestion of manure and plant tissue waste in a dedicated AD plant to produce a digestate;
- burning the biogas to produce energy to provide power on the farm or to export to the National Grid;
- sorting, screening, cutting, shredding, pulverising and chipping the waste to help the AD process.

Not included:

8 There is a composting industry code of practice which can be found on the Organics Recycling Group website.

- dispose of the biogas by flaring except if appliances burning biogas break down or when they are being maintained;
- dispose of the biogas by venting it to the atmosphere;
- compost the manure and plant tissue (exemption will need to be registered under T23 – Aerobic Composting);
- spread the digestate produced under this exemption (see related exemptions).

Types of waste are listed in the table below.

Waste code	Type of waste
020103, 020107, 170506, 200201	Plant tissue waste
020106	Horse and farmyard manure, slurry only
020199	Fully biodegradable animal bedding

Quantity of waste that can be stored or treated is up to 1,250 cubic metres of waste at any one time.

This storage limit does not include manure and slurry pits where these materials have been produced on the same farm and are being stored before being treated in the digester.

If manure and slurry is imported from other farms and stored before it is fed into an AD plant, the storage of this waste is included within the 1,250 cubic metre limit.

When manure and slurry is mixed with plant tissue waste the 1,250 cubic metre limit will include the storage of plant tissue waste, material in the digester and the storage of the resulting digestate.

Conditions

To comply then:

- the waste must be in the digester for at least 28 days;
- collect and burn the biogas produced by the AD process to produce energy;
- use a net rated thermal input of less than 0.4MW on the AD plant biogas burner;
- have a combined net rated thermal input of less than 0.4MW if there is more than one burner associated with the AD plant.

Anaerobic digestion is different from composting as the AD process involves de-composting waste in an atmosphere with no oxygen and it requires a purpose-built plant. Once the AD process is completed it produces a digestate.

If the only waste feedstock to the anaerobic digestion process is farmyard manure or slurry, the Environment Agency will accept that the digestate can be spread on agricultural land as a fertiliser or soil conditioner without being regulated as waste. It will still need to comply with the code of good agricultural practice and any requirements for Nitrate Vulnerable Zones.

Related exemptions

T23 – aerobic composting and associated prior treatment.

T25 – anaerobic digestion at premises not used for agriculture and burning of resulting biogas.

U10 – spreading waste to benefit agricultural land.

U11 – spreading waste on non-agricultural land.

T25: Anaerobic Digestion at Premises not Used for Agriculture and Burning Resulting Biogas

7.75 This exemption allows the treatment of food and other biodegradable waste by anaerobic digestion to produce a digestate, which can be used to benefit land. The gas produced (biogas) must be used for generating energy.

Types of activity include:

- a business or organisation, such as a hotel, prison or hospital using a small anaerobic digestion plant for their kitchen waste producing digestate for use on the gardens and biogas to generate electricity;
- sorting, screening, cutting, shredding, pulverising and chipping the waste to help the AD process.

Not included:

- aerobically treat the waste (see related exemptions);
- release the gas produced into the air;
- dispose of the biogas by flaring except if appliances burning biogas breakdown or are being routinely maintained;
- treat hazardous waste;
- treat waste that is an animal by-product without an appropriate authorisation from Animal Health.

Types of waste are listed in the table below.

Waste code	Type of waste
020103, 020107, 170506, 200201	Plant tissue waste
020106	Horse and farmyard manure only
200101	Paper and cardboard
200108	Biodegradable kitchen and canteen waste
020199	Fully biodegradable animal bedding
020202	Animal tissue waste
020501, 020601	Materials unsuitable for consumption or processing
200302	Biodegradable waste from markets only

Quantity of waste that can be stored or treated is up to 50 cubic metres of waste at any one time.

Conditions Must:

- collect and burn any gas produced by the anaerobic digester in an appliance;
- use an appliance with a net rated thermal input of less than 0.4MW. If there is more than one appliance, the total net rated thermal input must be less than 0.4MW (for example four 0.1 MW appliances);
- treat the waste within the digester for at least 28 days.

Associated prior treatment in relation to this exemption means screening, chipping, shredding, cutting, pulverising or sorting waste for the purposes of anaerobic digestion.

Animal By-Products Regulations – animal by-products can include 020202 animal tissue waste or catering waste from kitchens and restaurants (200108 biodegradable kitchen and canteen waste).

Approval may also be required from the Animal Health Veterinary Laboratory Agency (AHVLA). This is most likely if the intention is to supply compost to other users or use the compost at another place.

Farmed animals (includes pet sheep, cattle, pigs or poultry) must be excluded from the area where the composting takes place and where the compost will be used.

Related Exemptions

T23 – aerobic composting and associated prior treatment.

T24 – anaerobic digestion at premises used for agriculture and burning of resulting biogas.

U10 – spreading waste to benefit agricultural land.

U11 – spreading waste on non-agricultural land.

T26: Treatment of Kitchen Waste in a Wormery

7.76 This exemption allows very small-scale treatment of waste from kitchens using a wormery to produce compost. The compost can be spread on land to add nutrients or to improve the structure of the soil.

Type of activity includes:

- composting kitchen waste from an office canteen, along with a small amount of shredded cardboard, to provide a drier fibrous element.

Not included:

- treat waste not listed under this exemption.

Types of wastes listed in the table below.

Waste code	Type of waste
200101	Paper and cardboard
200108	Biodegradable kitchen and canteen waste

Quantity of waste that can be treated is up to six tonnes of waste over any 12-month period.

Conditions

The treatment must result in a stable, sanitised vermicompost that can be applied to land.

Animal By-Products Regulations – animal by-products can include catering waste from kitchens and restaurants (200108 biodegradable kitchen and canteen waste).

Approval may also be required from the Animal Health Veterinary Laboratory Agency (AHVLA). This is particularly if the intention is to supply compost to other users or use the compost at another place.

Farmed animals (including pet sheep, cattle, pigs or poultry) must be excluded from the area where the composting takes place and where the compost will be used.

Related exemptions

T23 – aerobic composting and associated prior treatment.

T25 – anaerobic digestion at premises not used for agriculture and burning of resulting biogas.

U10 – spreading waste to benefit agricultural land.

U11 – spreading waste on non-agricultural land.

T27: Treatment of Sheep Dip Using Organophosphate-degrading Enzyme

7.77 This exemption allows organophosphate sheep dip to be treated with an approved organophosphate-degrading enzyme.

This activity is to treat organophosphate sheep dip with an approved enzyme at the place where the sheep dip is used.

Not included:

- bringing back used and waste sheep dip from other places to be treated on a farm;
- allowing others to bring their waste sheep dip to another sheep farm to be treated in that sheep bath.

Mobile sheep dippers must not register this exemption if they are transporting waste sheep dip back to their premises to be treated.

Type of waste are listed in the table below.

Waste code	Type of waste
020109	Organophosphate sheep dip only

Quantity of waste that can be stored or treated is up to a maximum of 8,000 litres of sheep dip over a 24-hour period only at the place it was produced.

Conditions Must:

- treat the sheep dip only in accordance with the enzyme manufacturer's instructions;
- carry out the enzyme treatment in a secure container inside a drain pen or in a secure sheep dip bath;
- carry out the treatment and storage of the sheep dip so that no sheep dip (treated or untreated) can get into drains, watercourses or onto land.

Register this exemption

This exemption will need to be registered.

Related permits

Disposal of the sheep dip that has been treated with an approved enzyme by discharging it to land, will require an environmental permit.

T28: Sort and Denature Controlled Drugs for Disposal

7.78 This exemption allows pharmacies and other similar places to comply with the requirements of the Misuse of Drugs Regulations 2001, SI 2001/3998 by denaturing controlled drugs (making them unsuitable for consumption).

The activity includes that a pharmacy or veterinary surgery is required to denature controlled drugs before disposing of them.

Not included:

- any waste drugs that are hazardous waste;
- controlled drugs that are waste at any other place than where the waste was produced.

Types of waste are listed in the table below.

Waste code	Type of waste
180109	Medicines from natal care, diagnosis, treatment or prevention of disease in humans
180208	Medicines from research, diagnosis, treatment or prevention of disease involving animals
200132	Medicines separately collected as municipal waste

Quantity of waste:

- store or treat up to one cubic metre of waste at any one time;
- store waste for up to six months.

Conditions

Under this exemption only the sorting or denaturing of controlled drugs before disposing of them can be carried out.

Other returned or out of date medicines can be stored under the Non Waste Framework Directive exemption for temporary storage of waste at a collection centre. This does not need to be registered as the Non Waste Framework Directive exemption.

T29: Treatment of Non-Hazardous Pesticide Washings by Carbon Filtration for Disposal

7.79 This exemption allows the treatment of non-hazardous pesticide washings before disposing of them to land. Pesticide washings treated this way are less likely to damage the environment when discharged to land compared to untreated pesticide washings.

This activity allows the use of a sealed carbon filtration unit, such as a sentinel unit, to treat surplus spray and pesticide washings produced on the farm or in landscaping or other uses.

Not included:

- the importing of pesticide washings from other places or farms to treat in the plant;
- the treatment of hazardous pesticide washings.

Types of waste are listed in the table below.

Waste code	Type of waste
020109, 160509	Non-hazardous pesticide washings only

Quantity of waste that can be stored or treated is up to 8,000 litres of waste over any 24-hour period.

Conditions Must:

- treat and store waste in the same place that the pesticides were used;
- carry out treatment of the pesticide washings in a sealed unit with sealed drainage.

During treatment and storage, no pesticide washings (treated or untreated) can get into drains, watercourses or onto land and cause pollution.

Related exemptions

Pesticide washings can be treated by diluting in a biobed or biofilter under T32 – treatment of waste in a biobed or biofilter.

This exemption will need to be registered.

Related permits

If the disposal of the treated pesticides is by discharging them to land, then an application will need to be made for an environmental permit.

T30: Recover Silver

7.80 This exemption allows the recovery of silver from waste produced in connection with printing or photographic processes.

Types of activity include:

- extracting silver from photographic films or paper to use again;
- using an on-site contained unit to remove silver from fixer and developer solutions.

Not included:

- mixing hazardous waste with other hazardous or non-hazardous waste.

Types of waste are listed in the table below.

Waste code	Type of waste
090106*	Waste containing silver from on-site treatment of photographic waste
090107	Photographic film or paper containing silver or silver compounds

*May be hazardous waste

Quantity of waste that can be treated or stored is up to 1,000 litres of waste at any one time.

The treatment and storage of the waste is carried out in an area with sealed drainage.

This exemption will need to be registered with the Environment Agency.

T31: Recover Monopropylene Glycol from Aircraft Anti-freeze Fluids

7.81 This exemption allows airports to treat waste anti-freeze to recover monopropylene glycol.

This activity is the filtering and distilling of aircraft anti-freeze fluids that have been collected following de-icing of aircrafts to recover monopropylene glycol.

Not included:

- the treatment of waste antifreeze fluids for disposal to, for example, a sewer;
- the treatment of hazardous waste.

Type of waste is listed in the table below.

Waste code	Type of waste
160115	Antifreeze fluids not containing hazardous substances.

Quantity of waste that can be stored or treated is up to 250 cubic metres of waste over any seven-day period.

Conditions

The waste must be:

- treated in a location with sealed drainage;
- stored with secondary containment;
- treated at the airport where it was produced.

Anti-freeze is highly toxic to the water environment and there are controls on storing and treating waste antifreeze to prevent it being released into the environment.

Register this exemption

This exemption will need to be registered.

T32: Treatment of Waste in a Biobed or Biofilter

7.82 This exemption allows the treatment of non-hazardous pesticide washings in a biobed or biofilter. The treated washings can be re-used and the biobed material (biomix) can then be spread on land (under an exemption) for agricultural or ecological benefit.

This activity is the treatment of non-hazardous washings in a lined biobed or above-ground biofilter.

Not included:

- the treatment of pesticide washings that are hazardous waste;
- allowing others to bring their pesticide washings to be treated at another biobed or biofilter;
- agricultural contractors cannot bring back pesticide washings they have produced while working on other farms.

Types of waste are listed in the table below.

Waste code	Type of waste
020109, 160509	Non-hazardous pesticide washings only

Quantity of waste that can be stored or treated is up to 15,000 litres of dilute pesticide washings in a 12-month period.

Conditions

The treatment can only take place where the pesticide is used and the washings are produced.

The lined biobed or biofilter must be sited in a secure place and it must have an impermeable lining.

The biobed or biofilter must be located at least:

- 10 metres from a watercourse;
- 50 metres from a spring, well or borehole not used to supply water for producing domestic food;
- 250 metres from a borehole used to supply water for producing domestic food.

The lined biobed or biofilter must not be sited in an area that is designated as a groundwater Source Protection Zone 1. If the location is within a Source Protection Zone 1, the Environment Agency should be consulted to see if site specific circumstances can be agreed.

Using the treated washings

Biobeds and biofilters are classed as waste recovery systems, which means that where possible, the treated water should be re-used. Possible re-use options include irrigation, sprayer washing or as the carrier for pre-crop total herbicide applications.

If the treated water is used for irrigation, it must be applied:

- to a vegetated area;
- to ground that is not frozen or water logged;
- at least 10 metres away from any surface water;
- at least 50 metres from any spring, well or borehole not used for domestic or food production;
- at least 250 metres from a borehole used to supply domestic or food production.

Related exemptions

Dilute pesticide washings can be treated in a sentinel unit if they are registered and comply with exemption:

T29 – treating non-hazardous pesticide washings by carbon filtration for disposal.

Biofilter and biobed material can be spread on to land under exemption:

U10 – spreading waste to benefit agricultural land.

Biofilter and biobed material can also be spread on land under exemption:

U11 – spreading waste to benefit non-agricultural land.

T33: Recover Central Heating Oil by Filtration

7.83 This exemption allows the filtering of central heating oil so that it can be re-used.

This activity includes the filtering of central heating oil from disused tanks so that the oil can be re-used

Not included:

- the treatment of any other types of oil by filtration;
- the treatment of central heating oil for disposal;
- the mixing of hazardous waste with other hazardous or non-hazardous waste.

Types of waste are listed in the table below.

Waste code	Type of waste
130701*	Central heating oil only

*hazardous waste

Quantity of waste that can be treated is up to 400 litres of central heating oil over any seven-day period.

Conditions

The waste must be:

- stored with secondary containment;
- treated in a location with sealed drainage.

The exemption will need to be registered with the Environment Agency.

Definitions

7.84 'End-of-life vehicle (ELV)' – an ELV is any type of motor vehicle that is waste. Waste is anything discarded, intended to be discarded or there is a requirement to be discarded. This includes metal sent for recycling or re-use.

'Depolluted end-of-life vehicle' – a waste vehicle from which all hazardous substances or components, including fuels, lubricating and hydraulic oils, coolants, batteries, air conditioning fluids and air bags have been removed.

'Sealed drainage' – means a drainage system with an impermeable surface that ensures that:

- no liquid will run off the surface apart from via the system;
- except where they may be lawfully discharged, all liquids entering the system are collected in a sealed sump.

'Drain pen' – an impermeable area draining back to the sheep dip bath where newly-dipped sheep are held while they continue to dip.

'Secure' – a container, lagoon or other place is secure in relation to waste kept in it if:

- all reasonable precautions are taken to make sure the waste can't escape;
- members of the public cannot gain access to the waste.

'Place of production' – the place where waste was originally produced.

'Secure' – a container, lagoon or other place is secure if:

- all reasonable precautions are taken to make sure the waste cannot escape;
- members of the public cannot gain access to the waste.

'Secure' – a container, lagoon or other place is secure in relation to waste kept in it if:

- all reasonable precautions are taken to make sure the waste can't escape;
- members of the public can't gain access to the waste.

'Sealed drainage' – a drainage system with an impermeable surface that makes sure that:

- no liquid will run off the surface apart from via the system;
- except where they may be lawfully discharged, all liquids entering the system are collected in a sealed sump.

'Weatherproof covering' – may, in some circumstances, simply involve a lid or cover over a container, but, in others, a roofed shelter or building may be more appropriate.

Disposing

D1: depositing waste from dredging inland waters

7.85 This exemption allows the depositing of dredging spoil (dredgings) on the banks of the waters it was dredged from and to treat it by screening and removing water.

Types of activity include when:

- a contractor is involved in clearing silt from sections of a river and deposits the dredgings on the banks of the river;
- a contractor is involved in clearing silt and plant matter from sections of a canal. The contractor screens the dredgings to remove litter and then deposits the dredgings to let the water drain away. When the water has drained, the dredgings are taken away for spreading on land under an environmental permit.

Not included:

- the depositing of dredgings on the bank of a different inland water or deposit dredgings from any other waters – this is to prevent the spread of contamination or invasive non-native species, and to keep outside the requirements of the Landfill Directive;

- the depositing of dredgings that are hazardous waste – this is also to keep outside the requirements of the Landfill Directive;
- the treatment of dredgings other than by screening or removing water.

Types of waste are listed in the table below.

Waste code	Type of waste
170506	Dredging spoil not containing hazardous substances

Over any 12 month period, then up to 50 cubic metres of dredgings can be deposited or treated for each metre length of land on which waste is deposited.

D1: key conditions

The waste must be deposited as close as possible to where it was dredged from.

The waste must be deposited either:

- on the bank of the waters from where it was dredged;
- or on land next to the water it was dredged from (the dredgings must be removed from the waterway and deposited mechanically in one operation).

This means that it can't deposit onto a bank and then move it further away by the same or another machine.

A river or canal may be an example of a place, which can be regarded as a 'linear network' for the purpose of registering this exemption. Guidance on the meaning of place.

Non-hazardous dredgings that are deposited alongside the waterway from where they are dredged are excluded from the Landfill Directive. This can apply to dredgings placed into a lagoon.

As well as registering this exemption, it may also need to comply with other legislation. In particular, this may apply if there is also a U1 exemption.

This could include:

- planning permission;
- flood defence consent;
- if the work involves raising levels or building in the floodplain or near an ordinary watercourse, then the local flood authority needs to be contacted.

D1 related exemptions

Dredgings can be used in construction under:

U1: use of waste in construction.

Plant matter that is strimmed along the banks of the river can be spread where it has cut it under:

U13: spreading plant matter to provide benefits.

Dredgings produced from creating or maintaining habitats, ditches or ponds can be spread within parks, gardens, fields and forests only under:

- U10: spreading waste to benefit agricultural land.
- U11: spreading waste to benefit non-agricultural land.

D1 exemption will need to be registered.

Related permits

A permit will need to be applied for if:

- there is a requirement to treat or deposit more waste than this exemption allows;
- the lagoon is not excluded from the Landfill Directive.

and there is a requirement to spread dredgings to land or use dredgings in construction but cannot comply with the limits and conditions of those exemptions listed under related exemptions.

D2: Depositing Waste from a Railway Sanitary Convenience

7.86 Some older rolling stock, for example forming part of a vintage steam train, has not been or cannot be fitted with appropriate collection facilities. This exemption allows this rolling stock to deposit sanitary waste onto the track.

Latest standards for train design require collection of waste from the toilet but still allow deposit of waste water from hand washbasins and kitchen sinks.

This exemption allows the discharge of waste from a train sink or sanitary convenience onto a railway track.

Types of activity that cannot be carried out:

- discharge waste from a sanitary convenience from any other type of vehicle such as a bus;
- discharge any other types of waste onto a railway track;
- bringing in sanitary waste from trains operating on a different railway line.

Types of waste are listed in the table below.

Waste code	Type of waste
200399	Waste from railway sanitary conveniences only

There is no limit on the number of discharges that can be made, but there can only be a deposit of up to 25 litres of waste per discharge.

More than one railway operating company can register this exemption at the same place at the same time.

D2: Key conditions

For the purposes of this exemption, waste from a railway sanitary convenience means waste from any toilet or sink located in a railway carriage for carrying passengers.

A railway is an example of a place, which can be regarded as a 'linear network' for the purpose of registering this exemption. Guidance on the meaning of place.

D2 Related exemptions

Waste can be deposited from a portable toilet under:

D3: depositing waste from a portable sanitary convenience.

D3: Depositing Waste from a Portable Sanitary Convenience

7.87 This exemption allows the burial of waste from a portable toilet to avoid small quantities having to be transported long distances to sewage treatment works.

The activity includes a small group of boy scouts camping at a remote site for a night with portable toilets, and intending to bury the waste from the toilets.

Types of activity that cannot be carried out:

- deposit waste from a non-portable sanitary convenience;
- deposit waste by spreading or tipping it onto land;
- bring toilet waste in from other places to deposit on land;
- bury cesspit waste.

Type of waste that can be deposited is listed in the table below.

Waste code	Type of waste
200399	Waste from portable sanitary conveniences only

Up to one cubic metre of waste can be deposited in any 12-month period.

D3: Key conditions

Waste must not be deposited within 10 metres of any watercourse or 50 metres of any spring, well or borehole.

Waste can only be buried at the place of production – this means the place where the toilet is used.

In most cases it is better to dispose of the waste to a foul sewer at sites specially designed to do this. Most permanent campsites and caravan sites usually have a chemical toilet disposal point.

D3 Related exemptions

Waste from a railway sanitary convenience can be deposited under:

D2: depositing waste from a railway sanitary convenience.

D4: Depositing Agricultural Waste Consisting of Plant Tissue under a Plant Health Notice

7.88 This exemption allows the depositing of diseased crops, where they were grown, when a Plant Health Notice has been issued. This is to reduce the risk of spreading plant diseases or pests.

This activity includes disposing of diseased or spoilt crops on a farm if it has been issued with a Plant Health Notice.

Types of activity that cannot be carried out:

- disposal of spoilt crops that are not the subject of a Plant Health Notice;
- importing crops from other farms that are also subject to a Plant Health Notice;
- allowing landscape gardeners and tree surgeons to bring back wood and plant tissue from their jobs and deposit them at a farm.

Types of waste are listed in the table below.

Waste code	Type of waste
020103	Plant tissue waste

Where waste is placed in a pile, each pile of waste must be no more than 250 tonnes.

D4: Key conditions

Diseased crops must not be deposited:

- within 10 metres of any watercourse, including rivers and streams, or within 50 metres of any spring, well or borehole;
- on land that is frozen, covered with snow or waterlogged;
- on land that has been frozen for 12 hours or more in the 24 hours before it is required to be spread.

Piles of waste cannot be placed immediately next to each other.

These conditions are to reduce the risk of polluting watercourses or groundwater.

The deposit can only take place at the farm where the crops were grown.

D4: Related exemptions

T6: treating waste wood and waste plant matter by chipping, shredding, cutting or pulverising.

T23: aerobic composting and associated prior treatment.

T24: anaerobic digestion at premises used for agriculture and burning of resulting biogas.

T25: anaerobic digestion at premises not used for agriculture and burning of resulting biogas.

U10: spreading waste to benefit agricultural land.

U12: using mulch.

U13: spreading plant matter to provide benefit.

These exemptions cannot be used if the Plant Health Notice requires that the plant tissue is destroyed.

D5: Depositing Waste Samples for Testing or Analysis

7.89 This exemption allows the depositing and storage of waste samples, where these are required to comply with or enforce specified regulations, or before testing and analysis for research.

Types of activity include:

- a waste management company deposits and stores a sample of waste for testing, to make sure it has the correct description of the waste for Duty of Care purposes;
- waste is collected and stored before being analysed to find out the range of constituents making up a waste stream and its potential use.

Field research cannot be carried out under this exemption, as it only permits storing and depositing waste before it is tested or analysed.

Types of waste that can be deposited

The waste codes are those listed in the List of Wastes (LoW) Regulations.

Type of waste

Only samples of waste collected as listed under key conditions.

Quantity of waste that can be deposited

- deposit or store up to 10 tonnes of waste at any one time.
- waste can be stored for up to 12 months or longer if ordered by a court.

D5: Key conditions

Samples of waste collected in the following circumstances may be deposited or stored under this exemption before testing or analysis.

Exercising power under:

- Chapter 12 of the Radioactive Substances Act 1993;
- Chapter 40 of the Control of Pollution Act 1974;
- Section 5 of the Control of Pollution (Amendment) Act 1989 (as amended by s 37 of the Clean Neighbourhoods and Environment Act 2005);
- The Environmental Protection Act 1990;
- The Water Resources Act 1991;
- Chapter 56 of the Water Industry Act 1991;

- The Waste Electrical and Electronic Equipment Regulations 2006, SI 2006/3289 (as amended);
- The Producer Responsibility Obligations (Packaging Waste) Regulations 2007, SI 2007/871;
- Regulation 88 of the Waste Batteries and Accumulators Regulations 2009, SI 2009/890;
- by or on behalf of the holder of an environmental permit;
- by or on behalf of a person carrying on in relation to the waste;
- a waste operation covered by an exemption;
- a waste operation carried out under a licence issued under the Food and Environment Protection Act 1985 (or not needing a licence by an order made under that Act);
- by or on behalf of the owner or occupier of the land from which the samples are taken;
- by or on behalf of any person to whom s 34(1) or 34(1A) of the Environmental Protection Act 1990 (as amended) applies in connection with that person's duties under that section;
- by or on behalf of any person to whom the Transfrontier Shipment of Waste Regulations 2007, SI 2007/1711 apply in connection with that person's duties under those regulations

for the purposes of research.

D6: Disposal by Incineration

7.90 This exemption allows the disposal of small amounts of specific waste that have been produced on the site, in an incinerator.

This activity includes a joinery firm producing off-cuts of clean, untreated wood waste, which it burns in a purpose-built incinerator with a design capacity of less than 50 kilograms per hour.

Types of activity that cannot be carried out:

- incinerate someone else's waste or take waste to another place for burning;
- burn other types of waste than those listed below;
- burn treated wood waste, for example treated pallets, painted wooden door frames;
- use this exemption to burn waste as a fuel for producing heat or power, see related exemptions;
- use this exemption to burn waste in the open, see related exemptions.

Types of waste that can be burned:

To meet this exemption, the wastes have to be specifically excluded from the Industrial Emissions Directive. The relevant excluded wastes are:

- vegetable waste from agriculture and forestry;
- wood waste (except wood waste treated with preservatives or coatings);
- cork waste;
- animal carcasses.

Waste streams that can be burnt are listed in the table below.

Waste code	Type of waste
020103	Plant tissue waste (from agriculture, horticulture, aquaculture and fishing)
020107	Untreated wood from forestry only
030101	Waste bark and cork
030105	Untreated sawdust and wood shavings only
030301	Waste bark and wood
150103	Untreated wooden packaging only
170201	Untreated wood

Waste can be burnt in an incinerator that has a capacity of less than 50 kilograms per hour and a net rated thermal input of less than 0.4 megawatt.

Up to 5 tonnes of waste can be stored at any one time before burning.

D6: Key conditions

If there is more than one appliance, the total net rated thermal input must be less than 0.4MW (for example, 4 × 0.1MW appliances).

Only the person who produced the waste can burn the waste in an incinerator.

Animal by-products

If only the incineration of animal carcasses is to take place then the exemption does not have to be registered. Instead, approval will be needed from Defra's Animal Health section, which deals with the Animal By-Product Regulations.

If animal carcasses are being incinerated along with other wastes allowed here, they will need to register this exemption as well as obtaining the relevant approval from Animal Health.

D6 Related exemptions

Waste can be burnt in a boiler to produce heat and power under exemption U4: burning waste as fuel in a small appliance.

Untreated waste wood from joinery can be chipped under exemption T6: treating waste wood and waste plant matter by chipping, shredding, cutting or pulverizing.

Then use the wood chip for animal bedding under exemption U8: using waste for a specified purpose.

Plant tissue can be composted under exemption T23: aerobic composting.

Untreated wood waste can be burnt, for example hedge cuttings, in the open where it was produced under exemption D7: burning waste in the open.

D7: Burning Waste in the Open

7.91 This exemption allows the burning of plant tissue and untreated wood waste from joinery or manufacturing in the open air.

Types of activity that can be carried out include a landscape gardener who has trimmed hedges and branches and wants to burn them on a bonfire at the same place

Not included:

- the burning of other types of waste other than those listed below;
- the bringing in of waste from elsewhere to burn;
- the burning of treated wood waste or wood waste coming from any source other than listed;
- using this exemption to burn waste in an incinerator for disposal (see related exemptions);
- using this exemption to burn waste in a boiler to produce heat and power (see related exemptions).

Householders burning their own garden waste do not need to register any waste exemptions.

Types of waste are listed in the table below.

Waste code	Type of waste
020103, 020107, 200201	Plant tissue
030105	Sawdust, shavings and cuttings from untreated wood only
030301	Waste bark and wood

Quantity of waste:

- burn up to 10 tonnes of waste in any 24-hour period;
- store up to 20 tonnes of waste at any one time;
- store waste for up to six months before burning, to allow certain wood waste to dry out.

D7: Key conditions

The burning must take place on open land, not in an incinerator or a building.

The burning must take place only at the place where the waste is produced.

D7 Related exemptions

Untreated wood waste can be burned in a boiler to produce heat and power under exemption U4: burning waste as fuel in a small appliance.

Plant tissue can be composted under exemption T23: aerobic composting.

Untreated waste wood from joinery can be chipped under exemption T6: treating waste wood and waste plant matter by chipping, shredding, cutting or pulverising, then use the wood chip for animal bedding under exemption U8: using waste for a specified purpose.

These options above are preferable to burning in the open air because they result in recovering the waste rather than disposing of it.

Waste in an incinerator can be burned to dispose of the waste rather than using it as a fuel under D6: disposal by incineration.

D8: Burning Waste at a Port Under a Plant Health Notice

7.92 This exemption only allows the burning of plant tissue waste and wood packaging and packing material waste, at a port, when a Plant Health Notice has been served to prevent the spread of plant diseases.

The types of activity that cannot be carried out:

- the burning of waste that is not the subject of a Plant Health Notice;
- the burning of waste at a place other than the port where the waste was unloaded;
- the burning of waste other than those listed below under 'Types of waste that can be disposed of';
- the burning of treated wood waste, for example treated pallets;
- using this exemption to burn waste in a boiler to produce heat and power;
- using this exemption to burn plant tissue waste in the open other than at a port.

The wastes that can be disposed of are listed in the table below.

Waste code	Type of waste
020103, 020107, 020304, 200201	Plant tissue only
150103, 200138	Wood used to wedge or support parts of cargo, including packing material, spacers and pallets only

Quantity of waste that can be disposed of:

Up to 10 tonnes of waste can be burned in any 24-hour period.

D8: Key conditions

The burning must be required under a Plant Health Notice.

The burning can take place only at the port where the waste is unloaded.

Storing Waste Exemptions

S1: Storing Waste in Secure Containers

7.93 This exemption allows the storage of specific waste in secure containers at a different place to where the waste was produced, before the waste is transported to another site to be recovered.

Types of activity include:

- the storing of waste oil in a tank at an intermediate site before it is moved to another site to be recovered;
- the storing and bulking-up of plastic packaging in a container at a place before it is moved to another site to be recovered.

Types of activity that cannot be carried out:

- the treatment of waste under this exemption;
- the storage of waste destined to be disposed of;
- the storage of mixed waste;
- the mixing of hazardous with non-hazardous waste.

Types of waste are listed in the table below.

Waste code	Type of waste
130109 * to 130113 *, 130204 * to 130208 *, 130401 * to 130403 *, 130701 *	waste oils
150101, 200101	cartons
150102, 200139	plastics and plastic packaging
150104, 200140	cans and foil only
150101, 200101	paper and cardboard
150107, 200102	glass
150109, 200110, 200111	textiles and clothes
150202 *	absorbents, filter materials (including oil filters not otherwise specified), wiping cloths, protective clothing contaminated by hazardous substances
150203	absorbents, filter materials, wiping cloths, protective clothing not contaminated by hazardous substances
160107 *	oil filters

*These are hazardous waste.

Up to 400 cubic metres each of the following waste can be stored at any one time:

- cartons;
- plastics and plastic packaging;
- cans and foils;
- paper and cardboard;
- glass;
- textiles and clothes.

Storage of up to three cubic metres each of the following waste at any one time:

- waste oils;
- hazardous absorbents, filter material, wiping cloths and protective clothing contaminated by hazardous substances;
- non-hazardous absorbents, filter materials, wiping cloths and protective clothing;
- oil filters.

Waste can be stored in up to 20 containers at any one time but the total amount of waste stored must not exceed the maximum allowed under this exemption.

Conditions

The key conditions are:

- no waste is stored for longer than 12 months;
- the containers must be secure;
- the person storing the waste must own the containers or have the owner's consent;
- waste oils and oil filters must be stored with secondary containment;
- the waste is being stored before it is recovered elsewhere;
- each waste type is stored separately.

Related exemptions

T1: Cleaning, washing, spraying or coating relevant waste.

T2: Recovering textiles.

T4: Preparatory treatments (baling, sorting, shredding etc).

S2: Storing waste in a secure place.

S2: Storing Waste in a Secure Place

7.94 This exemption allows the storage of specific waste at a secure place that is different to where the waste was produced, before the waste is transported to another site to be recovered.

Types of activity include:

- storing construction and demolition waste that can be re-used without treatment, at an intermediate site before it is moved to another place to be re-used;

- storing olive pulp and marble chips at a port before it is moved to another place to be recovered.

Types of activity that cannot be carried out:

- the treatment of waste under this exemption. See related exemptions;
- the storage of waste destined to be disposed of. See related permits;
- the storage of mixed waste;
- the mixing of hazardous with non-hazardous waste;
- registering this exemption to increase the storage capacity at the place where another treatment exemption such as T1, T4, T11 or a U activity is carried out.

Types of waste are listed in the table below.

Waste code	Type of waste	Storage limit	Period (months)
161002	aqueous paint related waste only	1,000 litres	6
160601 *, 160602 *, 160603 *, 160604, 160605, 200133 *, 200134	batteries	10 tonnes	6
150104, 200140	cans and foil only	500 tonnes	12
140601 *	CFCs, HCFCs and HFCs	18 tonnes	6
170101, 170102, 170103, 170107, 170202, 170203, 170401 to 170407, 170504, 170604, 170802	construction and demolition waste (non hazardous) capable of being used in existing state only	100 tonnes	12
200125	edible oil and fat only	5,000 tonnes	12
100207 *	electric arc furnace dust only	2,500 tonnes	3
020104	farm plastics (non-packaging only)	500 tonnes	12
070213, 150101, 150102, 150105, 200139	Food and drink cartons	500 tonnes	12
101112, 150107, 160120, 170202, 191205, 200102	glass	5,000 tonnes	12
020102	mammalian protein only	60,000 tonnes	12
020202	mammalian tallow only	45,000 tonnes	12

Waste code	**Type of waste**	**Storage limit**	**Period (months)**
200307	mattresses only	5 tonnes	3
010408, 191209	marble chips only	5,000 tonnes	12
020304	olive pulp/pellet only	5,000 tonnes	3
080111 *, 080112, 200127 *, 200128	paints (excluding specialist and industrial paints, wood preservatives, aerosol and spray paints, inks, adhesives and resins) pending reuse as paints only	10,000 litres	6
030307, 030308, 150101, 191201, 200101	paper and cardboard (excluding food and drink cartons) only	15,000 tonnes	12
090107, 090108	photographic films and papers	50 tonnes	12
070213, 120105, 150102, 160119, 191204, 200139	plastic	500 tonnes	12
100101	poultry litter ash only	3,000 tonnes	12
080318, 150102, 160216, 200139	printer cartridges only	5,000 units	6
170301 *, 170302, 170504	road planings, waste road chippings, road sub-base only	500 tonnes	12
020110, 160117, 160118, 170401, 170402, 170403, 170404, 170405, 170406, 170407, 170411, 191202, 191203	scrap metal	15,000 tonnes	6
090110, 090111 *, 090112	single use cameras	400 cubic metres	6
020401, 020399	soils from cleaning fruit and veg only	100 tonnes	6

Waste code	Type of waste	Storage limit	Period (months)
100316, 100405 *, 100504, 100511, 100604, 100811, 100899	solder metal, skimmings, ashes and residues	100 tonnes	3
140602 *, 140603 *, 200113 *	solvents and solvent mixtures	5 cubic metres	6
100101, 100102, 100105, 100115	synthetic gypsum and pulverised fuel ash only	2,500 tonnes	3
040222, 150109, 191208, 200110, 200111	textiles and clothes	1,000 tonnes	12
160103, 191204	tyres, tyre chip and crumb	40 tonnes	3
110113 *, 120301 *, 160708 *	waste cleaning solution containing 2% sodium metasilicate and 1 to 2% waste oil only	3 tonnes	3
160211 *, 160213 *, 160214, 160216, 200121 *, 200123 *, 200135 *, 200136	WEEE	400 cubic metres	6
030301, 150102, 150103, 200138	wine bottle corks only	500 tonnes	12
030105, 170201, 170204 *, 191206 *, 191207, 200137 *, 200138	Wood, including telegraph poles and railway sleepers (hazardous and non hazardous)	100 tonnes	12

*These are hazardous waste.

Conditions

- each type of waste must be stored separately;
- the storage must take place at a secure place;
- the storage of the waste is pending its recovery elsewhere;
- each waste shall not be stored for longer than the specified time period.

The additional specific conditions apply to some specified types of waste which are listed in the table below.

Aqueous paint related waste only	the waste is stored in a container
	the waste is stored with secondary containment
Batteries:	the waste is stored in a container
	the storage place has sealed drainage
CFCs, HCFCs and HFCs:	the waste is stored in a container
	the waste is stored with secondary containment
Edible oil and fat only:	the waste is stored in a container
	the waste is stored with secondary containment
Electric arc furnace dust only:	the waste is stored indoors
	the waste is stored at a dock before being exported or after being imported
	the waste must arrive at the storage place in bags and must be stored there in bags or in drums
Farm plastics (non packaging):	the waste is stored indoors
Glass:	the storage place has sealed drainage
Mammalian protein only:	the waste is stored indoors
Mammalian tallow only:	the waste is stored indoors
Mattresses only:	the waste is stored indoors
Olive pulp/pellet:	the storage place has sealed drainage
	the waste is stored with secondary containment
	the waste is stored at a dock before being exported or after being imported
Paints (excluding specialist and industrial paints, wood preservatives, aerosol and spray paints, inks, adhesives and resins) pending reuse as paints:	the waste is stored in a container
	the waste is stored with secondary containment
Paper and cardboard (excluding food and drink cartons):	the waste is stored in a baled form, in a container or indoors

	within the quantity limit specified and despite this additional specific condition, up to 1,000 tonnes may be stored outdoors as long as it is stored in an enclosure designed and maintained to prevent litter escaping
Photographic films and papers:	the waste is stored in a baled form, in a container or indoors
Poultry litter ash only:	the waste is stored indoors
	the waste is stored at a dock before being exported or after being imported
Printer cartridges only:	the waste is stored indoors
Scrap metal:	the storage place has sealed drainage
	the waste is stored at a dock before being exported or after being imported
Solder metal, skimmings, ashes and residues:	the waste is stored in bags or in drums
Solvents and solvent mixtures:	the waste is stored in a container
	the waste is stored with secondary containment
Synthetic gypsum and pulverized fuel ash only:	the waste is stored indoors
	the waste is stored at a dock before being exported or after being imported
	the waste must arrive at the storage place in bags and must be stored there in bags or in drums
Tyres, tyre chip and crumb:	the total quantity of waste stored together doesn't exceed 10 tonnes; this is to reduce the risk of fire spreading
Waste cleaning solution containing 2% sodium metasilicate and 1% to 2% waste oil only:	the waste is stored in a container
	the waste is stored with secondary containment

WEEE:	must comply with the requirements in paragraph 1 of Annex III to the WEEE Directive; this means storage on an impermeable surface and with weatherproof covering where there is storage of WEEE containing hazardous materials or fluids and where storing whole appliances that are to be reused

Related exemptions

T1: Cleaning, washing, spraying or coating relevant waste.

T2: Recovering textiles.

T4: Preparatory treatments (baling, sorting, shredding and so on).

S1: Storing waste in secure containers.

The exemption will need to be registered.

Related permits

An environmental permit is needed if there is a requirement to store:

- waste at an intermediate site for treatment or use not covered in related exemptions or under a permit for recovery;
- mixed waste at an intermediate site for treatment;
- waste prior to disposal.

S3: Storing Sludge

7.95 This exemption allows the storage of sewage sludge at a place where it is to be used in accordance with the Sludge (Use in Agriculture) Regulations 1989, SI 1989/1263.

Types of activity include:

- storing sewage sludge at a farm before it is spread on land;
- storing septic tank sludge before it is spread on land.

Types of activity that cannot be carried out:

- store waste within a zone defined by a 50-day travel time for groundwater to reach a groundwater abstraction that is used to supply water for homes or producing food (this is known as a Source Protection Zone 1);
- store waste under this exemption where it is to be used at another place;
- store waste where it is not going to be used in accordance with the Sludge (Use in Agriculture) Regulations 1989;
- store waste within 0.3 metres of the top of an open storage container or within 0.75 metres of the top of an earthbank tank or lagoon.

Types of waste are listed in the table below.

Waste code	Type of waste
190805	residual sludge from sewage plants treating domestic or urban waste waters
200304	residual sludge from septic tanks and other similar installations for the treatment of sewage only

Quantity of waste:

- store up to 1,250 tonnes of waste at any one time;
- store waste for up to 12 months before using it.

Conditions

The waste must be stored at least:

- 10 metres from any watercourse;
- 50 metres from any spring or well, or from any borehole not used to supply water for homes or producing food;
- 250 metres from any borehole used to supply water for homes or producing food.

The waste must be stored in a secure location at the place where it is to be used.

Sludge must be used in accordance with the Sludge (Use in Agriculture) Regulations 1989.

SCOTLAND

Introduction

7.96 Licensing in Scotland is covered by The Pollution Prevention and Control (Scotland) Regulations 2000, SSI 2000/323 and the Waste Management Licensing (Scotland) Regulations 2011, SSI 2011/228.

Exemptions are covered in SSI 2011/228, regs 16 (Exclusion of activities under other control regimes from waste management licensing), 17 (Exemptions from waste management licensing), 18 (Exemptions from waste management licensing: controlled activities), 20 (Register of exempt activities), 21 (Register of exempt activities: registration of information), 22 (Register of exempt activities: registration of information), and 23 (register of exempt activities: requirements in respect of recovery and storage of scrap metal or waste motor vehicles). Also Sch 1 (Activities exempt from waste management licensing).

The Waste Management Licensing (Scotland) Regulations 2011, SI 2011/288

7.97 These Regulations consolidate the waste management licensing and exemption system currently contained in the Waste Management Licensing Regulations 1994, SI 1994/1056 as amended. They also make certain changes to that system (including the introduction of new exemptions from the requirement for licensing and the adjustment of various existing exemptions) and contain provision for the purpose of implementing Directive 2008/98/EC on waste ('the Directive').

Regulation 2 contains definitions, including transposition of relevant definitions contained in the Directive. Regulation 3 makes provision in connection with determining whether an applicant for a waste management licence is a fit and proper person. Regulations 4–7 contain procedural provision in relation to appeals under ss 43 and 66 of the Environmental Protection Act 1990 ('the 1990 Act').

Regulations 8 and 9 make provision in relation to the contents of public registers maintained under s 64 of the 1990 Act. Regulation 10 defines what is to be treated as mobile plant for the purposes of waste management licensing. Regulations 11–13 require the inclusion of particular types of condition in site licences dealing with activities involving waste electrical and electronic equipment and batteries. They provide continued transposition of requirements of Directive 2002/96/EC on waste electrical and electronic equipment (WEEE) and Directive 2006/66/EC on batteries and accumulators and waste batteries and accumulators.

Regulation 14 prohibits the imposition in waste management licences of conditions for the purpose of securing the health of persons at work. Regulation 15 makes provision, pursuant to Article 21 of the Directive, as to conditions to be included in a licence which relates to waste oil.

Regulations 16 and 17 exempt certain activities from the requirement to have a waste management licence. Regulation 16 disapplies s 33(1) of the 1990 Act in respect of certain waste activities controlled by other systems. Regulation 17 disapplies s 33(1)(a) and (b) of the 1990 Act in the case of the activities set out in Sch 1, subject to various conditions including compliance with Directives relating to protection of groundwater and water policy. References to benefit to agriculture or ecological improvement in Sch 1 are to be construed in accordance with Sch 2. Regulation 17 and Sch 1 contain the detail of the exemptions envisaged by Articles 24 and 25 of the Directive. Regulation 18 deals with the interaction between this exemption regime and the separate regulatory requirements relating to controlled activities affecting the water environment. Regulations 19–28 provide a system of registration for exempt activities relating to the recovery and disposal of waste, including special provision for specific exemptions (such as a requirement for annual renewal of registration on payment of a fee) and the imposition of criminal liability for breach of registration obligations.

Regulation 29 and Sch 4 contain provisions transposing aspects of the Directive. Schedule 4 Pt I requires specified public authorities to perform their functions under specified enactments with the objectives set out in the Directive. It also sets out requirements relating to the content of permits relating to waste activities, modification of enactments to facilitate compliance with requirements of the Directive, registration of waste collectors and transporters required to be registered by the Directive who would not otherwise be subject to any registration requirement, and the inspection of and keeping of records by establishments or undertakings carrying out waste activities. Parts II and III of Sch 4 provide illustrative lists of disposal and recovery operations, in connection with the definitions of those operations contained in reg 2.

Regulation 30 and Sch 5 require the registration of waste brokers and dealers from 1 April 2011. Schedule 5 makes provision as to the keeping of registers, and in respect of applications for registration, the duration and revocation of registration and appeals.

Regulations 31–33 contain transitional provisions and savings. Regulation 34 revokes the Waste Management Licensing Regulations 1994, SI 1994/1056, which are replaced by these Regulations, and gives effect to the minor and consequential amendments contained in Sch 6.

The Waste Management Licensing (Scotland) Amendment Regulations 2016, SI 2016/40

7.98 These Regulations amend the Waste Management Licensing (Scotland) Regulations 2011, SI 2011/288 ('the 2011 Regulations'). Regulation 2 amends Sch 1 to the 2011 Regulations, which provides for activities which are exempt from the requirement to have a waste management licence.

Regulation 2(2) removes an exemption relating to the baling, sorting or shredding of a limited quantity of waste tyres with a view to their recovery or reuse. Regulation 2(3) removes an exemption relating to the storage of a limited number of waste tyres in a secure place.

Regulation 3 makes transitional provisions for persons carrying out activities under one of the exemptions being removed by reg 2 when these Regulations come into force.

Activities which meet the requirements of reg 17 are exempt from licensing. Although an activity may be exempt from waste management licensing, it is subject to statutory controls to prevent environmental pollution and harm to human health.

Controls under reg 19 relate to the registration of exempt activities. The controls under reg 17 relate to certain conditions being met in order for the activity to be exempt, which includes the requirement that the activity must not endanger human health or harm the environment by:

- Presenting a risk to water, air, soil, plants or animals;

- Causing nuisance through noise or odours; or
- Adversely affecting the countryside or places of special interest.

Schedule 1 of the 2011 Regulations divide exemptions into Simple Exemptions and Complex Exemptions.

Types of Exemption

Simple Exemptions

7.99

- cleaning or coating waste packaging, containers and textiles;
- burning waste as a fuel in an exempt appliance;
- burning waste oil as a fuel in an engine;
- storing specified waste securely, eg waste hydraulic oils;
- treating specific waste to recover materials;
- burning waste in an exempt incinerator where it was produced;
- storing waste electrical and electronic equipment (WEEE) for recovery at a place other than where it was produced.

Complex Exemptions

7.100

- reclaiming and improving land;
- storing and spreading sludge;
- treating some waste at a water treatment works;
- composting;
- using waste in construction;
- disposing of pesticides;
- burning waste wood at docks;
- storing WEEE for recovery;
- repairing and refurbishing WEEE.

Simple Exemptions (by para no)

7.101

1 The use of waste glass in glass manufacture and production;
2 Operation of scrap metal furnace;
3 Burning of waste as fuel;
4 Cleaning, washing, spraying or coating of waste packaging and containers;
5 Burning waste as fuel in an appliance and recovery of fuel from waste vegetable oils;
6 Burning Waste Oil as a fuel in specified engines;

8(1)The storage and spreading of sludge;
11 Bailing, sorting, shredding etc of specified wastes;
13 Manufacture of specified goods from specified wastes;
14 The manufacture of finished goods and repair/refurbishment of waste goods;
15 Beneficial use of waste without further treatment;
16 The screening of specified wastes;
17 Storage of specified wastes in a secure place;
18 Secure storage of specified wastes on any premises;
20 Cleaning waste textiles for recovery;
21 Chipping etc waste plant matter;
22 Recovery of silver;
24 Size reduction of bricks, tiles or concrete;
25 The deposit of dredging wastes;
26 The recovery/disposal of waste at the place of production as an integral part of the production;
27* Bailing, compacting, crushing or shredding waste at the place it was produced;
28 The use of autoclaves to sterilise waste;
29 The disposal of waste at the place of production by burning in an incinerator;
30 Burning plant tissue waste on land in the open;
31 Discharge of sanitary waste from a passenger train;
32 Burial of sanitary waste from a removable receptacle;
33 Keeping or deposit of peat excavation waste;
35 The deposit of excavated material from a borehole or other excavation;
36 Temporary storage of specified waste at a harbour;
37 Burial of a dead domestic pet;
38 The deposit or storage of samples of waste for testing or analysis;
39 Secure storage of medical, nursing or veterinary wastes;
40 Secure storage of non-liquid waste other than at the place of production;
41 Temporary storage of waste at the place of production;
43 Crushing of waste gas discharge lamps;
44 Heating metal in an authorised furnace;
48 The storage of WEEE pending recovery elsewhere.

Complex Exemptions (by para no)

7.102

7 The treatment of land for agricultural benefit or ecological improvement;
8(2)The storage and spreading of sludge;
9 The reclamation or improvement of land;
10 Reception and treatment of specified waste at a Water Treatment Works;
12 Composting;
19 Waste for construction and other 'relevant work';
42 The disposal of waste pesticide solutions or washings in a lined biobed;
45 The recovery of scrap metal or the dismantling of motor vehicles;
46 The burning of plant tissue or wood at a dock;
47 Repair and/or refurbishment of WEEE;

49 The treatment of waste organophosphate sheep dip;
50 The mixing of pig or poultry ash for treatment of land;
51 The anaerobic digestion of agricultural benefit or ecological improvement.

Complex and Simple Exemptions (by para no)

Paragraph 1: The Use of Waste Glass in Glass Manufacture and Production

7.103 This exemption is for total quantity of waste glass used that does not exceed 600,000 tonnes in any period of 12 months.

The storage of the waste must be at the place the activity is carried out, and is intended for the use in this exemption.

This needs to be registered with SEPA, but no fee is required.

Paragraph 2: Operation of a Scrap Metal Furnace

7.104 This exemption applies to the operation of a scrap metal furnace, which should have a design capacity of less than 25 tonnes. The exemption also covers the loading or unloading of a furnace.

The storage at the place where the furnace is located is also covered by the exemption.

This exemption needs to be registered with SEPA, but no fee is required.

The registration of this exemption is not time limited but it can be de-registered by SEPA if it is not managed properly and no longer fulfills the criteria requirements of the exemption or if the activities cause pollution or harm to health. If the activity does comply with the exemption criteria the registration will not have to be renewed except where there is a significant change to the activity.

Paragraph 3: Burning of Waste as a Fuel

7.105 This exemption applies to:

- burning as a fuel, under a permit granted under the 2000 Regulations, of:
  - straw, poultry litter or wood;
  - waste oil listed in the table below (including waste oil which is special waste); or
  - solid fuel which has been manufactured from waste by a process involving the application of heat, to the extent that it is or forms part of an activity within Part B of any Section of Part 1 of Schedule 1 to the 2000 Regulations.

The waste needs to be securely stored on the premises on which it is to be burned, other than waste oil, which is intended to be burned as mentioned in that sub paragraph and the feeding of such wastes into an appliance in which they are to be so burned;

The secure storage of waste oil listed in the table below (including waste oil which is special waste) at the place where it is produced. This is for a period not exceeding 12 months if the waste oil is covered by the exemption and the total volume of that oil does not at any time exceed 23,000 litres.

Burning tyres as a fuel, under an authorisation granted under a permit granted under the 2000 Regulations, if it is or forms part of an activity within Part B of Section 1.1 of Part 1 of Schedule 1 to the 2000 Regulations, and the shredding and feeding of tyres into an appliance in which they are to be burned.

Tyres should be securely stored on the premises where they are to be burned, they should be stored separately and they should not be stored for longer than 12 months and the number of tyres should not exceed 1,000.

Codes*	Types of waste
Waste hydraulic oils (13 01)	
13 01 09**	Mineral based chlorinated hydraulic oils
13 01 10**	Mineral based non chlorinated hydraulic oils
13 01 11**	Synthetic hydraulic oils
13 01 12**	Readily biodegradable hydraulic oils
13 01 13**	Other hydraulic oils
Waste engine, gear and lubricating oils (13 02)	
13 02 04**	Mineral-based chlorinated engine, gear and lubricating oils
13 02 05**	Mineral-based non chlorinated engine, gear and lubricating oils
13 02 06**	Synthetic engine, gear and lubricating oils
13 02 07**	Readily biodegradable engine, gear and lubricating oils
13 02 08**	Other engine, gear and lubricating oils
Waste insulating and heat transmission oils (13 03)	
13 03 06**	Mineral based chlorinated insulating and heat transmission oils other than those mentioned in 13 03 01 (insulating or heat transmission oils containing PCBs)
13 03 07**	Mineral based non-chlorinated insulating and heat transmission oils
13 03 08**	Synthetic insulating and heat transmission oils
13 03 09**	Readily biodegradable insulating and heat transmission oils
13 03 10**	Other insulating and heat transmission oils
Bilge oils (13 04)	
13 04 01**	Bilge oils from inland navigation
13 04 02**	Bilge oils from jetty sewers
13 04 03**	Bilge oils from other navigation

Codes*	Types of waste
Waste of liquid fuels (13 07)	
13 07 01**	Fuel oil and diesel
13 07 02**	Petrol
13 07 03**	Other fuels (including mixtures)

* Codes referred to in the European Waste Catalogue.
** Includes special waste.

The exemption needs to be registered but no fee is required.

The registration of this exemption is not time limited but it can be de-registered by SEPA if it is not managed properly and no longer fulfils the criteria requirements of the exemption or if the activities cause pollution or harm to health. If the activity does comply with the exemption criteria the registration will not have to be renewed except where there is a significant change to the activity.

Paragraph 4: Cleaning, Washing, Spraying or Coating of Waste Packaging and Containers

7.106 This exemption applies to the cleaning, washing, spraying or coating of waste consisting of packaging or containers so that it or they can be re-used if the total quantity of such waste so dealt with at any place does not exceed 1,000 tonnes in any period of seven days.

Not included:

- the total quantity of such waste stored at that place exceeds 1,000 tonnes; or
- more than one tonne of metal containers used for the transport or storage of any chemical are dealt with in any period of seven days.

This exemption needs to be registered with SEPA. No fee is required.

The registration of this exemption is not time limited but it can be de-registered by SEPA if it is not managed properly and no longer fulfils the criteria requirements of the exemption or if the activities cause pollution or harm to health. If the activity does comply with the exemption criteria the registration will not have to be renewed except where there is a significant change to the activity.

Paragraph 5: Burning Waste as a Fuel in an Appliance and Recovery of Fuel from Waste Vegetable Oils

7.107 This exemption applies to the burning of waste as a fuel in a plant, which is an excluded plant for the purposes of Section 5.1 of Part 1 of Schedule 1 to the 2000 Regulations.

The secure storage of waste intended to be submitted to such burning:

(a) the plant is not located in an air quality management area designated under s 83 of the 1995 Act; and
(b) the waste is stored and burned in accordance with the requisite planning permission (if any).

The recovery, from waste vegetable oils (within category 20 01 25 of the European Waste Catalogue), of fuel for use in an engine of an aircraft, hovercraft, mechanically propelled vehicle, railway locomotive, ship or other vessel, or in appliances described in sub paragraph (1).

The secure storage of:

- such waste vegetable oil prior to the carrying out of the fuel recovery process, and
- waste produced by the fuel recovery process.

    (a) The fuel recovery process is conducted in equipment manufactured for this purpose;
    (b) the total quantity of waste (including waste produced by the fuel recovery process) stored at any one time does not exceed 23,000 litres;
    (c) no waste is stored for longer than 12 months;
    (d) waste produced by the fuel recovery process is stored at the place where it is produced;
    (e) accurate records (including information about sources, volumes and destinations) are maintained by the holder and processor of both the waste vegetable oil and the recovered fuel; and
    (f) the fuel recovery process is not carried out for commercial purposes on an industrial scale, and in any case no more than 200 tonnes of recovered fuel is produced by the fuel recovery process in any calendar year.

What this means:

Paragraph 5(1) permits wastes to be used as a fuel in an appliance or appliances that are exempt appliances under the Pollution Prevention and Control (Scotland) Regulations 2000, SI 2000/323.

Paragraph 5(2) allows storage of the wastes that are to be used as a fuel.

Paragraph 5(3) provides that burning cannot be carried out in a designated Air Quality Management Zone.

Registration requirements:

Needs to be registered with SEPA but no fee is required. This exemption can be registered online. However, establishments and undertakings can still use the paper form should they wish. This form should be sent to the Registrar at the local SEPA area office (addresses are given on the form).

The registration of this exemption is not time limited and does not require renewal. However, it can be removed from the register by SEPA if the activity no longer fulfils the criteria of the exemption or if the activities cause pollution or harm to health.

Some specific wastes can be used to generate heat or power in an appliance or appliances with a throughput of up to 50kg/hr. Activities with a throughput of 50kg/hr or greater require a PPC Permit.

Only the burning of 'biomass' wastes as defined in the Pollution Prevention and Control (Scotland) Regulations 2012, SI 2012/360 is allowed. The waste must burn cleanly and not pollute or cause a nuisance.

Paragraph 5 exemption applies to the burning of waste as a fuel and not for disposal. Disposing of a limited amount of wood, bark and plant matter by burning in the open is permitted under the Paragraph 30 exemption.

Storage should be in the place or places that are identified when in the registered exemption and must be secure so that waste cannot escape and access to it is controlled.

Examples of activities suitable for registration under the Paragraph 5 (1) exemption:

A joinery workshop burning particle board off-cuts for space heating. Particleboard (MDF, OSB and chipboard) where they have not been treated with coatings or preservatives containing heavy metals or halogenated organic compounds fall within the definition of 'biomass' and so are suitable for use as fuel under Paragraph 5(1).

Examples of activities not suitable for registration under the Paragraph 5(1) exemption:

- A waste transfer station separates out combustible wastes and burns them in a small skip – the burning of waste for disposal is not permitted by this exemption.
- A garage collects and stores waste oil, which it filters and feeds into a small waste oil burner for heating – the burning of waste oil is not covered by the Paragraph 5 exemption. Waste oil must only be burned in specialist incinerators, which comply with Chapter IV of the Industrial Emissions Directive.
- Company burns confidential, paper waste in an appliance that provides hot water for a building – The burning of waste paper is not covered by the Paragraph 5 exemption. Waste paper must only be burned in specialist incinerators, which comply with Chapter IV of the Industrial Emissions Directive.

Paragraph 6: Burning Waste Oil as a Fuel in Specified Engines

7.108 This exemption applies to the burning waste oil listed in the table below (including waste oil which is special waste) as a fuel in an engine of an aircraft, hovercraft, mechanically propelled vehicle, railway locomotive, ship or other vessel if the total amount burned of such waste does not exceed 2,500 litres an hour in any one engine.

The secure storage of waste oil (including waste oil which is special waste) intended to be so burned if the total quantity of oil stored does not at any time exceed 23,000 litres.

Codes*	Types of waste
Waste hydraulic oils (13 01)	
13 01 09**	Mineral based chlorinated hydraulic oils
13 01 10**	Mineral based non-chlorinated hydraulic oils
13 01 11**	Synthetic hydraulic oils
13 01 12**	Readily biodegradable hydraulic oils
13 01 13**	Other hydraulic oils
Waste engine, gear and lubricating oils (13 02)	
13 02 04**	Mineral-based chlorinated engine, gear and lubricating oils
13 02 05**	Mineral-based non chlorinated engine, gear and lubricating oils
13 02 06**	Synthetic engine, gear and lubricating oils
13 02 07**	Readily biodegradable engine, gear and lubricating oils
13 02 08**	Other engine, gear and lubricating oils
Waste insulating and heat transmission oils (13 03)	
13 03 06**	Mineral based chlorinated insulating and heat transmission oils other than those mentioned in 13 03 01 (insulating or heat transmission oils containing PCBs)
13 03 07**	Mineral based non-chlorinated insulating and heat transmission oils
13 03 08**	Synthetic insulating and heat transmission oils
13 03 09**	Readily biodegradable insulating and heat transmission oils
13 03 10**	Other insulating and heat transmission oils
Bilge oils (13 04)	
13 04 01**	Bilge oils from inland navigation
13 04 02**	Bilge oils from jetty sewers
13 04 03**	Bilge oils from other navigation
Waste of liquid fuels (13 07)	
13 07 01**	Fuel oil and diesel
13 07 02**	Petrol
13 07 03**	Other fuels (including mixtures)

* Codes referred to in the European Waste Catalogue.
** Includes special waste.

The exemption is required to be registered with SEPA but no fee is required.

The registration of this exemption is not time limited but it can be de-registered by SEPA if it is not managed properly and no longer fulfills the criteria requirements of the exemption or if the activities cause pollution or harm to health. If the activity does comply with the exemption criteria the registration will not have to be renewed except where there is a significant change to the activity.

Paragraph 7: The Treatment of Land for Agricultural Benefit or Ecological

7.109 This exemption applies to the treatment with any of the wastes listed in the table below of land used for agriculture where this treatment results in benefit to agriculture or ecological improvement.

The treatment with any of the wastes listed in Part I of the table below of—

- operational land of a railway, light railway, Scottish Water or British Waterways; or
- land which is a forest, woodland, park, garden, verge, landscaped area, sports ground, recreation ground, churchyard or cemetery; where the land in question is not used for agriculture and such treatment results in ecological improvement.

Codes*	Types of waste
PART I Wastes from agriculture, horticulture, aquaculture, forestry, hunting and fishing (02 01)	
02 01 03	Plant-tissue waste
Wastes from sugar processing (02 04)	
02 04 01	Soil from cleaning and washing beet
Wastes from wood processing and the production of panels and furniture (03 01)	
03 01 01	Waste bark and cork
03 01 05	Sawdust shavings, cuttings, wood, particle board***
Wastes from pulp, paper and cardboard production and processing (03 03)	
03 03 01	Waste bark and wood (including virgin pulp)
Soil (including excavated soil from contaminated sites), stones and dredging spoil (17 05)	
17 05 04	Soil and stones**
17 05 06	Dredging spoil***
Wastes from aerobic treatment of solid wastes (19 05)	
19 05 03	Off-specification compost consisting only of biodegradable waste***

Codes*	Types of waste
Wastes from anaerobic treatment of waste (19 06)	
19 06 04	Digestate consisting only of biodegradable waste***
19 06 06	Digestate from anaerobic treatment of animal and vegetable waste
Garden and park wastes (including cemetery waste) (20 02)	
20 02 01	Biodegradable waste
20 02 02	Soil and stones
PART II Wastes from agriculture, horticulture, aquaculture, forestry, hunting and fishing (02 01)	
02 01 99	Straw, wood or paper-based bedding waste, slurry or dirty water from stables, zoos, animal parks or livestock markets, animal faeces, urine and manure***
Wastes from the preparation and processing of meat, fish and other foods of animal origin (02 02)	
02 02 03	Materials unsuitable for consumption or processing consisting of blood and gut contents from abattoirs, poultry preparation plants or fish preparation plants; wash waters and sludges from abattoirs, poultry preparation plants or fish preparation plants; and shells from shellfish processing***
02 03	Wastes from fruit, vegetables, cereals, edible oils, cocoa, coffee, tea and tobacco preparation and processing; conserve production; yeast and yeast extract production, molasses preparation and fermentation – all wastes under this category
02 04	Wastes from sugar processing – all wastes under this category
02 05	Wastes from production of dairy products***
02 06	Wastes from the baking and confectionary industry – all wastes under this category
02 07	Wastes from the production of alcoholic and non-alcoholic beverages (except coffee, tea and cocoa) – all wastes under this category
Wastes from pulp, paper and cardboard production and processing (03 03)	
03 03 09	Lime mud waste
03 03 11	Sludges from on-site effluent treatment plants treating only virgin paper wastes which contain no inks***
03 03 99	De-inked paper sludge from paper recycling, paper crumble derived from virgin pulp which contains no inks***
Wastes from the leather and fur industry (04 01)	
04 01 07	Sludges, in particular from on-site effluent treatment free of chromium

Codes*	Types of waste
Wastes from the textile industry (04 02)	
04 02 10	Organic matter from natural products (for example grease, wax)
04 02 15	Wastes from finishing other than those containing organic solvents
04 02 20	Sludges from on-site effluent treatment**
04 02 21	Wastes from unprocessed textile fibres
04 02 22	Wastes from processed textile fibres
Wastes from the manufacture, formulation, supply and use (MFSU) of acids (06 01)	
06 01 99	Gypsum***
Wastes from the manufacture, formulation, supply and use of fine chemicals and chemical products not otherwise specified (07 07)	
07 07 12	Sludges from on site effluent treatment other than those containing dangerous substances
Wastes from power stations and other combustion plants (except wastes from waste management facilities, off site waste water treatment plants and the preparation of water intended for human consumption and water for industrial use) (10 01)	
10 01 99	Gypsum***
Wastes from manufacture of cement, lime and plaster and articles and products made from them (10 13)	
10 13 04	Gypsum***
Soil (including excavated soil from contaminated sites), stones and dredging spoil (17 05)	
17 05 06	Dredging spoil***
Wastes from anaerobic treatment of waste (19 06)	
19 06 03	Liquor consisting only of biodegradable waste***
19 06 05	Liquor from anaerobic treatment of animal and vegetable waste
Wastes from the preparation of water intended for human consumption or water for industrial use (19 09)	
19 09 02	Sludges from water clarification

* Codes referred to in the European Waste Catalogue.
** Wastes containing dangerous substances are not included.
*** The wastes listed do not include all of the wastes specified in the European Waste Catalogue under the code referred to.

The secure storage, at the place where it is to be used, should not be more than 1250 tonnes of waste provided that:

(a) the waste is stored at a distance of not less than—

- 10 metres from any inland or coastal waters;
- 50 metres from any well, borehole or similar work sunk into underground strata for the purpose of any water supply other than a domestic water supply; or
- 250 metres from any well, borehole or similar work sunk into underground strata for the purpose of a domestic water supply; and

(b) no waste is stored so that volume stored exceeds 90% of the available capacity of the container or lagoon where the storage takes place.
(c) no more than 250 tonnes or, in the case of dredgings from inland waters, 5,000 tonnes of waste per hectare are used on the land in any period of 12 months;
(d) the land to be treated by the waste does not exceed 50 hectares;
(e) the waste to be used has not been mixed with any material which does not itself provide benefit to agriculture or ecological improvement; and
(f) in the case of materials within category 02 02 03 of the European Waste Catalogue, as listed in Part II of Table 2 above, the waste is treated in accordance with the Animal By-Products (Scotland) Regulations 2003, SSI 2003/411 and Regulation (EC) No 1774/2002 of the European Parliament and of the Council of 3rd October 2002 laying down health rules concerning animal by-products not intended for human consumption.

(5) In this paragraph 'domestic water supply' means a supply of water—

(a) for such domestic purposes as consist in or include, cooking, drinking, food preparation or washing; or
(b) for any of these domestic purposes to premises in which food is produced.

Registration

Must be notified to SEPA and a fee applies. Registration 'expires' after 12 months unless a renewal notice has been received and registered by SEPA. A renewal notice should be submitted to SEPA at least 21 days before the registration is due to 'expire'.

Paragraph 8: The Storage and Spreading of Sludge

7.110 8. —(1) The storage in a secure container or lagoon (or, in the case of dewatered sludge, in a secure place) of sludge which is intended to be used

(a) in accordance with the 1989 Regulations; or
(b) for land treatment in reliance upon the exemption conferred by sub paragraph (2).

(2) The treatment with sludge of land which is not agricultural land within the meaning of the 1989 Regulations if–

(a) it results in ecological improvement, or in the case of the treatment of land used for non-food crops not grown in short term rotation with food crops, it enhances the growth of those crops; and
(b) it does not cause the concentration in the soil of any of the elements listed in column 1 of the soil table set out in Schedule 2 to the 1989 Regulations to exceed the limit specified in the corresponding entry in column 2 of the table.

(3) Sub paragraph (1) applies only if the following conditions are satisfied:–

(a) the sludge is stored at the place where it is to be used;
(b) the sludge is stored at a distance of not less than–

 (i) 10 metres from any inland or coastal waters;
 (ii) 50 metres from any well, borehole or similar work sunk into underground strata for the purpose of any water supply other than a domestic water supply; and
 (iii) 250 metres from any well, borehole or similar work sunk into underground strata for the purpose of a domestic water supply;

(c) no sludge is stored to the extent that the volume stored exceeds 90% of the available capacity of the container or lagoon where the storage takes place; and
(d) no sludge is stored for longer than six months.

(4) For the purpose of this paragraph, enhancement of crop growth or ecological improvement shall be construed in accordance with Schedule 2 to, and any guidance issued under regulation 17(11) of, these Regulations.

(5) In this paragraph–

(a) 'the 1989 Regulations' means the Sludge (Use in Agriculture) Regulations 1989 ('the 1989 Regulations');
(b) 'domestic water supply' has the meaning given by paragraph 7(5); and
(c) 'sludge' has the meaning given by regulation 2(1) of the 1989 Regulations and 'used', in relation to sludge, has the meaning given by that regulation.

Registration requirements:

8(1) – Storage of sludge

Needs to be registered with SEPA but no fee is required. This exemption can be registered online and does not have to be renewed.

8(2) – Spreading of sludge

Must be notified to SEPA using the appropriate form and enclosing the correct fee.

Registration 'expires' after 12 months unless a renewal notice has been received and registered by SEPA. A renewal notice should be submitted to SEPA at least 21 days before the registration is due to 'expire'.

Notification forms:

8(1) – Storage of sludge

This exemption can be registered online, however establishments and undertakings can still use the paper form should they wish. This form should be sent to the Registrar at the local SEPA area office.

8(2) – Spreading of sludge

- Use WMX-FM8(2) – Form to register or renew a paragraph 8(2) exemption.

Guidance:

- WMX-TG8 – Technical guidance note for Paragraph 8 exemptions.

Further information on agricultural benefit and ecological improvement can be found in Sch 2 of the Waste Management Licensing (Scotland) Regulations 2011, SSI 2011/228.

SEPA has published guidance on the application of organic materials to selected habitats for restoration or improvement. It contains information on target topsoil characteristics, typical application rates for a range of organic materials and a decision support checklist for their use in soil formation or habitat enhancement.

- Guidance on suitable organic material applications for land restoration and improvement.

Paragraph 9 – The Reclamation or Improvement of Land

7.111 This exemption applies to:

- the treatment of land (including the restoration of quarries) with any of the wastes listed in Part I of the table below;
- the treatment of land (including the restoration of quarries) with any of the wastes listed in Part II of the table below where such treatment results in benefit to agriculture or ecological improvement;
- the secure storage for the waste is at the place where it is to be used and for a period not exceeding six months.

Codes*	Types of waste
PART I Wastes from physical and chemical processing of non-metalliferous minerals (01 04)	
01 04 08	Waste gravel and crushed rocks**
01 04 09	Waste sand and clays
Wastes from sugar processing (02 04)	
02 04 01	Soil from cleaning and washing beet

Codes*	Types of waste
Wastes from power stations and other combustion plants (except wastes from waste management facilities, off-site waste water treatment plants and the preparation of water intended for human consumption and water for industrial use) (10 01)	
10 01 02	Pulverised fuel ash***
Wastes from manufacture of ceramic goods, bricks, tiles and construction products (10 12)	
10 12 08	Waste ceramics, bricks, tiles and construction products (after thermal processing)
Wastes from manufacture of cement, lime and plaster and articles and products made from them (10 13)	
10 13 14	Waste concrete and concrete sludge
Concrete, bricks, tiles and ceramics (17 01)	
17 01 01	Concrete
17 01 02	Bricks
17 01 03	Tiles and ceramics
17 01 07	Mixtures of concrete, bricks, tiles and ceramics**
Soil (including excavated soil from contaminated sites), stones and dredging spoil (17 05)	
17 05 08	Track ballast**
Wastes arising from waste water treatment plants not otherwise specified (19 08)	
19 08 99	Wastes not otherwise specified (specifically stone filter media)
Wastes from the preparation of water intended for human consumption or water for industrial use (19 09)	
19 09 99	Wastes not otherwise specified (specifically slow sand filter sand)
Wastes from the mechanical treatment of waste (for example. sorting, crushing, compacting, pelletising) not otherwise specified (19 12)	
19 12 09	Minerals (for example sand, stones)
Wastes from soil and groundwater remediation (19 13)	
19 13 02	Solid waste from soil remediation**
Garden and park wastes (including cemetery waste) (20 02)	
20 02 02	Soil and stones

Codes*	Types of waste
PART II Wastes from pulp, paper and cardboard production and processing (03 03)	
03 03 05	De-inked sludges from paper recycling
03 03 07	De-inked paper pulp from paper recycling***
03 03 09	Lime mud waste
Wastes from the manufacture, formulation, supply and use of fine chemicals and chemical products not otherwise specified (07 07)	
07 07 12	Sludges from on-site effluent treatment other than those containing dangerous substances
Soil (including excavated soil from contaminated sites), stones and dredging spoil (17 05)	
17 05 04	Soil and stones**
17 05 06	Dredging spoil**
Wastes from aerobic treatment of solid wastes (19 05)	
19 05 03	Off-specification compost consisting only of biodegradable waste***
Wastes from waste water treatment plants not otherwise specified (19 08)	
19 08 05	Sludges from treatment of urban waste water
19 08 99	Wastes not otherwise specified (specifically stone filter media)
Wastes from the preparation of water intended for human consumption or water for industrial use (19 09)	
19 09 02	Sludges from water clarification
19 09 99	Wastes not otherwise specified (specifically slow sand filter sand)
Wastes from soil and groundwater remediation (19 13)	
19 13 04	Sludges from soil remediation**

* Codes referred to in the European Waste Catalogue.
** Wastes containing dangerous substances are not included.
*** The wastes listed do not include all of the wastes specified in the European Waste Catalogue under the code referred to.

This does not apply to the use of waste at a site designed or adapted for the final disposal of waste by landfill at any time when such disposal is the subject of a waste management licence or a permit granted under reg 7 of the 2000 Regulations.

SSI 2000/228, Sch 1, para 9(3):

Sub paragraph (1) applies only where–

(a) the waste is used for the purpose of reclamation, restoration or improvement of land, which has been subject to industrial or other man made development;
(b) the waste is suitable for use for the purposes mentioned in sub paragraph (a);
(c) the waste is used in accordance with the requisite planning permission (if any);
(d) the waste is used to a depth not exceeding the lesser of 2 metres or the final cross sections shown on the plan submitted under regulation 18(8) of these Regulations; and
(e) the waste used does not exceed 20,000 cubic metres per hectare.

Registration requirements:

Registration must be made to SEPA. Registration 'expires' after 12 months unless a renewal notice has been received and registered by SEPA. A renewal notice should be submitted to SEPA at least 21 days before the registration is due to 'expire'.

Paragraph 10: Reception and Treatment of Specified Waste at a Water Treatment Works

7.112 The exemption is for the reception and treatment within the curtilage of a water treatment works of any of the wastes listed in the table below. The secure storage within the curtilage of a water treatment works of waste intended to be treated for recovery is reliant on the exemption above.

Codes*	Types of waste
Waste from waste water treatment plants not otherwise specified (19 08)	
19 08 01	Screenings
19 08 05	Sludges from treatment of urban waste water
Wastes from the preparation of water intended for human consumption or water for industrial use (19 09)	
19 09 02	Sludges from water clarification
19 09 03	Sludges from decarbonation
19 09 06	Solutions and sludges from regeneration of ion exchangers
Other municipal wastes (20 03)	
20 03 04	Septic tank sludge
20 03 99	Municipal waste not otherwise specified

* Codes referred to in the European Waste Catalogue.

The total quantity of waste brought to a sewage treatment works for the purposes of a recovery operation under this exemption should not exceed 100,000 cubic metres in any period of 12 months.

The area where a treatment or recovery operation takes place shall have an impermeable pavement capable of containing any spillage of waste received and connected to a drainage system with impermeable components which does not leak and which will ensure that—

- no liquid will run off the pavement otherwise than via that system; and
- except where they may be lawfully discharged, all liquids entering the system are collected in a sealed sump.

Must be registered with SEPA using the appropriate form and enclosing the correct fee. Registration 'expires' after 12 months unless a renewal notice has been received and registered by SEPA. A renewal notice should be submitted to SEPA at least 21 days before the registration is due to 'expire'.

Paragraph 11: Baling, sorting, shredding etc of specified wastes

7.113 This exemption is for the waste listed in column 1 of the table below and for the activities in column 2 of the table below where:

- the activity is carried on with a view to the recovery or re-use of the waste (whether or not by the person carrying on the activity listed in column 2 of that Table); and
- the total quantity of any particular kind of waste dealt with at that place does not in any period of seven days exceed the limit specified in relation to that kind of waste in column 3 of the table below.

Kind of waste	Activities	Maximum total quantity (tonnes per week)
Waste paper or cardboard	Baling, sorting or shredding	3,000
Waste textiles	Baling, sorting or shredding	100
Waste plastic	Baling, sorting, shredding, densifying or washing	100
Waste glass	Sorting, crushing or washing	1,000
Waste steel cans, aluminium cans or aluminium foil	Sorting, crushing, pulverising, shredding, compacting or baling	100
Waste food or drink cartons	Sorting, crushing, pulverising, shredding, compacting or baling	100
Waste wood	Baling, compacting, sorting or shredding	1,000
Waste paint	Sorting, bulking or mixing	100
Waste batteries*	Sorting	10

* includes special waste.

The exemption needs to be registered with SEPA but no fee is required.

The registration of this exemption is not time limited but it can be de-registered by SEPA if it is not managed properly and no longer fulfils the criteria requirements of the exemption or if the activities cause pollution or harm to health. If the activity does comply with the exemption criteria the registration will not have to be renewed except where there is a significant change to the activity.

Paragraph 12: Composting

7.114 This exemption applies to

(1)(a) the composting biodegradable waste of a type listed in the table below, other than for the purpose of cultivating mushrooms;

(b) chipping, shredding, cutting, pulverising, screening or storing such waste for the purpose of composting at the place where the composting activity is, or is to be, carried out; and

(c) screening any such waste, which has been composted.

Codes*	Types of waste
Wastes from agriculture, horticulture, aquaculture, forestry, hunting and fishing (02 01)	
02 01 03	Plant-tissue waste
02 01 06	Animal faeces, urine and manure (including spoiled straw)***
02 01 07	Wastes from forestry
02 01 99	Straw, wood or paper-based bedding waste, slurry or dirty water from stables, zoos, animal parks or livestock markets***
Wastes from fruit, vegetables, cereals, edible oils, cocoa, coffee, tea and tobacco preparation and processing; conserve production; yeast and yeast extract production, molasses preparation and fermentation (02 03)	
02 03 01	Sludges from washing, cleaning, peeling, centrifuging and separation
Wastes from wood processing and the production of panels and furniture (03 01)	
03 01 01	Waste bark and cork
03 01 05	Sawdust, shavings, cuttings, wood**
Wastes from pulp, paper and cardboard production and processing (03 03)	
03 03 01	Waste bark and wood
Wastes from the textile industry (04 02)	
04 02 10	Organic matter from natural products (eg grease, wax)

Codes*	Types of waste
Packaging (including separately collected municipal packaging waste) (15 01)	
15 01 01	Paper and cardboard packaging
15 01 03	Wooden packaging
15 01 09	Textile packaging
Wastes from aerobic treatment of solid wastes (19 05)	
19 05 03	Off-specification compost consisting only of biodegradable waste***
Separately collected fractions of municipal wastes (household waste and similar commercial, industrial and institutional wastes) (20 01)	
20 01 01	Paper and cardboard
20 01 08	Biodegradable kitchen and canteen waste
20 01 10	Clothes
20 01 11	Textiles
Garden and park waste (including cemetery waste) (20 02)	
20 02 01	Biodegradable waste
Other municipal wastes (20 03)	
20 03 02	Botanical waste from markets***

* Codes referred to in the European Waste Catalogue.
** Wastes containing dangerous substances are not included.
*** The wastes listed do not include all of the wastes specified in the European Waste Catalogue under the code referred to.

(2) Composting any of the wastes listed in the first column of the table below for the purpose of cultivating mushrooms at the place where the waste is produced or where the compost is to be used, or at any other place occupied by the person producing the waste or using the compost, where the quantity of waste being composted at any one time does not exceed 2,500 tonnes.

(3) The total quantity of waste treated and stored under sub-paragraph (1) at any time shall not exceed, in relation to the composting activities referred to in the second column of the table below, the quantity shown in the third column of that Table.

Code and type of waste (as referred to in Table 3B)*	**Composting activity**	**Maximum total quantity (including storage and treatment)**
Plant tissue waste 02 01 03	Open air windrow composting without containment** and without any impermeable pavement or sealed drainage	400 Tonnes
Wastes from forestry 02 01 07		
Waste bark and cork 03 01 01		
Biodegradable waste from gardens and park wastes (including cemetery waste 20 02 01)		
All above wastes namely 02 01 03, 02 01 07, 03 01 01, 20 02 01 plus 02 01 06, 15 01 01, 15 01 09, 20 01 10 and 20 01 11	Open air windrow composting without containment** on an impermeable pavement with sealed drainage	400 Tonnes
Any type of waste referred to in Table 3B	Composting with containment** of waste	400 Tonnes
Plant tissue waste 02 01 03	Open air windrow composting by a farm business of wastes arising within that business as defined for the purposes of the Integrated Administration and Control System for farm support without containment** and without any impermeable pavement or sealed drainage	1,000 Tonnes
Wastes from forestry 02 01 07		
Waste bark and cork 03 01 01		
Biodegradable waste from gardens and park wastes (including cemetery waste) 20 02 01		
All above wastes namely 02 01 03, 02 01 07, 03 01 01, 20 02 01 plus 02 01 06, 15 01 01, 15 01 09, 20 01 10 and 20 01 11	Open air composting by a farm business of wastes arising within that business as defined for the purposes of the Integrated Administration and Control System for farm support without containment** on an impermeable pavement with sealed drainage	1,000 Tonnes

Code and type of waste (as referred to in Table 3B)*	Composting activity	Maximum total quantity (including storage and treatment)
Any type of waste referred to in Table 3B	Composting by a farm business of wastes arising within that business as defined for the purposes of the Integrated Administration and Control System for farm support with containment** of waste	1,000 Tonnes

* Codes referred to in the European Waste Catalogue.
** Containment in this context means composting in a vessel, in a sealed building or by some other process equally effective to produce a wholly contained and controlled environment.

The secure storage of biodegradable waste in this exemption which is to be at a place other than the place where the composting is or is to be carried out, where:

- the waste is stored for a period not exceeding 48 hours and is to be taken directly from the storage site to that place;
- such storage is part of a scheme for the collection of waste from groups of premises; and
- the total quantity of waste so stored does not exceed 10 tonnes.

Definitions

'composting' means the autothermic and thermophilic biological decomposition and stabilisation of biodegradable waste under controlled aerobic conditions that result in a stable sanitised material that can be applied to land for the benefit of agriculture, horticulture or ecological improvement; and the reference to 'treated' waste includes waste which is undergoing maturation.

Code and type of waste (as referred to in Table 3B)*	**Composting activity**	**Maximum total quantity (including storage and treatment)**
Plant tissue waste 02 01 03	Open air windrow composting without containment** and without any impermeable pavement or sealed drainage	400 Tonnes
Wastes from forestry 02 01 07		
Waste bark and cork 03 01 01		
Biodegradable waste from gardens and park wastes (including cemetery waste 20 02 01)		
All above wastes namely 02 01 03, 02 01 07, 03 01 01, 20 02 01 plus 02 01 06, 15 01 01, 15 01 09, 20 01 10 and 20 01 11	Open air windrow composting without containment** on an impermeable pavement with sealed drainage	400 Tonnes
Any type of waste referred to in Table 6	Composting with containment** of waste	400 Tonnes
Plant tissue waste 02 01 03	Open air windrow composting by a farm business of wastes arising within that business as defined for the purposes of the Integrated Administration and Control System for farm support without containment** and without any impermeable pavement or sealed drainage	1,000 Tonnes
Wastes from forestry 02 01 07		
Waste bark and cork 03 01 01		
Biodegradable waste from gardens and park wastes (including cemetery waste) 20 02 01		
All above wastes namely 02 01 03, 02 01 07, 03 01 01, 20 02 01 plus 02 01 06, 15 01 01, 15 01 09, 20 01 10 and 20 01 11	Open air composting by a farm business of wastes arising within that business as defined for the purposes of the Integrated Administration and Control System for farm support without containment** on an impermeable pavement with sealed drainage	1,000 Tonnes

Code and type of waste (as referred to in Table 3B)*	Composting activity	Maximum total quantity (including storage and treatment)
Any type of waste referred to in Table 6	Composting by a farm business of wastes arising within that business as defined for the purposes of the Integrated Administration and Control System for farm support with containment** of waste	1,000 Tonnes
Biodegradable kitchen and canteen waste 20 01 08	Composting in compliance with regulation 15 of the Animal By-Products (Scotland) Regulations 2003 Link to Schedule 1 Part 1 requirements detailed by regulation 15 above	400 Tonnes

* Codes referred to in the European Waste Catalogue.
** Containment in this context means composting in a vessel, in a sealed building or by some other process equally effective to produce a wholly contained and controlled environment.

The secure storage of biodegradable waste, which is to be composted at a place other than the place where such composting is or is to be carried out, where:

- the waste is stored for a period not exceeding 48 hours and is to be taken directly from the storage site to that place;
- such storage is part of a scheme for the collection of waste from groups of premises; and
- the total quantity of waste so stored does not exceed 10 tonnes.

Definition

'composting' means the autothermic and thermophilic biological decomposition and stabilisation of biodegradable waste under controlled aerobic conditions that result in a stable sanitised material that can be applied to land for the benefit of agriculture, horticulture or ecological improvement; and

'treated' waste includes waste, which is undergoing maturation.

'The Integrated Administration and Control System for farm support' has the same meaning given by paragraph 7(7).

Registration is required and 'expires' after 12 months unless a renewal notice has been received and registered by SEPA. A renewal notice should be submitted to SEPA at least 21 days before the registration is due to 'expire'.

Paragraph 13: Manufacture of Specified Goods from Specified Wastes

7.115 This exemption covers the manufacture from:

- waste which arises from demolition or construction work or tunneling or other excavations; or
- waste which consists of ash, slag, clinker, rock, wood, bark, paper, straw, crushed glass or gypsum, digestate consisting only of biodegradable waste or compost consisting only of biodegradable waste, of timber products, straw board, plasterboard, bricks, blocks, roadstone, soil, soil substitutes or aggregate.

The treatment of waste soil or rock or digestate which consists only of biodegradable waste which, when treated, is to be spread on land under paragraph 7 or 9, if—

- it is carried out at the place where the waste is produced or the treated product is to be spread; and
- the total amount treated at that place in any day does not exceed 100 tonnes.

Waste should be stored at the place where the activity is to be carried out and the total quantity of waste stored at that place at any time does not exceed:

- in the case of the manufacture of roadstone from road planings, 50,000 tonnes; or
- in any other case, 20,000 tonnes.

Definitions:

'digestate' means a stable, sanitised material converted from the inputs to the process of controlled decomposition of waste under managed conditions where free oxygen is absent, at temperatures suitable for naturally occurring mesophilic or thermophilic anaerobe and facultative anaerobe bacteria species;

'compost' means a stable, sanitised material resulting from the autothermic and thermophilic biological decomposition and stabilisation of biodegradable waste in controlled aerobic conditions.

Paragraph 13(1)(a) and(b) permits a limited range of wastes to be used in the manufacture of certain materials. Demolition, construction, tunneling and excavation wastes or waste ash, slag, clinker, rock, wood, bark, paper, straw and gypsum in the manufacture of timber products, straw board, plasterboard, brick roadstone and aggregate can be used provided the waste used is non-hazardous.

Paragraph 13(2) allows up to 100 tonnes per day of waste soil or rock to be treated in order to produce materials that will be fit for use under either Paragraph 7 or Paragraph 9 land spreading exemptions. The waste must be processed at the place where it originated or at the place where it will be spread. A separate registration is required to spread the wastes under Paragraph 7 or 9.

Paragraph 13(3) allows 20,000 tonnes of wastes which are to be used in the ways described in Paragraph 13 (1), (2), to be stored at the place where they will be used. In the case of road planings, which are to be used in the manufacture of roadstone, this is increased to 50,000 tonnes.

Needs to be registered with SEPA but no fee is required.

The registration of this exemption is not time limited but it can be de-registered by SEPA if it is not managed properly and no longer fulfils the criteria requirements of the exemption or if the activities cause pollution or harm to health. If the activity does comply with the exemption criteria the registration will not have to be renewed except where there is a significant change to the activity.

Paragraph 14: The Manufacture of Finished Goods and Repair/ Refurbishment of Waste Goods

7.116 This exemption covers the manufacture of finished goods from any of the following kinds of waste – waste metal, plastic, glass, ceramics, rubber, textiles, wood, paper or cardboard.

The repair or refurbishment of waste goods is covered if the activity is carried out to the re-use of those goods for their original purpose.

The storage of the goods should be at the place where the manufacturing, repairing or refurbishment activity is to be carried out; and

the total quantity of the waste stored at that place at any time does not exceed 15,000 tonnes.

Needs to be registered with SEPA but no fee is required.

The registration of this exemption is not time limited but it can be de-registered by SEPA if it is not managed properly and no longer fulfils the criteria requirements of the exemption or if the activities cause pollution or harm to health. If the activity does comply with the exemption criteria the registration will not have to be renewed except where there is a significant change to the activity.

Paragraph 15: Beneficial Use of Waste Without Further Treatment

7.117 The beneficial use of waste if:

it is put to that use without further treatment; and

that use of the waste does not involve its disposal.

The storage of waste, which uses this exemption, does so providing the storage does not amount to disposal of the waste.

This exemption does not apply to the use or storage of waste if that activity is covered by the following exemptions:

Paragraph 7: The treatment of land for agricultural benefit or ecological improvement;

Paragraph 8: The storage and spreading of sludge;

Paragraph 9: The reclamation or improvement of land;

Paragraph 19: Waste for construction and other 'relevant work';

Paragraph 25: The deposit of dredging wastes;

or would have been covered except for any condition or limitation that the exemption is subject to.

Needs to be registered with SEPA but no fee is required, but it can be de-registered by SEPA if it is not managed properly and no longer fulfills the criteria requirements of the exemption or if the activities cause pollution or harm to health. If the activity does comply with the exemption criteria the registration will not have to be renewed except where there is a significant change to the activity.

Paragraph 16: The Screening of Specified Wastes

7.118 This exemption applies to the dry screening of any of the wastes listed in the table below if:

- the waste is screened at the place where it is produced; and no more than 100 tonnes of waste is screened at that place in any one day.

Codes*	Types of waste
Concrete, bricks, tiles and ceramics (17 01)	
17 01 01	Concrete
17 01 02	Bricks
17 01 03	Tiles and ceramics
17 01 07	Mixtures of concrete, bricks, tiles and ceramics other than those mentioned in 17 01 06
Soil (including excavated soil from contaminated sites), stones and dredging spoil (17 05)	
17 05 04	Soils and stones other than those mentioned in 17 05 03
17 05 06	Dredging spoil other than those mentioned in 17 05 05
17 05 08	Track ballast other than those mentioned in 17 05 07

* Codes referred to in European Waste Catalogue

This exemption needs to be registered with SEPA. The registration of this exemption is not time limited but it can be de-registered by SEPA if it is not managed properly and no longer fulfils the criteria requirements of the exemption or if the activities cause pollution or harm to health. If the activity does comply with the exemption criteria the registration will not have to be renewed except where there is a significant change to the activity.

Paragraph 17: Storage of specified wastes in a secure place

7.119 This exemption applies to the storage in a secure place on any premises of waste of a kind described in column 1 of the table below if:

the total quantity of that kind of waste stored on those premises at any time does not exceed the quantity specified in column 2 of the table in relation to that kind of waste.

The waste is to be re-used, or used for the purposes of:

- an activity described in paragraph 11 (baling, sorting, shredding etc. of specified wastes); or
- any other recovery operation.

In the case of any waste batteries, storage takes place in accordance with the treatment conditions.

Each kind of waste listed in the Table stored on the premises is kept separately; and

No waste is stored on the premises for longer than 12 months.

Kind of waste	Maximum total quantity
Waste paper or cardboard	15,000 tonnes
Waste textiles	1,000 tonnes
Waste plastics	500 tonnes
Waste glass	5,000 tonnes
Waste steel cans, aluminium cans or aluminium foil	500 tonnes
Waste food or drink cartons	500 tonnes
Waste paint within Codes* 20 01 27** and 20 01 28	25 cubic metres
Waste wood	1,000 tonnes
Waste articles which are to be used for construction work which are capable of being so used in their existing state	100 tonnes
Solvents within Codes* 14 06 02**, 14 06 03** and 20 01 13**	5 cubic metres
Batteries within Codes* 20 01 33** and 20 01 34	5 cubic metres
Refrigerants and halons within Code* 14 06 01** (chlorofluorocarbons, HCFC and HFC)	18 tonnes
Waste mammalian protein	60,000 tonnes
Waste mammalian tallow	45,000 tonnes

* Codes referred to in the European Waste Catalogue.
** Includes special waste.

Definitions

'mammalian protein' means proteinaceous material, and

'mammalian tallow' means fat, which in each case is derived from the whole or part of any dead mammal by a process of crushing, cooking or grinding.

Needs to be registered with SEPA but no fee is required. Where waste is only being stored at the place of production (unlikely), the activity does not have to be registered.

This exemption needs to be registered with SEPA. The registration of this exemption is not time limited but it can be de-registered by SEPA if it is not managed properly and no longer fulfils the criteria requirements of the exemption or if the activities cause pollution or harm to health. If the activity does comply with the exemption criteria the registration will not have to be renewed except where there is a significant change to the activity.

Paragraph 18: Secure Storage of Specified Wastes on Any Premises

7.120 This applies to the secure storage of any waste described in paragraph 17 (Storage of specified wastes on any premises) secure storage on any premises except for waste solvents (within categories 14 06 02, 14 06 03 and 20 01 13 of the European Waste Catalogue), refrigerants or halons and waste oil or waste vegetable oil listed in Table 10 (including oil which is special waste).

Providing that:

- the storage capacity of the container or containers does not exceed 400 cubic metres in total;
- in the case of waste oil or waste vegetable oil listed in the table below, the storage capacity of any container or containers used for its storage does not exceed three cubic metres in total, and provision is made to prevent oil escaping into the ground or a drain;
- there are no more than 20 containers on those premises;
- the waste will be re-used, or used for the purposes of:
    - any activity described in paragraph 11(baling, sorting, shredding etc of specified wastes) carried on at those premises; or
    - any other recovery activity;
- in the case of any waste batteries, storage takes place in accordance with the treatment conditions.
- each kind of waste stored on the premises is separately stored;
- no waste is stored on the premises for longer than 12 months; and
- the person storing the waste is the owner of the container or containers or has the consent of the owner of the container or containers to use them for that purpose.

Codes*	Types of waste
Waste hydraulic oils (13 01)	
13 01 09**	Mineral-based chlorinated hydraulic oils
13 01 10**	Mineral based non-chlorinated hydraulic oils
13 01 11**	Synthetic hydraulic oils
13 01 12**	Readily biodegradable hydraulic oils
13 01 13**	Other hydraulic oils
Waste engine, gear and lubricating oils (13 02)	
13 02 04**	Mineral-based chlorinated engine, gear and lubricating oils
13 02 05**	Mineral-based non-chlorinated engine, gear and lubricating oils
13 02 06**	Synthetic engine, gear and lubricating oils
13 02 07**	Readily biodegradable engine, gear and lubricating oils
13 02 08**	Other engine, gear and lubricating oils
Waste insulating and heat transmission oils (13 03)	
13 03 06**	Mineral-based chlorinated insulating and heat transmission oils other than those mentioned in 13 03 01 (insulating or heat transmission oils containing PCBs)
13 03 07**	Mineral-based non-chlorinated insulating and heat transmission oils
13 03 08**	Synthetic insulating and heat transmission oils
13 03 09**	Readily biodegradable insulating and heat transmission oils
13 03 10**	Other insulating and heat transmission oils
Bilge oils (13 04)	
13 04 01**	Bilge oils from inland navigation
13 04 02**	Bilge oils from jetty sewers
13 04 03**	Bilge oils from other navigation
Waste of liquid fuels (13 07)	
13 07 01**	Fuel oil and diesel
13 07 02**	Petrol
13 07 03**	Other fuels (including mixtures)
Separately collected fractions (except 15 01) (20 01)	
20 01 25	Edible oil and fat

* Codes referred to in the European Waste Catalogue.
** Includes special waste.

This exemption needs to be registered with SEPA but no fee is required. Where waste is only being stored at the place of production (unlikely), the activity does not have to be registered. The registration of this exemption is not time limited but it can be de-registered by SEPA if it is not managed properly and no longer fulfils the criteria requirements of the exemption or if the activities cause pollution or harm to health. If the activity does comply with the exemption criteria the registration will not have to be renewed except where there is a significant change to the activity.

Paragraph 19: Waste for Construction and Other 'Relevant Work'

7.121 This applies to the storage on a site of any of the wastes listed in the table below, if:

- the waste in question is suitable for use for the purposes of relevant work which will be carried on at the site;
- in the case of waste which is not produced on the site, it is not stored there for longer than six months and no more than 50,000 tonnes of the waste is stored on the site at any time; and
- in the case of waste tyre bales, no more than 100 tonnes of the waste is stored on the site at any time.

Codes*	Types of waste
Wastes from physical and chemical processing of non-metalliferous minerals (01 04)	
01 04 08	Waste gravel and crushed rocks**
01 04 09	Waste sand and clays
Wastes from agriculture, horticulture, aquaculture, forestry, hunting and fishing (02 01)	
02 01 99	Clean shells***
Wastes from sugar processing (02 04)	
02 04 01	Soil from cleaning and washing beet
Wastes from power stations and other combustion plants (except wastes from waste management facilities, off-site waste water treatment plants and the preparation of water intended for human consumption and water for industrial use) (10 01)	
10 01 02	Pulverised fuel ash***
10 01 99	Gypsum***
Wastes from the iron and steel industry (10 02)	
10 02 02	Unprocessed slag
Wastes from casting of ferrous pieces (10 09)	
10 09 03	Furnace slag

Codes*	Types of waste
Waste from casting of non ferrous pieces (10 10)	
10 10 03	Furnace slag
Wastes from manufacture of ceramic goods, bricks, tiles and construction products (10 12)	
10 12 08	Waste ceramics, bricks, tiles and construction products (after thermal processing)
Wastes from manufacture of cement, lime and plaster and articles and products made from them (10 13)	
10 13 14	Waste concrete and concrete sludge
End-of-life vehicles from different means of transport (including off-road machinery) and wastes from dismantling of end-of-life vehicles and vehicle maintenance (except 13, 14, 16 06 and 16 08) (16 01)	
16 01 03	Waste tyre bales
Concrete, bricks, tiles and ceramics (17 01)	
17 01 01	Concrete
17 01 02	Bricks
17 01 03	Tiles and ceramics
17 01 07	Mixtures of concrete, bricks, tiles and ceramics**
Bituminous mixtures, coal tar and tarred products (17 03)	
17 03 02	Road base and road planings***
Soil (including excavated soil from contaminated sites), stones and dredging spoil (17 05)	
17 05 04	Soil and stones***
17 05 06	Dredging spoil**
17 05 08	Track ballast**
Wastes from incineration or pyrolysis of waste (19 01)	
19 01 12	Bottom ash and slag**
Wastes from aerobic treatment of solid wastes (19 05)	
19 05 03	Off-specification compost
Wastes from anaerobic treatment of wastes (19 06)	
19 06 04	Digestate consisting only of biodegradable waste***
19 06 06	Digestate from anaerobic treatment of animal and vegetable waste

Codes*	Types of waste
Wastes from the mechanical treatment of waste (for example sorting, crushing, compacting pelletising not otherwise specified) (19 12)	
19 12 09	Minerals (for example sand, stones)
Wastes from soil and groundwater remediation (19 13)	
19 13 02	Solid waste from soil remediation**
Garden and park wastes (including cemetery waste) (20 02)	
20 02 02	Soil and stones

* Codes referred to in the European Waste Catalogue.
** Wastes containing dangerous substances are not included.
*** The wastes listed do not include all of the wastes specified in the European Waste Catalogue under the code referred to.

The waste is used to a depth not exceeding the dimensions of the final cross sections shown on the plan submitted under regulation 25(2) or 26(2) of these Regulations; and

in the case of clean shells (within category 02 01 99 of the European Waste Catalogue), the shells from shellfish processing are treated in accordance with the Animal By-Products (Scotland) Regulations 2003, SSI 2003/411 and Regulation (EC) No 1774/2002 of the European Parliament and of the Council of 3rd October 2002 laying down health rules concerning animal by-products not intended for human consumption.

Definition

'relevant work' means the use of waste:

(a) for the construction, maintenance or improvement of:

- a building, road, railway, airport, dock or other transport facility;
- recreational facilities; or
- drainage; or

(b) for engineering works relating to or adjacent to any part of the water environment, but does not include work involving land reclamation.

'drainage' is a reference to drainage works within the meaning of the Land Drainage (Scotland) Act 1958.

This must be registered with SEPA and the relevant fee paid.

Registration 'expires' after 12 months unless a renewal notice has been received and registered by SEPA. A renewal notice should be submitted to SEPA at least 21 days before the registration is due to 'expire'.

Where an exempt activity relates to the storage, treatment or recovery of waste then operating out-with the parameters specified for that activity may indicate that a waste disposal activity is taking place. In these circumstances, the activity would no longer be exempt and the activity may have to cease or remove some or all of the waste or apply for an appropriate authorisation.

Paragraph 20: Cleaning Waste Textiles for Recovery

7.122 This applies to the laundering or otherwise cleaning of waste textiles with a view to their recovery or re-use.

The storage of waste textiles at the place where they are to be so laundered or cleaned.

This needs to be registered with SEPA but no fee is required.

The registration of this exemption is not time limited but it can be de-registered by SEPA if it is not managed properly and no longer fulfils the criteria requirements of the exemption or if the activities cause pollution or harm to health. If the activity does comply with the exemption criteria the registration will not have to be renewed except where there is a significant change to the activity.

Examples of activities that are not suitable for registration under the Paragraph 20 exemption:

An establishment collects waste textiles at its depot, which are sorted and despatched elsewhere. The sorting activity does not qualify for a Paragraph 20 exemption. It may be permissible under Paragraph 11 (baling, sorting, shredding etc, of specified wastes).

Examples of activities that are suitable for registration under the Paragraph 20 exemption.

A company collects discarded textiles from household waste sites, which are sorted and where they are suitable for re-sale, are cleaned.

The company would need to register a Paragraph 20 exemption at the place where the cleaning and storing is undertaken. The sorting of the textiles should also be registered under the paragraph 11 exemption.

Paragraph 21: Chipping etc Waste Plant Matter

7.123 This applies to the chipping, shredding, cutting or pulverising waste plant matter (including wood or bark), or sorting and baling sawdust or wood shavings, on any premises if:

- those activities are carried on for the purposes of recovery or re-use; and
- no more than 1,000 tonnes of such waste are dealt with on those premises in any period of seven days.

The total quantity of waste stored at those premises at any time does not exceed 1,000 tonnes.

This needs to be registered with SEPA but no fee is required.

The registration of this exemption is not time limited but it can be de-registered by SEPA if it is not managed properly and no longer fulfils the criteria requirements of the exemption or if the activities cause pollution or harm to health. If the activity does comply with the exemption criteria the registration will not have to be renewed except where there is a significant change to the activity.

Paragraph 22: Recovery of Silver

7.124 This applies to the recovery at any premises of silver from waste produced in connection with printing or photographic processing if no more than 50,000 litres of such waste are dealt with on those premises in any day.

This needs to be registered with SEPA but no fee is required.

The registration of this exemption is not time limited but it can be de-registered by SEPA if it is not managed properly and no longer fulfils the criteria requirements of the exemption or if the activities cause pollution or harm to health. If the activity does comply with the exemption criteria the registration will not have to be renewed except where there is a significant change to the activity.

Paragraph 23: The Recovery of Animal By-Product Wastes

7.125 This applies to the recovery of waste consisting of animal by-products at a collection centre in accordance with an authorisation under reg 27 of the Animal By-Products (Scotland) Regulations 2003, SSI 2003/411('the 2003 Regulations') if the total quantity of waste being recovered at that collection centre at any time does not exceed 50 tonnes.

The waste should be securely stored and no waste is to be stored for longer than 12 months.

Definition

'animal by-products' and 'collection centre' have the same meaning as in the Community Regulation as defined in reg 2(1) of the 2003 Regulations.

This exemption does not have to be registered with SEPA. It should be registered with Animal Health (previously known as the State Veterinary Service) an executive agency of the Department of Environment, Food and Rural Affairs (Defra).

Paragraph 24: Size Reduction of Bricks, Tiles or Concrete

7.126 This applies to the crushing, grinding or other size reduction of waste bricks, tiles or concrete, under an authorisation granted under Pt I of the Environmental Protection Act 1990, to the extent that it is or forms part of a process within para (c) of Part B of Section 3.4 (other mineral processes) of Sch 1 to the 1991 Regulations or under a permit under the 2000 Regulations (SSI 2000/323), to the extent that it is or forms part of an activity within paragraph (a) of Part B of Section 3.5 (other mineral activities) of Pt 1 of Sch 1 to the 2000 Regulations.

Where the crushing, grinding or other size reduction is carried on other than at the place where the waste is produced, the exemption only applies if those activities are carried on with a view to recovery or re-use of the waste.

The quantity of such waste to be stored does not at any time exceed 20,000 tonnes.

This needs to be registered but no fee is required.

The registration of this exemption is not time limited but it can be de-registered by SEPA if it is not managed properly and no longer fulfils the criteria requirements of the exemption or if the activities cause pollution or harm to health. If the activity does comply with the exemption criteria the registration will not have to be renewed except where there is a significant change to the activity.

Paragraph 25: The Deposit of Dredging Wastes

7.127 This applies to the deposit of waste arising from dredging inland waters or public SUD[9] systems, or from clearing plant matter from inland waters or public SUD systems, if either:

- the waste is deposited along the bank or towpath of the waters, or on land adjacent to the public SUD system, where the dredging or clearing takes place; or
- the waste is deposited along the bank or towpath of any inland waters, or on land adjacent to a public SUD system, so as to result in benefit to agriculture or ecological improvement.

The total amount of waste deposited along the bank or towpath, or on land adjacent to the public SUD system must not at any time exceed 50 tonnes for each metre of the bank, towpath or land along or on which it is deposited.

This does not apply to waste deposited in a container or lagoon.

Where the waste is deposited along the bank or towpath of the waters, or on land adjacent to the public SUD system, where the dredging or clearing takes place only applies to the deposit of waste by an establishment or undertaking where the waste deposited is the establishment or undertaking's own waste.

The treatment by screening or dewatering of such waste:

On the bank or towpath of the waters, or on land adjacent to the public SUD system, where either the dredging or clearing takes place or the waste is to be deposited, prior to its being deposited in reliance upon the exemption conferred by the foregoing provisions of this paragraph;

9 SUD system – sustainable urban drainage systems.

On the bank or towpath of the waters, or on land adjacent to the public SUD system, where the dredging or clearing takes place, or at a place where the waste is to be spread, prior to its being spread in reliance upon the exemption conferred by paragraph 7(1)[10] or (2)[11]; or

In the case of waste from dredging, on the bank or towpath of the waters, or on land adjacent to the public SUD system, where the dredging takes place, or at a place where the waste is to be spread, prior to its being spread in reliance upon the exemption conferred by paragraph 9(1)[12].

Definition

'Public SUD system' has the same meaning as in s 59(1) of the Sewage (Scotland) Act 1968.

This needs to be registered with SEPA but no fee is required. The registration of this exemption is not time limited but it can be de-registered by SEPA if it is not managed properly and no longer fulfils the criteria requirements of the exemption or if the activities cause pollution or harm to health. If the activity does comply with the exemption criteria the registration will not have to be renewed except where there is a significant change to the activity.

Paragraph 26: The Recovery/Disposal of Waste at the Place of Production as Part of the Production Process

7.128 This applies to the recovery or disposal of waste, at the place where it is produced, as an integral part of the process that produces it.

The storage, at the place where it is produced, of waste which is intended to be so recovered or disposed of but does not apply to the final disposal of waste by deposit in or on land.

This needs to be registered with SEPA but no fee is required. The registration of this exemption is not time limited but it can be de-registered by SEPA if it is not managed properly and no longer fulfils the criteria requirements of the exemption or if the activities cause pollution or harm to health. If the activity

10 Paragraph 7(1) The treatment with any of the wastes listed in Table 2 of land used for agriculture where such treatment results in benefit to agriculture or ecological improvement.

11 Paragraph 7(2) The treatment with any of the wastes listed in Part I of Table 2 of—
- (a) operational land of a railway, light railway, Scottish Water or British Waterways; or
- (b) land which is a forest, woodland, park, garden, verge, landscaped area, sports ground, recreation ground, churchyard or cemetery; where the land in question is not used for agriculture and such treatment results in ecological improvement.

12 9. —(1) Subject to the following provisions of this paragraph–
- (a) the treatment of land (including the restoration of quarries) with any of the wastes listed in Part I of Table 3;
- (b) the treatment of land (including the restoration of quarries) with any of the wastes listed in Part II of that Table where such treatment results in benefit to agriculture or ecological improvement;
- (c) the secure storage, at the place where it is to be used and for a period not exceeding 6 months, of waste intended to be used in reliance upon the exemption conferred by paragraph (a) or (b).

does comply with the exemption criteria the registration will not have to be renewed except where there is a significant change to the activity.

Paragraph 27: Baling, Compacting, Crushing or Shredding Waste at the Place it was Produced

7.129 This applies to the baling, compacting, crushing, shredding or pulverising of waste at the place where it is produced.

The storage, at the place where it is produced, of waste which is to be submitted to any of those operations.

As the activity does not involve the recovery or disposal of waste, this exemption does not have to be registered with SEPA. The registration of this exemption is not time limited but it can be de-registered by SEPA if it is not managed properly and no longer fulfils the criteria requirements of the exemption or if the activities cause pollution or harm to health. If the activity does comply with the exemption criteria the registration will not have to be renewed except where there is a significant change to the activity.

Paragraph 28: The Use of Autoclaves to Sterilise Waste

7.130 This applies to the use of autoclaves to sterilise waste, including special waste, as listed in the table below, at the place where the waste is produced, if:

- the autoclaves have a capacity of no more than three cubic metres;
- the autoclaves are accredited by the Medicine and Healthcare Products Regulatory Agency;
- no more than 100 tonnes of waste is so sterilised at any one place in any one calendar month; and
- where the waste is special waste, it is so sterilised as part of, or as a preliminary to, a recovery operation.

Codes*	Types of waste
Wastes from natal care, diagnosis, treatment or prevention of disease in humans (18 01)	
18 01 01	Sharps (except 18 01 03)
18 01 02	Body parts and organs including blood bags and blood preserves (except 18 01 03)
18 01 03**	Wastes whose collection and disposal is subject to special requirements in order to prevent infection
18 01 04	Wastes whose collection and disposal is not subject to special requirements in order to prevent infection (for example, dressings, plaster casts, linen, disposable clothing, diapers)
18 01 06**	Chemicals consisting of or containing dangerous substances
18 01 07	Chemicals other than those mentioned in 18 01 06

Codes*	Types of waste
18 01 08**	Cytotoxic and cytostatic medicines
18 01 09	Medicines other than those mentioned in 18 01 08
18 01 10**	Amalgam waste from dental care

* Codes referred to in the European Waste Catalogue.
** Includes special waste.

This needs to be registered with SEPA but no fee is required. Where waste is being stored at the place of production, the activity does not have to be registered. The registration of this exemption is not time limited but it can be de-registered by SEPA if it is not managed properly and no longer fulfils the criteria requirements of the exemption or if the activities cause pollution or harm to health. If the activity does comply with the exemption criteria the registration will not have to be renewed except where there is a significant change to the activity.

Paragraph 29: The Disposal of Waste at the Place of Production by Burning in an Incinerator

7.131 This applies to the disposal of waste at the place where it is produced, by the person producing it, by burning it in an incinerator:

- which is an exempt incinerator for the purposes of Section 5.1 (incineration) of Sch 1 to the 1991 Regulations (SSI 1991/472, or which is an excluded plant for the purposes of Section 5.1 of Part 1 of Schedule 1 to the 2000 Regulations (SSI 2000/323) with a capacity of less than 50 kg per hour; or
- which is not an incineration plant, a co-incineration plant or an excluded plant for the purposes of Section 5.1 of Part 1 of Schedule 1 to the 2000 Regulations.

The secure storage at that place of any such waste intended to be submitted to such burning.

This needs to be registered with SEPA but no fee is required.

SEPA's current position on the use of drum incinerators is as follows:

1. Drum incinerators and waste exemptions can still be used providing they do not cause pollution or a public nuisance.
2. SEPA is obliged to register the use of drum incinerators under para 29 of Sch 1 of the Regulations.
3. However, where agricultural waste is burned in a drum incinerator in a manner likely to cause pollution of the environment or harm to human health an offence may be committed under s 33(1)(c) of the Environmental Protection Act 1990 in contravention of the relevant objectives. SEPA will investigate such complaints.

4. There is also potential for legal action by local authorities under the Clean Air Act 1993 if dark or black smoke is produced and under Pt III of the Environmental Protection Act 1990 if there is statutory nuisance.
5. The Prevention of Environmental Pollution from Agricultural Activities Code of Good Practice (the PEPFAA Code) advises against the burning of plastics, rubber, tyres and other wastes which may cause pollution or nuisance.
6. There is a risk of harm to human health from noxious fumes if burning substances such as certain plastics.
7. The disposal of ash from a drum incinerator must be to a suitably licensed facility. As farm landfills require Pollution Prevention and Control (Scotland) Regulations 2012, SSI 2012/360 landfill permits, on-site disposal is unlikely to be an option and off-site disposal will be required.
8. Farm Quality Assurance schemes often discourage burning on farms.

The Waste (Scotland) Regulations 2012, SSI 2012/148 require that all organisations, businesses etc have a duty to take all reasonable steps to recycle the five key waste streams (cardboard, paper, glass, plastic and metal). It is recognised that limited recycling services may be available in some areas, however where recycling services do exist the key waste streams should be recycled.

For these reasons, SEPA advises against the use of drum incinerators. Farmers should be advised to look into off-farm recovery and disposal routes (eg, through local authority commercial collections).

The registration of this exemption is not time limited but it can be de-registered by SEPA if it is not managed properly and no longer fulfils the criteria requirements of the exemption or if the activities cause pollution or harm to health. If the activity does comply with the exemption criteria the registration will not have to be renewed except where there is a significant change to the activity.

Paragraph 30: Burning Plant Tissue Waste on Land in the Open

7.132 This applies to burning waste on land in the open if:

- the waste consists of plant tissue;
- it is agricultural waste or it is produced on land which is operational land of a railway, light railway, tramway, British Waterways, or which is a forest, woodland, park, garden, verge, landscaped area, sports ground, recreation ground, bank of an inland waterway, churchyard or cemetery, or it is produced on other land as a result of demolition work;
- it is burned on the land where it is produced; and
- the total quantity burned in any period of 24 hours does not exceed 10 tonnes.

It only applies to the burning of waste by an establishment or undertaking where the waste burned is the establishment or undertaking's own waste.

The burning of waste consisting of plant tissue waste, in pursuance of a notice given under art 32 of the Plant Health (Scotland) Order 2005, SSI 2005/613.

The total quantity of waste burned in any period of 24 hours shall not exceed 10 tonnes.

The storage of waste before burning on the land where it is to be burned in reliance upon the exemption conferred by sub paragraph (1) or (3).

The treatment of land for the benefit of agriculture or ecological improvement by incorporation into soil of ash from waste which is burned on that land in reliance upon the exemption conferred by sub paragraph (1) or (3).

This needs to be registered with SEPA but no fee is required. The registration of this exemption is not time limited but it can be de-registered by SEPA if it is not managed properly and no longer fulfils the criteria requirements of the exemption or if the activities cause pollution or harm to health. If the activity does comply with the exemption criteria the registration will not have to be renewed except where there is a significant change to the activity.

Up to 10 tonnes per day of plant tissue waste can be burned in the open provided the wastes are the ones you have generated.

Plant tissue waste is intended as a very broad description of all materials of plant origin such as wood and vegetation and may include engineered materials of this type provided they are untreated.

The exemption allows the burning of untreated wood, bark and plant matter provided it does not cause pollution or a nuisance. The exemption does not allow any other waste type to be burnt in the open. Paragraph 30(3) NOTE: If the activity is carried out in a way likely to cause a nuisance or the activity involves burning in a designated Air Quality Management Area, this exemption will not cover the activity. If the Local Authority inform SEPA that a nuisance has been caused by the activity or that the exemption is in an Air Quality Management Area, SEPA can remove the activity from the register.

Paragraph 31: Discharge of Sanitary Waste from a Passenger Train

7.133 This applies to the discharge of waste onto the track of a railway from a sanitary convenience or sink forming part of a vehicle used for the carriage of passengers on the railway.

The discharge should not exceed 25 litres.

This needs to be registered with SEPA but no fee is required.

The registration of this exemption is not time limited but it can be de-registered by SEPA if it is not managed properly and no longer fulfils the criteria requirements of the exemption or if the activities cause pollution or harm to health. If the activity does comply with the exemption criteria the registration will not have to be renewed except where there is a significant change to the activity.

Paragraph 32: Burial of Sanitary Waste from a Removable Receptacle

7.134 This applies to the burial on premises of waste arising from the use on those premises of a sanitary convenience, which is equipped with a removable receptacle.

The total amount buried in any period of 12 months does not exceed five cubic metres.

This needs to be registered with SEPA but no fee is required. The registration of this exemption is not time limited but it can be de-registered by SEPA if it is not managed properly and no longer fulfils the criteria requirements of the exemption or if the activities cause pollution or harm to health. If the activity does comply with the exemption criteria the registration will not have to be renewed except where there is a significant change to the activity.

Paragraph 33: Keeping or Deposit of Peat Excavation Waste

7.135 This applies to the keeping or deposit of waste consisting of excavated materials arising from peat working at the place where that activity takes place.

The exemption only applies to the keeping or deposit of waste by an establishment or undertaking where the waste kept or deposited is the establishment or undertaking's own waste.

This needs to be registered with SEPA but no fee is required. The registration of this exemption is not time limited but it can be de-registered by SEPA if it is not managed properly and no longer fulfils the criteria requirements of the exemption or if the activities cause pollution or harm to health. If the activity does comply with the exemption criteria the registration will not have to be renewed except where there is a significant change to the activity

Paragraph 34: Keeping or Deposit of Spent Railway Ballast

7.136 This applies to the keeping or deposit of spent railway ballast on land at the place where it is produced if the land is the operational land of a railway, light railway or tramway.

The total amount kept or deposited at that place does not exceed 10 tonnes for each metre of track from which the ballast derives.

It only applies to the keeping or deposit of waste by an establishment or undertaking where the waste kept or deposited is the establishment or undertaking's own waste.

It needs to be registered with SEPA but no fee is required. The registration of this exemption is not time limited but it can be de-registered by SEPA if it is not managed properly and no longer fulfils the criteria requirements of the exemption or if the activities cause pollution or harm to health. If the activity does comply with the exemption criteria the registration will not have to be renewed except where there is a significant change to the activity.

Paragraph 35: The Deposit of Excavated Material from a Borehole or Other Excavation

7.137 This applies to the deposit of waste consisting of excavated material from a borehole or other excavation made for the purpose of mineral exploration if:

- it is deposited in or on land at the place where it is excavated; and
- the total quantity of waste so deposited during any period of 24 months does not exceed 45,000 cubic metres per hectare.

This only applies if:

- the drilling of the borehole or the making of any other excavation is development which is permitted by Class 53 or 54 of Sch 1 to the Town and Country Planning (General Permitted Development) (Scotland) Order 1992, SI 1992/223; and
- the conditions subject to which the development is permitted are observed.

The expressions used in this paragraph, which are also used in the Town and Country Planning (General Permitted Development) (Scotland) Order 1992 shall have the same meaning as in that Order.

This needs to be registered with SEPA but no fee is required. The registration of this exemption is not time limited but it can be de-registered by SEPA if it is not managed properly and no longer fulfils the criteria requirements of the exemption or if the activities cause pollution or harm to health. If the activity does comply with the exemption criteria the registration will not have to be renewed except where there is a significant change to the activity.

Paragraph 36: Temporary Storage of Specified Waste at a Harbour

7.138 This applies to the temporary secure storage of specified waste within a harbour area in accordance with the Merchant Shipping and Fishing Vessels (Port Waste Reception Facilities) Regulations 2003, SI 2003/1809 ('the 2003 Regulations'), where such storage is incidental to the collection or transport of the waste.

This applies to waste other than tank washings (including such special waste as is listed in the table below), only if—

- the amount of waste so stored within a harbour area at any time does not exceed 20 cubic metres for each ship from which waste has been landed; and
- no waste is so stored for more than seven days.

This applies to waste, which consists of tank washings (including such special waste as is listed in the second table below), but only if—

- the amount of tank washings consisting of dirty ballast so stored within a harbour area at any time does not exceed 30% of the total deadweight of the ships from which such washings have been landed; and

- the amount of tank washings consisting of waste mixtures containing oil so stored within a harbour area at any time does not exceed 1% of the total deadweight of the ships from which such washings have been landed.

Definition

'harbour area' has the same meaning as in the Dangerous Substances in Harbour Areas Regulations 1987;

'ship' has the same meaning as in the 2003 Regulations;

'tank washings' means waste residues from the tanks (other than the fuel tanks) or holds of a ship or waste arising from the cleaning of such tanks or holds.

Codes*	Types of waste
Absorbents, filter materials, wiping cloths and protective clothing (15 02)	
15 02 02**	Absorbents, filter materials (including oil filters not otherwise specified), wiping cloths, protective clothing contaminated by dangerous substances
Wastes from electrical and electronic equipment (16 02)	
16 02 09**	Transformers and capacitors containing PCBs
16 02 10**	Discarded equipment containing or contaminated by PCBs other than those mentioned in 16 02 09
16 02 11**	Discarded equipment containing chlorofluorocarbons, HCFC, HFC
16 02 12**	Discarded equipment containing free asbestos
16 02 13**	Discarded equipment containing hazardous components other than those mentioned in 16 02 09 to 16 02 12
16 02 15**	Hazardous components removed from discarded equipment
Waste explosives (16 04)	
16 04 03**	Other waste explosives
Gases in pressure containers and discarded chemicals (16 05)	
16 05 04**	Gases in pressure containers (including halons) containing dangerous substances)
Batteries and accumulators (16 06)	
16 06 01**	Lead batteries
16 06 02**	Ni-Cd batteries
16 06 03**	Mercury-containing batteries
16 06 06**	Separately collected electrolyte from batteries and accumulators

Codes*	Types of waste
Wastes from natal care, diagnosis, treatment or prevention of disease in humans (18 01)	
18 01 03**	Wastes whose collection and disposal is subject to special requirements in order to prevent infection
18 01 06**	Chemicals consisting of or containing dangerous substances
18 01 08**	Cytotoxic and cytostatic medicines
Separately collected fractions (except 15 01) (20 01)	
20 01 13**	Solvents
20 01 21**	Fluorescent tubes and other mercury containing waste
20 01 23**	Discarded equipment containing chlorofluorocarbons
20 01 26**	Oil and fat other than those mentioned in 20 01 25
20 01 27**	Paint, inks, adhesives and resins containing dangerous substances
20 01 29**	Detergents containing dangerous substances
20 01 31**	Cytotoxic and cytostatic medicines
20 01 33**	Batteries and accumulators included in 16 06 01, 16 06 02 or 16 06 03 and unsorted batteries and accumulators containing these batteries
20 01 35**	Discarded electrical and electronic equipment other than those mentioned in 20 01 12 and 20 01 23 containing hazardous components
20 01 37**	Wood containing dangerous substances

* Codes referred to in the European Waste Catalogue.
** Includes special waste.

Codes*	Types of waste
Waste hydraulic oils (13 01)	
13 01 09**	Mineral-based chlorinated hydraulic oils
13 01 10**	Mineral-based non-chlorinated hydraulic oils
13 01 11**	Synthetic hydraulic oils
13 01 12**	Readily biodegradable hydraulic oils
13 01 13**	Other hydraulic oils
Waste engine, gear and lubricating oils (13 02)	
13 02 04**	Mineral-based chlorinated engine, gear and lubricating oils
13 02 05**	Mineral-based non-chlorinated engine, gear and lubricating oils

Codes*	Types of waste
13 02 06**	Synthetic engine, gear and lubricating oils
13 02 07**	Readily biodegradable engine, gear and lubricating oils
13 02 08**	Other engine, gear and lubricating oils
Waste insulating and heat transmission oils (13 03)	
13 03 06**	Mineral-based chlorinated insulating and heat transmission oils other than those mentioned in 13 03 01 (insulating or heat transmission oils containing PCBs)
13 03 07**	Mineral-based non-chlorinated insulating and heat transmission oils
13 03 08**	Synthetic insulating and heat transmission oils
13 03 09**	Readily biodegradable insulating and heat transmission oils
13 03 10**	Other insulating and heat transmission oils
Bilge oils (13 04)	
13 04 01**	Bilge oils from inland navigation
13 04 02**	Bilge oils from jetty sewers
13 04 03**	Bilge oils from other navigation
Waste of liquid fuels (13 07)	
13 07 01**	Fuel oil and diesel
13 07 02**	Petrol
13 07 03**	Other fuels (including mixtures)
Oil wastes not otherwise specified (13 08)	
13 08 01**	Desalter sludges or emulsions
13 08 02**	Other emulsions
13 08 99**	Wastes not otherwise specified
Wastes from transport tank, storage tank and barrel cleaning (except 05 and 13) (16 07)	
16 07 08**	Wastes containing oil
16 07 09**	Wastes containing other dangerous substances
Aqueous liquid wastes destined for off-site treatment (16 10)	
16 10 01**	Aqueous liquid wastes containing dangerous substances
16 10 03**	Aqueous concentrates containing dangerous substances

* Codes referred to in the European Waste Catalogue.
** Includes special waste.

This needs to be registered with SEPA but no fee is required. Where waste is being stored at the place of production, the activity does not have to be registered. The registration of this exemption is not time limited but it can be de-registered by SEPA if it is not managed properly and no longer fulfils the criteria requirements of the exemption or if the activities cause pollution or harm to health. If the activity does comply with the exemption criteria the registration will not have to be renewed except where there is a significant change to the activity.

Paragraph 37: Burial of a Dead Domestic Pet

7.139 This applies to the burial of a dead domestic pet in the garden of a domestic property where the pet lived.

This does not apply if:

- the dead domestic pet may prove hazardous to anyone who may come into contact with it; or
- the burial is carried out by an establishment or undertaking and the pet did not die at the property.

As this activity is more than likely to be carried out by an individual rather than an establishment or undertaking, this exemption does not have to be registered with SEPA. The registration of this exemption is not time limited but it can be de-registered by SEPA if it is not managed properly and no longer fulfils the criteria requirements of the exemption or if the activities cause pollution or harm to health. If the activity does comply with the exemption criteria the registration will not have to be renewed except where there is a significant change to the activity.

Paragraph 38: The Deposit or Storage of Samples of Waste for Testing or Analysis

7.140 This applies to the deposit or storage of samples of waste, including samples of waste which are special waste (including the temporary storage of WEEE pending its recovery), which are being or are to be subjected to testing and analysis, at any place where they are being or are to be tested or analysed, if the samples do not exceed 10 tonnes and are taken:

- in the exercise of any power under the Radioactive Substances Act 1993, the Sewerage (Scotland) Act 1968, the Water Environment and Water Services (Scotland) Act 2003, the Environmental Protection Act 1990, the Environment Act 1995 or the Water Environment (Controlled Activities) (Scotland) Regulations 2005, SSI 2005/348.
- by or on behalf of the holder of a waste management licence in pursuance of the conditions of that licence;
- by or on behalf of a person carrying on in relation to the waste an activity described in this Schedule or in reg 16(1);
- by or on behalf of the owner or occupier of the land from which the samples are taken;

- by or on behalf of any person to whom s 34 applies in connection with that person's duties under that section; or
- for the purposes of research.

This needs to be registered with SEPA but no fee is required. Where waste is being stored at the place of production, the activity does not have to be registered. The registration of this exemption is not time limited but it can be de-registered by SEPA if it is not managed properly and no longer fulfils the criteria requirements of the exemption or if the activities cause pollution or harm to health. If the activity does comply with the exemption criteria the registration will not have to be renewed except where there is a significant change to the activity.

Paragraph 39: Secure Storage of Medical, Nursing or Veterinary Wastes

7.141 This applies to the secure storage at a pharmacy or at the premises of a medical, nursing or veterinary practice or a needle exchange facility, of the wastes listed in the table below (including those which are special waste) which have been returned to those places from care services, households or by individuals if:

- the total quantity of such returned waste at the pharmacy or premises does not at any time exceed 10 cubic metres;
- the total amount of such waste returned in any 24-hour period does not exceed five kilograms or, as the case may be, five litres; and
- any waste so returned to the pharmacy or premises is not stored there for longer than three months.

The secure storage at the premises of a medical, nursing or veterinary practice of the wastes listed in the table below (including special waste) produced in carrying on that practice if:

- in the case of liquid waste, it is stored in a secure container and the total volume of that waste does not at any time exceed 23,000 litres;
- in any other case, either:
  - it is stored in a secure container and the total volume of that waste does not at any time exceed 80 cubic metres; or
  - it is stored in a secure place and the total volume of that waste does not at any time exceed 50 cubic metres; and
- no such waste is stored at those premises for longer than one year.

In the case of any waste batteries, secure storage must take place in accordance with the treatment conditions.

Definition

'care services' has the same meaning as in s 2 of the Regulation of Care (Scotland) Act 2001.

Codes*	Types of waste
Wastes from electrical and electronic equipment (16 02)	
16 02 09**	Transformers and capacitors containing PCBs
16 02 10**	Discarded equipment containing or contaminated by PCBs other than those mentioned in 16 02 09
16 02 11**	Discarded equipment containing chlorofluorocarbons, HCFC, HFC
16 02 12**	Discarded equipment containing free asbestos
16 02 13**	Discarded equipment containing hazardous components other than those mentioned in 16 02 09 to 16 02 12
16 02 14	Discarded equipment other than those mentioned in 16 02 09 to 16 02 13
16 02 15**	Hazardous components removed from discarded equipment
16 02 16	Components removed from discarded equipment other than those mentioned in 16 02 15
Batteries and accumulators (16 06)	
16 06 01**	Lead batteries
16 06 02**	Ni-Cd batteries
16 06 03**	Mercury-containing batteries
16 06 04	Alkaline batteries (except 16 06 03)
16 06 05	Other batteries and accumulators
Wastes from natal care, diagnosis, treatment or prevention of disease in humans (18 01)	
18 01 01	Sharps (except 18 01 03 – waste whose collection and disposal is subject to special requirements in order to prevent infection)
18 01 02	Body parts and organs including blood bags and blood preserves (except 18 01 03)
18 01 03**	Wastes whose collection and disposal is subject to special requirements in order to prevent infection
18 01 04	Wastes whose collection and disposal is not subject to special requirements in order to prevent infection (for example dressings, plaster casts, linen, disposable clothing, diapers)
18 01 07	Chemicals other than those mentioned in 18 01 06 (chemicals consisting of or containing dangerous substances)
18 01 08**	Cytotoxic and cytostatic medicines
18 01 09	Medicines other than those mentioned in 18 01 08

Codes*	Types of waste
Wastes from research, diagnosis, treatment or prevention of disease involving animals (18 02)	
18 02 01	Sharps (except 18 02 02 – wastes whose collection and disposal is subject to special requirements in order to prevent infection)
18 02 02**	Wastes whose collection and disposal is subject to special requirements in order to prevent infection
18 02 03	Wastes whose collection and disposal is not subject to special requirements in order to prevent infection
18 02 06	Chemicals other than those mentioned in 18 02 05 (chemicals consisting of or containing dangerous substances)
18 02 07**	Cytotoxic and cytostatic medicines
18 02 08	Medicines other than those mentioned in 18 02 07
Separately collected fractions (20 01)	
20 01 31**	Cytotoxic and cytostatic medicines
20 01 32	Medicines other than those mentioned in 20 01 31 (cytotoxic and cytostatic medicines)
20 01 33**	Batteries and accumulators included in 16 06 01, 16 06 02 or 16 06 03 and unsorted batteries and accumulators containing these batteries
20 01 34	Batteries and accumulators other than those mentioned in 20 01 33
20 01 35**	Discarded electrical and electronic equipment other than those mentioned in 20 01 21 and 20 01 23 containing hazardous components
20 01 36	Discarded electrical and electronic equipment other than those mentioned in 20 01 21, 20 01 23 and 20 01 35
20 01 99	Other fractions not otherwise specified

* Codes referred to in the European Waste Catalogue.
** Includes special waste.

This needs to be registered with SEPA but no fee is required. Where waste is being stored at the place of production, the activity does not have to be registered. The registration of this exemption is not time limited but it can be de-registered by SEPA if it is not managed properly and no longer fulfils the criteria requirements of the exemption or if the activities cause pollution or harm to health. If the activity does comply with the exemption criteria the registration will not have to be renewed except where there is a significant change to the activity.

Paragraph 40: Secure Storage of Non-Liquid Waste Other Than at the Place of Production

7.142 This applies to the secure storage of non-liquid waste, including WEEE, at any place other than the premises where it is produced if:

- the amount of waste stored does not at any time exceed 50 cubic metres in total and is not kept for longer than three months;
- the person storing the waste is the owner of the place where it is stored or has the consent of the owner;
- the place where it is stored is not a site designed or adapted for the reception of waste with a view to its being disposed of or recovered elsewhere; and
- such storage is incidental to the collection or transport of the waste.

It does not apply to the storage of waste at a place designed or adapted for the recovery of scrap metal or the dismantling of waste motor vehicles.

The temporary storage of scrap rails on operational land of a railway, light railway or tramway if the total quantity of that waste in any one place does not at any time exceed 10 tonnes and the storage is incidental to the collection or transport of the scrap rails.

As the activity does not involve the recovery or disposal of waste, this exemption does not have to be registered with SEPA. The registration of this exemption is not time limited but it can be de-registered by SEPA if it is not managed properly and no longer fulfils the criteria requirements of the exemption or if the activities cause pollution or harm to health. If the activity does comply with the exemption criteria the registration will not have to be renewed except where there is a significant change to the activity.

Paragraph 41: Temporary Storage of Waste at the Place of Production

7.143 This applies to the temporary storage of waste, including WEEE, pending its collection, on the site where it is produced.

This does not apply to the storage of waste at a place designed or adapted for the recovery of scrap metal or the dismantling of waste motor vehicles.

It will only apply if such temporary storage conforms with the obligations and the minimum technical requirements described, respectively, in para 1 to each of Pts 1 and 2 of the Schedule to the End-of-Life Vehicles (Storage and Treatment) (Scotland) Regulations 2003, SSI 2003/593.

This will apply to special waste if:

- it is stored on the site for no longer than 12 months;
- in the case of liquid waste, it is stored in a secure container and the total volume of that waste does not at any time exceed 23,000 litres; and
- in any other case, either:
    - it is stored in a secure container and the total volume of that waste does not at any time exceed 80 cubic metres; or

- it is stored in a secure place and the total volume of that waste does not at any time exceed 50 cubic metres.

As the activity does not involve the recovery or disposal of waste, this exemption does not have to be registered with SEPA. The registration of this exemption is not time limited but it can be de-registered by SEPA if it is not managed properly and no longer fulfils the criteria requirements of the exemption or if the activities cause pollution or harm to health. If the activity does comply with the exemption criteria the registration will not have to be renewed except where there is a significant change to the activity.

Paragraph 42: The Disposal of Waste Pesticide Solutions or Washings in a Lined Biobed

7.144 This applies to the disposal of waste consisting of pesticide solution or washings in a lined biobed at the place where the waste is produced, provided that no more than 8000 litres of such waste is so disposed of in any 12-month period.

Secure storage of the waste needs to be at the place where it is produced, prior to its disposal in accordance provided that no more than 1500 litres of such waste is stored at any one time.

Applies to the recovery or re-use of liquid residue from a lined biobed at the place where the liquid residue is produced.

The treatment of land by spreading with biobed material from a lined biobed where:

- the biobed material has been securely stored for at least 12 months prior to being spread on the land; and
- that treatment results in benefit to agriculture or environmental improvement.

The secure storage, at the place where it is produced, of biobed material intended to be used:

- no more than 10 tonnes of biobed material is stored at any one time; and
- no waste is stored for longer than three years.

This must be notified to SEPA and the appropriate fee paid. Registration 'expires' after 12 months unless a renewal notice has been received and registered by SEPA. A renewal notice should be submitted to SEPA at least 21 days before the registration is due to 'expire'.

Paragraph 43: Crushing of Waste Gas Discharge Lamps

7.145 This applies to the crushing of waste gas discharge lamps (including fluorescent tubes which are special waste within category 20 01 21 of the European Waste Catalogue) for the purposes of volume reduction prior to collection, where the material is intended for recovery or re-use.

The storage of such lamps:

- prior to crushing; and
- after crushing but prior to collection.

It applies only where:

- the crushing operation is carried out in equipment designed for the purpose of volume reduction prior to collection;
- the crushing operation is carried out solely for that purpose;
- the mercury concentration in emissions does not exceed 50 micrograms/m3;
- the total quantity of lamps processed in any period of 24 hours does not exceed three tonnes;
- the storage of the lamps prior to crushing is under weather-proof covering in compliance with paragraph 1 of Annex III to the WEEE Directive; and
- the storage of the lamps after crushing is in a secure container.

This needs to be registered with SEPA but no fee is required. The registration of this exemption is not time limited but it can be de-registered by SEPA if it is not managed properly and no longer fulfils the criteria requirements of the exemption or if the activities cause pollution or harm to health. If the activity does comply with the exemption criteria the registration will not have to be renewed except where there is a significant change to the activity.

Paragraph 44: Heating Metal in an Authorised Furnace

7.146 This applies to the heating of iron, steel or any ferrous-alloy, non-ferrous metal or non-ferrous metal alloy, in one or more furnaces or other appliances the primary combustion chambers of which have in aggregate a net rated thermal input of less than 0.2 megawatts, for the purpose of removing grease, oil or any other non-metallic contaminant.

This does not apply to the removal by heat of plastic or rubber covering from scrap cable or of any asbestos contaminant.

In the case of a process involving the heating of iron, steel or any ferrous-alloy, it does not apply if that process is related to a process described in any of paragraphs (a)–(h), or (j)–(m), of Part A or paragraphs (a)–(c), or (e)–(f), of Part B of Section 2.1 of Sch 1 to the 1991 Regulations (SSI 1991/472) or an activity described in Section 2.1 (other than in paragraph (d) of Part B) of Part 1 of Sch 1 to the 2000 Regulations (SSI 2000/323).

In the case of a process involving the heating of any non-ferrous metal or non ferrous metal alloy, sub paragraph (1) does not apply if that process is related to a process described in any of paragraphs (a)–(g), or (i)–(k), of Part A of Section 2.2 of Sch 1 to the 1991 Regulations or an activity described in Part A of Section 2.2 of Part 1 of Sch 1 to the 2000 Regulations.

The secure storage of waste intended to be submitted to heating to which sub-paragraph (1) applies if the waste or, as the case may be, any container in which the waste is stored, is stored on an impermeable pavement which is provided with a sealed drainage system.

Definition

'net rated thermal input' means the rate at which fuel can be burned at the maximum continuous rating of the appliance multiplied by the net calorific value of the fuel and expressed as megawatts thermal.

'ferrous alloy' means an alloy of which iron is the largest constituent, or equal to the largest constituent, by weight, whether or not that alloy also has a non-ferrous metal content greater than any percentage specified in Section 2.2 of Sch 1 to the 1991 Regulations and 'non-ferrous metal alloy' shall be construed accordingly.

It needs to be registered with SEPA but no fee is required. The registration of this exemption is not time limited but it can be de-registered by SEPA if it is not managed properly and no longer fulfils the criteria requirements of the exemption or if the activities cause pollution or harm to health. If the activity does comply with the exemption criteria the registration will not have to be renewed except where there is a significant change to the activity.

Paragraph 45: The Recovery of Scrap Metal or the Dismantling of Motor Vehicles

7.147 (1) This applies to the carrying on, at any secure place designed or adapted for the recovery of scrap metal or the dismantling of depolluted motor vehicles, in respect of a kind of waste described in column 1 of the table below, of any of the activities specified in column 2 of the table below in relation to that kind of waste if:

- the total quantity of any particular kind of waste so dealt with at that place does not in any period of seven days exceed the limit specified in relation to that kind of waste in column 3 of the table below;
- the activity is carried on with a view to the recovery of the waste (whether or not by the person carrying on the activity listed in the table below);
- every part of that place upon which the activity is carried out is surfaced with an impermeable pavement provided with a sealed drainage system; and
- the plant or equipment used in carrying on the activity is maintained in reasonable working order.

Kind of waste	Activities	Seven-day limit
Ferrous metals (within category 16 01 17 of the European Waste Catalogue) or ferrous alloys in metallic non-dispersible form (but not turnings, shavings or chippings of those metals or alloys)	Sorting; grading; baling; shearing by manual feed; compacting; crushing; cutting by hand-held equipment	8,000 tonnes

Kind of waste	Activities	Seven-day limit
The following non-ferrous metals (within category 16 01 18 of the European Waste Catalogue), namely copper, aluminium, nickel, lead, tin, tungsten, cobalt, molybdenum, vanadium, chromium, titanium, zirconium, manganese or zinc, or non-ferrous alloys, in metallic non dispersible form, of any of those metals (but not turnings, shavings or chippings of those metals or alloys)	Sorting; grading; baling; shearing by manual feed; compacting; crushing; cutting by hand-held equipment	400 tonnes
Turnings, shavings or chippings of any of the metals or alloys listed in either of the above categories	Sorting; grading; baling; shearing by manual feed; compacting; crushing; cutting by hand-held equipment	300 tonnes
De-polluted motor vehicles	Dismantling, rebuilding, restoring or reconditioning	40 vehicles
Lead acid motor vehicles batteries (within category 16 06 01 of the European Waste Catalogue) (including those whose contents are special waste) not forming part of, nor contained in, a motor vehicle	Sorting	20 tonnes

(2) The storage, at any place designed or adapted for the recovery of scrap metal or the dismantling of waste motor vehicles, of waste of a kind listed in column 1 of the table above, if:

- the waste is to be submitted to any of the activities specified in the table above in relation to that kind of waste, or to a recycling or reclamation operation authorised by a waste management licence or an authorisation granted under Pt I of the 1990 Act or a permit granted under the 2000 Regulations (SSI 2000/323);
- the total quantity of waste of that kind stored at that place does not exceed the maximum total quantity specified in column 2 of the table above in relation to that kind of waste;
- no waste is stored at that place for longer than 12 months;
- each kind of waste is either stored separately or is kept in separate containers, but in a case where a consignment consisting of more than one

kind of waste is delivered to that place it may be stored unseparated at that place pending sorting for no longer than two months;

- in the case of waste, which is liquid or consists of motor vehicle batteries, it is stored in a secure container;
- in the case of waste motor vehicles, they are, where appropriate, stored on an impermeable pavement;
- subject to sub paragraph (f), the waste or, as the case may be, any container in which it is stored, is stored on an impermeable pavement which is provided with a sealed drainage system; and
- the height of any pile or stack of waste does not exceed five metres.

Kind of waste	Maximum total quantity
Ferrous metals (within category 16 01 17 of the European Waste Catalogue) or ferrous alloys in metallic non-dispersible form (but not turnings, shavings or chippings of those metals or alloys)	50,000 tonnes
The following non-ferrous metals (within category 16 01 18 of the European Waste Catalogue), namely copper, aluminium, nickel, lead, tin, tungsten, cobalt, molybdenum, vanadium, chromium, titanium, zirconium, manganese or zinc, or non-ferrous alloys, in metallic non-dispersible form, of any of those metals (but not turnings, shavings or chippings of those metals or alloys)	1,500 tonnes
Turnings, shavings or chippings of any of the metals or alloys listed in either of the above categories	1,000 tonnes
Motor vehicles, stored where appropriate on an impermeable pavement	1000 vehicles
Lead acid motor vehicle batteries (within category 16 06 01 of the European Waste Catalogue) (including those whose contents are special waste) not forming part of, nor contained in, a motor vehicle	40 tonnes

(3) This only applies to the carrying on of an activity at a place if:

- the person responsible for the management of that place:
  - has established administrative arrangements to ensure that–
    - waste accepted at that place is of a kind listed in Table 4G or, as the case may be, Table 4H; and
    - no waste is accepted at that place in such a quantity as would cause there to be a breach of any of the terms and conditions of the exemption; and
  - carries out a monthly audit to confirm compliance with the terms and conditions of the exemption;

- the records required by para 14 of Pt I of Sch 4 are kept in such a form as to show, for each month, the total quantity of each kind of waste recovered during that month at that place, and details of the total quantity of each kind of waste recovered at that place during the preceding 12 months are sent annually to the appropriate registration authority with the charge referred to in paragraph (d);
- an up to date plan of that place containing the details referred to in reg 18(6)(c)(i)–(iv) is sent annually to the appropriate registration authority with the annual fee referred to in paragraph (d); and
- the charge prescribed for the purpose by a charging scheme under s 41 of the Environment Act 1995 is paid in respect of that place to the appropriate registration authority by the due date which shall be ascertained in accordance with the following:

(4) For the purposes of ascertaining the due date in any year for payment of the fee referred in sub-paragraph (3)(d) in respect of any place–

(a) the appropriate registration authority shall serve notice in accordance with the following provisions of this sub paragraph on the establishment or undertaking from which notice has been received by the authority under reg 18(6) in respect of that place;
(b) a notice required by paragraph (a) shall be served not later than one month before the anniversary of the date when the notice, plan and fee referred to in reg 18(6) were received by the authority in respect of that place and shall specify–

 (i) the amount of the payment due,
 (ii) the method of payment,
 (iii) the date of such anniversary,
 (iv) that payment is due on that date or, if later, upon the day falling one month after the date of the notice, and
 (v) the effect of payment not being made by the date on which it is due, and the due date for payment of the annual fee for that year by that establishment or undertaking in respect of that place shall be the date specified for payment in the notice.

(5) The temporary storage of waste (in this sub paragraph referred to as 'the non-scrap waste'), pending its collection, at a place designed or adapted for the recovery of scrap metal or the dismantling of waste motor vehicles if–

(a) the non scrap waste is not of a kind described in Table 4H;
(b) the non-scrap waste was delivered to that place as part of a consignment of waste of which–

 (i) at least 70% by weight was waste consisting of waste motor vehicles; or
 (ii) at least 95% by weight was waste of any kind described in Table 4H other than waste motor vehicles, and is capable of being separated from that waste by sorting or hand dismantling;

(c) the non-scrap waste is stored at that place for no longer than three months;

(d) in a case where the non-scrap waste is liquid, it is stored in a secure container; and
(e) the non scrap waste or, as the case may be, the container in which the non-scrap waste is stored, is stored on an impermeable pavement which is provided with a sealed drainage system.

(6) In Table 4G, ;'shearing' means the cold cutting of metal by purpose-made shears, and 'depolluted' in relation to waste motor vehicles has the meaning given in reg 2 of the End-of-Life Vehicles (Storage and Treatment) (Scotland) Regulations 2003.

(7) For the purposes of this paragraph and paragraph 44, 'sealed drainage system', in relation to an impermeable pavement, means a drainage system with impermeable components which does not leak and which will ensure that–

(a) no liquid will run off the pavement otherwise than via the system; and
(b) except where they may be lawfully discharged, all liquids entering the system are collected in a sealed sump.

What this means:

This paragraph authorises the storage of metals but in relation to End-of-Life Vehicles (ELVs) it can only cover the storage of depolluted vehicles. A maximum of 1000 vehicles may be stored for a period of 12 months (only 40 vehicles can be accepted to these sites in a seven-day period). These exemptions are mostly used by operators who also hold a waste management licence and use exempt yards as storage for de-polluted shells and non-oily parts for resale.

This must be notified to SEPA and the correct fee paid and an up-to-date plan of the site must be enclosed with the application. The plan must show the following:

- the boundaries of the site;
- the locations within the site at which the exempt activity is to be carried on;
- the location and specifications of any such impermeable pavements, drainage systems or hard standings; and
- the location of any secure containers.

There is no time limit on registration but an inspection of the site must be carried out as part of the assessment process. Registration lasts for 12 months unless a renewal notice has been received and registered by SEPA. SEPA must notify the establishment before the renewal date. A renewal form, new plan and fee must be submitted every year for the exemption to remain valid.

If the renewal is not received within two months of the renewal date, the registration can be cancelled and removed from the register.

Paragraph 46: The Burning of Plant Tissue or Wood at a Dock

7.148 This applies to the burning at a dock of waste consisting of:

- plant tissue waste, in pursuance of a notice given under art 32 of the Plant Health (Scotland) Order 2005, SSI 2005/613; or

- wood of any kind used to wedge or support parts of cargo, including packing material, spacers and pallets, in pursuance of a notice given under art 31 of the Plant Health (Forestry) Order 2005, SI 2005/2517.

The storage of waste at the dock where it was unloaded.

The total quantity of waste stored or burned in any period of 24 hours, shall not exceed 15 tonnes.

Any waste burned shall be burned on a hard standing, within a secure location at the dock where it was unloaded.

This must be notified to SEPA and the correct fee paid. This activity does not have to be notified to SEPA 21 days before it is due to commence, due to the immediate requirement to burn the waste. Registration 'expires' after 12 months unless a renewal notice has been received and registered by SEPA. A renewal notice should be submitted to SEPA at least 21 days before the registration is due to 'expire'.

Paragraph 47: Repair and/or Refurbishment of WEEE

7.149 This applies to the repair or refurbishment or both at a secure place of WEEE as described in column 1 of the table below, together with any incidental storage and other incidental activities in relation to that type of WEEE if:

- the activity is carried on with a view to the reuse of the WEEE for its original purpose;
- best available treatment, recovery and recycling techniques are used when carrying out the activity;
- in respect of a description of waste contained in column 1 of the table below, the total quantity of WEEE so dealt with at that place does not exceed either the storage limit specified in column 2 of the table below or the treatment limit specified in column 3 in relation to that type of WEEE;
- the technical requirements specified in Annex III of the WEEE Directive are met;
- prior to the treatment of WEEE and in relation to any waste components from WEEE, such WEEE or components are stored in such a manner that their environmentally sound reuse or recycling is not hindered; and
- no WEEE is stored at that place for longer than 12 months.

Codes and type of waste*	Storage limits**	Treatment limits
16 02 14 (WEEE other than those mentioned in 16 02 09 to 16 02 13)	50 cubic metres	two tonnes/day
20 01 36 (WEEE other than those mentioned in 20 01 21 and 20 01 23 and 20 01 35)	50 cubic metres	two tonnes/day

* Codes referred to in the European Waste Catalogue.

** In total either awaiting repair or refurbishment or stored following such treatment.

The carrying on, at any secure place in respect of WEEE that is special waste as described in the table above, the treatment activities of repair or refurbishment or both together with any incidental storage and other incidental activities, but not including the de-gassing and capture of ozone depleting substances in relation to that type of WEEE if:

- the activity is carried on primarily with a view to the reuse of the WEEE for its original purpose;
- best available treatment, recovery and recycling techniques are used when carrying out the activity;
- in respect of a description of waste contained in a particular row to the table below, the total quantity of WEEE so dealt with at that place does not exceed either the storage limit specified in column 2, or the treatment limit specified in column 3 of that Table for the types of WEEE in that row;
- the technical requirements specified in Annex III of the WEEE Directive are met;
- the other requirements specified in column 4 of that Table for that type of WEEE are met;
- prior to the treatment of WEEE and in relation to any waste components from WEEE, such WEEE or components are stored in such a manner that their environmentally sound reuse or recycling is not hindered; and
- no WEEE is stored at that place for longer than 12 months.

Codes and type of waste*	**Storage limits****	**Treatment limits**	**Other requirements**
16 02 11 (WEEE containing chlorofluorocarbons, HCFC or HFC) and 20 01 23 (WEEE containing chlorofluorocarbons)	80 cubic metres	five tonnes/day	Stored in a manner that will prevent the release of the CFCs, HCFCs or HFCs
Televisions and computer monitors containing cathode ray tubes falling within 16 02 13 (WEEE containing hazardous components other than those mentioned in 16 02 09 to 16 02 12) and 20 01 35 (WEEE other than those mentioned in 20 01 21 and 20 01 23 containing hazardous components)	80 cubic metres	five tonnes/day	

* Codes referred to in the European Waste Catalogue.
** In total either awaiting repair or refurbishment or stored following such treatment.

Only applies to the carrying out of an activity at a place if the person responsible for the management of that place has established administrative arrangements to ensure that:

- WEEE accepted at that place is of a type set out in column 1 of either of the tables immediately above, as the case may be; and
- no waste is accepted at that place in such a quantity as would cause there to be a breach of any of the terms and conditions of the exemption.

The storage and treatment limits set out in columns 2 and 3 of the tables above are overall limits that apply to all waste falling within the 6 digit code or codes specified in column 1 of those Tables.

This must be notified to SEPA and enclosing the appropriate fee. SEPA has 21 days to register or reject the activity otherwise it is automatically placed on the register. Registration lasts for 12 months unless a renewal notice has been received and registered by SEPA. SEPA must notify the establishment before the renewal date.

Paragraph 48: The Storage of WEEE Pending Recovery Elsewhere

7.150 This applies to the storage at any secure place of WEEE of a type listed in column 1 of the table below if:

- the WEEE is stored for the purpose of its recovery elsewhere;
- the total quantity of any particular type of WEEE stored at the site at any time does not exceed the storage limit specified in column 2 of the Table in relation to that type of WEEE;
- any particular type of WEEE is stored for no longer than the duration limits specified in column 3 of the Table;
- the other relevant storage requirements specified in column 4 of the Table are met; and
- the WEEE is stored in such a manner that its environmentally sound reuse or recycling is not hindered.

Codes and type of waste*	Maximum quantity	Maximum duration	Type of containment
16 02 14 (WEEE other than those mentioned in 16 02 09 to 16 02 12) and 20 01 36 (WEEE other than those mentioned in 20 01 21, 20 01 23 and 20 01 35)	80 cubic metres	three months	Impermeable surface within a secure store; weatherproof covering of stored WEEE
16 02 14 (WEEE other than those mentioned in 16 02 09 to 16 02 12) and 20 01 35 (WEEE other than those mentioned in 20 01 21 and 20 01 23 containing hazardous components)	80 cubic metres	three months	Appropriate leak proof containers providing an impermeable surface

* Codes referred to in the European Waste Catalogue.

The secure storage of WEEE of a type listed in column 1 of the table above, if–

- the WEEE is stored for the purpose of its recovery elsewhere;
- the total quantity of any particular type of WEEE stored at the site at any time does not exceed the storage limit specified in column 2 of the Table;
- any particular type of WEEE is stored for no longer than the duration limits specified in column 3 of the Table;
- the relevant type of containment and other relevant requirements specified in columns 4 and 5 of the Table are met;
- the WEEE is stored in such a manner that its environmentally sound reuse or recycling is not hindered.

Codes and type of waste*	Maximum quantity	Maximum duration	Type of containment	Other requirements
16 02 11 (WEEE containing chlorofluorocarbons, HCFC or HFC) and 20 01 23 (WEEE containing chlorofluorocarbons)	80 cubic metres	three months	Impermeable surface within a secure store; weatherproof covering of stored WEEE	Stored in a manner that will prevent the release of the CFC, HCFC and HFC; overall height of any stack shall not exceed two units or 3.5 metres, whichever is the lower
16 02 13 WEEE containing hazardous components other than those mentioned in 16 02 09 to 16 2 12) and 20 01 35 (WEEE other than those mentioned in 20 01 21 and 20 01 23 containing hazardous components)	80 cubic metres	three months	Impermeable surface within a secure store; weatherproof covering of stored WEEE	
20 01 21 (fluorescent tubes and other mercury-containing waste)	50 cubic metres	three months	Appropriate leak proof containers; weatherproof covering	Stored in such a way that the glass is not broken

* Codes referred to in the European Waste Catalogue.

The activity of storage shall be taken to include the incidental sorting of waste of that type.

The storage and treatment limits set out in columns 2 and 3 of the tables immediately above are overall limits that apply to all waste falling within the 6 digit code or codes specified in column 1 of those Tables.

This needs to be registered with SEPA but no fee is required. Where waste is being stored at the place of production, the activity does not have to be registered. The registration of this exemption is not time limited but it can be de-registered by SEPA if it is not managed properly and no longer fulfils the criteria requirements of the exemption or if the activities cause pollution or harm to health. If the activity does comply with the exemption criteria the registration will not have to be renewed except where there is a significant change to the activity.

Paragraph 49: The Treatment of Waste Organophosphate Sheep Dip with an Enzyme Preparation

7.151 This applies to the treatment of waste organophosphate sheep dip (including such waste as is special waste) with an enzyme preparation, provided that no more than two tonnes of such waste is treated in any one day.

This must be notified to SEPA and the appropriate fee paid. Registration 'expires' after 12 months unless a renewal notice has been received and registered by SEPA. A renewal notice should be submitted to SEPA at least 21 days before the registration is due to 'expire'.

Paragraph 50: The Mixing of Pig or Poultry Ash with Manure for the Treatment of Land

7.152 This applies to the mixing of ash from the incineration of pig or poultry carcasses at its place of production with manure for the treatment of land.

The treatment of land used for agriculture with agricultural waste resulting in benefit to agriculture or ecological improvement, if:

- the waste consists of ash from the incineration of pig or poultry carcasses, or such ash mixed with manure relying on the exemption;
- the land is at the place of production of the ash;
- the land is at least:
  - 10 metres from any inland or coastal waters;
  - 50 metres from any well, borehole or similar work sunk into underground strata for the purpose of any water supply other than a domestic water supply; and
  - 250 metres from any well, borehole or similar work sunk into underground strata for the purpose of any domestic water supply;
- at the start of and throughout the treatment:
  - the land has not been frozen for 12 or more hours during the preceding 24 hours; and
  - the land is not waterlogged, flooded or snow-covered;

- the activity is carried out in accordance with any requirement imposed by or under the Action Programme for Nitrate Vulnerable Zones (Scotland) Regulations 2008, SSI 2008/298;
- where the waste consists only of ash from the incineration of pig or poultry carcasses, it is incorporated into the soil as soon as possible; and
- the total quantity of ash incorporated into the soil does not exceed 150 kilograms per hectare in any period of 12 months.

The secure storage of not more than 100 tonnes of waste intended to be used for the treatment of land.

Definition

'Domestic water supply' has the meaning given by para 7(5).

(5) In this paragraph 'domestic water supply' means a supply of water–

(a) for such domestic purposes as consist in or include, cooking, drinking, food preparation or washing; or
(b) for any of these domestic purposes to premises in which food is produced.

This must be notified to SEPA and the appropriate fee enclosed. Registration 'expires' after 12 months unless a renewal notice has been received and registered by SEPA. A renewal notice should be submitted to SEPA at least 21 days before the registration is due to 'expire'.

Paragraph 51: The Anaerobic Digestion of Agricultural or Distillery Waste

7.153 This applies to the anaerobic digestion of biodegradable waste which is agricultural waste or waste from a distillery.

Definition

'anaerobic digestion' means the process of controlled decomposition of waste under managed conditions:

- where free oxygen is absent;
- at temperatures suitable for naturally occurring mesophilic or thermophilic anaerobe and facultative anaerobe bacteria species; and
- where the inputs to the process are converted to a methane rich biogas for use in an energy recovery process and to a stable sanitised material, the application of which material to land results in benefit to agriculture or horticulture or ecological improvement.

This must be notified to SEPA and the appropriate fee paid. Registration 'expires' after 12 months unless a renewal notice has been received and registered by SEPA. A renewal notice should be submitted to SEPA at least 21 days before the registration is due to 'expire'.

NORTHERN IRELAND

Introduction

7.154 Environmental Licensing and Permitting is covered by the Waste Management Licensing Regulations (Northern Ireland) 2003, NISR 2003/493 and the Pollution Prevention and Control Regulations (Northern Ireland) 2003, NISR 2003/46.

There are a number of waste management licensing exemptions.

- The use and storage of waste glass in a process under an IPC authorisation – paragraph 1 exemption.
- The use and storage of waste glass under a Pollution Prevention Control (PPC) permit – paragraph 2 exemption.
- Operation of a scrap metal furnace under an IPC authorisation – paragraph 3 exemption.
- Operation of a scrap metal furnace under a Pollution Prevention Control (PPC) permit – paragraph 4 exemption.
- Burning of waste as a fuel – paragraph 5 exemption.
- Cleaning, washing, spraying or coating of waste packaging or containers – paragraph 6 exemption.
- Burning of waste as a fuel in an appliance with a net rated thermal input of less than 0.4 megawatts – paragraph 7 exemption.
- Burning waste oil as a fuel – paragraph 8 exemption.
- Land treatment for agricultural benefit or ecological improvement – paragraph 9 exemption.
- The spreading, or storage and spreading of sludge on land which is not land used for agriculture – paragraph 10(a) exemption.
- The storage and spreading of sludge – paragraph 10(b) exemption.
- The reclamation, restoration or improvement of land – paragraph 11 exemption.
- The treatment of waste with a view to recovery or reuse – paragraph 12 exemption.
- The composting and storage of biodegradable waste – paragraph 13 exemption.
- The manufacture of products from waste, the treatment of waste soil or rock which is to be spread on land under paragraph 9 or 11, and the storage of waste which is to be submitted to the preceding activities – paragraph 14 exemption.
- The manufacture of finished goods from waste metal, plastic, glass, ceramics, rubber, textiles, wood, paper or cardboard, and the storage of those wastes – paragraph 15 exemption.
- The beneficial use of waste – paragraph 16 exemption.
- The storage of wastes for the purposes of recovery or re-use – paragraph 17 exemption.
- The storage of waste – paragraph 18 exemption.
- The use and/or storage of waste for the purpose of construction, maintenance or improvement of relevant work – paragraph 19 exemption.

- The laundering or otherwise cleaning of waste textiles with a view to recovery or re-use – paragraph 20 exemption.
- The chipping, shredding, cutting or pulverising of waste plant matter – paragraph 21 exemption.
- The recovery of silver from waste produced in connection with printing or photographic processing – paragraph 22 exemption.
- The recovery of waste consisting of animal by-products – paragraph 23 exemption.
- The crushing, grinding or other size reduction of waste bricks, tiles or concrete – paragraph 24 exemption.
- The deposit of waste arising from dredging inland waters or from clearing plant matter from inland waters – paragraph 25 exemption.
- The recovery of waste as an integral part of the process that produces it – paragraph 26 exemption.
- The baling, compacting, crushing, shredding or pulverising of waste at the place of production – paragraph 27 exemption.
- The storage of returned goods pending recovery or disposal – paragraph 28 exemption.
- The disposal of specified wastes at the place of production in an incinerator – paragraph 29 exemption.
- The burning of waste on land in the open – paragraph 30 exemption.
- The discharge of sanitary waste from a passenger train – paragraph 31 exemption.
- The burial on premises of sanitary waste arising from the use on those premises of a sanitary convenience which is equipped with a removable receptacle – paragraph 32 exemption.
- The keeping or deposit of waste arising from peat working – paragraph 33 exemption.
- The keeping or deposit of spent railway ballast – paragraph 34 exemption.
- The deposit of borehole or excavated waste – paragraph 35 exemption.
- The temporary storage of garbage or tank washings from ships at a port – paragraph 36 exemption.
- The burial of a dead domestic pet in the garden of a domestic property where the pet lived – paragraph 37 exemption.
- The storage or deposit of waste samples to be tested or analysed – paragraph 38 exemption.
- The storage of waste medicines at a pharmacy or storage of waste at the premises of a medical, nursing or veterinary practice – paragraph 39 exemption.
- The storage of non-liquid waste including waste electrical and electronic equipment (WEEE) at any site other than the premises where it is produced – paragraph 40 exemption.
- The temporary storage of waste, including WEEE, pending its collection, at the place where it is produced – paragraph 41 exemption.
- The waste management transitional arrangements for the treatment, keeping or disposal at any premises, of scrap metal or waste motor vehicles, which are to be dismantled pending the issue of a licence – paragraph 42 exemption.

- The waste management transitional arrangements to cover (up to 19 December 2004) the treatment, keeping or disposal of waste – paragraph 43 exemption.
- The heating of metals in a furnace or other combustion chamber for the purpose of removing grease, oil or any other non-metallic contaminant – paragraph 44 exemption.
- The recovery of scrap metal or the dismantling of de-polluted end-of-life vehicles – paragraph 45 exemption.
- The recovery and storage of waste from a water treatment works or a sewage treatment works – paragraph 46 exemption.
- The burning of waste at a dock – paragraph 47 exemption.
- The repair, refurbishment, and storage of waste electrical and electronic equipment (WEEE) – paragraph 49 exemption.
- The secure storage of waste electrical and electronic equipment (WEEE) – paragraph 50 exemption.
- The crushing of waste gas discharge lamps – paragraph 51 exemption.
- The secure storage of waste batteries and accumulators – paragraph 52 exemption.

The Waste Management Licensing Regulations (Northern Ireland) 2003, NISR 2003/493[13]

7.155 These Regulations make provision related to the bringing into operation of the waste management licensing system under Pt II of the Waste and Contaminated Land (Northern Ireland) Order 1997, NI Order in Council 1997/278 ('the 1997 Order') and for the purpose of implementing certain Council Directives relating to waste.

Regulations 16 (Exclusion of activities under the control regimes from waste management licensing) deals with exclusions from waste management licensing and 17 (Exemptions form waste management licensing) deal with exemptions. Schedule 2 deals with the specifics of exemptions from waste management licensing.

The Regulations have also been subject to a number of amendments.

The Waste Management Licensing (Amendment) Regulations (Northern Ireland) 2011, NISR 2011/403[14]

7.156 These Regulations amend the Waste Management Licensing Regulations (Northern Ireland) 2003, NISR 2003/493 to provide for an exemption for the land spreading of sludges from on-site effluent treatment of waste from abattoirs

13 NISR 2003/493.
14 NISR 2011/403.

and meat and fish processing plants (European Waste Catalogue reference 02 02 04) following 'conventional treatment' or 'enhanced treatment'.

The Waste Management Licensing (Amendment) Regulations (Northern Ireland) 2014, NISR 2014/137

7.157 These Regulations amend the Waste Management Licensing Regulations (Northern Ireland) 2003, NISR 2003/493 by substituting definitions from Directive 2008/98/EC of the European Parliament and of the Council on Waste ([2008] OJ No L312, 22/11/2008, p 3) (the Waste Framework Directive) and inserting a reference to Directive 2012/19/EU of the European Parliament and of the Council on Waste Electrical and Electronic Equipment ([2012] OJ No L197, 24.7.2012, p 38) (the WEEE Directive).

They substitute conditions of site licences for the storage and treatment of Waste Electrical and Electronic Equipment ('WEEE') together with exemptions from the permit requirements of the WEEE Directive.

Regulation 2 substitutes the definitions as reflected in Article 3 of the Waste Framework Directive.

Regulation 3 substitutes the conditions of site licences for the storage or treatment of WEEE in accordance with Article 8 and Annexes VII and VIII of the WEEE Directive.

Regulation 4 substitutes the technical requirements specified in Annex VIII of the WEEE Directive in relation to exemptions from waste management licensing.

Copies of a full Regulatory Impact Assessment in relation to these Regulations can be obtained from Environmental Policy Division, 44-58 May Street, Town Parks, Belfast BT1 4JJ.

The Waste Management Licensing (Amendment No 2) Regulations (Northern Ireland) 2014, NISR 2014/253[15]

7.158 These Regulations implement Regulation (EU) No 493/2012 of 11 June 2012 laying down, pursuant to Directive 2006/66/EC of the European Parliament and of the Council, detailed rules regarding the calculation of recycling efficiencies of the recycling process for waste batteries and accumulators.

Regulation 2 implements Article 3 of the Commission Regulation in relation to the methods of calculation of recycling efficiency and for the purposes of Article 3(4) of the Regulation the Department is the competent authority.

15 NISR 2014/253.

The Waste Management Licensing (Amendment) Regulations (Northern Ireland) 2015, NISR 2015/301[16]

7.159 These Regulations amend regs 3 to 5 of and Sch 1 to the Waste Management Licensing Regulations (Northern Ireland) 2003, NISR 2003/493 (the 2003 Regulations) and prescribe the accepted technical competencies for the purpose of determining if a person can be deemed a fit and proper person to hold a waste management licence.

Regulation 2(2) updates reg 3 of the 2003 Regulations and introduces the requirement for existing operators to obtain up-to-date certification within 24 months of the operational date of these Regulations. It also introduces a requirement for operators to renew their certification every 24 months.

Regulation 2(3) revokes reg 4 of the 2003 Regulations, which was transitional and is outdated and no longer required.

Regulations 2(4) and 2(5) amend reg 5 of the 2003 Regulations and reduce the maximum amount of time allowed to acquire technical certification from two years to 12 months.

Regulation 2(6) replaces Sch 1 to the 2003 Regulations with an updated list of Operator Competence Certificates, which takes account of new waste technologies.

The Waste Management Licensing (Amendment No 2) Regulations (Northern Ireland) 2015, NISR 2015/386[17]

7.160 These Regulations amend reg 2 of and Sch 1 to the Controlled Waste (Registration of Carriers and Seizure of Vehicles) Regulations (Northern Ireland) 1999, NISR 1999/362 (the 1999 Regulations). They also amend reg 2 of the Waste Management Licensing Regulations (Northern Ireland) 2003, NISR 2003/493 (the 2003 Regulations) which prescribe offences which must be taken into account by the Department when determining if a person can be deemed a fit and proper person to hold a waste management licence. The Regulations also introduce a new Sch 6 to the 2003 Regulations, to provide an updated list of offences that includes environmental, financial and other relevant offences.

In particular, reg 2 amends the 1999 Regulations.

Regulation 3(2) substitutes reg 2 of the 2003 Regulations with a new reg 2.

Regulation 3(3) inserts a new Sch 6 to the 2003 Regulations.

16 NISR 2015/301.
17 NISR 2015/386.

The Waste Management Licensing (Amendment) Regulations (Northern Ireland) 2016, NISR 2016/96[18]

7.161 These Regulations amend the Waste Management Licensing Regulations (Northern Ireland) 2003, NISR 2003/493 (the 2003 Regulations).

Regulation 3 amends paras 12, 15, 19, 27 and 45 of Pt 1 of Sch 2 to the 2003 Regulations.

Regulation 3(4) revokes para 17 of Pt 1 of Sch 2 to the 2003 Regulations.

Regulation 3(5) revokes para 18 of Pt 1 of Sch 2 to the 2003 Regulations.

The Use and Storage of Waste Glass in a Process Under an IPC Authorisation – paragraph 1

7.162 A paragraph 1 exemption allows the use, under an authorisation granted under the Industrial Pollution Control Order, of waste glass as part of a process within Part B of section 3.5 (glass manufacture and production) of Sch 1 of The Industrial Pollution Control (Prescribed Processes and Substances) Regulations (Northern Ireland) 1998, NISR 1998/28[19].

Quantity

The total quantity of waste glass used in the process must not exceed 600,000 tonnes in any period of 12 months.

Storage

This must be at the place where the manufacturing process is carried on. There is no limit to the amount of waste glass that can be stored.

Additional conditions

There are additional conditions in that the process must be authorised under an authorisation granted under the Industrial Pollution Control (Northern Ireland) Order of 1997 for waste glass as part of a process which includes processes authorised by Part B of section 3.5 (glass manufacture and production) of

18 NISR 2016/ 96.

19 NISR 1998/28, Sch1 Pt B Section 3.5

Part B

(a) The manufacture of glass at any location where the person concerned has the capacity to make 5000 tonnes or more of glass in any 12 month period, and any process involving the use of glass which is carried on at any such location in conjunction with its manufacture.

(b) The manufacture of glass where the use of lead or any lead compound is involved.

(c) The making of any glass product where lead or any lead compound has been used in the manufacture of glass except—

 (i) the making of products from lead glass blanks;

 (ii) the melting, or mixing with another substance, of glass manufactured elsewhere to produce articles such as ornaments or road paint;

(d) Polishing or etching glass or glass products in the course of any manufacturing process if—

 (i) hydrofluoric acid is used; or

 (ii) hydrogen fluoride may be released into the air.

(e) The manufacture of glass frit or enamel frit and its use in any process where that process is related to its manufacture if not falling within Part A of this Section.

Sch 1 of the Industrial Pollution Control (Prescribed Processes and Substances) Regulations (Northern Ireland) 1998 as:

- manufacture of glass at any location with the capacity to make 5000 tonnes or more of glass in any 12-month period and any process involving the use of glass, which is carried out at the location in conjunction with its manufacture.
- manufacture of glass where the use of lead or any lead compound is involved.

The Use and Storage of Waste Glass Under a Pollution Prevention Control (PPC) Permit – paragraph 2

7.163 Paragraph 2 allows the use under a permit, of waste glass as part of an activity within Part B of section 3.3 (the manufacture of glass and glass fibre) of Pt 1, Sch 1 to the Pollution Prevention and Control Regulations (Northern Ireland) 2003, NISR 2003/46.

Quantity

The total quantity of waste glass used in the activity must not exceed 600,000 tonnes in any period of 12 months.

Storage

This must be at the place where the activity is carried out.

Additional conditions

The activity must be permitted under Sch 1, Pt 1, Section 3.3 Part B of The Pollution Prevention and Control Regulations (Northern Ireland) 2003, which allows for:

- Manufacturing glass at any location where the person concerned has the capacity to make 5,000 tonnes or more of glass in any period of 12 months, and any activity involving the use of glass, which is carried out at any such location in conjunction with its manufacture.
- Manufacturing glass where the use of lead or any lead compound is involved.
- Manufacturing any glass product where lead or any lead compound has been used in the manufacture of the glass except–

    making products from lead glass blanks; or
    melting, or mixing with another substance, glass manufactured elsewhere to produce articles such as ornaments or road paint.

- Polishing or etching glass or glass products in the course of any manufacturing activity if–

    hydrofluoric acid is used; or
    hydrogen fluoride may be released into the air.

- Manufacturing glass frit or enamel frit and its use in any activity where that activity is related to its manufacture.

Part B processes are permitted by the Industrial Pollution and Radiochemical Inspectorate of the Northern Ireland Environment Agency (NIEA).

Operation of a Scrap Metal Furnace Under an IPC Authorisation – paragraph 3

7.164 Paragraph 3[20] allows the operation, under an authorisation granted under the Industrial Pollution Control Order, of a scrap metal furnace with a designed holding capacity of less than 25 tonnes and the loading or unloading of the furnace in connection with its operation.

Quantity

There is no limit to the amount of scrap metal that can be processed, however the designated holding capacity of the furnace must not exceed 25 tonnes.

Storage and location

Storage of the scrap metal must be at the site where the furnace is located (but not in cases where that place is used for carrying on business as a scrap metal dealer). There is no limit to the amount of waste that can be stored.

Additional conditions

The operation must be authorised by an authorisation granted under the Industrial Pollution Control Order to the extent that it is or forms part of a process within paragraphs (b), (e) or (f) of Part B[21], or (a), (b) or (d) of Part C[22]

20 (1) The operation, under an authorisation granted under the Industrial Pollution Control Order, of a scrap metal furnace with a designed holding capacity of less than 25 tonnes to the extent that it is or forms part of a process within paragraphs (b), (e) or (f) or Part B or paragraphs (a), (b) or (d) of Part C of section 2.1 (iron and steel), or paragraphs (a), (b) or (c) of Part B or paragraphs (a) or (b) of Part C of section 2.2 (non-ferrous metals), of Schedule 1 to the 1988 Regulations.

(2) The loading or unloading of such a furnace in connection with its operation in a manner covered by the exemption conferred by sub-paragraph (1).

(3) The storage, at the place where such a furnace is located (but not in cases where that place is used for carrying on business as a scrap metal dealer), of scrap metal intended to be submitted to an operation covered by the exemption conferred by sub-paragraph (1).

21 (b) Any process for the refining or making of iron, steel or any ferrous alloy in which air or oxygen or both are used, if related to a process described in this Part of this Section.

(e) Any foundry process (including ancillary foundry operations such as the manufacture and recovery of moulds, the reclamation of sand, fettling, grinding and shot-blasting) if related to another process described in this Part of this Section.

(f) Any process involving the casting of iron, steel or any ferrous alloy from deliveries of 50 tonnes or more at one time of molten metal.

22 (a) Making, melting or refining iron, steel or any ferrous alloy in an electric arc furnace, a cupola, crucible furnace, reverberatory furnace, rotary furnace, induction furnace or resistance furnace with a designed holding capacity of less than 5 tonnes.

(b) Any process for the refining or making of iron, steel or any ferrous alloy in which air or oxygen or both are used, if related to a process described in this Part of this Section.

(d) Any such process as is described in paragraph (*i*) of Part A or paragraph (*d*) of Part B above, if not falling within those paragraphs; but a process does not fall within this paragraph if—

(i) it is a process for heating iron, steel or any ferrous alloy in one or more furnaces or other appliances the primary combustion chambers of which have in aggregate a net rated thermal input of less than 0.2 megawatts;

(ii) it does not involve the removal by heat of plastic or rubber covering from scrap cable or of any asbestos contaminant; and

(iii) it is not related to any other process described in this Part of this Section.

of section 2.1 (iron and steel) or paragraphs (a), (b) or (c) of Part B[23], or (a) or (b) of Part C[24] of section 2.2 (non-ferrous metals), of Part I of Schedule 1 to the Pollution Prevention and Control Regulations (Northern Ireland) 2003, NISR 2003/46.

Part B processes are authorised by the Industrial Pollution and Radiochemical Inspectorate of the Northern Ireland Environment Agency.

Part C processes are authorised by the local District Council.

The relevant IPC authorisation must be submitted with the paragraph 3 exemption application.

Operation of a Scrap Metal Furnace Under a Pollution Prevention Control (PPC) Permit – paragraph 4

7.165 Paragraph 4 allows the operation, under a Pollution Prevention and Control Regulations (Northern Ireland) 2003, NISR 2003/46 permit, of a scrap metal furnace with a holding capacity of less than 25 tonnes and the loading or unloading of the furnace in connection with its operation.

Quantity

The amount of waste processed will be determined by the scrap metal furnace with a designated holding capacity of less than 25 tonnes.

Storage

Storage of the scrap metal must be at the site where the furnace is located (but not in cases where that place is used for carrying on business as a scrap metal dealer).

Additional conditions

The operation must be authorised by an authorisation granted under the Pollution Prevention and Control Regulations (Northern Ireland) 2003 to the

23 (a) Any process not described in Part A of this Section for making or melting any non-ferrous metal or non-ferrous alloy (other than tin or any alloy which in molten form contains 50 per cent. or more by weight of tin) where the designed holding capacity of molten metal is 0.5 tonnes or more (together with any additional refining).
(b) The fusion of calcined bauxite for the production of artificial corundum.
(c) Melting zinc or a zinc alloy in conjunction with a galvanising process.

24 (a) The making or melting of any non-ferrous metal or non-ferrous metal alloy (other than tin or any alloy which, in molten form, contains 50 per cent. or more by weight of tin) where the designed holding capacity of molten metal is less than 0.5 tonnes (together with any incidental refining).
(b) Any such process as is described in paragraph (*i*) of Part A above, if not related to another process described in that Part; but a process does not fall within this paragraph if—
(i) it involves the use of one or more furnaces or other appliances the primary combustion chambers of which have in aggregate a net rated thermal input of less than 0.2 megawatts; and
(ii) it does not involve the removal by heat of plastic or rubber covering from scrap cable or of any asbestos contaminant.

extent that it is or forms part of an activity within paragraph (a), (b) or (d) of Part C[25] of Section 2.1 (ferrous metals), or (a) of Part B[26] or paragraph (a) or (b) of Part C[27] of section 2.2 (non-ferrous metals), of Part I of Schedule 1 to the above regulations.

Part B processes are permitted by the Industrial Pollution and Radiochemical Inspectorate of the Northern Ireland Environment Agency (NIEA). An application form is available from our Integrated Pollution Prevention and Control (IPPC) section.

Part C processes are permitted by the local District Council.

Burning of waste as a fuel – paragraph 5

7.166 Paragraph 5 allows burning of waste as a fuel at a rate of less than one tonne per hour to the extent that it is or forms part of a process within Part B or Part C of any section of Sch 1 to the Industrial Pollution Control (Prescribed Processes and Substances) (Northern Ireland) Regulations 1998, NISR 1998/28, or an activity within Part B or Part C of any Section of Pt 1 of Sch 1 to the Pollution Prevention and Control Regulations (Northern Ireland) 2003, NISR 2003/46 and the feeding of such waste into an appliance in which it is to be so burned.

25 (a) Making, melting or refining iron, steel or any ferrous alloy in an electric arc furnace, a cupola, crucible furnace, reverberatory furnace, rotary furnace, induction furnace or resistance furnace with a designed holding capacity of less than 5 tonnes.
(b) Any process for the refining or making of iron, steel or any ferrous alloy in which air or oxygen or both are used, if related to a process described in this Part of this Section.
(d) Any such process as is described in paragraph (*i*) of Part A or paragraph (*d*) of Part B above, if not falling within those paragraphs; but a process does not fall within this paragraph if—
(i) it is a process for heating iron, steel or any ferrous alloy in one or more furnaces or other appliances the primary combustion chambers of which have in aggregate a net rated thermal input of less than 0.2 megawatts;
(ii) it does not involve the removal by heat of plastic or rubber covering from scrap cable or of any asbestos contaminant; and
(iii) it is not related to any other process described in this Part of this Section.

26 (a) Any process not described in Part A of this Section for making or melting any non-ferrous metal or non-ferrous alloy (other than tin or any alloy which in molten form contains 50 per cent. or more by weight of tin) where the designed holding capacity of molten metal is 0.5 tonnes or more (together with any additional refining).

27 (a) The making or melting of any non-ferrous metal or non-ferrous metal alloy (other than tin or any alloy which, in molten form, contains 50 per cent. or more by weight of tin) where the designed holding capacity of molten metal is less than 0.5 tonnes (together with any incidental refining).
(b) Any such process as is described in paragraph (*i*) of Part A above, if not related to another process described in that Part; but a process does not fall within this paragraph if—
(i) it involves the use of one or more furnaces or other appliances the primary combustion chambers of which have in aggregate a net rated thermal input of less than 0.2 megawatts; and
(ii) it does not involve the removal by heat of plastic or rubber covering from scrap cable or of any asbestos contaminant.

Permitted waste streams

- straw (02 01 03), poultry litter (02 01 06) or wood (02 01 07); or
- solid fuel which has been manufactured from waste by a process involving the application of heat (19 12 10).

Quantity

Waste must be burned at a rate of less than one tonne per hour.

Storage and location

No more than 25 tonnes of waste may be stored on the premises at any one time. No waste may be stored for longer than one year.

Additional conditions

The operation may only be carried on under an authorisation granted under the Industrial Pollution Order or a permit under the Pollution Prevention and Control Regulations (Northern Ireland) 2003 to the extent that it forms part of a process within Part B or Part C of any section of Sch 1 to the 1998 Regulations, or an activity within Part B or Part C of any Section of Part 1 of Sch 1 to the 2003 Regulations.

The relevant IPC authorisation or PPC permit must be submitted with the paragraph 5 exemption application.

Cleaning, washing, spraying or coating of waste packaging or containers – paragraph 6 exemption.

Burning of waste as a fuel in an appliance with a net rated thermal input of less than 0.4 megawatts – paragraph 7 exemption.

Burning of Waste as a Fuel in an Appliance with a Net Rated Thermal Input of Less than 0.4 Megawatts – paragraph 7

7.167 A paragraph 7 exemption allows the burning of waste as a fuel in an appliance or appliances that have an aggregate net thermal input of less than 0.4 megawatts.

Any waste is permitted

Quantity

The quantity of waste to be burned is determined by the designated holding capacity of the appliance or appliances, which have a net thermal input of less than 0.4 megawatts.

Storage and location

Waste must be securely stored on the premises where it is to be burned.

Definition

'net rated thermal input', means the rate at which fuel can be burned at the maximum continuous rating of the appliance multiplied by the net calorific value of the fuel expressed as megawatts thermal.

Burning Waste Oil as a Fuel – paragraph 8

7.168 A paragraph 8 exemption allows the burning of waste oil as a fuel in an engine of an aircraft, hovercraft, mechanically propelled vehicle, railway locomotive, ship or other vessel.

Permitted waste

Waste oil

Quantity

The total amount of waste oil burned in one hour in a single engine must not exceed 2,500 litres.

Storage

Storage must be in a secure container.

Additional conditions

For this particular exemption, the following information is required to be submitted in accordance with Pt II of Sch 2 of the Waste Management Licensing Regulations (Northern Ireland) 2003, NISR 2003/493 (as amended):

- location, name and address and six figure Ordinance Survey Irish grid reference of the place where any waste is to be stored;
- quantity of waste oil to be stored at any one time;
- construction and capacity details of storage tanks and their bunds;
- type of vehicle in which the waste will be used.

Land Treatment for Agricultural Benefit or Ecological Improvement – paragraph 9

7.169 A paragraph 9 exemption allows certain wastes to be spread on land used for agriculture where the spreading will result in benefit to agriculture or ecological improvement or on land not used for agriculture where the spreading will result in ecological improvement.

Permitted waste:

Wastes identified by European Waste Catalogue (EWC) Codes in Column 1 of Table 3, Sch 2, Pt I of the Waste Management Licensing Regulations (Northern Ireland) 2003, NISR 2003/493 (as amended) and referred to in Column 2 of that Table may be spread on land used for agriculture where the spreading results in benefit to agriculture or ecological improvement.

Wastes listed in Part I of Table 3 and identified by EWC Code in Column 1 of Table 3 and referred to in Column 2 of that Table may be spread on operational land of a railway, light railway, or any Government Department; or land which is a forest, woodland, park, garden, verge, bank of an inland waterway, landscaped area, sports ground, recreation ground, churchyard or cemetery; where the land in question is not used for agriculture where the spreading results in ecological improvement.

Quantity

Where waste is to be spread for agricultural benefit the quantity shall be determined by the crop requirement and shall be based on the soil and waste analyses.

No more than 250 tonnes, or in the case of dredgings from inland waters, 5,000 tonnes of waste per hectare can be spread in any 12-month period.

Storage

No more than 1000 cubic metres of waste can be stored and storage must be at the place where it is to be spread. Waste must be stored in a secure container or lagoon or place.

Additional conditions

The spreading of the waste must result in benefit to agriculture or ecological improvement and a certificate must be submitted describing how the spreading activity will achieve this. The certificate must be prepared by or based on advice from a person with appropriate technical or professional expertise.

Benefit to agriculture shall be assessed by reference to whether the spreading will result in an improvement to the soil for the purpose of growing crops.

Ecological improvement shall be assessed by reference to the extent to which wildlife habitats, which might otherwise deteriorate, are maintained or supported. The creation of a new habitat or the restoration of an old habitat shall be considered an improvement.

Benefit to agriculture and ecological improvement shall be subject to the criteria given in Sch 2, Pt III of the Waste Management Licensing (Northern Ireland) Regulations 2003 (as amended).

A certificate of benefit to agriculture or ecological improvement must be submitted with the paragraph 9 exemption application. Where the waste to be spread involves animal by-products, it must have been treated in accordance with the Animal By-Products Regulations (Northern Ireland) 2003, NISR 2003/495.

Where the Nitrates Action Programme Regulations (Northern Ireland) 2006, NISR 2006/489 apply, treatment shall be carried out in accordance with those Regulations.

The following information contained in Pt II of the Schedule is required for the registration for an exemption under paragraph 9:

1. The application shall include the following particulars–

   (a) a description of the waste to be used, its physical form and the process from which it arose;
   (b) a description of where and how the waste will be stored pending its use;
   (c) a description of the land which is to be treated with the waste, including:

   - the farm survey number;
   - client reference number;
   - field number (if applicable);
   - area available for treatment;
   - the soil;
   - crops grown or to be grown in the soil; and
   - all wastes used to treat the land in the previous six months;

   (d) the method and intended date of treatment, the quantities of waste to be used and the rate of application;
   (e) details of the benefit to agriculture or ecological improvement to be expected from the treatment, including any analysis of the waste or the soil, and management plans;
   (f) the location of any waterway within 15 metres of the land on which the waste is to be spread.

2. The application shall be accompanied by the following documents–

   (a) an analysis of the wastes to be used;
   (b) an analysis of the soil which is to be treated with the wastes;
   (c) an assessment of the risk of pollution caused by the use;
   (d) a certificate describing how the treatment will result in benefit to agriculture or the ecological improvement, which shall be prepared or be based upon advice from a person with the appropriate technical or professional expertise.

3. Wastes shall be analysed in relation to the following parameters–

   (a) percentage dry solids content;
   (b) pH;
   (c) conductivity (mS);
   (d) total content of major nutrients and readily plant available ammonium-nitrogen with the results expressed on a fresh weight basis;
   (e) biochemical oxygen demand; and
   (f) in relation to the types of waste that are listed in the first column of Table 12 in Part III of Schedule 2, the parameters ticked in the remaining columns of the Table.

Definition:

'Operational Land' has the meaning given by art 2 of the Planning (General Development) Order (Northern Ireland) 1996.

The Spreading, or Storage and Spreading of Sludge on Land which is not Land Used for Agriculture – paragraph 10(a)

7.170 Paragraph 10(a) allows the storage and spreading of sewage sludge on land which is not land used for agriculture within the meaning of the Sludge (Use in Agriculture) Regulations (Northern Ireland) 1990, SR 1990/245 if the spreading results in ecological improvement or, in the case of spreading on land used for non-food crops not grown in short rotation with food crops, it results in benefit to agriculture.

Permitted waste

Sewage sludge

Quantity

The quantity of waste spread will be determined by the crop requirement and be based on the soil and waste analyses where the spreading will be carried out in accordance with the Sludge (Use in Agriculture) Regulations (Northern Ireland) 1990, SR 1990/245.

Storage

If sludge is to be stored prior to spreading, it must be stored at the place where it is to be spread and must not be stored for a period of more than six months.

Storage must be in a secure container or lagoon or, in the case of dewatered sludge, in a secure place on land.

Additional conditions

The spreading of sewage sludge must result in benefit to agriculture or ecological improvement and a certificate must be submitted describing how the spreading activity will achieve this. The certificate must be prepared by or based on advice from a person with appropriate technical or professional expertise.

Benefit to agriculture shall be assessed by reference to whether the spreading will result in an improvement to the soil for the purpose of growing non-food crops not grown in short rotation with food crops.

Ecological improvement shall be assessed by reference to the extent to which wildlife habitats, which might otherwise deteriorate, are maintained or supported. The creation of a new habitat or the restoration of an old habitat shall be considered an improvement.

The following information contained in Pt II of Sch 2 of the Waste Management Licensing Regulations (Northern Ireland) 2003, NISR 2003/493 (as amended) is required for the registration of an exemption under paragraph 10:

- the names, addresses and telephone numbers and if applicable, the fax numbers and e-mail addresses of the establishment or undertaking that is to store or spread the sludge and the establishment or undertaking supplying the sludge;

- the quantity of sludge to be stored or spread and its origin;
- information on how the sludge has been treated and where it has been treated;
- the location in which the sludge is to be stored or spread, including a description of any container or lagoon to be used;
- the locations of any buildings, public rights of way, abstraction points or surface waters above ordinance datum which are situated within 400 metres of the boundaries of that place;
- where sludge is to be spread, the application shall be accompanied by a certificate describing how the activity will result in benefit to agriculture or ecological improvement, which shall be prepared by or based on advice from a person with appropriate technical or professional expertise.

What the legislation says

The full wording of the paragraph 10 exemption can be found in Schedule 2 of the Waste Management Licensing Regulations (Northern Ireland) 2003.

For storage and spreading on non-agricultural land application should be made for a paragraph 10(a) exemption.

For storage of sludge to allow the spreading on agricultural land under the Sludge (Use in Agriculture) Regulations (Northern Ireland) 1990, paragraph 10(b) exemption should be applied for.

Relevant objectives in relation to the disposal or recovery of waste are given in para 4 of Sch 3 to the Regulations.

The Storage and Spreading of Sludge – paragraph 10(b)

7.171 A paragraph 10(b) exemption allows the storage of sewage sludge, which is to be used in accordance with the Sludge (Use in Agriculture) Regulations (Northern Ireland) 1990, SR 1990/245.

Permitted waste

Sewage sludge.

Quantity

The quantity of waste to be stored will be dependent on the crop requirement and be based on the soil and waste analyses.

Storage

Sludge must be stored at the place where it is to be spread and must not be stored for a period of more than six months.

Storage must be in a secure container or lagoon or, in the case of dewatered sludge, in a secure place on land.

The following information contained in Pt II of Sch 2 of the Waste Management Licensing Regulations (Northern Ireland) 2003, NISR 2003/493 (as amended) is required for the registration of an exemption under paragraph 10:

- the names, addresses and telephone numbers and if applicable, the fax numbers and e-mail addresses of the establishment or undertaking that is to store the sludge and the establishment or undertaking supplying the sludge;
- the quantity of sludge to be stored and its origin;
- information on how the sludge has been treated and where it has been treated;
- the location in which the sludge is to be stored or spread, including a description of any container or lagoon to be used;
- the locations of any buildings, public rights of way, abstraction points or surface waters above ordinance datum which are situated within 400 metres of the boundaries of that place.

For storage and spreading on non-agricultural land then the paragraph 10(a) exemption applies.

The Reclamation, Restoration or Improvement of Land – paragraph 11

7.172 A paragraph 11 exemption allows the spreading and storage of waste for the purpose of reclaiming, restoring, or improving land which has been subject to industrial or other manmade development where the use to which that land could be put would be improved by the spreading.

Permitted waste

Wastes in Pt I of Table 4, Sch 2, Pt I of the Waste Management Licensing Regulations (Northern Ireland) 2003, NISR 2003/493, (as amended), may be spread on any land. Wastes from Pt II of Table 4 may be spread on land where the activity results in benefit to agriculture or ecological improvement.

Quantity

The activity must be carried out in accordance with any planning permission (where such permission is required) and the quantity of waste used will be determined by planning permission requirements.

The relevant planning permission, relevant planning drawings and cross-sections must be submitted with the paragraph 11 exemption application.

The depth of waste spread must not exceed the lesser of two metres or the final cross sections shown on any plan or drawings required to be submitted in support of the exemption application.

The waste spread must not exceed 20,000 cubic metres per hectare.

Storage

The waste must be securely stored for a period not exceeding six months and storage must be at the place where it is to be spread.

Additional conditions

Where any of the wastes listed in Pt II of Table 4 are to be spread, the notice shall be accompanied by a certificate describing how the activity will result in benefit to agriculture or ecological improvement. The certificate shall be prepared by or based on advice from a person with appropriate technical or professional expertise.

Benefit to agriculture shall be assessed by reference to whether the spreading will result in an improvement to the soil for the purpose of growing crops or grazing.

Ecological improvement shall be assessed by reference to the extent to which wildlife habitats, which might otherwise deteriorate, are maintained or supported. The creation of a new habitat or the restoration of an old habitat shall be considered an improvement.

Benefit to agriculture and ecological improvement shall be subject to the criteria given in Sch 2, Pt III of the Waste Management Licensing (Northern Ireland) Regulations 2003 (as amended).

In areas where the Nitrates Action Programme Regulations (Northern Ireland) 2006[28] apply the spreading shall be carried out in accordance with those Regulations.

The following information contained in Pt II of Sch 2 is required for the registration for an exemption under paragraph 11:

- where less than 2,500 cubic metres of waste are to be spread in any one year, a description of the spreading, the type and quantity of waste to be spread and the location of the spreading;
- where 2,500 or more cubic metres of waste are to be spread–
    - the total quantity of waste to be spread;
    - the type of waste to be spread, identified by reference to the descriptions in the second column of Table 4;
    - the name, address and telephone number of all relevant landowners and the location of their land where the waste is to be spread or stored;
    - a plan of the spreading with cross-sections showing the proposed final levels of the land affected by the spreading;
    - the intended start and completion date of the spreading and any related storage.

The full wording of the paragraph 11 exemption can be found in Sch 2 to the Waste Management Licensing Regulations (Northern Ireland) 2003. These Regulations have been amended a number of times.

28 NISR 2006/489.

The Treatment of Waste with a View to Recovery or Re-use – paragraph 12

7.173 A paragraph 12 exemption allows the treatment of specific wastes with a view to recovery or reuse.

Permitted waste streams

The following kinds of waste, activities and limits are permitted:

Kind of waste	Activities	Limit (tonnes per week)
Paper or Cardboard	Baling, sorting or shredding	3,000
Textiles	Baling, sorting or shredding	100
Plastic	Baling, sorting or shredding, densifying or washing	100
Glass	Sorting, crushing or washing	1,000
Steel cans, aluminium cans or aluminium foil	Sorting, crushing, pulverising, shredding, compacting or baling	100
Food or drink cartons	Sorting, crushing, pulverising, shredding, compacting or baling	100

Storage

Must be at the place where the treatment activity takes place and must be in accordance with the conditions described in paragraph 17 of the Waste Management Licensing Regulations (Northern Ireland) 2003, NISR 2003/493.

Additional conditions

This activity must be carried out for the purpose of recovery or reuse of the waste.

The Composting and Storage of Biodegradable Waste – paragraph 13

7.174 A paragraph 13 exemption allows the composting and/or the storage of certain biodegradable wastes that can be applied to land for the benefit of agriculture or ecological improvement. Paragraph 13 also allows for the chipping, shredding, cutting or pulverising of waste for the purpose of composting at the place where the composting activity is to be carried out, and the screening of waste, which has been composted.

Permitted waste

The waste types are listed in Table 6, Sch 2 of the Waste Management Licensing Regulations (Northern Ireland) 2003, NISR 2003/493 (as amended).

Quantity

The total quantity of waste treated or stored, at the place where composting is to be carried out, must not exceed 200 tonnes at any one time. The 200 tonne limit includes all wastes onsite prior to treatment.

The quantity of waste stored, at a place other than the place where composting is to be carried out, must not exceed 10 tonnes.

Storage

All waste must be stored securely on an impermeable pavement.

Waste may be stored for a period not exceeding 48 hours at a place other than the place where the composting activity is to be carried out. Such storage must be part of a scheme for the collection of waste from groups of premises and the waste must be taken directly from the place of storage to the place where the composting activity is to be carried out.

Additional conditions

In the case of biodegradable waste containing animal by-products, including gut contents, the waste must have been treated in accordance with the Animal By-Products Regulations (Northern Ireland) 2003[29].

Interpretation

- 'Composting' means the autothermic and thermophilic biological decomposition and stabilisation of biodegradable waste under controlled conditions that are aerobic or anaerobic and results in a stable sanitised material that can be applied to land for the benefit of agriculture or ecological improvement.
- 'Impermeable pavement' means a pavement in the area where a waste recovery or storage operation takes place which is capable of containing any spillage of waste received and is connected to a drainage system with impermeable components, which does not leak, and which will ensure that:
  - no liquid will run off the pavement otherwise than via the drainage system; and
  - except where they may be lawfully discharged, all liquids entering the drainage system are collected in a sealed sump.

29 NISR 2003/495.

The Manufacture of Products from Waste, the Treatment of Waste Soil or Rock which is to be Spread on Land under paragraph 9 or 11, and the Storage of Waste which is to be Submitted to the Preceding Activities – paragraph 14

7.175 A paragraph 14 exemption allows for the manufacture of timber products, straw board, plasterboard, bricks, blocks, road stone, or aggregate, from waste; the manufacture of soil, or soil substitutes, from waste; the treatment of waste soil or rock which, when treated, is to be spread on land under paragraph 9 or 11; or the storage of waste which is to be subjected to any of the preceding activities.

Permitted Waste

Waste arising from demolition or construction work, tunneling or other excavations; or waste, which consists of ash, slag, clinker, rock, wood, bark, paper, straw and gypsum. Waste soil or rock, which when treated, is to be spread on land under a paragraph 9 or 11 exemption.

Storage and location

Waste must be stored at the place where the activity is to be carried on and the total quantity of waste stored must not exceed 20,000 tonnes.

Additional conditions

The manufacture of soil or soil substitutes must be carried out at the place where either the waste is produced or the manufactured product is to be applied to land under paragraph 9 or 11.

The total amount of soil, or soil substitutes, manufactured at that place on any day must not exceed 500 tonnes.

The treatment of waste soil or rock must be carried out at the place where the waste is produced or the treated product is to be spread.

The total amount of waste soil or rock treated at that place on any day must not exceed 100 tonnes.

Documentary evidence must be provided to demonstrate that the manufacturing activity will result in a product.

The Manufacture of Finished Goods from Waste Metal, Plastic, Glass, Ceramics, Rubber, Textiles, Wood, Paper or Cardboard, and the Storage of those Wastes – paragraph 15

7.176 A paragraph 15 exemption allows the manufacture of specified waste into finished goods or the storage of waste for the manufacture of finished goods.

Permitted waste

Metal, plastic, glass, ceramics, rubber, textiles, wood, paper or cardboard.

Quantity

The total amount of any particular kind of waste stored at any time must not exceed 15,000 tonnes.

Storage and location

Waste must be stored at the place of the manufacturing activity.

The Beneficial Use of Waste – paragraph 16

7.177 A paragraph 16 exemption allows the beneficial use of waste if it is put to that use without further treatment and that use of the waste does not involve its disposal.

Permitted waste

Any waste that can be put to use without further treatment.

Quantity

The quantity will be determined by the activity for which the waste can be used beneficially without further treatment.

Storage and location

Secure storage must be provided on the premises where the waste is to be used insofar as that storage does not amount to disposal of the waste.

Additional conditions

This paragraph does not apply to the use or storage of waste if that activity is covered by an exemption conferred by paragraph 9, 10, 11, 19 or 25, or would be so covered but for any condition or limitation to which that exemption is subject by virtue of any provision contained in the paragraph by which that exemption is conferred.

What the legislation says

The full wording of the paragraph 16 exemption can be found in Sch 2 of the Waste Management Licensing Regulations (Northern Ireland) 2003, NISR 2003/493.

Relevant objectives in relation to the disposal or recovery of waste are given in para 4 of Sch 3 to the Regulations.

The Storage of Wastes for the Purposes of Recovery or Re-use – paragraph 17

7.178 The paragraph 17 exemption allows the secure storage of particular waste types where the waste is to be reused, or used for the purposes of an activity described in paragraph 12 of the Waste Management Licensing Regulations (Northern Ireland) 2003, NISR 2003/493 or, for any other recovery operation.

Permitted waste:

Kind of waste	**Maximum total quantity**
Waste paper or cardboard	15,000 tonnes
Waste textiles	1,000 tonnes
Waste plastics	500 tonnes
Waste glass	5,000 tonnes
Waste steel cans, aluminium cans or aluminium foil	500 tonnes
Waste food or drink cartons	500 tonnes
Waste articles which are to be used for construction work, and can be so used in their existing state	100 tonnes
Solvents	5 cubic metres
Refrigerants and halons	18 tonnes
Tyres	250 tyres
Waste mammalian protein	100 tonnes
Waste mammalian tallow	100 tonnes

Quantity

The total quantity of waste to be stored at any one time must not exceed the maximum total quantities stated in Table 7 of Sch 2, Pt 1 of the Waste Management Licensing Regulations (Northern Ireland) 2003 (see the table above)

Storage and location

Each kind of waste must be stored separately. The waste can be stored for no more than 12 months. Storage must be in a secure place and can be on any premises.

Additional conditions

The stored waste must be recovered or reused by an activity described in paragraph 12 of the Regulations, or in any other recovery operation.

The Storage of Waste – paragraph 18

7.179 The paragraph 18 exemption allows the storage on any premises (in a secure container or containers) of permitted waste.

Permitted waste:

- Waste paper and cardboard;
- Waste textiles;
- Waste plastics;
- Waste glass;
- Waste steel cans, aluminium cans or aluminium foil;
- Waste food or drink cartons;
- Waste articles which are to be used for construction work which are capable of being so used in their existing state;
- Waste tyres;
- Waste mammalian protein;
- Waste mammalian tallow;
- Waste oil.

Quantity

There must be no more than 20 containers on the premises and the storage capacity of the container or containers must not exceed 400 cubic metres in total.

In the case of waste oil, the storage capacity of any container or containers used must not exceed three cubic metres in total.

Storage and location

Storage can be on any premises. Each kind of waste must be kept separately.

The waste must be in a secure container or containers and the person storing the waste must be the owner of the container(s) or have the consent of the owner.

Waste cannot be stored for longer than 12 months.

Where waste oil is being stored, provision must be made to prevent oil escaping into the ground or a drain.

Additional conditions

The waste must be stored with a view to being reused, or used for the purposes of any activity described in paragraph 12, Sch 2 of the Waste Management Licensing Regulations (Northern Ireland) 2003, NISR 2003/493 (as amended) or any other recovery activity.

The Use and/or Storage of Waste for the Purpose of Construction, Maintenance or Improvement of Relevant Work – paragraph 19

7.180 A paragraph 19 exemption allows specified wastes to be used and stored for the purpose of relevant work.

Permitted waste:

The waste types that can be used are listed in Table 8, Sch 2 of the Waste Management Licensing Regulations (Northern Ireland) 2003, NISR 2003/493 (as amended).

Quantity

The waste must be used in accordance with the planning permission granted (if required) and must not exceed the dimensions of the final cross-sections shown on any plan required to be submitted with the application.

Storage and location

Waste may be stored on a site where relevant work is taking place or is to take place and can be stored for a period not exceeding three months. No more than 20,000 tonnes of waste consisting of road planings and roadbase, which are to be used for the purposes of construction, maintenance or improvement, to be carried on elsewhere, may be stored.

Additional conditions

The waste must be suitable for the purposes of the relevant work.

The waste must be used in accordance with any planning permission where such permission is required.

Definition:

- 'relevant work' means work for the purposes of construction, maintenance or improvement of:
  - a building, road, railway, airport, dock or other transport facility;
  - recreational facilities;
  - drainage;
  - a waterway;

  but does not include any work involving land reclamation.

The Laundering or Otherwise Cleaning of Waste Textiles with a View to Recovery or Re-use – paragraph 20

7.181 A paragraph 20 exemption allows the laundering or otherwise cleaning of waste textiles with a view to their recovery or reuse.

Permitted waste

Waste textiles

Storage and location

Storage must be at the place where the waste is to be laundered or cleaned.

The Chipping, Shredding, Cutting or Pulverising Waste Plant Matter – paragraph 21

7.182 A paragraph 21 exemption allows the chipping, shredding, cutting or pulverising of waste plant matter, or the sorting and baling of sawdust or wood shavings.

Permitted waste:

- Waste plant matter (including wood or bark);
- Sawdust or wood shavings.

Quantity

No more than 1,000 tonnes of waste may be dealt with in any period of seven days.

Storage and location

Storage of the waste must be at the site where the activity is to be carried out. The total amount stored must not at any time exceed 1000 tonnes.

Additional conditions

The activities must be carried on for the purpose of recovery or re-use.

The Recovery of Silver from Waste Produced in Connection with Printing or Photographic Processing – paragraph 22

7.183 A paragraph 22 exemption allows the recovery of silver from wastes produced in connection with printing or photographic processing. Guidance and application forms can be found below.

Permitted waste

Wastes produced as a result of printing or photographic processing.

Quantity

No more than 50,000 litres of waste can be dealt with in any day.

Storage and location

The storage must be at the premises where the recovery operation takes place.

The Recovery of Waste Consisting of Animal By-Products – paragraph 23

7.184 A paragraph 23 exemption allows the recovery of waste consisting of animal by-products at one of the collection centres in accordance with an authorisation under reg 27 of the Animal By-Products Regulations (Northern Ireland) 2003, NISR 2003/495.

Permitted waste

Animal by-products.

Storage and location

Waste may be stored in a secure location for up to 12 months.

Quantity

The total quantity of waste being recovered at the authorised collection centre at any time must not exceed 10 tonnes.

Additional conditions

The activity must be authorised under reg 27 of the Animal By-Products Regulations (Northern Ireland) 2003.

Interpretation

- 'Animal By-Products' has the meaning given by Article 2 of the Community Regulation (EC) No.1774/2002 laying down health rules on animal by-products not intended for human consumption.
- 'Collection centres' means premises collecting and treating certain animal by-products intended to be used for the feeding of the animals specified in Article 23(2)(c) of Regulation (EC) No.1774/2002 laying down health rules concerning animal by-products not intended for human consumption.

Animals specified are:

- zoo animals;
- circus animals;
- reptiles and birds of prey other than zoo or circus animals;
- fur animals;
- wild animals the meat of which is not destined for human consumption;
- dogs from recognised kennels or packs of hounds; and
- maggots for fishing bait.

The Crushing, Grinding or Other Size Reduction of Waste Bricks, Tiles or Concrete – paragraph 24

7.185 The paragraph 24 exemption allows the crushing, grinding, or other size reduction of waste bricks, tiles and concrete, under a permit granted under the Pollution Prevention and Control Regulations (Northern Ireland) 2003, NISR 2003/46.

Permitted waste

- Bricks;
- Tiles;
- Concrete.

Quantity

No more than 20,000 tonnes of waste can be stored prior to size reduction at any one time.

Storage and location

Waste must be stored at the place where the activity is to take place.

Additional conditions

Where the activity is carried on otherwise than at the place where the waste is produced it must be carried on with a view to recovery or reuse of the waste.

For new and current applications the operation must be permitted under a permit granted under the Pollution Prevention and Control Regulations (Northern Ireland) 2003 (PPC Regs), to the extent that it is or forms part of an activity within para (a) of Part B or para (b) of Part C of section 3.5 (other mineral activities) of Pt 1 of Sch 1 to the 2003 Regulations.

Part B processes are authorised by the Industrial Pollution and Radiochemical Inspectorate of the Northern Ireland Environment Agency.

Part C processes are authorised by the local District Council. Application must be made to the local district council.

The relevant PPC permit must be submitted with the paragraph 24 exemption application.

The Deposit of Waste Arising from Dredging Inland Waters or from Clearing Plant Matter from Inland Waters – paragraph 25

7.186 A paragraph 25 exemption allows the deposit and, treatment by screening or de-watering of waste arising from dredging or from clearing plant matter from inland waters.

Permitted waste

Inland water dredgings or plant matter.

Quantity

The total amount of waste deposited along the bank or towpath on any day must not exceed 50 tonnes for each metre of the bank or towpath along which it is deposited.

Location

The deposit of the waste must be along the bank or towpath of the waters where the dredging or clearing takes place; or

Along the bank or towpath of any inland waters so as to result in benefit to agriculture or ecological improvement.

Additional conditions

The deposit of waste does not apply to waste deposited in a container or lagoon.

The deposit of waste arising from dredging or clearing plant matter from inland waters only applies to an establishment or undertaking where the waste deposited is the establishment or undertaking's own waste.

The treatment by screening or dewatering of the waste must be on the bank or towpath of the waters where either the dredging or clearing takes place or the waste is to be deposited; or

On the bank or towpath of the waters where the dredging or clearing takes place, or at the place where the waste is to be spread, prior to its being spread in accordance with a paragraph 9(1) or (2) exemption; or

In the case of waste from dredging, on the bank or towpath of the waters where the dredging takes place, or at the place where the waste is to be spread, prior to its being spread in accordance with a paragraph 11(1) exemption.

The Recovery of Waste as an Integral Part of the Process that Produces it – paragraph 26

7.187 A paragraph 26 exemption allows the recovery, or disposal of waste at the place where it is produced, as an integral part of the process that produces it. It also allows the storage at the place where it is produced, of waste, which is intended to be so recovered or disposed of. It does not allow the recovery or disposal of waste at the place where it is produced where the final disposal of the waste is by deposit in or on land.

Waste stream(s) permitted

Any waste which can be recovered, at the place where it is produced, as an integral part of the process that produces it.

Quantity

The amount produced as a result of the process.

Storage

Storage must be at the place where the waste is produced.

Additional conditions

This exemption does not apply to the final disposal of the waste by deposit in or on land.

What the legislation says

The full wording of the paragraph 26 exemption can be found in Sch 2 of the Waste Management Licensing Regulations (Northern Ireland) 2003, NISR 2003/493.

Relevant objectives in relation to the disposal or recovery of waste are given in para 4 of Sch 3 to the Regulations.

The Baling, Compacting, Crushing, Shredding or Pulverising of Waste at the Place of Production – paragraph 27

7.188 A paragraph 27 exemption allows the baling, compacting, crushing, shredding or pulverising of waste at the place where it is produced.

Storage and location

Waste, which is to be submitted to any of the above treatments, may be stored temporarily, at the site where it was produced.

What the legislation says

The full wording of the paragraph 27 exemption can be found in Sch 2 of the Waste Management Licensing Regulations (Northern Ireland) 2003, NISR 2003/493.

Relevant objectives in relation to the disposal or recovery of waste are given in para 4 of Sch 3 to the Regulations.

The Storage of Returned Goods Pending Recovery or Disposal – paragraph 28

7.189 A paragraph 28 exemption allows the storage of returned goods that are waste, pending recovery or disposal, by their manufacturer, distributor or retailer.

Waste stream(s) permitted

Any waste.

Storage and location

Waste may be stored at the place of the manufacturer, distributor or retailer of the waste for a period not exceeding one month.

Quantity

There is no limit to the amount of waste that can be stored.

What the legislation says

The full wording of the paragraph 28 exemption can be found in Sch 2 of the Waste Management Licensing Regulations (Northern Ireland) 2003, NISR 2003/493.

Relevant objectives in relation to the disposal or recovery of waste are given in para 4 of Sch 3 to the Regulations.

The Disposal of Specified Wastes at the Place of Production in an Incinerator – paragraph 29

7.190 A paragraph 29 exemption allows the disposal of waste at the place where it is produced, by the person producing it, by burning it in an incinerator, which has a net rated thermal input of less than 0.3 megawatt if the total amount of waste burned does not exceed 50 kg per hour.

Waste stream(s) permitted

Clean wood, straw, vegetable waste and animal carcasses.

Quantity

The total amount of waste burned must not exceed 50 kg per hour in an incinerator, which has a net rated thermal input of less than 0.3 megawatts.

Storage and location

Waste must be securely stored at the place of production.

Interpretation

- 'Net rated thermal input', means the rate at which fuel can be burned at the maximum continuous rating of the appliance multiplied by the net calorific value of the fuel and expressed as megawatts thermal.
- Wastes not controlled by the European Council Directive 2000/76/EC (WID).

What the legislation says

The full wording of the paragraph 29 exemption can be found in Sch 2 of the Waste Management Licensing Regulations (Northern Ireland) 2003, NISR 2003/493.

Relevant objectives in relation to the disposal or recovery of waste are given in para 4 of Sch 3 to the Regulations.

The Burning of Waste on Land in the Open – paragraph 30

7.191 A paragraph 30 exemption allows the burning of plant tissue waste on land in the open and the storage of this waste on land where it is to be burned pending the burning of it.

Waste stream(s) permitted

Plant tissue

Quantity

The total quantity of waste burned in any period of 24 hours must not exceed 10 tonnes.

Storage and location

Storage must be on the land where the waste is intended to be burned.

Additional conditions

- waste must be burned on the land where it is produced;
- the waste must be produced on land which is operational land for a railway, light railway, tramway, or any Government Department; or which is a forest; woodland, park, garden, verge, landscaped area, sports ground, recreation ground, bank of an inland .waterway, churchyard or cemetery or on other land as a result of demolition work
- waste must only be burned by an establishment or undertaking where it is the establishment or undertaking's own waste.

Interpretation

- 'Plant tissue' means tissue from a plant and includes any linseed residues, oil-seed rape residues, field beans harvested dry residues, peas harvested dry residues, cereal straw or cereal stubble.
  'Plant' has the same meaning as in art 2 of the Plant Health Order (Northern Ireland) 2006 and means a living plant (including a fungus, tree or shrub), or a living part of a plant (including a living part of a fungus, tree or shrub), at any stage of growth; and living parts of a plant shall include: fruit or seed, vegetables, other than those preserved by deep freezing, tubers, corms, bulbs or rhizomes, cut flowers, branches with or without foliage, a plant, tree or shrub that has been cut and which retains any foliage, leaves or foliage, a plant, tree or shrub in tissue culture, live pollen, and bud wood, cuttings or scions (a detached shoot or twig containing buds from a woody plant, used in grafting).
  Plant Health Order (Northern Ireland) 2006, NISR 2006/82.
- 'Operational Land' has the meaning given by art 2 of the Planning (General Development) Order (Northern Ireland) 1993.

What the legislation says

The full wording of the paragraph 30 exemption can be found in Sch 2 of the Waste Management Licensing Regulations (Northern Ireland) 2003, NISR 2003/493.

Relevant objectives in relation to the disposal or recovery of waste are given in para 4 of Sch 3 to the Regulations.

The Discharge of Sanitary Waste from a Passenger Train – paragraph 31

7.192 A paragraph 31 exemption allows the discharge of waste onto the track from a sanitary convenience or sink forming part of a vehicle used for the carriage of passengers on the railway.

Waste stream(s) permitted

Waste from a sanitary convenience or sink.

Quantity

No more than 25 litres can be discharged.

What the legislation says

The full wording of the paragraph 31 exemption can be found in Sch 2 of the Waste Management Licensing Regulations (Northern Ireland) 2003, NISR 2003/493.

Relevant objectives in relation to the disposal or recovery of waste are given in para 4 of Sch 3 to the Regulations.

The Burial on Premises of Sanitary Waste Arising from the Use on those Premises of a Sanitary Convenience which is Equipped with a Removable Receptacle – paragraph 32

7.193 A paragraph 32 exemption allows the burial on premises of waste arising from the use on those premises of a sanitary convenience.

Waste stream(s) permitted

Sanitary waste

Quantity

No more than five cubic metres can be buried in any period of 12 months.

Location

Burial must be on the premises where the sanitary waste arises.

Additional conditions

The sanitary convenience must be equipped with a removable receptacle and the total amount buried in any period of 12 months must not exceed five cubic metres.

What the legislation says

The full wording of the paragraph 32 exemption can be found in Sch 2 of the Waste Management Licensing Regulations (Northern Ireland) 2003, NISR 2003/493.

Relevant objectives in relation to the disposal or recovery of waste are given in para 4 of Sch 3 to the Regulations.

The Keeping or Deposit of Waste Arising from Peat Working – paragraph 33

7.194 A paragraph 33 exemption allows the keeping or deposit of waste consisting of excavated materials arising from peat working.

Waste stream(s) permitted

Excavated materials arising from peat working.

Location

The keeping or deposit of the waste must be at the place where the peat working is carried out.

Additional conditions

This exemption only applies to the keeping or deposit of waste by an establishment or undertaking where the waste kept or deposited is the establishment or undertaking's own waste.

What the legislation says

The full wording of the paragraph 33 exemption can be found in Sch 2 of the Waste Management Licensing Regulations (Northern Ireland) 2003, NISR 2003/493.

Relevant objectives in relation to the disposal or recovery of waste are given in para 4 of Sch 3 to the Regulations.

The Keeping or Deposit of Spent Railway Ballast –paragraph 34

7.195 A paragraph 34 exemption allows the keeping or deposit on land where it is produced of spent ballast if the land is operational land of a railway, light railway, or tramway.

Waste stream(s) permitted

Waste railway ballast.

Location

Keeping or deposit of ballast must be on the land where it is produced.

Quantity

The total amount of waste kept or deposited at that place must not exceed 10 tonnes for each metre of track from which the ballast derives.

Additional conditions

The exemption only applies to the keeping or deposit of waste by an establishment or undertaking where the waste kept or deposited is the establishment or undertaking's own waste.

Interpretation

- 'Operational Land' has the meaning given by art 2 of the Planning (General Development) Order (Northern Ireland) 1993.

What the legislation says

The full wording of the paragraph 34 exemption can be found in Sch 2 of the Waste Management Licensing Regulations (Northern Ireland) 2003, NISR 2003/493.

Relevant objectives in relation to the disposal or recovery of waste are given in para 4 of Sch 3 to the Regulations.

The Deposit of Borehole or Excavated Waste –paragraph 35

7.196 A paragraph 35 exemption allows the deposit of waste consisting of excavated material from a borehole or other excavation made for the purpose of mineral exploration.

Waste stream(s) permitted

Excavated materials from a borehole or other excavation made for the purpose of mineral exploration.

Location

The deposit of the waste must be in or on the land at the place where it is excavated.

Additional conditions

The total quantity of waste deposited over any period of 24 months must not exceed 45,000 cubic metres per hectare.

Paragraph 35 only applies if–

- the drilling of the borehole or the making of any other excavation is development for which planning permission is granted by art 3 of, and Sch 1 Pt 16 to, the Planning (General Development) Order (Northern Ireland) 1993; and
- the conditions subject to which the development is permitted are observed.

Interpretation

Expressions used in this paragraph which are also used in the Planning (General Development) Order (Northern Ireland) 1993 shall have the same meaning as in that Order.

What the legislation says

The full wording of the paragraph 35 exemption can be found in Sch 2 of the Waste Management Licensing Regulations (Northern Ireland) 2003, NISR 2003/493.

Relevant objectives in relation to the disposal or recovery of waste are given in para 4 of Sch 3 to the Regulations.

The Temporary Storage of Garbage or Tank Washings from Ships at a Port – paragraph 36

7.197 A paragraph 36 exemption allows the temporary storage of waste, either garbage or tank washings, at a harbour or terminal, from ships that normally use that harbour or terminal.

Waste stream(s) permitted

Garbage or tank washings from ships.

Quantity

The amount of garbage landed for storage must not exceed 20 cubic metres from each ship.

Storage and location

Garbage must be stored for no more than seven days.

Tank washings being stored which consist of dirty ballast must not exceed more than 30% of the total deadweight of the ships from which they are landed.

Tank washings being stored which contain oil must not exceed 1% of the total deadweight of the ships from which they are landed.

Additional conditions

The waste reception facility must be subject to and in accordance with the Merchant Shipping and Fishing Vessels (Port Waste Reception Facilities) Regulations 2003, SI 2003/1809.

Interpretation

- 'Garbage' means all kinds of victual, domestic and operational waste excluding fresh fish and parts thereof, generated during the normal operation of the ship and liable to be disposed of continuously or periodically, except sewage originating from ships.
- 'Washings' means waste residues from the tanks (other than the fuel tanks) or holds of a ship or waste arising from the cleaning of such tanks or holds.
- 'Ship' means a vessel of any type whatsoever operating in the marine environment including submersible craft, floating craft and any structure, which is a fixed or floating platform.

What the legislation says

The full wording of the paragraph 36 exemption can be found in Sch 2 of the Waste Management Licensing Regulations (Northern Ireland) 2003, NISR 2003/493.

Relevant objectives in relation to the disposal or recovery of waste are given in para 4 of Sch 3 to the Regulations.

The Burial of a Dead Domestic Pet in the Garden of a Domestic Property where the Pet Lived – paragraph 37

7.198 A paragraph 37 exemption allows the burial of a dead domestic pet in the garden of the domestic property where the pet lived.

Waste stream(s) permitted

Dead domestic pet.

Location

The garden of the domestic property where the pet lived.

Additional conditions

The burial of a dead domestic pet is not allowed if it may prove hazardous to anyone who may come into contact with it.

The burial is not allowed to be carried out by an establishment or undertaking.

What the legislation says

The full wording of the paragraph 37 exemption can be found in Sch 2 of the Waste Management Licensing Regulations (Northern Ireland) 2003, NISR 2003/493.

Relevant objectives in relation to the disposal or recovery of waste are given in para 4 of Sch 3 to the Regulations.

The Storage or Deposit of Waste Samples to be Tested or Analysed – paragraph 38

7.199 The steps that have to be taken to apply for a paragraph 38 exemption which allows the deposit or storage of samples of waste which are being or are to be subjected to testing and analysis are set out below.

Exempted activity

Paragraph 38 allows the deposit or storage of samples of waste, which are being or are to be subjected to testing and analysis if the samples are taken:

- in the exercise of any power under –

  The Radioactive Substances Act 1993;

  The Pollution Control and Local Government (Northern Ireland) Order 1978, NIOC 1978/1049 (The 1978 Order);

  The Waste and Contaminated Land (Northern Ireland) Order 1997, NIOC 1997/2778 (The 1997 Order);

  The Water (Northern Ireland) Order 1999, NIOC 1999/662;

  The Industrial Pollution Control (Northern Ireland) Order 1997, NIOC 1997/2777;

  The Environment (Northern Ireland) Order 2002, NIOC 2002/3153; or

- by or on behalf of the holder of a Waste Management Licence, a disposal licence or a resolution in pursuance of the conditions of that licence or resolution; or
- by or on behalf of a person carrying on in relation to the waste an activity described in Pt 1 of Sch 2, or reg 16(1), of the Waste Management Licensing Regulations (Northern Ireland) 2003, NISR 2003/493
- by or on behalf of the owner or occupier of the land from which the samples were taken;
- by or on behalf of any person to whom:
  Article 5 of The 1997 Order applies in connection with his duties under that article; or

the Special Waste Regulations (Northern Ireland) 1998, NISR 1998/289 applies;

- for the purpose of research.

Waste stream(s) permitted

Waste samples, which are being or are to be tested or analysed.

Quantity

There is no limit on the quantity of samples being deposited or stored.

What the legislation says

The full wording of the paragraph 38 exemption can be found in Sch 2 of the Waste Management Licensing Regulations (Northern Ireland) 2003, NISR 2003/493.

Relevant objectives in relation to the disposal or recovery of waste are given in para 4 of Sch 3 to the Regulations.

The Storage of Waste Medicines at a Pharmacy or Storage of Waste at the Premises of a Medical, Nursing or Veterinary Practice – paragraph 39

7.200 A paragraph 39 exemption allows the secure storage at a pharmacy pending their disposal elsewhere of waste medicines which have been returned to the pharmacy from households or individuals. Paragraph 39 also allows the secure storage at the premises of a medical, nursing, or veterinary practice of waste produced in carrying on that practice.

Waste stream(s) permitted

Waste medicines returned from households or individuals to a pharmacy.

Medical, nursing, or veterinary practice waste produced in carrying on that practice.

Quantity

The total quantity of returned waste medicines at the pharmacy must not exceed five cubic metres at any one time.

The total quantity of waste stored at the premises of a medical, nursing, or veterinary practice must not at any time exceed five cubic metres.

Storage and location

Returned waste medicines at a pharmacy must not be stored for a period exceeding six months.

Wastes from medical, nursing and veterinary practices must be stored at the premises and must not be stored for a period exceeding three months.

What the legislation says

The full wording of the paragraph 39 exemption can be found in Sch 2 of the Waste Management Licensing Regulations (Northern Ireland) 2003, NISR 2003/493.

Relevant objectives in relation to the disposal or recovery of waste are given in para 4 of Sch 3 to the Regulations.

The Storage of Non-Liquid Waste Including Waste Electrical and Electronic Equipment (WEEE) at Any Site Other Than the Premises Where it is Produced – paragraph 40

7.201 A paragraph 40 exemption allows non-liquid waste, including WEEE, to be stored pending collection or transport, at a site where it was not produced.

Waste stream(s) permitted

Non-liquid waste, including WEEE.

Quantity

The amount stored must not at any time exceed 50 cubic metres in total.

Up to 10 tonnes of scrap rails can be stored on operational land of a railway, light railway or tramway, pending their collection or transport.

Storage and location

Storage must be in a secure container or containers.

Waste must not be stored for a period exceeding three months.

The person storing the waste must either own the storage container(s) or have the owner's consent.

The storage container(s) must not be on a site that also receives waste for disposal or recovery elsewhere. This excludes storage at sites commonly known as transfer stations. This exemption is to apply at sites where the transfer of waste is merely ancillary to the operations of the site.

A transfer station is a facility designed or adapted for the purpose of transferring waste.

Additional conditions

This exemption does not allow the storage of waste at a place used for the dismantling of waste motor vehicles or scrap metal recovery.

Under Annexe III of the WEEE Directive, sites for storage (including temporary storage) of WEEE prior to their treatment must have impermeable surfaces and waterproof covering for appropriate areas.

Interpretation

- Waste electrical and electronic equipment or 'WEEE' means electrical or electronic equipment which is waste within the meaning of Article 1(a) of Directive 75/442/EEC, including all components, subassemblies and consumables which are part of the product at the time of discarding.

- Operational land has the meaning given by Article 2 of the Planning (General Development) Order (Northern Ireland) 1993.

What the legislation says

The full wording of the paragraph 40 exemption can be found in Sch 2 of the Waste Management Licensing Regulations (Northern Ireland) 2003, NISR 2003/493.

Relevant objectives in relation to the disposal or recovery of waste are given in para 4 of Sch 3 to the Regulations.

The Temporary Storage of Waste, Including WEEE, Pending its Collection, at the Place Where it is Produced – paragraph 41

7.202 A paragraph 41 exemption allows the temporary storage of waste, including WEEE, at the place where it is produced, pending its collection. This paragraph does not apply to the storage of waste at a place used for the recovery of scrap metal or the dismantling of waste motor vehicles.

Waste stream(s) permitted

Any waste which is produced at that location.

Storage

This paragraph shall apply to special waste if:

- the waste is stored on the site for no more than 12 months;
- in the case of liquid waste, it is stored in a secure container and the total volume of that waste does not at any time exceed 23,000 litres; and
- in any other case–
    - the waste is stored in a secure container and the total volume of that waste does not exceed 80 cubic metres; or
    - the waste is stored in a secure place and the total volume of that waste does not at any time exceed 50 cubic metres.

Interpretation

- Waste electrical and electronic equipment or 'WEEE' means electrical or electronic equipment which is waste within the meaning of Article 1(a) of Directive 75/442/EEC, including all components, subassemblies and consumables which are part of the product at the time of discarding.
- Special waste has the meaning given by reg 2 of the Special Waste Regulations (Northern Ireland) 1998, NISR 1998/289, except that it does not include radioactive waste within the meaning of the Radioactive Substances Act 1993.

What the legislation says

The full wording of the paragraph 41 exemption can be found in Sch 2 of the Waste Management Licensing Regulations (Northern Ireland) 2003, NISR 2003/493.

Relevant objectives in relation to the disposal or recovery of waste are given in para 4 of Sch 3 to the Regulations.

The Waste Management Transitional Arrangements for the Treatment, Keeping or Disposal at any Premises, of Scrap Metal or Waste Motor Vehicles, which are to be Dismantled Pending the Issue of a Licence – paragraph 42

7.203 A paragraph 42 exemption allows the treatment, keeping or disposal by any person at any premises of scrap metal or waste motor vehicles, which are to be dismantled.

Waste stream(s) permitted

Scrap metal and waste motor vehicles.

Additional conditions

Paragraph 42 allows the treatment, keeping or disposal by any person at any premises of scrap metal or waste motor vehicles, which are to be dismantled if:

- he was carrying out the activity in question before 19 December 2003; and
- he had applied before that date for a waste disposal licence under Part II of the Pollution Control and Local Government Order (NI) 1978, authorising the activities described, and the application is pending on 19 December 2003.

The exemption will cease to have effect in relation to the carrying on of the activities at those premises on the date on which the licence applied for is granted.

If the application for a waste disposal licence is (or is deemed to be) rejected, the exemption will expire on the date on which:

- The period for appealing expires without an appeal being made; or
- Any appeal is withdrawn or finally determined.

This exemption is only suitable for an applicant who had applied for a waste disposal licence for his premises prior to 19 December 2003. New applicants will be required to apply for either a Waste Management Licence, or a paragraph 45 exemption, depending upon the activity they wish to undertake.

What the legislation says

The full wording of the paragraph 42 exemption can be found in Sch 2 of the Waste Management Licensing Regulations (Northern Ireland) 2003,

NISR 2003/493. These Regulations have been amended a number of times. A document containing subsequent legislative updates can be found on the NIEA website.

The Waste Management Transitional Arrangements to Cover (up to 19 December 2004) the Treatment, Keeping or Disposal of Waste – paragraph 43

7.204 A paragraph 43 exemption allows the treatment, keeping or disposal by any person, at any premises, of waste, by that person, if they were carrying on that activity at those premises before 19 December 2003 and, before that date no disposal licence was required under Part ll of the Pollution Control and Local Government Order (Northern Ireland) 1978 for that activity.

Waste stream(s) permitted

Waste types which were previously handled in the carrying out of the treatment, keeping or disposal activity before 19 December 2003.

Storage and location

Waste may be stored at the premises where the activity was carried out before 19 December 2003.

Additional conditions

After 19 December 2004 the exemption conferred in relation to an activity carried on by a person at any premises shall cease to have effect unless, on or before that date, the operator applies for a waste management licence in relation to the activity in question.

Where the operator makes an application for a waste management licence, the permission to carry on the activity under paragraph 43 shall continue to have effect in relation to the activity, until the date on which the licence applied for is granted or, if the application is (or is deemed to be) rejected, until the period for appealing expires without an appeal being made, or any appeal is withdrawn or finally determined.

The amount of waste being handled for the purpose of the treatment, keeping or disposal activity must comply with the relevant objectives under para 4 of Sch 3 to the Waste Management Licensing Regulations (Northern Ireland) 2003, NISR 2003/493 (as amended).

What the legislation says

The full wording of the paragraph 43 exemption can be found in Sch 2 of the Waste Management Licensing Regulations (Northern Ireland) 2003.

Relevant objectives in relation to the disposal or recovery of waste are given in para 4 of Sch 3 to the Regulations.

The Heating of Metals in a Furnace or Other Combustion Chamber for the Purpose of Removing Grease, Oil or Any Other Non-Metallic Contaminant – paragraph 44

7.205 A paragraph 44 exemption allows the heating of iron, steel or any ferrous-alloy, non-ferrous metal or non-ferrous metal alloy, in one or more furnaces or other appliances, the primary combustion chambers of which have in aggregate a net rated thermal input of less than 0.2 megawatts, for the purpose of removing grease, oil or any other non-metallic contaminant.

Waste stream(s) permitted

Iron, steel or any ferrous-alloy, non-ferrous metal or non-ferrous metal alloy.

Storage and location

Waste must be securely stored at the premises where the heating is to take place. All waste, including waste stored in containers must be stored on an impermeable pavement.

Additional conditions

In the case of a process involving the heating of iron, steel or any ferrous-alloy, the exemption does not apply if that process is related to a process described in any of paragraphs (a) to (h), or (j) to (l), of Part A or paragraphs (a), (b), (e) or (f) of Part B of section 2.1 of Sch 1 to the Industrial Pollution Control (Prescribed Processes and Substances) Regulations (Northern Ireland) 1998, NISR 1998/28; or

Paragraphs (a), (b) or (c) of Part C of section 2.1 of Schedule 1 to the Industrial Pollution Control (Prescribed Processes and Substances) Regulations 1998 or,

An activity described in section 2.1 (other than in paragraph (d) of Part C) of Part I of Schedule 1 to the Pollution Prevention and Control Regulations (Northern Ireland) 2003.

In the case of a process involving the heating of any non-ferrous metal or non-ferrous metal alloy, sub-paragraph (1) does not apply if that process is related to a process described in any of paragraphs (a) to (g), or (j) or (k), of Part A of section 2.2 of Schedule 1 to the Industrial Pollution Control (Prescribed Processes and Substances) Regulations 1998; or

an activity described in Part A of section 2.2 of Part I of Schedule 1 to the Pollution Prevention and Control Regulations (Northern Ireland) 2003.

This exemption does not apply to the removal by heat of plastic or rubber covering from scrap cable or of any asbestos contaminant.

Interpretation

- 'ferrous alloy' means an alloy of which iron is the largest constituent, or equal to the largest constituent, by weight, whether or not that alloy also has a non-ferrous metal content greater than any percentage specified in section 2.2 of Sch 1 to the 1998 Regulations, or as the case may be, section 2.2 of Sch 1 to the 2003 Regulation and 'non-ferrous metal alloy' shall be construed accordingly.

- 'net rated thermal input' means the rate at which fuel can be burned at the maximum continuous rating of the appliance multiplied by the net calorific value of the fuel and expressed as megawatts thermal.

What the legislation says

The full wording of the paragraph 44 exemption can be found in Sch 2 of the Waste Management Licensing Regulations (Northern Ireland) 2003, NISR 2003/493.

Relevant objectives in relation to the disposal or recovery of waste are given in para 4 of Sch 3 to the Regulations.

The Recovery of Scrap Metal or the Dismantling of De-Polluted End-of-Life Vehicles – paragraph 45

7.206 A paragraph 45 exemption allows the recovery of scrap metal or the dismantling of de-polluted end-of-life vehicles and storage of these wastes at any secure place. It also allows the temporary storage of non-scrap waste, which has been received as part of a consignment of scrap metal or de-polluted end-of-life vehicles.

Waste stream(s) permitted

Scrap metal and de-polluted end of life vehicles specified in Tables 9 and 10 of Schedule 2, Part 1 of the Regulations:

Kind of waste	Activities	Seven-day limit (tonnes)	Maximum total storage quantity (tonnes)	Maximum storage period (months)
Ferrous metals or ferrous alloys in metallic non-dispersible form (but not turnings, shavings or chippings of those metals or alloys)	Sorting; grading; baling; shearing by manual feed; compacting; crushing; cutting by hand-held equipment	8,000	50,000	12

Kind of waste	Activities	Seven-day limit (tonnes)	Maximum total storage quantity (tonnes)	Maximum storage period (months)
The following non-ferrous metals, namely copper, aluminium, nickel, lead, tin, tungsten, cobalt, molybdenum, vanadium, chromium, titanium, zirconium, manganese or zinc, or non-ferrous alloys, in metallic non-dispersible form, of any of those metals (but not turnings, shavings or chippings of those metals or alloys)	Sorting; grading; baling; shearing by manual feed; compacting; crushing; cutting by hand-held equipment	400	1,500	12
Turnings, shavings or chippings of any of the metals or alloys listed in either of the above categories	Sorting; grading; baling; shearing by manual feed; compacting; crushing; cutting by hand-held equipment	300	1,000	12
Depolluted end of life vehicles and their components	Dismantling and selling depolluted components	seven-day limit not applicable	40 vehicles	12

Storage and location

The storage must be for waste which is to be submitted to any of the activities specified in the above table or to a recycling or reclamation operation authorised by a waste management licence or, an authorisation under the Industrial Pollution Control Order or a permit under the 2003 Regulations.

Each kind of waste must be either stored separately or kept in separate containers, (but in the case where a consignment of more than one kind of waste is delivered to that place, it may be stored un-separated at that place pending sorting, for a period not exceeding two months).

No waste may be stored at the site for a period exceeding 12 months.

Liquid waste must be stored in a secure container.

De-polluted vehicles must be stored on a hardstanding or an impermeable pavement.

The height of any pile or stack being stored must not exceed five metres.

Additional conditions

Only depolluted end-of-life vehicles are permitted on the site.

The activity must be carried on with a view to the recovery of the waste (whether or not by the person carrying on the activity).

Every part of the site, except where depolluted vehicles are stored, must be surfaced with an impermeable pavement.

The plant or equipment used in carrying on the activity must be maintained in reasonable working order.

The paragraph 45 exemption can only apply to the carrying on of an activity at a place if the person responsible for the management of that place has established administrative arrangements to ensure that:–

- no waste must be accepted in such a quantity as would breach any of the terms and conditions of the paragraph 45 exemption;
- records are maintained which show, for each month, the total quantity of each kind of waste recovered during the preceding 12 months.

Storage of non-scrap waste

The temporary storage, pending its collection, of non-scrap waste, at a secure place used for the recovery of scrap metal or the dismantling of depolluted waste vehicles where the waste was received as part of a consignment eg, glass, seat-belts etc.

Waste stream(s) permitted

Non-scrap waste received as part of a consignment of scrap metal and non-depolluted vehicles where the non-scrap waste is not of a kind listed above (Tables 9 and 10).

Storage and location

The non-scrap waste must not be stored at that place for more than three months.

Liquid non-scrap waste must be stored in a secure container.

Non-scrap waste or the container in which the non-scrap waste is stored must be stored on an impermeable pavement.

Additional conditions

The non-scrap waste must be received as part of a consignment of waste of which:

- at least 70% by weight was waste consisting of waste motor vehicles; or
- at least 95% by weight was waste of any kind described in Tables 9 and 10 other than waste motor vehicles, and is capable of being separated from that waste by sorting or hand dismantling.

Interpretation

- 'shearing' means the cold cutting of metal by purpose-made shears.

What the legislation says

The full wording of the paragraph 45 exemption can be found in Sch 2 of the Waste Management Licensing Regulations (Northern Ireland) 2003, NISR 2003/493 (as amended).

Relevant objectives in relation to the disposal or recovery of waste are given in para 4 of Sch 3 to the Regulations.

The Recovery and Storage of Waste from a Water Treatment Works or a Sewage Treatment Works – paragraph 46

7.207 A paragraph 46 exemption allows the recovery and storage of waste from within the curtilage of a water treatment works and waste from within the curtilage of a sewage treatment works.

Waste stream(s) permitted

Waste from Water Treatment and Waste from Sewage Treatment (Table 11. Part 1, Sch 2 of the Waste Management Licensing Regulations (Northern Ireland) 2003, NISR 2003/493 (as amended).

Storage and location

Storage must be carried out on an impermeable pavement.

Quantity

The maximum quantity of wastes permitted to be accepted in any period of 12 months from the preparation of water intended for human consumption or water from industrial use at a Water Treatment Works, must not exceed 10,000 cubic metres.

The maximum quantity of other municipal wastes or waste from wastewater treatment plants permitted to be accepted at a Waste Water Treatment Works in any period of 12 months must not exceed 100,000 cubic metres.

What the legislation says

The full wording of the paragraph 46 exemption can be found in Sch 2 of The Waste Management Licensing Regulations (Northern Ireland) 2003, NISR 2003/493.

Relevant objectives in relation to the disposal or recovery of waste are given in para 4 of Sch 3 to the Regulations.

The Burning of Waste at a Dock – paragraph 47

7.208 A paragraph 47 exemption allows the burning at a dock of waste consisting of plant tissue or wood of any kind used to wedge or support parts of cargo, including packing material, spacers and pallets. The activity is carried out in pursuance of a notice given under art 22 of the Plant Health Order (Northern Ireland) 1993.

Waste stream(s) permitted

Plant tissue waste and wood of any kind used to wedge or support parts of cargo, including packing material, spacers and pallets.

Quantity

The total quantity of waste stored or burned in any period of 24 hours must not exceed 15 tonnes.

Storage and location

Storage of the waste must be at the dock where it was unloaded and where it is intended to be burned.

Additional conditions

The burning at a dock of plant tissue waste must be in pursuance of a notice given by the Department of Agriculture and Rural Development (DARD) under art 22 of the Plant Health Order (Northern Ireland) 2006, NISR 2006/82 (as amended).

The burning at a dock of wood of any kind used to wedge or support parts of cargo, including packing material, spacers and pallets must be in pursuance of a notice given by DARD under Article 20 of the Plant Health (Wood and Bark) Order (Northern Ireland) 2006, NISR 2006/66 (as amended).

In accordance with reg 18 of the Waste Management Licensing Regulations (Northern Ireland) 2003, NISR 2003/493, the burning activity may be carried out as soon as required under the notice issued by DARD and the information notified to the Northern Ireland Environment Agency thereafter as soon as is practicable.

Interpretation

- 'Plant tissue' means tissue from a plant and includes any linseed residues, oil-seed rape residues, field beans harvested dry residues, peas harvested dry residues, cereal straw or cereal stubble.
- 'Plant' has the same meaning as in Article 2 of the Plant Health Order (Northern Ireland) 2006 and means a living plant (including a fungus, tree or shrub), or a living part of a plant (including a living part of a fungus, tree or shrub), at any stage of growth; and living parts of a plant shall include: fruit or seed, vegetables, other than those preserved by deep freezing, tubers, corms, bulbs or rhizomes, cut flowers, branches with or without foliage, a plant, tree or shrub that has been cut and which retains any foliage, leaves or foliage, a plant, tree or shrub in tissue culture, live pollen, and bud wood, cuttings or scions.
- Plant Health Order (Northern Ireland) 2006

What the legislation says

The full wording of the paragraph 47 exemption can be found in Sch 2 of the Waste Management Licensing Regulations (Northern Ireland) 2003, NISR 2003/493.

Relevant objectives in relation to the disposal or recovery of waste are given in para 4 of Sch 3 to the Regulations.

The Repair, Refurbishment, and Storage of Waste Electrical and Electronic Equipment (WEEE) – paragraph 49

7.209 A paragraph 49 exemption allows the treatment and storage of Waste Electrical and Electronic Equipment (WEEE) for the purpose of its repair or refurbishment.

Waste stream(s) permitted

(a) Waste electronic and electrical equipment specified in Table 11A of Schedule 2, Part 1 of the Waste Management Licensing Regulations (Northern Ireland) 2003 as follows:

- EWC 16 02 14, (WEEE other than those mentioned in EWC 16 02 09 to 16 02 13);
- EWC 20 01 36, (WEEE other than those mentioned in EWC 20 01 21, 20 01 23 and 20 01 35).

(b) Waste electronic and electrical equipment specified in Table 11B of Schedule 2, Part 1 of the Regulations as follows:

- EWC 16 02 11*, (WEEE containing chlorofluorocarbons, HCFC, HFC);
- EWC 20 01 23*, (WEEE containing chlorofluorocarbons).

(c) Waste electronic and electrical equipment specified in Table 11B of Schedule 2, Part 1 of the Regulations as follows:

- Television and computer monitors containing cathode ray tubes falling within EWC 16 02 13* (WEEE containing hazardous components other than those mentioned in 16 02 09 to 16 02 12) and EWC 20 01 35* (WEEE other than those mentioned in 20 01 21 and 20 01 23 containing hazardous components).

Annex 1B of the WEEE Directive provides a list of products which fall under each category. The list is not exhaustive but shows examples of the type of products included in the broader categories.

An asterisk against an EWC code indicates that this waste is hazardous. If hazardous waste is handled under this exemption there may be additional obligations under the Hazardous Waste Regulations.

Quantity

The maximum quantity of waste that can be treated is five tonnes per day.

Storage

WEEE must be securely stored at the place where it is intended to be repaired or refurbished.

The total quantity of any particular kind of WEEE stored must not exceed 80 cubic metres.

The WEEE must be stored in such a manner that is environmentally sound and reuse or recycling is not hindered.

No WEEE can be stored at that place for more than 12 months.

The waste must be stored and treated in a manner that will prevent the release of CFCs, HCFCs, or HFCs.

Additional conditions

All WEEE must be stored on an impermeable pavement with a weatherproof covering.

The activity must be carried on with a view to the reuse of the WEEE for its original purpose.

Best available treatment, recovery and recycling techniques must be used when carrying out the activity.

The technical requirements specified in Annex III of the WEEE Directive must be met.

Interpretation

- 'Waste electrical and electronic equipment' or 'WEEE' means electrical or electronic equipment which is waste within the meaning of Article 1(a) of Directive 75/442/EEC, including all components, subassemblies and consumables which are part of the product at the time of discarding;

- 'Impermeable pavement' means a pavement in the area where a waste recovery or storage operation takes place and which is capable of containing any spillage of waste received and is connected to a drainage system with impermeable components, which does not leak and which will ensure that:
  - no liquid will run off the pavement otherwise than via the drainage system; and
  - except where they may be lawfully discharged, all liquids entering the drainage system are collected in a sealed sump.

What the legislation says

The full wording of the paragraph 49 exemption can be found in Sch 2 of the Waste Management Licensing Regulations (Northern Ireland) 2003.

Relevant objectives in relation to the disposal or recovery of waste are given in para 4 of Sch 3 to the Regulations.

The Secure Storage of Waste Electrical and Electronic Equipment (WEEE) – paragraph 50

7.210 A paragraph 50 exemption allows the secure storage of Waste Electrical and Electronic Equipment (WEEE) for the purpose of its recovery elsewhere. The exemption also allows for the incidental sorting of WEEE.

Waste stream(s) permitted

Waste electronic and electrical equipment (WEEE) specified in Tables 11C and 11D of Schedule 2, Part 1 of the Waste Management Licensing Regulations (Northern Ireland) 2003, NISR 2003/493 (as amended) as follows:

- EWC 16 02 14, (WEEE other than those mentioned in EWC 16 02 09 to 16 02 12);
- EWC 20 01 36, (WEEE other than those mentioned in EWC 20 01 21, 20 01 23 and 20 01 35);
- EWC 16 02 11 (WEEE containing chloroflurocarbons, HCFC or HFC);
- EWC 20 01 23 (WEEE containing chloroflurocarbons);
- EWC 16 02 13 (WEEE containing hazardous components other than those mentioned in 16 02 09 to 16 02 12);
- EWC 20 01 35 (WEEE other than those mentioned in 20 01 21 and 20 01 23 containing hazardous components);
- EWC 20 01 21 (fluorescent tubes and other mercury-containing waste).

Annex 1B of the WEEE Directive provides a list of products which fall under each category. The list is not exhaustive but shows examples of the type of products included in the broader categories.

Quantity

For waste other than fluorescent tubes and other mercury containing lamps the maximum quantity that can be stored is 80 cubic metres.

For fluorescent tubes and other mercury containing lamps the maximum quantity that can be stored is 50 cubic metres.

Storage

Waste may be stored for a maximum period of three months.

Waste other than fluorescent tubes and other mercury containing lamps must be stored on an impermeable surface and have weatherproof covering.

Fluorescent tubes and other mercury containing waste must be stored in appropriate secure containers with weatherproof covering, and in such a way that the glass is not broken.

WEEE must be stored in such a manner that its environmentally sound reuse or recycling is not hindered.

Waste containing chlorofluorocarbons must be stored in a manner that will prevent the release of the CFC, HCFC and HFC. The number of units in any stack (eg fridges and freezers) must not exceed two and the overall height of any stack shall not exceed 3.5m.

Additional conditions

WEEE must be stored for the purposes of recovery elsewhere.

All WEEE other than fluorescent tubes and other mercury containing lamps must be stored on an impermeable pavement with a weatherproof covering. The impermeable surface must direct all liquid to run off to a drainage system with impermeable components, which does not leak, and where all liquids entering the drainage system are stored in a sealed sump.

Interpretation

- 'Waste electrical and electronic equipment' or 'WEEE' means electrical or electronic equipment which is waste within the meaning of Article 1(a) of Directive 75/442/EEC, including all components, subassemblies and consumables which are part of the product at the time of discarding.
- 'impermeable pavement' means a pavement in the area where a waste recovery or storage operation takes place and which is capable of containing any spillage of waste received and is connected to a drainage system with impermeable components, which does not leak and which will ensure that:
  - no liquid will run off the pavement otherwise than via the drainage system; and
  - except where they may be lawfully discharged, all liquids entering the drainage system are collected in a sealed sump.

What the legislation says

The full wording of the paragraph 50 exemption can be found in Sch 2 of the Waste Management Licensing Regulations (Northern Ireland) 2003, NISR 2003/493.

Relevant objectives in relation to the disposal or recovery of waste are given in para 4 of Sch 3 to the Regulations.

The Crushing of Waste Gas Discharge Lamps – paragraph 51

7.211 A paragraph 51 exemption allows the crushing of waste gas discharge lamps (including florescent tubes EWC 20 01 21*) for the purposes of volume reduction prior to collection, where the material is intended for recovery or reuse. It also allows the secure storage of the lamps prior to crushing, or after crushing prior to collection.

Waste stream(s) permitted

Waste gas discharge lamps.

Quantity

The total quantity of lamps processed in any period of 24 hours must not exceed three tonnes.

Storage and location

Lamps must be stored under weatherproof covering prior to crushing or after crushing but prior to collection. After crushing, the lamps must be stored in a secure container.

Additional conditions

The activity must be carried out for the purpose of recovery or reuse, but not for disposal.

The mercury concentration in emissions must not exceed 50 micrograms per cubic metre.

What the legislation says

The full wording of the paragraph 51 exemption can be found in Sch 2 of the Waste Management Licensing Regulations (Northern Ireland) 2003, NISR 2003/493.

Relevant objectives in relation to the disposal or recovery of waste are given in para 4 of Sch 3 to the Regulations.

The Secure Storage of Waste Batteries and Accumulators – paragraph 52

7.212 A paragraph 52 exemption allows the secure storage of waste batteries or accumulators at a place other than the place of production, pending their recovery.

Waste stream(s) permitted

Waste batteries or accumulators specified in Table 11E of Sch 2, Pt 1 of the Waste Management Licensing Regulations (Northern Ireland) 2003, NISR 2003/493(as amended) as follows:

- EWC 16 06 01*, lead batteries;

- EWC 16 06 02*, Ni-Cd batteries;
- EWC 16 06 04, alkaline batteries except 16 06 03, mercury-containing batteries;
- EWC 20 01 33*, batteries and accumulators included in 16 06 01, 16 06 02 or 16 06 03 and unsorted batteries and accumulators containing these batteries;
- EWC 20 01 34, batteries and accumulators other than those mentioned in 20 01 33.

Quantity

The maximum quantity of automotive batteries and accumulators that can be stored at any one time is five tonnes.

The maximum quantity of portable batteries and accumulators that can be stored at any one time is 10 tonnes.

Storage

The storage site must have a sealed drainage system and all storage must take place on an impermeable pavement, under weatherproof covering.

The duration of storage must not be longer than six months.

Interpretation

- 'Waste battery or accumulator' means any battery or accumulator, which is waste within the meaning of Article 1(1)(a) of Directive 2006//12/EC of the European Parliament and of the Council on waste.
- 'Battery or accumulator' means any source of electrical energy generated by direct conversion of chemical energy and consisting of one or more primary battery cells (non-rechargeable) or consisting of one or more secondary battery cells (rechargeable).
- 'Automotive battery or accumulator' means any battery or accumulator used for automotive starter, lighting or ignition power.
- 'Portable battery or accumulator' means any battery, accumulator or battery pack which is sealed; can be hand-carried by an average person without difficultyis neither an automotive battery or accumulator nor an industrial battery or accumulator.
- 'Impermeable pavement' means a pavement in the area where a waste recovery or storage operation takes place and which is capable of containing any spillage of waste received and is connected to a drainage system with impermeable components, which does not leak and which will ensure that:
  - no liquid will run off the pavement otherwise than via the drainage system;
  - except where they may be lawfully discharged, all liquids entering the drainage system are collected in a sealed sump.

An asterisk against an EWC code indicates that the waste is hazardous. If hazardous waste is handled under this exemption there may be additional obligations under the Hazardous Waste Regulations.

What the legislation says

The full wording of the paragraph 52 exemption can be found in Sch 2 of the Waste Management Licensing Regulations (Northern Ireland) 2003, NISR 2003/493.

Relevant objectives in relation to the disposal or recovery of waste are given in para 4 of Sch 3 to the Regulations.

Chapter 8

Special/Hazardous Waste

INTRODUCTION

8.1 The Special Waste Regulations 1996[1] were enacted on 1 September 1996, causing a change in the manner by which toxic waste is defined and controlled. For England and Wales these have been replaced by the Hazardous Waste (England and Wales) Regulations 2005, SI 2005/894 and for Northern Ireland the Hazardous Waste (Northern Ireland) Regulations 2005, NISR 2005/300 and the term 'hazardous waste' is used for Special Waste. In Scotland the term 'special waste' is still used.

Waste is defined as 'hazardous waste' (in England, Wales and Northern Ireland) or 'special waste (in Scotland) if it is classified as 'hazardous' in the European Waste Catalogue (or list of wastes). Generally waste is hazardous/special if it, or the materials or substances it contains, are harmful to human health or environment.

This chapter looks at the application of the Regulations in each of these areas in some detail and there is a table which shows how the Regulations control hazardous/special waste by topic.

Waste legislation is constantly changing and so it is anticipated that further amendments of these Regulations will occur.

ENGLAND, WALES AND NORTHERN IRELAND

8.2 The Hazardous Waste (England and Wales) Regulations 2005, SI 2005/894 and the Hazardous Waste (Northern Ireland) Regulations 2005, NISR 2005/300 are very similar in the way they address the issue of dealing with hazardous waste. This section looks at the two sets of Regulations and where there are differences, makes this clear – otherwise the text should be read as applying to both.

1 Special Waste Regulations 1996, SI 1996/972.

Environment Agency Cost Recovery

8.3 The Environment Agency recovers the costs it incurs carrying out its functions in relation to hazardous waste by means of a charge imposed for issuing consignment notes, which are needed in order to move waste in accordance with the Special Waste Regulations.

These Regulations will not rely on the Environment Agency issuing consignment notes. The Agency will need to recover its costs by a different route and this in turn means that payments are required in different circumstances. However, whilst some producers may face fee increases, those who make more than two consignments of hazardous waste will be able to benefit from reduced costs.

The Regulations also require notification of premises at which waste is produced or removed. This notification requirement will be subject to a charge. Producers will notify the premises at which the waste is produced. If waste is removed from other premises, the person responsible for the removal (the consignor) will notify.

The Regulations also now require a quarterly return from consignees (ie, those who receive the waste) and from those producers who dispose on the site of production. The Environment Agency intends to impose a charge in both cases, with consignees given a right to re-claim any payment from the consignor who sent them the waste. In practice, consignees will pass on any charges in any event but structuring the charging regime in this way is administratively more efficient and results in lower costs to the Agency and hence lower charges overall. Again a transitional charging scheme is put in place until a scheme can be made.

The Regulations implement the Hazardous Waste Directive (91/689/EEC) ([1991] OJ L 377, 31/12/91 p 0020–0026).

Wales

8.4 With the exception of Pt 11[2] the National Assembly for Wales will make its own regulations for hazardous waste in Wales. The changes in Pt 11 rely on s 156 of the Environmental Protecion Act 1990 (the 1990 Act) to broaden the offence in s 33 of the 1990 Act by including a new definition of hazardous waste and to include a new power to list waste which displays hazardous properties. Section 156 has not been transferred to Wales and it has therefore been necessary for those provisions to extend to Wales. The definition of hazardous waste inserted into the 1990 Act differs for Wales because it is not possible to refer to the equivalent Regulations in Wales until they are made. The intention is that the equivalent National Assembly for Wales regulations will come into force on the same day and there will then be an opportunity

2 Hazardous Waste (England and Wales) Regulations 2005, SI 2005/894, Part 11 Replacement of Controls on Special Waste.

to bring the definitions into line. If the National Assembly for Wales do not make the bring forward regulations governing hazardous waste in Wales for any reason, it will be necessary to revoke these provisions in relation to Wales before they come into force.

European Waste Catalogue (EWC) Codes

8.5 Any controlled waste that is collected in the UK must under the Duty of Care, be covered by a Waste Transfer Note or, for hazardous/special waste a consignment note.

A requirement of these documents is to describe the waste and this will require a written description that will vary according to the waste carrier, but it will always require an EWC code. This is a list of wastes that has been produced by the EU to standardise the description of waste across Europe and to enable waste to be recorded, monitored and controlled more effectively. The EWC codes have been transposed into UK law by the List of Wastes Regulations (SI 2005/895, SI 2005/1820, NISR 2005/301).

Technical Guidance on waste classification has been provided by the relevant agencies for England, Northern Ireland, Scotland and Wales[3].

Mixing of Hazardous Waste

8.6 There is a general prohibition on the mixing of different types of hazardous waste or mixing of hazardous waste with non-hazardous waste in the Hazardous Waste Directive. This provision, which applied to those who dispose of, recover, collect or transport hazardous waste, is extended to those producing hazardous waste (mixing will still be allowed where it is being undertaken in accordance with a permit). As the source of hazardous waste, producers are the first place where wastes could be mixed. Mixing of different hazardous wastes is generally discouraged, as there are unknown risks associated with the resulting waste. The types of reactions that can occur when mixing such wastes are complex and the hazard characteristics of the resulting waste are not always easy to ascertain.

The mixing of hazardous waste with any other materials, whether they are hazardous or not, is also undesirable for a number of reasons. In the first instance it may dilute the hazardous waste to below threshold value levels, therefore avoiding hazardous waste controls. Secondly, such mixing could in some cases increase the volumes of hazardous waste arising, by making the non-hazardous waste, hazardous. For example, fluorescent tubes disposed in a

3 Waste Classification – Guidance on the classification and assessment of waste (1 edn 2015) Technical Guidance WM3 Adapted from the third edition of technical guidance WM2 LIT 10121 available at: https://www.gov.uk/government/uploads/system/uploads/attachment_data/file/427077/LIT_10121.pdf.

skip full of otherwise non-hazardous commercial waste. The presence of these tubes would result in the whole skip full of waste being classified as hazardous waste.

Notification of Premises

8.7 There is a general requirement for hazardous waste producers to notify the Environment Agency of sites where hazardous waste is produced and this attracts an annual 'site notification' fee.

As some waste producers will only produce minimal amounts of hazardous waste and these may be on a sporadic basis and whilst these wastes must be subject to the control system there is little environmental benefit to be gained from the notification of such sites. The Hazardous Waste Regulations therefore exempt certain small businesses from the requirement to notify. The following premises may use this exemption: offices and shops, small agricultural businesses, caravans, camp sites, houseboats, residential homes, premises forming part of a university, school or other educational establishment or a hospital or nursing home, doctors' surgeries, dental surgeries, health clinics, prisons, public meeting halls, and premises used by charities.

Multiple Collections – Modification of Carrier Rounds

8.8 Small loads of similar wastes, such as waste oils from garages, can be collected in small loads, as small loads from many producers in the same vehicle. The Hazardous Waste Regulations modified procedures for loads collected on carrier rounds (called 'multiple collections'). To meet the Hazardous Waste Directive requirements, information must be obtained about each load collected, including details about each producer. Using the multiple collection procedure, each load collected will be required to have a unique consignment code and a separate consignment note for multiple collections will obtain all the information required on individual loads whilst also providing summary information for carriers and consignees. Charging for loads on multiple collections also means that each load will be subject to a waste movement fee (although this fee will be lower than the fee for single waste movements). In order to simplify administrative procedures, and avoid excessive paperwork, the Regulations include a schedule for recording loads on multiple collections.

Coding Standards

8.9 It is the duty of the Agency from time to time to designate and at all times to maintain in force, a designation of a standard (in these Regulations referred to as a 'coding standard'), making provision for the composition of consignment codes.

The coding standard must enable each consignment of hazardous waste to be given a unique consignment code.

The standard makes provision for different codes for consignment notes completed in relation to hazardous waste removed by pipeline reg 41[4], where piping is continuous for more than one quarter.

The standard shall make provision for consignment codes to consist of letters, numbers or symbols, or any combination of letters, numbers and symbol.

NORTHERN IRELAND

8.10 The Hazardous Waste Regulations (Northern Ireland) 2005[5] and the List of Waste Regulations (Northern Ireland) 2005[6] came into force on 16 July 2005. They revoked the Special Waste Regulations (Northern Ireland) 1998[7]. These Regulations were introduced to implement the provisions of European Commission (EC) Directive 91/689/EEC as amended and Commission Decision 2000/532/EC.

They aim to provide a more effective system of control over special waste from the time of its production to the time of its final destination for disposal or recovery. They also enable waste holders to comply with the statutory 'Duty of Care' in relation to waste.

These Regulations deal with the assignment of these codes to hazardous waste and the relevant fees charged for doing so, the completion of consignment notes, the pre-notification of a consignment and also the keeping of registers and site records. Details of offences and penalties are also provided.

SCOTLAND

8.11 In Scotland, the Special Waste Regulations 1996[8] as amended set out procedures to be followed when disposing of, or carrying and receiving special waste. The Special Waste Amendment (Scotland) Regulations 2004, SI 2004/112 amended the Pollution Prevention and Control (Scotland) Regulations 2000, SI 2000/323.

It included the following definitions[9]:

4 SI 2005/894, Part 6 Documents to be completed for consignments, reg 41 Removal of wastes by pipeline.
5 The Hazardous Waste Regulations (Northern Ireland) 2005, NISR 2000/300. NISR 2005/300.
6 The List of Waste Regulations (Northern Ireland) 2005, NISR 2005/301.
7 Special Waste Regulations (Northern Ireland) 1998, NISR 1998/289.
8 The Special Waste Amendment (Scotland) Regulations 2004, SSI 2004/112.
9 Regulation 3(2).

'Hazardous Waste' unless the context requires otherwise means any waste, which is special waste to which the Special Waste Regulations 1996 apply.

'Waste' means, unless the context otherwise requires, anything that–

(i) is waste for the purposes of the Waste Directive; and

(ii) is not excluded from the scope of the Directive by Article 2[10] of the Directive.

The regulations set out procedures for the disposing of, carrying and receiving special waste – they are the principal piece of legislation covering special waste arising in Scotland.

ENGLAND AND WALES REGULATIONS

The Hazardous Waste (England and Wales) Regulations 2005, SI 2005/894

8.12 The Regulations were made to implement the Hazardous Waste Directive (91/689/EEC). The Regulations replaced controls imposed by the special waste regime set out in Pt 2 of the Environmental Protection Act 1990 and the Special Waste Regulations 1996, SI 1996/972. They partly address an unfavourable judgment made by the ECJ in relation to the transposition of the Directive (Case C-431/02)[11] as the domestic provisions did not dovetail adequately with those in the Directive.

The Regulations also introduced the requirement for notification of premises at which waste is produced or removed. This requirement has been revoked in the Hazardous Waste (England and Wales) (Amendment) Regulations 2016, SI 2016/336.

10 75/442/EEC Article 2
  1 Without prejudice to this Directive, Member States may adopt specific rules for particular categories of waste.
  2 The following shall be excluded from the scope of this Directive:
    (a) radioactive waste;
    (b) waste resulting from prospecting, extraction, treatment and storage of mineral resources and the working of quarries;
    (c) animal carcasses and the following agricultural waste : faecal matter and other substances used in farming;
    (d) waste waters, with the exception of waste in liquid form;
    (e) gaseous effluents emitted into the atmosphere;
    (f) waste covered by specific Community rules.

11 *European Commission v United Kingdom* (Case C-431/02)
By failing to adopt all the measures necessary to comply with its obligations under Articles 1(4) and (5), 2(1), (2) and (4), 3(1) to (4), 4(1) to (3) and 5(2) of Council Directive (EEC) 91-689 (on hazardous waste), the United Kingdom had failed to fulfil its obligations under that directive and under the EC Treaty.

WALES

The National Assembly for Wales (Transfer of Functions) (No 2) Order 2006, SI 2006/3334

8.13 This Order makes provision for the transfer of a number of powers to the National Assembly for Wales. In relation to s 156 of the Environmental Protection Act 1990, the National Assembly has the power to give effect to Community law obligations with respect to waste management under s 2(2)[12] of the European Communities Act 1972 (ECA 1972).

The Hazardous Waste (Wales) Regulations 2005, SI 2005/1806[13]

Provision of Information to the National Assembly (Reg 60)

8.14 The Agency must inform the Assembly each year of any changes in the following information for every establishment or undertaking which carries out disposal or recovery of hazardous waste principally on behalf of third parties and which is likely to form part of the integrated network referred to in Article 5 of the Waste Directive:

- name and address;
- the method used to treat waste; and
- the types and quantities of waste, which can be treated.

The Agency must provide the information required by paragraph (1) in the format provided for pursuant to the fourth paragraph of Article 8(3) of the Hazardous Waste Directive.

12 ECA 1973

(2) Subject to Schedule 2 to this Act, at any time after its passing Her Majesty may by Order in Council, and any designated Minister or department may [**F2** by order, rules, regulations or scheme] , make provision—

(a) for the purpose of implementing any [**F3** EU obligation] of the United Kingdom, or enabling any such obligation to be implemented, or of enabling any rights enjoyed or to be enjoyed by the United Kingdom under or by virtue of the Treaties to be exercised; or

(b) for the purpose of dealing with matters arising out of or related to any such obligation or rights or the coming into force, or the operation from time to time, of subsection (1) above;

and in the exercise of any statutory power or duty, including any power to give directions or to legislate by means of orders, rules, regulations or other subordinate instrument, the person entrusted with the power or duty may have regard to the [**F4** objects of the EU] and to any such obligation or rights as aforesaid.

In this subsection 'designated Minister or department' means such Minister of the Crown or government department as may from time to time be designated by Order in Council in relation to any matter or for any purpose, but subject to such restrictions or conditions (if any) as may be specified by the Order in Council.

13 The Hazardous Waste (Wales) Regulations 2005, SI 2005/1806.

The Hazardous Waste (Miscellaneous Amendments) (Wales) Regulations 2015, SI 2015/1417

8.15 These Regulations amend certain statutory instruments, which are concerned with hazardous waste and refer to the List of Waste Regulations 2005 (SI 2005/895, SI 2005/1820, NISR 2005/301).

They make amendments to the following:

- Landfill Allowances Scheme (Wales) Regulations 2004, SI 2004/1490;
- Hazardous Waste (Wales) Regulations 2005, SI 2005/1806;
- Environmental Permitting (England and Wales) Regulations 2010, SI 2010/675;
- Waste (England and Wales) Regulations 2011, SI 2011/988;
- Controlled Waste (England and Wales) Regulations 2012, SI 2012/811.

The Hazardous Waste (Wales) (Amendment) Regulations 2009, SI 2009/2861[14]

8.16 These Regulations are designated for the purposes of s 2(2) of the European Communities Act 1972[15] in relation to the prevention, reduction and elimination of pollution caused by waste and exercises of the powers conferred by s 2(2) of the Act, the Welsh ministers made these Regulations.

These amendment Regulations are almost the same in content as the Hazardous Waste (England and Wales) Regulations 2005, SI 2005/894. References to the amendments are given for each regulation including at reg 42 (duty of a consignee not accepting delivery) and amendment to reg 70.

These amendments only apply to Wales.

The Hazardous Waste (England and Wales) (Amendment) Regulations 2009, SI 2009/507[16]

8.17 This amends the Hazardous Waste (England and Wales) Regulations 2005, SI 2005/894 to:

- increase the qualifying limitation in reg 30 from 200kg to 500kg in any period of 12 months and extend the scope of the exemption from the requirement to notify from specified premises to all premises within the qualifying limitation; and
- clarify certain provisions of the 2005 Regulations.

14 The Hazardous Waste (Wales) (Amendment) Regulations 2009, SI 2009/2861.
15 Section 2(2) was amended by s 27(1) of the Legislative Regulatory Reform Act 2006.
16 The Hazardous Waste (England and Wales) (Amendment) Regulations 2009, SI 2009/507.

The Hazardous Waste (Miscellaneous Amendments) Regulations 2015, SI 2015/1360

8.18 These regulations update domestic legislation in light of amendments to two EU instruments establishing the classification for waste and hazardous waste and the replacement of an EU instrument concerning waste electrical equipment.

The classification system for waste and hazardous waste is established by two EU instruments: Directive 2008/98/EC of the European Parliament and of the Council on waste ('the Waste Framework Directive') (see Annex III) and Commission Decision 2000/532/EC establishing a list of wastes ('the EU List of Wastes') ([2000] OJ:L226/1/00) These instruments were amended in December 2014. In addition, a 2002 EU instrument concerning waste electrical equipment was replaced by a new instrument: Directive 2012/19/EU of the European Parliament and of the Council on waste electrical and electronic equipment ([2012] OJ L 197, 24.7.12, p 38–71).

Accordingly, provisions of the following domestic instruments require updating:

- the Environmental Protection Act 1990;
- the Hazardous Waste (England and Wales) Regulations 2005, SI 2005/894;
- the Batteries and Accumulators (Placing on the Market) Regulations 2008, SI 2008/2164;
- the Cremation (England and Wales) Regulations 2008, SI 2008/2841;
- the Waste Batteries and Accumulators Regulations 2009, SI 2009/890;
- the Environmental Permitting (England and Wales) Regulations 2010, SI 2010/675;
- the Waste (England and Wales) Regulations 2011, SI 2011/988;
- the Controlled Waste (England and Wales) Regulations 2012, SI 2012/811.

The Hazardous Waste (England and Wales) (Amendment) Regulations 2016, SI 2016/336[17]

8.19 This revokes Part 5 of the Hazardous Waste (England and Wales) Regulations 2005, SI 2005/894 ('the 2005 Regulations'), which requires the notification of premises at which hazardous waste, is produced at, collected at or removed from. There are also consequential amendments arising from the revocation of Part 5, and a correction to an error in reg 49(4) of the 2005 Regulations.

17 The Hazardous Waste (England and Wales) (Amendment) Regulations 2016, SI 2016/336.

The Hazardous Waste (England and Wales) Regulations 2005, SI 2005/894

8.20 Parts 1 to 3 of the Regulations define hazardous waste and how the Regulations apply to that waste. The definition of waste in reg 6 refers to the list of hazardous wastes set out in the List of Wastes (England) Regulations 2005[18] and the Lists of Wastes (Wales) Regulations 2005[19]. There is an exclusion from these controls for domestic waste which displays hazardous properties but not if it comprises asbestos waste or is collected separately. In both cases, the regulations do not impose obligations directly on householders.

Part 1 General

The Waste Directive and the Meaning of Waste (reg 2)

8.21 The Waste Directive is the Council Directive 75/442/EEC[20] as amended by 91/156/EEC[21], the Commission's decision 96/350/EC[22] and Regulation (EC) No 1882/2003.

In these Regulations, a reference to the Waste Directive conditions are those laid down in Article 4[23] of that Directive. Any waste should be recovered or disposed of without endangering human health and without using processes or methods which could harm the environment and in particular:

- without risk to water, air, soil and plants and animals;
- without causing a nuisance through noise or odours; and
- without adversely affecting the countryside or places of special interest.

The Hazardous Waste Directive (reg 3)

8.22 The Regulations make clear that the Hazardous Waste Directive (HWD) Annex 1, Annex 2 and Annex 3 are referenced in Sch 1, Sch 2 and Sch 3 of these Regulations.

18 The List of Wastes (England) Regulations 2005, SI 2005/895.
19 The List of Wastes (Wales) Regultions 2005, SI 2005/1820.
20 Council Directive 75/442/EEC 15 July 1975 on waste.
21 Council Directive 91/556/EEC 18 March 1991 amending Directive 75/442/EEC on waste.
22 96/350/EC Commission Decision 24 May 1996 adapting Annexes IIA and IIB of Council Directive 75/442/EEC on waste.
23 *Article 4*
Member States shall take the necessary measures to ensure that waste is recovered or disposed of without endangering human health and without using processes or methods which could harm the environment, and in particular:
- without risk to water, air, soil and plants and animals,
- without causing a nuisance through noise or odours,
- without adversely affecting the countryside or places of special interest.

Member States shall also take the necessary measures to prohibit the abandonment, dumping or uncontrolled disposal of waste.

List of Wastes (reg 4)

8.23 Any controlled waste that is collected in the UK must, under the Duty of Care (see **Chapter 4**) be covered by a Waste Transfer note, or for Hazardous Special Waste a consignment note. These documents require a description of waste, which will vary according to the waste carrier – but also a European Waste Catalogue (EWC) code. One of the requirements of The Lists of Wastes has been produced by the EU to standardise the description of waste across Europe and to enable waste to be recorded, monitored and controlled more effectively.

'The List of Wastes Decision' means Commission Decision 2000/532/EC of 3 May 2000 replacing Decision 94/3/EC establishing a list of wastes pursuant to Article 1(a) of Council Directive 75/442/EEC on waste and Council Decision 94/904/EC establishing a list of hazardous waste pursuant to Article 1(4) of Council Directive 91/689/EEC on hazardous waste, as amended by amendments thereto which have effect from time to time in relation to England pursuant to the List of Wastes Regulations.

'The List of Wastes Regulations' means the List of Wastes (England) Regulations 2005, SI 2005/895.

'The List of Wastes' means the list of wastes set out in the List of Wastes Decision as it is from time to time set out in the List of Wastes Regulations, being the list referred to in the first indent of Article 1(4) of the Hazardous Waste Directive drawn up on the basis of Annexes I and II, having one or more of the properties listed in Annex III, taking account of the origin and composition of the waste and, where necessary, limit values of concentration.

Within these Regulations 'listed as a waste' and 'listed as a hazardous waste' refers to that waste being listed as a waste, or as a hazardous waste, in the List of Wastes, in the case of a waste to which a limit value of concentration applies, then it will only be considered to be as a hazardous waste where the relevant limit value of concentration is satisfied.

Waste that is not listed as a hazardous waste in the List of Wastes, whether or not it is listed as a waste, and whether or not it is otherwise a hazardous waste pursuant to these Regulations.

General Interpretation (reg 5)

8.24 This section defines the terms used within the regulations.

The original definition of 'premises' is changed to[24]:

> '"premises" includes land and any ship and any other means of transport from which a mobile service is operated;'.

24 The Hazardous Waste (England and Wales) (Amendment) Regulations 2009, SI 2009/507, reg 3; the Hazardous Waste (Wales) (Amendment) Regulations 2005, SI 2005/1806, reg 3.

This clarifies that the term includes land.

The definitions of 'mobile service' and 'SIC' are omitted[25]. In the definition of premises the words 'and any other means' is changed to 'operated'.

In paragraph (3)(c) the word notification is omitted.

(3) (c) any requirement for a signature on a [...] consignment note, schedule of carriers or multiple collection consignment note, may be satisfied by an electronic signature incorporated into the document;

Part 2 Hazardous and Non-hazardous Waste

Hazardous Waste (reg 6)

8.25 Waste is hazardous if it is:

- Listed as a hazardous waste in the List of Wastes[26];
- Listed in regulations made under the Environmental Protection Act 1990, s 62A[27] or
- A specific load of waste, which under reg 8[28] is determined to be hazardous waste.

Non Hazardous Waste (reg 7)

8.26 A waste is non-hazardous if it is:

- a waste which is not a hazardous waste pursuant to reg 6 (Hazardous waste); or

25 The Hazardous Waste (England and Wales) (Amendment) Regulations 2016, SI 2016/336, reg 2(2)(a)(i) and (ii).

26 Wastes listed as hazardous in the list of Wastes are considered hazardous in Article 1.4 of the Hazardous Waste Directive.

27 **62A Lists of waste displaying hazardous properties**
(1) The Secretary of State shall by regulations list any controlled waste in England which—
(a) is not listed as a hazardous waste in the Hazardous Waste List; and
(b) appears to him to display any of the properties listed in Annex III to Council Directive 91/689/EEC.
(2) The National Assembly for Wales shall by regulations list any controlled waste in Wales which—
(a) is not listed as a hazardous waste in the Hazardous Waste List; and
(b) appears to it to display any of the properties listed in Annex III to Council Directive 91/689/EEC.
(3) In this section the Hazardous Waste List means the list referred to in the first indent of Article 1(4) of Council Directive 91/689/EEC.
(4) Regulations under subsection (2) shall be made by statutory instrument but section 161(2) shall not apply to regulations under that subsection.

28 The Hazardous Waste (England and Wales) Regulations 2005, SI 2005/894, reg 8 Specific waste to be treated as hazardous.

- a specific batch of waste, which is determined pursuant to reg 9 (Specific waste to be treated as non-hazardous) to be a non-hazardous waste.

Specific Waste to be treated as hazardous (reg 8)

8.27 The Regulations give the Secretary of State the power to decide that in exceptional cases, a specific batch of waste in England which:

- is not listed as a hazardous waste in the List of Wastes;
- is not listed in regulations made under s 62A(1) of the EPA 1990; or
- though of a type listed as a hazardous waste in the List of Wastes, is treated as non-hazardous pursuant to reg 9(2)

displays one or more of the hazardous properties, and then it will be treated as hazardous waste.

Where a batch of waste is produced in Wales, Scotland or Northern Ireland and has been determined by the Welsh Assembly Government, the Scottish Executive or the Northern Ireland Department of the Environment it will be treated as hazardous waste in England.

Specific waste to be treated as non-hazardous (reg 9)

8.28 This Regulation gives the Secretary of State the power to decide whether in exceptional cases and on the basis of documentary evidence provided by the holder and having regard to Annexes I, II and III and the limit values of concentration in the List of Wastes that a specific batch of waste in England which:

- is listed as hazardous waste in the List of Wastes;
- is listed in regulations made under s 62A(1) of the EPA1990; or
- though of a type not listed as a hazardous waste in the List of Wastes, is treated as hazardous pursuant to reg 8(2);

does not display any of the properties listed in Annex III to the Hazardous Waste Directive and accordingly that it shall be treated for all purposes as non-hazardous in England.

If a specific batch of waste produced in Wales, Scotland or Northern Ireland and listed as a hazardous waste in the List of Wastes and for which the Welsh Assembly Government, the Scottish Executive or the Northern Ireland Department of the Environment, as the case may be, is determined to be non-hazardous then it will be treated as non-hazardous in England.

Common provisions in Regulations 8 and 9 (reg 10)

8.29 Where a determination about hazardous and non-hazardous waste has been made under regs 8 (Specific waste to be treated as hazardous) and 9 (Specific waste to be treated as non-hazardous) then the Secretary of State may revoke this.

Before making revocation then the Secretary of State must consider whether is would be inappropriate taking into account the nature of the emergency or grave danger consult with:

- the requisite bodies;
- the holder of the specific batch of waste; and
- any other person appearing to the Secretary of State
  - to have an interest in the specific waste; or
  - to be otherwise directly affected by the determination.

The Secretary of State must then give notice of any determination made under reg 49 (Producers', holders' and consignors' records) or 50 (Carriers records) or revocation of such a determination to:

- the requisite bodies[29];
- the holder of the specific batch of waste concerned; and
- any person she has consulted who has an interest in a specific waste or directly affected by the determination.

The notice shall give reasons for the determination or revocation, as the case may be.

Requisite bodies (reg 11)

8.30 This lists the agencies and parties involved:

- The Agency;
- SEPA;
- The National Assembly of Wales;
- The Scottish Assembly;
- The Northern Ireland Department of the Environment;
- The Health and Safety Executive;
- The organisation appearing to the Secretary of State to be representative of the persons likely to be affected by the relevant determination or revocation of a determination as the case may be.

29 SI 2005/894, Pt 2, reg 11.
For the purposes of this Part, the 'requisite bodies' are—
(a) the Agency;
(b) SEPA;
(c) the National Assembly for Wales;
(d) the Scottish Executive;
(e) the Northern Ireland Department of the Environment;
(f) the Health and Safety Executive; and
(g) any organisation appearing to the Secretary of State to be representative of persons likely to be affected by the relevant determination or revocation of a determination, as the case may be.

Part 3 Application of Parts 2–11

General application of Parts 4–11 (reg 12)

8.31 Parts 4–11 apply to hazardous waste except:

For waste covered under regs 13 (application to asbestos waste) and 14 (application to separately collected fractions) these Regulations do not apply to domestic waste[30].

Nothing in Pt 6 of these Regulations (movement of hazardous waste) applies in relation to shipments of waste to which the provisions of Council Regulation 259/93/EEC, other than Title III of that Regulation, apply.

These Regulations apply to hazardous waste in England notwithstanding that the waste:

- was produced on or removed from premises in Scotland, Wales, Northern Ireland or Gibraltar; or
- is, or is to be, transported from premises in England to premises located in one of those places.

For the avoidance of doubt, in their application to:

- ships' waste, these Regulations apply to any ship;
- the internal waters and the territorial sea of the United Kingdom adjacent to England, these Regulations apply[31], to a consignment of waste transported in any ship.

in each case (whether the ship is a United Kingdom ship or otherwise and, if a United Kingdom ship, whether registered in England or otherwise).

Asbestos Waste (reg 13)[32]

8.32 These Regulations apply to asbestos waste whether or not it is domestic waste.

Nothing in these Regulations imposes obligations on an occupier of domestic premises in relation to asbestos waste produced on those premises.

A contractor who is engaged to carry out any works at domestic premises which produce asbestos waste or which relate to asbestos waste is to be treated as the producer of the asbestos waste, and, where the contractor does not engage another person as consignor, as the consignor of the asbestos waste.

30 Article 1(5) of the Hazardous Waste Directive made provision for specific rules to be made by the European Community taking into consideration the particular nature of domestic waste; no such rules have, at the date of making these Regulations, been adopted.

31 SI 2005/894, Pt 3, reg 12(3) without prejudice to paragraph (3).

32 The Hazardous Waste (England and Wales) (Amendments) Regulations 2009, SI 2009/507, reg 5; the Hazardous Waste (Wales) (Amendment) Regulations 2005, SI 2005/1806, reg 5.

Separated domestic fractions (reg 14)[33]

8.33 Regulation 14 was amended to clarify that the occupier of domestic premises is exempt from obligations relating to separated domestic fractions of hazardous waste. This also clarifies the restrictions on the mixing of hazardous waste in Pt 4 of the 2005 Regulations apply from the time that waste is either accepted for collection, disposal or recovery from domestic premises or accepted at a site for domestic waste when it is taken there by the occupiers of domestic premises.

This regulation applies to any separated domestic fraction, that is to say, hazardous waste which:

- is domestic waste; and
- has been separated from other domestic waste.

Nothing in these Regulations imposes obligations on an occupier of domestic premises in relation to separated domestic fractions, which have been produced at those premises.

Part 4 of these Regulations applies to separated domestic fractions from the time at which the waste is accepted for collection, disposal or recovery:

- from the domestic premises at which the waste was produced; or
- at a site for the reception of domestic waste to which the separated domestic fractions are taken by an occupier of domestic premises.

After separated domestic fractions have been removed from the premises at which the waste was produced and taken to other premises for collection, disposal or recovery, any establishment or undertaking which accepts the separated domestic fractions for collection, disposal or recovery must be treated from the time at which the waste is so accepted as the producer of the waste for the purposes of these Regulations.

Hazardous waste produced at a shop premises (reg 14A)[34]

8.34 This is new Regulation to extend the application of a provision.

Hazardous waste produced at shop premises by customers of the occupier shall be treated as being produced by the occupier for the purposes of these Regulations.

For the purposes of this regulation, 'shop premises' has the meaning given by s 1(3) of the Offices, Shops and Railway Premises Act 1963.

33 The Hazardous Waste(England and Wales) (Amendment) Regulations 2009 , SI 2009/507, reg 14.

34 The Hazardous Waste (England and Wales) (Amendment) Regulations 2009, SI 2009/507, reg 7; The Hazardous Waste (Wales) (Amendment) Regulations 2005, SI 2005/1806, reg 7.

Radioactive Waste (reg 15)

8.35 This regulation applies where radioactive waste[35] is exempt for the time being from the requirements of:

- s 13[36] (disposal of radioactive waste); or
- s 14[37] (accumulation of radioactive waste)

of the Radioactive Substances Act 1993 Act by or pursuang to s 15[38] of that Act; and

has one or more hazardous properties arising other than from its radioactive nature.

Agricultural waste (reg 16)

8.36 These Regulations do not apply to agricultural waste before 1 September 2006, but apply on and after that date to agricultural waste whenever it became waste. For the purpose of this regulation, 'agricultural waste' means waste from premises used for agriculture within the meaning of the Agriculture Act 1947[39].

Mines and quarries waste (reg 17)

8.37 These Regulations do not apply to waste from a mine or quarry before 1 September 2006, but apply on and after that date to such waste whenever it became waste.

Part 4 Mixing of Hazardous Waste

8.38 This bans the mixing of hazardous waste unless permitted as part of a disposal or recovery operation in accordance with the WFD. It also imposes

35 Radioactive Substances Act 1993, s 2
Meaning of 'radioactive waste'.
In this Act 'radioactive waste"'means waste which consists wholly or partly of—
(a) a substance or article which, if it were not waste, would be radioactive material, or
(b) a substance or article which has been contaminated in the course of the production, keeping or use of radioactive material, or by contact with or proximity to other waste falling within paragraph (a) or this paragraph.

36 Radioactive Substances Act 1993 Authorisation of disposal and accumulation of radioactive waste s 13 Disposal of radioactive waste.

37 Radioactive Substances Act 1993 Authorisation of disposal and accumulation of radioactive waste s 14 Accumulations of Radioactive waste.

38 Radioactive Substances Act 1993 Authorisation od disposal and accumulation of radioactive waste s 15 Further exemptions from ss 13 and 14.

39 Agriculture Act 1947, Ch 48 Part V s 109(3) 'agriculture' includes horticulture, fruit growing, seed growing, dairy farming and livestock breeding and keeping, the use of land as grazing land, meadow land, osier land, market gardens and nursery grounds, and the use of land for woodlands where that use is ancillary to the farming of land for other agricultural purposes, and "agricultural" shall be construed accordingly.

a duty to separate different categories of Hazardous Waste where this is technically feasible.

Meaning of mixing hazardous waste (reg 18)

8.39 For the purposes of these Regulations, hazardous waste of any description shall be considered to have been mixed if it has been mixed with—

- a different category of hazardous waste;
- a non-hazardous waste; or
- any other substance or material.

Prohibition on mixing hazardous waste without a permit (reg 19)

8.40 Any establishment that carries out disposal or recovery of hazardous waste or which produces, collects or transports hazardous waste will not mix any hazardous waste.

Duty to separate mixed wastes (reg 20)

8.41 The holder of hazardous waste which has been mixed other than in accordance with a waste permit or a registered exemption and separation is both:

- technically and economically feasible; and
- necessary in order to comply with the Waste Directive conditions.

Then the holder must make arrangements for separation of the waste to be carried out in accordance with a waste permit or registered exemption as soon as reasonably practicable.

In this Regulation 'separation' means separation of a waste from any other waste, substance or material with which it has been mixed.

Part 5 Notification of Premises – Revoked

8.42 This section has been revoked by the Hazardous Waste (England and Wales) (Amendment) Regulations 2016, SI 2016/336.

Part 6 Movement of Hazardous Waste

8.43 This Part requires that documents are completed whenever hazardous waste is removed from premises (including removal from ships and removal by pipeline). The various types of form are set out in Schs 4 to 6. This ensures that there is an accurate description of consignments of waste that accompanies them whenever they move. This is in addition to any requirements to ensure hazardous waste is properly packaged and labelled.

Producers, holders, carriers, consignors and consignees are all required to complete various parts of forms. If the consignee rejects the waste, suitable alternative arrangements must be made. Schedule 7 deals with cross-border transfers within the United Kingdom and Gibraltar.

Documents to be completed by consignments

Coding standard (reg 33)

8.44 It is the duty of the Agency from time to time to designate, and at all times to maintain in force a designation of a standard (in these Regulations referred to as a 'coding standard', making provision for the composition of consignment codes for the purposes of this Part.

The coding standard must enable each consignment of hazardous waste to be given a unique consignment code.

The standard shall make provision for different codes for consignment notes completed in relation to hazardous waste removed by pipeline pursuant to reg 41, where piping is continuous for more than one quarter.

The standard shall make provision for consignment codes to consist of letters, numbers or symbols, or any combination of letters, numbers and symbols.

Consignment Codes (reg 34)

8.45 (1) It is the duty of:

- the producer, in relation to:
  - a consignment of hazardous waste to be removed from premises at which the waste is produced (other than a ship);
  - hazardous waste to be removed by pipeline from premises (other than a ship) situated in a case to which regulation 41 applies; or
  - hazardous waste to be deposited within the curtilage of premises at which it was produced;
- the master of the ship, in relation to any hazardous waste removed from a ship in a harbour area (including waste accidentally spilled on land adjacent to the ship); and
- the consignor, in relation to any other consignment of hazardous waste,

to assign to the hazardous waste a unique code in accordance with the coding standard for the time being in force.

(2) The code assigned pursuant to paragraph (1) shall be the consignment code of the hazardous waste concerned for the purposes of these Regulations.

Documents to be completed for consignments[40]

8.46 Under the Hazardous Waste Regulations there is no requirement for pre-notification of a consignment, which eliminates a regulatory and administrative burden on companies consigning waste and provides greater flexibility on the timing of consignments to consignors. Waste movement notes will still need to accompany consignments although the obligation to submit them to the Agency has been removed. Copies must be kept by consignors and consignees for at least three years.

The audit trail of hazardous waste arisings will be monitored by the Environment Agency through the submission of quarterly reports by consignees. Charges for waste movements will be made, by the Environment Agency to the consignee, on the basis of these reports.

Completion of consignment note (reg 35)

8.47 (1) Where hazardous waste is removed from any premises:

(a) a consignment note shall be completed in accordance with paragraph (2), this regulation and the requirements of the relevant regulation if one of the following regulations applies:

- (i) regulation 36 (standard procedure);
- (ii) regulation 39 (removal of ship's waste to reception facilities);
- (iii) regulation 40 (removal of ship's waste other than to reception facilities);
- (iv) regulation 41(removal of waste by pipeline); or
- (v) where the consignment or any part thereof is rejected by the consignee, in accordance in each case with regulation 42 (Duty of consignee not accepting delivery) and 43 (Further consignment note for rejected consignment).

(b) if regulation 37 (Schedule of carriers) applies, a schedule of carriers shall be completed in accordance with that regulation and paragraph (3) of this regulation; or

(c) if regulation 38 (multiple collection) or regulation 44 (multiple collection procedure for rejected consignments) applies, a multiple collection consignment note shall be completed in accordance with the applicable regulation and paragraph (4).

(2) The form of consignment note set out in Schedule 4, or a form requiring the same information in substantially the same format, shall be used and shall be completed so as to contain (in the place indicated in the form) all the information provided for in that Schedule which is applicable to the case.

(3) The form of the schedule of carriers set out in Schedule 5 (General interpretation), or a form requiring the same information in substantially the same format, shall be used and shall be completed so as to contain (in the place

40 Waste is defined as anything for the purpose of the Waste Directive subject to reg 15 (reg 35).

indicated in the form) all the information provided for in that Schedule which is applicable to the case.

(4) The form of multiple collection consignment note set out in Schedule 6 (Hazardous waste), or a form requiring the same information in substantially the same format, shall be used and shall be completed so as to contain (in the place indicated in the form) all the information provided for in that Schedule which is applicable to the case.

(5) In this Part, a reference to a part of a consignment note, schedule of carriers or multiple collection consignment note by its number or description is a reference to that part as it is required to be so numbered or described as the case may be in the form in Schedule 4 (The List of wastes), 5 (General interpretation) or 6 (Hazardous waste) as the case may require.

Standard procedure (reg 36)

8.48 (1) This regulation applies in all cases where a consignment of hazardous waste is to be removed from premises except in cases to which any of regulations 38 to 41 apply.

(2) Before the consignment is removed:

(a) the hazardous waste producer, or holder, as the case may be, shall:

- (i) prepare a copy of the consignment note for each of the following: the hazardous waste producer or holder, where different from the consignor; the consignor; the carrier; and the consignee;
- (ii) complete Parts A and B on each copy; and
- (iii) give every copy to the carrier

(b) the carrier shall complete Part C on each copy and give every copy to the consignor
(c) the consignor shall:

- (i) complete Part D on each copy;
- (ii) where the hazardous waste producer or holder, as the case may be, is not the consignor, give one copy to him;
- (iii) retain one copy; and
- (iv) give every remaining copy to the carrier

(3) The carrier shall ensure that every copy which he has received:

- (a) travels with the consignment; and
- (b) is given to the consignee on delivery of the consignment.

(4) Subject to regulation 42 (Duty of consignee not accepting delivery), on receiving the consignment the consignee shall:

(a) complete Part E on both copies; and
(b) give one copy to the carrier.

Schedule of carriers (reg 37)

8.49 (1) This regulation applies in all cases, whether under regulation 36 (Standard procedure), or regulation 40 (Removal of ships' waste other than to reception facilities) where more than one carrier transports, or is to transport, the consignment.

(2) Before the consignment is removed:

(a) the consignor shall:

(i) prepare a copy of the schedule of carriers for the hazardous waste producer or holder (where different from the consignor), the consignor, every carrier and the consignee; and
(ii) give every copy to the first carrier;

(b) the first carrier shall ensure that every copy he has received travels with the consignment;
(c) on delivery of the consignment to each subsequent carrier:

(i) the previous carrier shall give the subsequent carrier every copy of the schedule which he has received;
(ii) the subsequent carrier shall complete the relevant certificate on each copy, give one to the previous carrier who shall retain it, and ensure that every remaining copy which he has received travels with the consignment; and

(d) when the consignment is delivered to a consignee, the carrier shall retain one copy of the carrier's schedule and give every remaining copy to the consignee.

(3) Where:

(a) arrangements for the transport of the consignment have not been made with all carriers intended to be involved before transport commences; or
(b) there is a change in any such arrangements after transport commences for any reason

Paragraph (2) applies as if the carrier in possession of the consignment when further arrangements are made, in the case of paragraph (a) or take effect, in the case of paragraph (b), were the consignor and the next carrier were the first carrier.

Multiple collections (reg 38)

Multiple collections – modification of carrier rounds

8.50 Small loads of similar wastes, such as waste oils from garages, can be collected in small loads as small loads from many producers in the same vehicle. The Hazardous Waste Regulations modified procedures for loads collected on carrier rounds (called 'multiple collections'). To meet the Hazardous Waste Directive requirements, information must be obtained about each load collected, including details about each producer. Using the multiple collection procedure, each load collected will be required to have a unique consignment code and a separate consignment note for multiple collections will obtain all the information required on individual loads whilst also providing summary information for carriers and consignees. Charging for loads on multiple collections also means that each load will be subject to a waste movement fee (although this fee will be lower than the fee for single waste movements). In order to simplify administrative procedures, and avoid excessive paper work, the Regulations include a schedule for recording loads on multiple collections.

(1) This regulation applies to a journey made by a single carrier, which meets the following conditions:

(a) the carrier collects more than one consignment of hazardous waste in the course of the journey;
(b) each consignment is collected from different premises (none of which is a ship);
(c) all the premises from which a collection is made are in England; and
(d) all consignments collected are transported by that carrier in the course of the journey to the same consignee, and a journey which meets these conditions is referred to in these Regulations as a 'multiple collection'.

(2) Where the carrier elects to apply the multiple collection procedure set out in this regulation to a multiple collection, the requirements of this regulation apply to the carrier, and to the producers, holders and consignors of the consignments collected in the course of the round.

(3) Before the first collection, the carrier shall:

(a) prepare two copies of the multiple consignment collection note, plus one copy for each hazardous waste producer or holder, as the case may be, from whom waste is to be collected during the round, and one copy for each consignor, in cases where the hazardous waste producer, or holder, as the case may be, is not the consignor; and
(b) complete Parts A and B on each copy.

(4) Before the removal of waste from each set of premises from which a collection is made:

(a) the producer, or holder, must complete the annex to the multiple collection consignment note on each copy;
(b) the consignor and carrier must sign their respective declarations to the annex to the multiple collection consignment note on each copy of the note; and
(c) the carrier must pass a completed copy to the producer or holder in each case (and where the producer or holder is not the consignor, the consignor).

(5) After collection of the last consignment but before delivery to the consignee, the carrier must complete the particulars for completion by the carrier in section C on both remaining copies of the consignment note.

(6) Subject to regulation 42 (Duty of the consignee not accepting delivery), on delivery of the waste:

(a) the carrier must pass to the consignee both remaining copies of the note;
(b) the consignee must complete the particulars for completion by the consignee in Section C and complete the certificate in part D of the note on both copies; and
(c) the consignee must return one copy of the completed note to the carrier.

Removal of ships' wastes to reception facilities (reg 39)

8.51 (1) This regulation applies where hazardous waste is removed from a ship (including excesses or spillage from loading or unloading, which have been accidentally spilled on land adjacent to the ship) in a harbour area:

(a) to reception facilities provided within that harbour area; or
(b) by pipeline to any such facilities provided outside a harbour area.

(2) Before the waste is removed from the ship the master of the ship shall:

(a) prepare two copies of the consignment note;
(b) complete Parts A, B and D on each copy;
(c) retain one copy; and
(d) give one copy to the operator of the facilities.

(3) Subject to regulation 42 (Duty of the consignee not accepting delivery), on receiving a consignment of hazardous waste the operator of the facilities shall complete Part E on the copy which he has received.

Removal of ships' wastes other than to reception facilities (reg 40)

8.52 (1) This regulation applies where hazardous waste is removed from a ship in a harbour area other than in a case to which regulation 39 (Removal of ships' wastes to reception facilities) applies.

(2) Before the consignment is removed:

(a) the master of the ship shall–

 (i) prepare three copies of the consignment note;
 (ii) complete Parts A and B on each copy; and
 (iii) give every copy to the carrier.

(b) the carrier shall complete Part C on each copy;
(c) the master of the ship shall–

 (i) complete Part D on each copy
 (ii) retain one copy; and
 (iii) give every remaining copy to the carrier.

(d) the carrier shall ensure that every copy which he has received–

 (i) accompanies the consignment; and
 (ii) is given to the consignee on delivery of the consignment.

(3) Subject to regulation 42 (Duty of the consignee not accepting delivery), on receiving the consignment the consignee shall–

(a) complete Part E on both copies; and
(b) give one copy to the carrier.

Removal of wastes by pipeline (reg 41)

8.53 (1) This regulation applies where hazardous waste is removed from any premises (other than a ship) on which it is produced or stored by pipeline.

(2) Before the waste is piped, and, where the piping is continuous, at the commencement of each quarter, the producer, or holder, as the case may be, shall–

(a) prepare one copy of the consignment note for each of the following: the producer or holder, as the case may be, where different from the consignor, the consignor and the consignee;
(b) complete Parts A, B and D on each copy, provided that in Part B3 the average flow rate shall be recorded, together with an estimate of the total volume of waste to be piped per week or calendar month.

(3) The consignee shall complete Part E on each copy with the following modifications–

(a) in Part E1, the reference to the date on which the waste was received shall be taken to be a reference to the last date on which waste was received by the consignee pursuant to that consignment note; and
(b) in Part E2, the particulars of the vehicle registration are not required.

(4) The consignee shall retain one copy and ensure that a copy is given to the consignor, and to the hazardous waste producer or holder, where different from the consignor.

Rejected consignments

Duty of consignee not accepting delivery (reg 42)

8.54 This has been amended to correct the reference to the part of the form of multiple consignment notes, which is the consignee's certificate. This also includes amendments to clarify references to Environmental Permitting (England and Wales) Regulations 2007, SI 2007/3538.

(1) This regulation and regulations 43 and 44 apply where the consignee does not accept delivery of a consignment of hazardous waste, whether wholly or in part.

(2) The requirements of regulation 36(4), 38(6)(b) and (c), 39(3) or 40(3) (which relate to the duties of the consignee on acceptance of the consignment) as the case may be, do not apply to the consignee in respect of a consignment, or part thereof, which has been rejected.

(3) If copies of the consignment note relating to a rejected consignment have been given to the consignee they shall:

(a) indicate, in the part headed "consignee's certificate"[41] of each copy that he receives that he does not accept the consignment, or part of the consignment, as the case may be, and the reasons why he does not accept the consignment or part;
(b) retain one copy give one copy to the carrier; and
(c) as soon as reasonably practicable, send a copy to the consignor, and (if different from the consignor, the producer or holder, where known).

(4) If no copy of the consignment note has been given to the consignee he shall:

(a) prepare a written explanation of his reasons for not accepting delivery, including such dctails of the consignment, the hazardous waste producer or holder, the consignor and the carrier as are known to him;
(b) give such written explanation to the carrier
(c) as soon as reasonably practicable, send one copy to the consignor, and (if different from the consignor) the producer or holder, where known; and
(d) retain a copy of his written explanation.

(5) On being informed that the consignee will not accept delivery of the consignment or part, the carrier shall:

(a) inform the Agency;
(b) seek instructions from the hazardous waste producer or holder; and
(c) take all reasonable steps to ensure those instructions are fulfilled (including completing any consignment note on their behalf).

(6) It is the duty of the hazardous waste producer or holder identified in the relevant part of the consignment note, as the case may be to:

(a) make arrangements as soon as reasonably practicable for the transfer of the rejected consignment or part to another specified consignee who holds a waste permit or carries on an exempt waste operation[42] for the recovery or disposal of the waste; and
(b) Forthwith:
    (i) give instructions to the carrier accordingly; and
    (ii) inform the Agency of the arrangements and instructions.

(7) If in any case within paragraph (6)(a) no alternative consignee can be found within five business days, the hazardous waste producer or holder identified in the relevant part of the consignment note shall make arrangements to return the waste to premises from which it was removed for its storage in accordance with the Waste Directive conditions until a suitable consignee can be found.

41 The Hazardous Waste (England and Wales) (Amendment) Regulations 2009, reg 13(a); The Hazardous Waste (Wales) (Amendment) Regulations 2009, reg 13(a).
42 The Hazardous Waste (England and Wales) (Amendment) Regulations 2009, SI 2009/507, reg 13(b); The Hazardous Waste (Wales) (Amendment) Regulations 2009, SI 2009/2861, reg 13(b).

In this regulation[43], 'exempt waste operation' has the meaning given in reg 5 of the Environmental Permitting (England and Wales) Regulations 2007[44].

Further consignment note for rejected consignment (reg 43)

8.55 (1) This regulation applies to the removal of any consignment following rejection by the consignee other than a case to which regulation 44 (Procedure for rejected multiple collection assignments) applies.

(2) Before the consignment or part is moved from the original place for delivery, the hazardous waste producer or holder identified in the relevant part of the original consignment note shall ensure that a copy of a new consignment note is prepared in respect of the rejected consignment or part for each of the following–

(a) the hazardous waste producer;
(b) where the hazardous waste producer or holder is not the consignor, the consignor;
(c) the carrier; and
(d) the new consignee.

(3) The producer or holder shall–

(a) complete Parts A and B on each copy of the note as follows:
  (i) the information required to complete Part A should be copied from the original consignment note, including the relevant consignment code to which the letter 'R' shall be added to the end;
  (ii) the new consignee's name and address (including the postcode) shall be entered in Part A4; and
  (iii) subject to R14, the relevant information from the original consignment note shall (where part of the load is rejected, in so far as it relates to that part) be copied to Part B; and

(b) give every copy to the carrier.

(4) Where the consignee who rejected the consignment or part states in his written explanation that the description of the waste in the original consignment note is incorrect, the producer or holder shall instead include an accurate description of the waste in the new consignment note.

(5) The carrier shall complete Part C on each copy.

(6) The producer or holder identified in the relevant part of the original consignment note shall–

(a) complete Part D on each copy;

43 The Hazardous Waste (England and Wales) (Amendment) Regulations 2009, SI 2009/507, reg 13(c); The Hazardous Waste (Wales) (Amendment) Regulations 2009, SI 2009/2861, reg 13(c).
44 SI 2007/3538.

(b) where the producer or holder is not the consignor, give one copy to the producer or holder as the case may be; and
(c) give every remaining copy to the carrier.

(7) The carrier shall ensure that every copy of the new consignment note which he has received–

(a) travels with the rejected consignment or part; and
(b) is given to the new consignee on delivery of the consignment or part.

(8) The new consignee shall–

(a) complete Part E on both copies of the new consignment note; and
(b) give one copy to the carrier.

(9) Where there is to be more than one carrier–

(a) in paragraphs (3)(b), (5) and (6)(c), references to the carrier shall be treated as references to the first carrier;
(b) in paragraphs (2)(c) and (7) references to the carrier shall be treated as references to each carrier;
(c) in paragraph (7)(b), in relation to a carrier who is not the final carrier, the reference to 'the new consignee' shall be treated as a reference to 'the subsequent carrier'; and
(d) in paragraph (8)(b) the reference to the carrier shall be treated as a reference to the final carrier.

Procedure for rejected multiple collection consignments (reg 44)

8.56 (1) Where two or more consignments comprising part of a multiple collection are rejected and are to be delivered to the same consignee, if the carrier elects to apply the multiple collection procedure set out in regulation 38 (Multiple Collections) to such a delivery, the following requirements shall apply—

(a) the carrier shall—

  (i) prepare two copies of the multiple consignment collection note, plus one copy for each hazardous waste producer or holder, as the case may be, from whose consignment has been rejected, and one copy for each consignor, in cases where the hazardous waste producer, or holder, as the case may be, is not the consignor; and
  (ii) complete Parts A and B on each copy

(b) before the removal of waste from the original delivery premises:

  (i) the producer, or holder, shall complete the annex to the multiple collection consignment note on each copy prepared by the carrier;
  (ii) the consignor and carrier must sign their respective declarations to the annex to the multiple consignment note on each copy of the note prepared by the carrier;

(iii) the carrier must pass a completed copy to the producer or holder in each case (and where the producer is not the consignor, to the consignor);

(c) on delivery of the waste to the new consignee—

(i) the carrier must complete the particulars for completion by the carrier in section C on every copy of the consignment note;
(ii) the carrier must pass to the consignee every copy of the note;
(iii) the consignee must complete the particulars for completion by the consignee in Section C and complete the certificate in part D of the note on every copy of the note; and
(iv) the consignee must return one copy of the completed note to the carrier.

Duty to deliver within time limit

Duty to deliver consignment promptly (reg 45)

8.57 It is the duty of the carrier to deliver the consignment to the consignee promptly and without undue delay.

Cross Border movements

Cross border movement of hazardous waste (reg 46)

8.58 Schedule 7 (Non-hazardous waste) has effect in connection with the mutual recognition of consignment notes and the cross-border movement of hazardous waste between England and Wales, Scotland, Northern Ireland and Gibraltar.

Part 7 Records and Returns

8.59 This part requires that producers, holders, carriers, consignors and consignees keep records and these must be kept for a minimum of three years except in the case of carriers where this is 12 months. Consignees are also required to provide the Environment Agency with a quarterly return setting out the consignments they have received during that period. Consignees may also be required to pay a fee to the Environment Agency but are given the right to recover from consignors who have sent the waste to them. They are also required to send a return to producers or holders who sent waste to them. Schedule 9 sets out a transitional scheme for fees.

Site records

Records of tipped (discharged) hazardous waste (reg 47)

8.60 (1) Any person who tips (discharges) hazardous waste (whether by way of disposal or storage) in or on any land shall record and identify the waste in accordance with the following requirements of this regulation and regulation 51(Registers and records: common provisions).

(2) A record shall include either:

(a) a site plan marked with a grid, or
(b) a site plan with overlays on which the deposits of the tipped (discharged) waste are shown in relation to the contours of the site.

(3) Records made under this regulation shall be kept in a register.

(4) Deposits shall be identified by reference to both:

(a) the relevant description and six digit code in the List of Wastes, together with a description of the composition of the waste; and
(b) the consignment note relating to such waste, save that where waste is disposed of within the curtilage of the premises at which it is produced the deposits shall be described by reference to the quarterly return made to the Agency by the hazardous waste producer under regulation 53[45].

(5) A person who is required to make or retain a register pursuant to this regulation shall:

(a) update the register as soon as reasonably practicable and in any event within 24 hours of the receipt, or deposit, as the case may be, of the waste;
(b) keep the register on the site where the tipping takes place; and
(c) retain the records—
    (i) for three years after deposit of the waste; or
    (ii) if he has a waste permit pursuant to which the site is operated, until that permit is surrendered or revoked.

(6) In reckoning any period of hours for the purposes of this regulation or regulation 48[46], only the days or hours of any business day shall be counted.

Records of disposal or recovery of hazardous waste by other means (reg 48)

8.61 This has been amended to require that a record of hazardous waste includes, as applicable, details of the disposal method in respect of waste in addition to the recovery method used.

(1) Any person who–

(a) disposes of hazardous waste in or on land (other than any disposal covered by regulation 47 (Records of tipped (discharged) hazardous waste);
(b) recovers hazardous waste in or on land; or
(c) receives hazardous waste at a transfer station;

45 SI 2005/894, Pt 7, reg 53 Consignee and self-disposal quarterly returns.
46 SI 2005/894, Pt 7 reg 48 Records of disposal or recovery of hazardous waste by other means.

(d) or both shall record and identify any hazardous waste received in accordance with the following requirements of this regulation and regulation 51 (Registers and records: common provisions).

(2) The waste shall be identified in the record by reference to the relevant description in the List of Wastes and six digit code, and the entry shall include a description of the composition of the waste.

(3) The record shall include–

(a) the quantity, nature and origin of any such waste;
(b) the relevant hazardous properties;
(c) the disposal or recovery method in respect of the waste by reference to the numbering and description applicable in accordance with Annex IIA or IIB of the Waste Directive[47]; and
(d) an inventory showing the specific location at which the waste is being held.

(4) Records made under this regulation shall be kept in a register.

(5) The register shall be updated as soon as reasonably practicable and in any event no later than 24 hours after:

(a) a consignment of hazardous waste is received;
(b) any recovery or disposal operations are carried out or any hazardous waste is placed in storage at the transfer station, as the case may be; and
(c) any hazardous waste is removed from the premises.

(6) A person who is required to make or retain records pursuant to this regulation shall:

(a) keep the register of the records on the site where the recovery operations take place or the transfer station is operated, as the case may be; and
(b) retain the records—
    (i) where the hazardous waste is fully recovered, or remains at a transfer station, as the case may be, until it leaves the site and for three years thereafter; or
    (ii) if he has a waste permit pursuant to which the site is operated, until that permit is surrendered or revoked.

Producer and transport records

Producers', holders' and consignors' records (reg 49)

8.62 This has been corrected so that the 2005 Regulations apply where the consignor of hazardous waste is different from the producer or the holder of hazardous waste.

47 The Hazardous Waste (England and Wales) (Amendment) Regulations 2009, SI 2009/507, reg 14; The Hazardous Waste (Wales) (Amendment) Regulations 2009, SI 2009/2861, reg 14.

(1) A producer or holder of hazardous waste, and where different from the producer or holder[48] a consignor of hazardous waste, shall keep a record of the quantity, nature, origin and, where relevant, the destination, frequency of collection, mode of transport and treatment method of the waste.

(2) Where the waste is transported, the duty in paragraph (1) includes a requirement to keep a record of particulars sufficient to identify the carrier.

(3) The producer, holder or consignor, as the case may be, shall preserve the records to be made pursuant to this regulation whilst he remains the holder of the waste and for at least three years afterwards commencing on the date on which the waste is transferred to another person.

(4) A producer, holder, dealer, broker or consignor required by paragraph (1) to keep records shall keep the records chronologically in a register[49].

(5) The register required to be kept and retained by a producer, holder or consignor under paragraph (4) of this regulation shall be kept:

(a) in relation to the register required to be kept by the producer or holder:

  (i) at the premises notified pursuant to regulation 24;
  (ii) if he no longer occupies those premises, at his principal place of business (or such other address as agreed with the Agency for that purpose); ~~or~~

  ~~o if no premises were notified in relation to the waste, at his principal place of business (or such other address as agreed with the Agency for that purpose).~~[50]

(b) the register required to be kept by a consignor other than the producer or holder shall be kept at his principal place of business.

~~If the producer or holder ceases to occupy the notified premises before the period referred to in paragraph (3) expires, he shall inform the Agency forthwith.~~[51]

Carrier's records (reg 50)

8.63 (1) An establishment or undertaking which transports hazardous waste shall keep a record of the quantity, nature, origin and, where relevant, the destination, frequency of collection, mode of transport and treatment method of the waste in accordance with the following requirements of this regulation.

48 The Hazardous Waste (England and Wales) (Amendment) Regulations 2009, SI 2009/507, reg 15.

49 The Hazardous Waste (England and Wales) (Amendment) Regulations 2016, SI 2016/336, reg 2(4).

50 The Hazardous Waste (England and Wales) (Amendment) Regulations 2016, SI 2016/336, reg 2(4)(b)(ii).

51 The Hazardous Waste (England and Wales) (Amendment) Regulations 2016, SI 2016/336, reg 2(4)(c).

(2) The establishment or undertaking shall keep the records made pursuant to this regulation for at least 12 months commencing on the date of delivery of the waste to its destination.

(3) The information required to be recorded in paragraph (1) shall be entered in a register and the register kept at the carrier's principal place of business.

Registers and records: common provisions (reg 51)

8.64 (1) The following provisions of this regulation apply in relation to registers required to be kept under regulations 47 to 50.

(2) A person who is required to keep a register shall enter in the register each copy received by them of–

(a) any consignment note (including multiple consignment notes and, where consignments are not accepted, the original note, a copy of any explanation of the reasons for rejection prepared pursuant to regulation 42 (Duty of consignee not accepting delivery) and the consignment note prepared pursuant to regulation 43 (Further consignment note for rejected consignment) or 44 (Procedure for rejected multiple collection assignments);
(b) any consignee's return to the producer, holder or consignor received pursuant to regulation 54 (Consignee's return to the producer, holder or consignor); and
(c) any carrier's schedule given to him pursuant to regulation 37 (Schedule of carriers).

(3) A person who is required to keep a register or retain records until his waste permit is surrendered or revoked shall send those records or that register to the Agency when the permit is surrendered or revoked.

(4) Every register kept or record made pursuant to regulation 15 (Radioactive waste) or 16 (Agricultural waste) of the 1996 Regulations, and every record made pursuant to regulation 13 (Asbestos waste) or 14 (Separately collected domestic fractions) of the Control of Pollution (Special Waste) Regulation 1980, shall:

(a) be kept with the register kept pursuant to regulations 47 (Records of tipped (discharged) hazardous waste) to 49 (Producers', holders' and consignors' records) by the person required to keep that register for so long as is mentioned in the relevant regulation; and
(b) be sent by that person together with that register if it is sent to the Agency.

Previous holder's right to information (reg 52)

8.65 (1) An establishment or undertaking to which hazardous waste is delivered for disposal or recovery, as the case may be, has the duty, owed to any previous holder of the waste, to supply to that previous holder on request documentary evidence that the disposal or recovery operation concerned has been carried out, indicating where applicable, the relevant entry listed in Annex IIA or Annex IIB, as the case may be, to the Waste Directive.

(2) Any request for information under this regulation shall be in writing and shall specify the period (which may not be less than seven days) within which the information is to be supplied.

Consignee and self-disposal quarterly returns (reg 53)

8.66 (1) Every consignee shall make a return, in these Regulations referred to as a consignee quarterly return, to the Agency of information relating to all consignments of hazardous waste received by him in any quarter in accordance with paragraph (4).

(2) The return shall include:

(a) rejected consignments;
(b) hazardous waste delivered by pipeline where regulation 41 applies; and
(c) in the case of multiple collections, each individual consignment collected.

(3) Where in any quarter hazardous waste is disposed of by depositing it within the curtilage of the premises at which it is produced, the producer shall make a return in respect of that quarter of information relating to the deposit to the Agency (including the consignment code), in accordance with paragraph (4).

(4) A return to be made pursuant to this regulation in respect of a quarter shall be made no later than the time specified in the right hand column of the table below in relation to the quarter specified in the left hand column:

(5) The Agency may prescribe a format for returns to be made under this regulation and, where for the time being a format is prescribed pursuant to this paragraph:

(a) the Agency shall publish the format on its website and in such other manner as it may consider appropriate for informing persons required to submit such returns of its content; and
(b) the Agency shall not be obliged to consider a return as properly made for the purposes of these Regulations unless it is made in that format, or in a format substantially to like effect.

(6) If the Agency prescribe a fee payable by a consignee by charging scheme made under section 41 of the Environment Act 1995 as a means of recovering its costs incurred in performing functions in relation to the consignments included on the consignee quarterly returns, a consignee may recover from a consignor any fees paid under these Regulations in relation to consignments sent by that consignor.

Consignee's return to the producer, holder or consignor (reg 54)

8.67 (1) Without prejudice to any duty under these Regulations on the part of the consignee to send any document or copy thereof to the producer, holder or consignor, a consignee shall send to a producer or holder identified in the relevant part of a consignment note:

(a) a return in a form corresponding to that set out in Schedule 8 or a form substantially to like effect within one month of the end of the quarter in which the waste concerned was accepted; or
(b) a copy of the consignment note together with a description of the method of disposal or recovery undertaken in relation to the waste, within one month of the end of the quarter in which the waste concerned was accepted.

(2) Where hazardous waste was delivered by pipeline in a case to which regulation 41 applies, paragragh (1) shall apply so that the return required in paragraph (1)(a) or the information required in paragraph (1)(b) shall be supplied within one month of the end of the quarter in which the waste concerned was piped.

(3) Where regulation 42 (Duty of consignee not accepting delivery) applies, this regulation does not apply to the consignment or part concerned.

Duties to supply information (reg 55)

8.68 (1) A person who is required to retain any record pursuant to any of the foregoing provisions of this Part shall, at any time during the period in which the record is required to be retained produce that record to the Agency or emergency services on request.

(2) A producer holder, previous holder, consignor, carrier or consignee of hazardous waste shall supply to the Agency on request such information as it may reasonably require for the purposes of performing its functions in connection with these Regulations and for the purposes of monitoring the production, movement, storage, treatment, recovery and disposal of hazardous waste.

(3) An establishment or undertaking to which hazardous waste is delivered for recovery or disposal, as the case may be, has the duty to supply to the Agency on request documentary evidence that the disposal or recovery operation concerned has been carried out, indicating where applicable, the relevant entry listed in Annex IIA or Annex IIB, as the case may be, of the Waste Directive.

(4) A person who is required to supply information to the Agency pursuant to this regulation shall supply that information in such form as the Agency may reasonably request.

(5) The power conferred in paragraph (4) includes power to require the production in a visible and legible documentary form of any information is held in electronic form.

(6) Any request for information under this regulation shall be in writing and shall specify the period within which the information is to be supplied.

Part 8 The Agency's Functions

8.69 This part sets out the Agency's functions – in particular that the Agency is required to inspect producers of hazardous waste periodically and to keep any records sent to it (as per Part 7) for a minimum of three years.

Inspections of hazardous waste producers (reg 56)

8.70 It is the duty of the Agency to carry out appropriate periodic inspections of hazardous waste producers.

Inspections of collection and transport operations (reg 57)

8.71 (1) In relation to the appropriate periodic inspections of collection and transport operations which are required to be conducted pursuant to Article 13 of the Waste Directive by the Agency, without prejudice to the generality of the requirement to conduct such inspections, it is the duty of the Agency insofar as the inspections relate to hazardous waste to conduct the inspections so that they cover more particularly the origin and destination of the hazardous waste.

(2) In paragraph (1) 'collection and transport operations' includes operations where the hazardous waste is transported following transfer between different carriers.

Agency to retain registers etc (reg 58)

8.72 The Agency shall retain registers and any accompanying records sent to it pursuant to regulation 51(3) for a period of not less than three years commencing with the receipt thereof by the Agency.

Fees (reg 59)

8.73 (1) Section 41 of the Environment Act 1995 (power to make schemes imposing charges) is amended as follows.

~~as a means of recovering costs incurred by it in performing functions conferred by regulations under section 62 of the 1990 Act (dangerous or intractable waste) each of the new Agencies may require the payment to it of such charges as may from time to time be prescribed;~~

(2) Subsection (1)(c) – as a means of recovering costs incurred by it in performing functions conferred by regulations made for the purpose of implementing Council Directive 91/689/EEC the Agency may require the payment to it of such charges as may from time to time be prescribed.

(3) Schedule 9 (Transitional Charges by the Agency) has effect to make provision in relation to fees to be charged by the Agency in connection with these Regulations until a charging scheme made under section 41 of the 1995 Act to recover the costs incurred by the Agency in performing functions under these Regulations takes effect.

Provision of information to the Secretary of State (reg 60)

8.74 (1) The Agency shall inform the Secretary of State each year of any changes in the following information for every establishment or undertaking which carries out disposal or recovery of hazardous waste principally on behalf of third parties and which is likely to form part of the integrated network referred to in Article 5 of the Waste Directive:

(a) name and address;
(b) the method used to treat waste; and
(c) the types and quantities of waste which can be treated.

(2) The Agency shall provide the information required by paragraph (1) in the format provided for pursuant to the fourth paragraph of Article 8(3) of the Hazardous Waste Directive[52].

Part 9 Emergencies and Grave Dangers

General (reg 61)

8.75 (1) This Part has effect for the purpose of making provision for an emergency or grave danger.

(2) For the purposes of these Regulations, an "emergency or grave danger" is a present or threatened situation arising from a substance or object which is, or which there are reasonable grounds to believe is, hazardous waste, and the situation constitutes a threat to the population or the environment in any place.

General duties on the holder in the event of an emergency or grave danger (reg 62)

8.76 (1) In cases of emergency or grave danger, a holder of hazardous waste which gives rise to an emergency or grave danger shall:

(a) take all lawful and reasonable steps to avert the emergency or grave danger; or

52 Hazardous Waste Directive 91/689/EEC Article 8(3). In addition, by 12 December 1994, the Member States shall send the Commission the following information for every establishment or undertaking which carries out disposal and/or recovery of hazardous waste principally on behalf of third parties and which is likely to form part of the integrated network referred to in Directive 75/442/EEC:
- name and address,
- the method used to treat waste,
- the types and quantities of waste, which can be treated.

Once a year, Member States shall inform the Commission of any changes in this information. The Commission shall make this information available on request to the competent authorities in the Member States.
The format in which this information will be supplied to the Commission shall be agreed upon in accordance with the procedure laid down in Article 18 of Directive 75/442/EEC.

(b) where it is not reasonably practicable to comply with paragraph (a), take all lawful and reasonable steps to mitigate the emergency or grave danger.

(2) For the purposes of paragraph (1) an act or omission may be considered lawful notwithstanding that it would, apart from this regulation, constitute a breach of these Regulations.

(3) Where the holder of hazardous waste knows or has reasonable grounds to believe that an emergency or grave danger has arisen, he shall notify the Agency as soon as reasonably practicable of the circumstances.

(4) Where the holder takes any step in compliance with paragraph (1) he shall, save where such step has completely averted the emergency or grave danger without breach of these Regulations, notify the Agency as soon as reasonably practicable.

For the purposes of paragraph (3), an emergency or grave danger shall not be considered to have been completely averted if any release has taken place of a substance or object which is, or which there are reasonable grounds to believe is, hazardous waste, whether or not the holder believes he has completely destroyed, retrieved or rendered harmless the substance or object.

In the event of notification in paragraphs (3) or (4) being made orally, the holder shall confirm in writing to the Agency the matters notified within one week of the oral notification.

For the purposes of this regulation a person shall not be released from any requirement imposed by this regulation merely because he has ceased to be the holder through the waste having left his possession or control due to release.

General duties of the Agency (reg 63)

8.77 The Agency shall exercise its functions (whether under these Regulations or otherwise) so as to take all reasonably practicable steps necessary or expedient to avert or mitigate an emergency or grave danger.

An authorised person shall in exercising functions in relation to an emergency or grave danger so exercise his powers under ss 108 and 109 of the 1995 Act as to take all reasonably practicable steps to avert or mitigate the emergency or grave danger.

Schedules

8.78 SCHEDULE 1 ANNEX I to the Hazardous Waste Directive

SCHEDULE 2 ANNEX II to the Hazardous Waste Directive

SCHEDULE 3 ANNEX III to the Hazardous Waste Directive

SCHEDULE 4 Form of Consignment Note[53]

53 The Hazardous Waste (England and Wales) (Amendment) Regulations 2016, SI 2016/336.

SCHEDULE 5 Schedule of Carriers

SCHEDULE 6 Form of Multiple Collection Consignment Note

SCHEDULE 7 UK Cross Boarder Movement of Hazardous Waste

SCHEDULE 8 Form of Consignee's Return to Producer or Holder

SCHEDULE 9 Transitional Charges by Agency

SCHEDULE 10 Form of Fixed Penalty Notices

SCHEDULE 11 Consequential Amendments

SCHEDULE 12 Transitional Provisions

NORTHERN IRELAND

The Hazardous Waste Regulations (Northern Ireland) 2005, NISR 2005/300[54]

8.79 The Hazardous Waste Northern Ireland Regulations 2005 set out the regime for the control and tracking of the movement of hazardous waste for the purpose of implementing the Hazardous Waste Directive[55] in Northern Ireland.

The Hazardous Waste (Amendment) Regulations (Northern Ireland) 2005, NISR 2005/461[56]

8.80 These Regulations amend regs 43 (Offences) and 45 (penalties) to refer to the appropriate Regulation, Part and maximum fine in the Hazardous Waste Regulations (Northern Ireland) 2005.

The Hazardous Waste (Amendment) Regulations (Northern Ireland) 2015, NISR 2015/238[57]

8.81 These Regulations were revoked by the Hazardous Waste (Amendment No 2) Regulations (Northern Ireland) 2015, SI 2015/288.

54 The Hazardous Waste Regulations (Northern Ireland) 2005, NISR 2005/300.
55 Directive 91/689/EC.
56 The Hazardous Waste (Amendment) Regulations (Northern Ireland) 2005, NISR 2005/461.
57 The Hazardous Waste (Amendment) Regulations (Northern Ireland) 2015, NISR 2015/238.

The Hazardous Waste (Amendment No 2) Regulations (Northern Ireland) 2015, NISR 2015/288

8.82 These Regulations are made under s 2(2) of and para 1A of Sch 2 to the European Communities Act 1972. They amend a number of pieces of Northern Ireland legislation to bring into effect the changes introduced by Commission Decision 2014/955/EU amending Decision 2000/532/EC (the Decision) on the list of waste pursuant to Directive 2008/98/EC of the European Parliament and of the Council. Commission Decision 2000/532/EC is directly binding as regards determination of waste that is to be considered as hazardous waste. The changes being introduced by way of the updated Decision came into effect on 1 June 2015.

Regulation 2 amends the Waste and Contaminated Land (Northern Ireland) Order 1997 to insert a definition of 'List of Wastes'.

Regulation 3 amends the Controlled Waste (Duty of Care) Regulations (Northern Ireland) 2002 to update the definition of the European Waste Catalogue to reflect the new requirements.

Regulation 4 provides for similar updating amendments to the Waste Management Licensing Regulations (Northern Ireland) 2003.

Regulation 5 amends the Hazardous Waste Regulations (Northern Ireland) 2005 to remove references to the List of Wastes Regulations (Northern Ireland) 2005 and Article 3 of the Decision and, where appropriate, replaces such references to directly apply the Decision, Article 7 of and Annex III to Directive 2008/98/EC.

Regulation 6 revokes the List of Wastes Regulations (Northern Ireland) 2005.

Regulation 7 revokes the Hazardous Waste (Amendment) Regulations (Northern Ireland) 2015.

Hazardous Waste Regulations (Northern Ireland) 2005, NISR 2005/300

Part 1 General

8.83 Parts 1–3 of the Regulations define hazardous waste and set out how the Regulations apply to that waste. The definition of hazardous waste in reg 6 refers to the list of hazardous wastes set out in the List of Wastes (Northern Ireland) Regulations 2005, NIRS 2005/301. Regulations 9–11 permit the Department of the Environment to determine specific wastes to be classed as hazardous or non-hazardous within Northern Ireland.

Hazardous and Non Hazardous Waste

8.84 The Waste Framework Directive[58] regulates, subject to the certain exclusions, the management of all types of waste. The Hazardous Waste

58 WFD 75/442/EEC.

Directive supplements the Waste Framework Directive by imposing additional requirements in relation to Directive waste, which displays certain hazardous properties. These requirements have previously been transposed by the Special Waste Regulations (Northern Ireland) 1998 through controls on 'special waste'. These Regulations repeal the Special Waste Regulations (Northern Ireland) 1998 and replace the term 'special waste' with 'hazardous waste'.

Meaning of Hazardous Waste (reg 6)

8.85 A waste is hazardous and subject to Regulation 10 (Specific waste to be treated as non-hazardous) if it is–

(a) Listed as a hazardous waste in the List of Wastes[59];

(b) any other waste which (being a waste considered)[60] to display one or more of the properties listed in Annex III) is a hazardous waste pursuant to regulation 8 (Determination that a waste is hazardous); and

(c) a specific waste which, though of a type indicated in the List of Wastes as non-hazardous, is determined pursuant to regulation 9 (Specific waste to be treated as hazardous) to be a hazardous waste;

and the term 'hazardous and cognate expressions' shall be construed accordingly.

Meaning of Non Hazardous Waste (reg 7)

8.86 A waste is non-hazardous if it is:

(a) a waste which is not a hazardous waste pursuant to regulation 6 (Meaning of hazardous waste) or

(b) a specific batch of waste which is determined pursuant to regulation 10 (Specific waste to be treated as non-hazardous) to be a non-hazardous waste

and the term ''non-hazardous and cognate expressions' shall be construed accordingly

Determination

Determination of waste that is hazardous (reg 8)

8.87 (1) Subject to the following provisions of this regulation, where it appears to the Department that, having regard to Annexes I, II and III and the

59 Wastes listed as hazardous in the List of Wastes are considered hazardous pursuant to the first indent of Article 1.4 of the Hazardous Waste Directive and the European Waste Catalogue.

60 A waste which a member state considers displays any of the properties in Annex III is a hazardous waste pursuant to the second indent of Article 1.4 of the Hazardous Waste Directive.

limit values of concentration in the List of Wastes, there are reasonable grounds to suspect that a waste of a type listed, or falling within a type listed, in the List of Wastes as non-hazardous displays one or more hazardous properties, it may determine that, in Northern Ireland, such waste is a hazardous waste.

(2) The Department may revoke a determination made under paragraph (1).

(3) Before making a determination under paragraph (1) or revoking a determination under paragraph (2) the Department shall consult the requisite bodies, except where it considers it inappropriate to do so in any case on account of the nature of an emergency or grave danger.

(4) The Department shall publish in such manner, as it considers appropriate for the purpose of informing persons likely to be affected thereby, a notice of any determination made under paragraph (1) or revocation under paragraph (2), and the notice shall include:

(a) The date and time at which the determination or revocation, as the case may be, is to take effect (which may be immediately upon publication); and
(b) a description of the waste sufficient to identify it,

and shall send a copy of the notice to the requisite bodies.

(5) The notice shall give reasons for the determination or revocation, as the case may be.

(6) Where waste enters Northern Ireland from England, Scotland or Wales, a determination in force under this regulation shall apply to it notwithstanding that a determination to like effect is not in force in respect of England, Scotland or Wales, as the case may be.

Specific waste to be treated as hazardous (reg 9)

8.88 Without prejudice to regulation 8 (Determination that waste is hazardous), the Department, having regard to Annexes I, II and III and the limit values of concentration in the List of Wastes, may determine, in exceptional cases, that a specific waste produced in Northern Ireland and indicated in the List of Wastes as non-hazardous displays one or more of the hazardous properties, and accordingly that that waste is a hazardous waste.

A specific waste produced in England, Wales or Scotland and indicated in the List of Wastes as non-hazardous and which is for the time being determined to be hazardous, pursuant to Article 3 of the List of Wastes Decision Article 7(3) of the Waste Directive[61] shall be treated for all purposes as hazardous waste in Northern Ireland.

61 The Hazardous Waste (Amendment) Regulations (Northern Ireland) 2015, NISR 2015/238, reg 5(5)(b).

Specific waste to be treated as non-hazardous (reg 10)

8.89 The Department may decide, in exceptional cases, on the basis of documentary evidence provided in an appropriate way by the holder, and having regard to Annexes I, II and III and ~~the limit values of concentration in~~[62] the List of Wastes, that a specific waste produced in Northern Ireland and indicated in the List as being hazardous does not display any of the properties listed in Annex III to the Hazardous Waste Directive.

A specific waste produced in England, Wales or Scotland and indicated in the List of Wastes as hazardous waste and which is for the time being determined to be non-hazardous pursuant to ~~Article 3 of the List of Wastes Decision~~ Article 7(2) of the Waste Directive[63] shall be treated for all purposes as non-hazardous in Northern Ireland.

Provisions Common in Regulations 9 and 10 (reg 11)

8.90 (1) The department may revoke determination made under regulation 9 or 10.

(2) Before the Department makes a determination as per egulation 9 and 10 or revoking such a determination – except where it considers it inappropriate to do so on account of the nature of any emergency or grave danger, consult:

(a) the requisite bodies;
(b) the holder of the specific waste; and
(c) any other person appearing to it—

   (i) to have an interest in the specific waste; or
   (ii) to be otherwise directly affected by the determination.

(3) The Department shall give notice of any determination or revocation made under regulation 9 (Specific waste to be treated as non-hazardous) or 10 (Specific waste to be treated as non-hazardous) to:

(a) the requisite bodies;
(b) the holder of the specific waste concerned;
(c) any person it has consulted pursuant to paragraph (2)(c).

(4) The notice shall give reasons for the determination or revocation, as the case may be.

Requisite Bodies (reg 12)

8.91 The 'requisite bodies' are—

(a) the Department of the Environment Food and Rural Affairs;
(b) the Environment Agency of England and Wales;

62 NISR 2015/238, reg 5(5)(a).
63 NISR 2015/238, reg 5(6).

(c) the Scottish Executive;
(d) the Scottish Environment Protection Agency;
(e) the National Assembly for Wales;
(f) the Health and Safety Executive for Northern Ireland; and
(g) any organisation appearing to the Department to be representative of persons likely to be affected by the relevant determination, direction or revocation of a determination or direction, as the case may be.

Part 3 Application

8.92 Part 3 of the Regulations includes an exemption from these controls for domestic waste, which displays hazardous properties but not if it comprises asbestos waste or is collected separately. In both cases, the Regulations do not impose obligations directly on householders.

Application of Regulations to domestic asbestos waste (reg 13)

8.93 (1) Subject to paragraphs (2) to (4) below and Regulation 14, these Regulations apply to hazardous waste including waste that was produced on or removed from premises in England, Scotland, Wales, or Gibraltar; or which is, or is to be, transported to a place in one of those jurisdictions, and to any consignment of waste transported on any ship in the internal waters and the territorial sea of the United Kingdom adjacent to Northern Ireland, whether or not the ship is registered in the United Kingdom, or if a United Kingdom registered ship, whether or not registered in Northern Ireland.

(2) Except as provided for in paragraphs (3) and (4) below these regulations apply to domestic asbestos waste, except in so far as the Regulations would, apart from this paragraph, impose obligations on a person to whom this paragraph applies.

(3) This paragraph applies to a person who is both the original producer of the domestic waste and who is either:

(a) a person who resides at the domestic premises at which the asbestos waste arises; or
(b) a person who is acting on behalf of such a person without reward.

(4) In the application of these Regulations to the asbestos waste:

(a) not being domestic waste; and
(b) produced in the course of any of the activities of construction, modification, repair and maintenance (including structural works) or demolition of domestic premises or any part thereof

(5) These Regulations operate in relation to a domestic occupier who has engaged any person, other than a person mentioned in paragraph (3)(b), for the carrying out of any such activity so as to treat that person as the producer of the domestic asbestos waste to the exclusion of the occupier.

Separately collected domestic fractions (reg 14)

8.94 (1) This regulation applies to separately collected domestic fractions, that is to say, hazardous waste which is:

(a) domestic waste; and
(b) collected from the premises on which it is produced separately from the collection of other waste from those premises.

(2) For the purposes of paragraph (1) hazardous waste may be considered to be collected separately from the collection of other waste notwithstanding that it is collected at the same time or on the same vehicle or both, provided that the hazardous waste is not mixed with the other waste.

(3) These Regulations apply to separately collected fractions, provided that:

(a) nothing in this regulation imposes any obligations in relation to such waste on any person other than an establishment or undertaking which collects, transports, recovers or disposes of such waste; and
(b) where pursuant to sub-paragraph (a) these Regulations do not apply to the person who, but for this sub-paragraph, would be treated as the consignor, the person who collects the waste shall be treated as the consignor.

Radioactive waste (reg 15)

8.95 (1) This regulation applies where radioactive waste within the meaning of section 2 of the Radioactive Substances Act 1993[64]

(a) is exempt for the time being from the requirements of:

 (i) section 13 (disposal of radioactive waste); or
 (ii) section 14 (accumulation of radioactive waste), of that Act by or
 (iii) section 15 of that Act; and

(b) has one or more hazardous properties arising other than from its radioactive nature.

(2) Notwithstanding the definition of waste[65] radioactive waste to which this regulation applies is treated as waste for the purposes of these Regulations, and accordingly it is treated as hazardous waste and these Regulations apply to that waste.

64 Radioactive Substances Act 1993. This Act extends to Northern Ireland (s 51(3).
65 NISR 2005/300, Pt 1, reg 3(1)(b).
'waste' means anything that:
- is waste for the purposes of the Waste Directive; and
- subject to regulation 8, is not excluded from the scope of that Directive by Article 2 of that Directive.

Agricultural waste (reg 16)

8.96 These Regulations do not apply to agricultural waste before 1 September 2006, but apply on and after that date to agricultural waste whenever it became waste.

For the purpose of this regulation, "agricultural waste" means waste from premises used for agriculture within the meaning of the Agricultural Act (Northern Ireland) 1949.

Mines and quarries waste (reg 17)

8.97 These Regulations do not apply to waste from a mine or quarry before 1 September 2006, but apply on and after that date to such waste whenever it became waste.

Part 4 Mixing of Hazardous Waste

8.98 Part 4 bans the mixing of hazardous waste during disposal or recovery operation or by anyone who produces, collects or transports hazardous waste unless such activity is permitted or authorised in accordance with the Waste Framework Directive. It also imposes a duty to separate different categories of Hazardous Waste where technically and economically feasible.

Meaning of mixing hazardous waste (reg 18)

8.99 For the purposes of these Regulations, hazardous waste of any description shall be considered to have been mixed if it has been mixed with—

- a different category of hazardous waste;
- a non-hazardous waste; or
- any substance or material other than hazardous waste of the same category.

Prohibition on mixing hazardous waste without a permit (reg 19)

8.100 (1) Subject to paragraphs (2) and (3), no establishment or undertaking which carries out the disposal or recovery of hazardous waste, or which produces, collects or transports hazardous waste, shall mix any hazardous waste.

(2) The prohibition in paragraph (1) shall not apply so as to prohibit a process by which waste is produced and which results in the production of mixed wastes, being a process other than one which mixes a waste with any other waste substance or material, resulting in:

(a) a change in the nature or composition of that waste; or
(b) the production of another waste.
(c) paragraph (1) above shall not apply to the extent that the mixing is authorised by and is in compliance with the conditions of a permit.

Duty to separate mixed wastes (reg 20)

8.101 This regulation applies to the holder where:

- the hazardous waste has been mixed other than under and in accordance with a waste permit or a registered exemption, whether by the holder or a previous holder; and
- separation is both—
  - technically and economically feasible; and
  - necessary in order to comply with the Waste Directive conditions.

The obligated holder must make arrangements for separation of the waste to be effected by an authorised person (which may be himself) as soon as reasonably practicable.

For the purposes of these Regulations:

- 'authorised person' in relation to any type of hazardous waste, means a person who is authorised (howsoever expressed) by a waste permit to separate hazardous waste of that type;
- 'obligated holder' means a person who operates an establishment or undertaking which carries out disposal or recovery of, or collects or transports, hazardous waste, whether or not that holder is also an authorised person;
- 'separation' means separation of a waste from any other waste, substance or material with which it has been mixed; and
- separation shall, in the absence of evidence to the contrary, be considered to be reasonably practicable when the waste arrives at, or could, without unreasonable delay, have arrived at, a site where separation is permitted by a waste permit.

Part 5 Movement of Hazardous Waste

8.102 Part 5 requires documents to be completed whenever hazardous waste is removed from premises, (which includes removal from ships and removal by pipeline). The Regulations require that all hazardous waste produced in Northern Ireland must be accompanied by the appropriate documentation for transfer within the United Kingdom. The various types of form are set out in Schs 4–7. The Regulations require an accurate description of consignments of waste accompanies them whenever they move and each consignment to be accompanied with a unique consignment code. This is in addition to any requirements to ensure hazardous waste is properly packaged and labelled (see in particular the Packaging (Essential Requirements) Regulations 2003, SI 2003/1941).

Producers, holders, carriers, consignors and consignees are all required to complete various parts of the forms. There is provision for a schedule of the carriers to monitor the waste chain and a copy of the completed Schedule of Carriers must be sent to the Department of the Environment. If the consignee rejects the waste, suitable alternative arrangements, including a new consignment note, must be made. Schedule 7 deals with cross border transfers within the United Kingdom and Gibraltar.

Packaging and Labelling

General Duty Related to Packaging and Labelling (reg 21)

8.103 (1) In the course of its collection, transport and temporary storage, the consignor shall ensure that hazardous waste is properly packaged and labelled in accordance with the relevant international and European Community standards in force.

(2) For the purposes of paragraph (1) the international and Community standards in force are taken to be:

(a) ADR, that is to say, the European Agreement concerning the International Carriage of Dangerous Goods by Road signed at Geneva on 30th September 1957;
(b) RID, that is to say, Annex I to Appendix B of the Convention concerning International Carriage by Rail (COTIF);
(c) IMDG, that is to say, the International Maritime Dangerous Goods Code;
(d) the Technical Instructions for the Safe Transport of Dangerous Goods by Air issued by the International Aviation Authority (ICAO); and
(e) the ADN, that is to say, the European Agreement concerning the International Carriage of Dangerous Goods by Inland Waterway.

Coding of consignments (reg 22)

8.104 (1) Subject to paragraph (3), the Department shall assign or supply forthwith to any person, on request, for the purpose of assigning to a consignment of hazardous waste or, where the consignment is one in a carrier's round, to the consignments in that round, a code unique to that consignment or round, as the case may be.

A code assigned or supplied in accordance with paragraph (1) may consist of letters, numbers or symbols, or any combination of letters, numbers and symbols, or a bar code, which enables the consignment or carrier's round, as the case may be, to be identified electronically.

(3) The Department need not assign or supply a code for a consignment or round until any fee required in respect of it under regulation 36 (Fees) has been paid.

Standard procedure (reg 23)

8.105 Except in a case to which regs 24 (cases in which pre-notification is not required), 27 (procedure where pre-notification is not required), 28 (removal of ships' wastes to reception facilities), 29 (Removal of ships' wastes other than to reception facilities) or 30 (removal of wastes by pipeline) applies, this regulation applies where a consignment of hazardous waste is to be removed from the premises at which it is being held, and any consignment of waste produced or held at premises in Northern Ireland (including rejected loads to which regulation 31 (duty of consignee not accepting consignment) applies) can only be removed under a consignment note in the form set out in Sch 4 to these regulations.

Before the consignment is removed:

- the producer, holder, or other person on his behalf as the case may be, shall prepare six copies of the consignment note, and, on each copy, and shall complete Parts A and B of the consignment note as set out in Schedule 4 to these regulations including reference to:
  - the relevant code;
  - the six digit code referable to the waste in accordance with the List of Waste Regulations (Northern Ireland) 2005; and
  - the postcode of the producer of the waste.
- the consignor shall ensure that one of those copies (on which Parts A and B have been completed and the relevant code entered) is furnished to the Department;
- the carrier shall complete Part C on each of the five remaining copies; and
- the consignor—
  - shall complete Part D on each of those copies;
  - where the hazardous waste producer or holder, as the case may be, is not the consignor, ensure that one of those copies is given to him;
  - shall retain one copy (on which Parts A to D have been completed and the relevant code entered); and
  - shall give the three remaining copies (on which Parts A to D have been completed and the relevant code entered) to the carrier.
- The carrier shall ensure that the copies which he has received—
  - travel with the consignment; and
  - are given to the consignee on delivery of the consignment.
- Subject to regulation 31 (duty of the consignee not accepting delivery of a consignment), on receiving the consignment the consignee shall—
  - complete Part E on the three copies of the consignment note given to him;
  - retain one copy;
  - give one copy to the carrier;
  - forthwith furnish one copy to the Department.
- The carrier shall retain the copy of the consignment note given to him by the consignee.

Cases in which pre-notification is not required (reg 24)

8.106 (1) For the purposes of regulation 25, except in a case to which regulation 27 applies, this regulation applies:

(a) subject to paragraph (2)(a), to the removal, from the premises at which it is being held, of each of the second and any subsequent consignment of hazardous waste in a succession of consignment of hazardous waste,
(b) subject to paragraph (2)(b), to the removal as a consignment of hazardous waste of a product or material for the purposes of the return by the person to whom the product or material had been supplied to the person who supplied it to him or who manufactured it,
(c) subject to paragraph (2)(c), to the removal of a consignment of hazardous waste where the consignor and the consignee are bodies corporate belonging to the same group,
(d) to the removal from a ship in a harbour area of a consignment of hazardous waste to a conveyance for transportation to a place outside that area, and
(e) to the removal of a consignment of hazardous waste which consists entirely of lead acid batteries.

(2) This regulation does not apply unless–

(a) in the case mentioned in paragraph (1)(a), in respect of each consignment—
    (i) the waste is of the same description as the waste in the first of the consignments in the succession;
    (ii) the consignor is the same person;
    (iii) the consignee is the same person;
    (iv) the premises from which the consignment is removed are the same;
    (v) the place to which the consignment is transported is the same; and
    (vi) the removal of the consignment takes place within one year of the removal of the first consignment in the succession;

(b) in the case mentioned in paragraph 1(2)(b), the person to whom the product or material was supplied is satisfied that, as supplied, the product or material fails to meet any specification which he expected it to meet:

(c) in the case mentioned in paragraph (1)(c), the removal is for either storage of waste pending any of the operations listed in Annex IIA to Council Directive 75/442/EEC[66], or storage of waste consisting of materials intended for submission to any operation listed in Annex IIB to that Directive, but, in either case, excluding temporary storage, pending collection, on the site where the waste is produced, and the consignee is the holder of a permit which authorises the relevant operation.

3 In paragraph (1)(c) "group", in relation to a body corporate, means that body corporate, any other body corporate, which is its holding company or

66 ([1975] O.J. No. L194, 25.7.75, p 39); as amended by Council Directive 91/156/EEC ([1991] O.J. No. L78, 26.3.91, p 32).

subsidiary, and any other body corporate that is a subsidiary of that holding company; and for these purposes:

'body corporate' does not include a corporation sole, but includes a company incorporated elsewhere than in Northern Ireland; and

'holding company' and 'subsidiary' have the meaning given by Article 4 of the Companies (Northern Ireland) Order 1986[67].

Procedure where pre-notification is not required (reg 25)

8.107 Paragraph (2), with the exception of sub-paragraph (b) and paragraphs (3) to (5) of regulation 23 shall apply in cases to which regulation 24 applies as if—

(a) 'five' were substituted for "six" in sub-paragraph (a) of paragraph (2) of regulation 23;
(b) references to the consignor were references—
    (i) in relation to the case mentioned in regulation 24(1)(b), to the person to whom the product or material was supplied; and
    (ii) in relation to the case mentioned in regulation 24(1)(d), to the master of the ship; and
(c) references to the consignee were references, in relation to the case mentioned in regulation 24(1)(b) to the person to whom the product or material is to be returned.

Schedule of Carriers (reg 26)

8.108 (1) This regulation applies in all cases (whether under regulation 23, or any of regulations 27 to 31) where more than one carrier transports, or is to transport, the consignment.

(2) Before the consignment is removed:

(a) the consignor shall ensure that—
    (i) a copy of the schedule of carriers in the form set out in Schedule 5 is prepared for the hazardous waste producer or holder (where different from the consignor), the consignor, every carrier and the consignee;
    (ii) all copies are given to the first carrier.
(b) the first carrier shall ensure that all copies travel with the consignment.
(c) on delivery of the consignment to each subsequent carrier—
    (i) the previous carrier shall give the subsequent carrier all copies of the schedule;
    (ii) the subsequent carrier shall complete the relevant certificate on each copy, give one to the previous carrier who shall retain it, and ensure that all remaining copies travel with the consignment.

67 SI 1986/1032 (N.I. 6) as amended by SI 1990/1504 (N.I. 10), Article 62.

(d) where the consignment is delivered to a consignee, the carrier shall retain one copy of the carriers schedule and give the remaining copies to the consignee.
(e) on receipt of completion of consignment, the consignee must forward forthwith a copy of the completed schedule of carriers to the Department.

(3) Where:

(a) arrangements for the transport of the consignment have not been made with all carriers intended to be involved before transport commences; or
(b) though such arrangements have been made, there is a change in the arrangements for any reason, paragraphs (1) and (2) apply as if the carrier in possession of the consignment when further arrangements are made, in the case of paragraph (3)(a), or take effect, in the case of paragraph (b), were the hazardous waste producer or holder and the next carrier were the first carrier.

(4) Where this regulation applies, other than in the case of a rejected consignment of hazardous waste under regulation 31, regulation 23 has effect as if:

(a) a reference to the hazardous waste being removed includes reference to its possession being transferred to the next carrier;
(b) in paragraph (2)(c) and (d), the reference to 'the carrier' were a reference to 'the first carrier';
(c) for the purposes of paragraph (3)(a) the reference to 'the carrier' in paragraph (3) were a reference to ''every carrier';
(d) for the purposes of paragraph (3)(b), the reference to 'the carrier' in paragraph (3) were a reference to 'the final carrier';
(e) in paragraph (4)(c) and (5) the reference to 'the carrier' was a reference to 'the final carrier'.

Carrier's rounds (reg 27)

8.109 (1) This regulation applies to a carrier's round or to a succession of such rounds by the same carrier starting and ending within a 12-month period in respect of which–

(a) every consignor is a person specified in the consignment note or in the schedule prepared in accordance with paragraph (2)(b)(iii) or whose particulars are notified in writing to the Department not less than 72 hours before the removal of the first waste on the carrier's round
(b) the premises from which the hazardous waste is removed are within Northern Ireland and are specified in the consignment note or in the schedule prepared in accordance with paragraph (2)(b)(iii) or notified in writing to the Department not less than 72 hours before the removal of the first waste on the carrier's round;
(c) the hazardous waste is of a description specified in the consignment note; and
(d) in the case of a single round other than a round that satisfies the requirements of regulation 36(2)(a), the time between the collection of the first consignment and the final consignment is no more than 24 hours.

(2) Before the first removal of waste, the carrier shall–

(a) on any carrier's round which is not in a succession or on the first round in such a succession, ensure that—

 (i) Parts A and B of the consignment note are completed and that the relevant code is entered;
 (ii) except where the hazardous waste to be collected on the carrier's round consists entirely of lead acid batteries, one copy of the consignment note is furnished to the Department.

(b) on every round—

 (i) prepare three copies of the consignment note in addition to one copy for each consignor from whom waste is to be collected during the round;
 (ii) complete on those copies Parts A and B, the carrier's particulars and particulars of transport in Part C, the code assigned or supplied under regulation 4 in respect of the round and, if it is a second or subsequent round, the code in respect of the first round; and
 (iii) ensure that four copies of a schedule are prepared in the form set out in Schedule 6 in addition to one consignor's copy for each site from which waste is to be collected during that round.

(3) In a case where waste of more than one description is specified in the consignment note, either:

(a) the schedule referred to in paragraph (2)(b)(iii) shall contain a separate entry for each description of waste to be collected from each consignor showing the description of waste to which that entry relates; or
(b) each entry in the schedule shall show the different descriptions of the waste to be collected and, for each such description, the quantity of the waste to be collected.

(4) The consignor shall, before the removal of waste from a site, complete on all the copies that part of the schedule indicated on it as for completion by him.

(5) The carrier shall ensure, before the removal of the waste, that—

(a) the part of the schedule indicated on it as for completion by him is completed on all the copies and includes a record of the time at which it is completed; and
(b) he has all copies of the schedule (on which the part to be completed by the consignor has been completed) except the copy to be retained by the consignor under paragraph (6).

(6) The consignor shall retain in respect of each site one copy of the consignment note and of that part of the schedule on which the parts to be completed by him and by the carrier have been completed.

(7) Before the removal of the last consignment of waste on the carrier's round, the carrier shall complete Part C on the three copies of the consignment note retained by him.

(8) The carrier shall ensure that the copies of the consignment note and of the schedule which he has received:

(a) travel with the waste to which they refer;
(b) are given to the consignee on delivery of the waste.

(9) Subject to regulation 31 (Duty of consignee not accepting a consignment), on receiving the waste collected on each round, the consignee shall:

(a) complete Part E on the three copies of the consignment note given to him;
(b) retain one copy of the consignment note and one copy of the schedule;
(c) give to the carrier a copy of the consignment note and a copy of the schedule; and
(d) forthwith furnish to the Department one copy of the consignment note and one copy of the form in Schedule 6.

(10) The carrier shall retain the copies given to him in accordance with paragraph 9(c).

Removal of ships' waste to reception facilities (reg 28)

8.110 (1) This regulation applies where hazardous waste (including excesses or spillage from loading or unloading, which have been accidentally spilled on land adjacent to the ship) is removed from any ship, whether or not a United Kingdom registered ship, in a harbour area to:

(a) reception facilities provided within that harbour area; or
(b) by pipeline to any such facilities provided outside a harbour area.

(2) Before the waste is removed from the ship:

(a) three copies of the consignment note shall be prepared and Parts A and B shall be completed and the relevant code entered on each of those copies;
(b) the operator of the facilities shall complete Part C on each of those copies; and
(c) the master of the ship—

    (i) shall ensure that Part D is completed on each of those copies;
    (ii) shall retain one copy (on which Parts A to D have been completed); and
    (iii) shall give the two remaining copies (on which Parts A to D have been completed) to the operator of the facilities.

(3) Subject to regulation 31, on receiving a consignment of hazardous waste the operator of the facilities shall:

(a) complete Part E on the copies of the consignment note which he has received;
(b) retain one copy; and
(c) forthwith furnish the other copy to the Department.

Removal of ships' wastes other than to reception facilities (reg 29)

8.111 (1) This regulation applies where hazardous waste is removed from any ship, whether or not a United Kingdom registered ship, in a harbour area other than in a case to which regulation 27 (Carrier's rounds) applies.

(2) Before the consignment is removed—

(a) the master of the ship shall ensure that—

- (i) four copies of the consignment note are prepared;
- (ii) on each of the copies Parts A and B are completed; and
- (iii) all copies are given to the carrier.

(b) the carrier shall complete Part C on each of the copies;
(c) the master of the ship shall—

- (i) ensure that Part D is completed on each of the copies;
- (ii) retain one copy; and
- (iii) give the remaining copies to the carrier.

(d) the carrier shall ensure that the copies which he has received—

- (i) travel with the consignment; and
- (ii) are given to the consignee on delivery of the consignment.

(3) Subject to regulation 31(Duty of a consignee not accepting a consignment), on receiving the consignment the consignee shall—

(a) complete Part E on both copies; and
(b) give one copy to the carrier;
(c) retain a copy;
(d) forthwith furnish the other copy to the Department.

Removal of wastes by pipeline (reg 30)

8.112 (1) This regulation applies where hazardous waste is removed from any premises (other than a ship) on which it is produced or stored by pipeline other than in a case to which regulation 28 (Removal of ships' wastes to reception facilities) or 29 (removal of ships' wastes other than to reception facilities) applies.

(2) Before the waste is first piped, and, where the piping is continuous and exceeds one quarter in duration, at the commencement of each further quarter the producer, or holder, as the case may be, shall ensure that:

(a) one copy of the consignment note is prepared for each of the following: the producer or holder, as the case may be, where different from the consignor, the consignor, the consignee and the Department;
(b) on each of the copies Parts A, B and D are completed, provided that in Part B5 the average flow rate shall be recorded, together with an estimate of the total volume of waste to be piped per week or calendar month.

(3) Part C shall not be completed.

(4) Subject to regulation 31, the consignee shall complete Part E on each copy of the note, the application of which is modified as follows—

(a) in Part E1, the reference to the date on which the waste was received shall be taken to be a reference to the last date on which waste was received by the consignee pursuant to that consignment note; and
(b) in Part E3, the particulars of the vehicle registration are not required.

(5) The consignee shall retain one copy of the consignment note and ensure that a copy is given to the consignor, to the hazardous waste producer or holder, where different from the consignor, and to the Department.

Duty of consignee not accepting delivery of a consignment (reg 31)

8.113 (1) This regulation applies where the consignee does not accept delivery of a consignment of hazardous waste in whole or in part.

(2) In a case to which this regulation applies the requirements of regulation 23(4), 27(9), 28(3), 29(3) or 30(4) (which relate to the duties of the consignee on acceptance of the consignment), as the case may be, shall not apply to the consignee in respect of the consignment, or part thereof, which has been rejected.

(3) If, in a case to which this regulation applies, copies of the consignment note have been given to the consignee he shall:

(a) indicate on Part E of each copy that he does not accept the consignment and the reasons why he does not accept the consignment;
(b) retain one copy;
(c) ensure that one copy, accompanied by one copy of any carrier's schedule given to him in accordance with regulation 27(Carrier's rounds), are furnished forthwith to the Department; and
(d) ensure that the other copy is returned to the carrier forthwith.

(4) If, in a case to which this regulation applies, no copies of the consignment note have been given to the consignee he shall ensure that a written explanation of his reasons for not accepting delivery, including such details of the consignment and of the carrier as are known to him, is furnished forthwith to the Department.

(5) In a case to which this regulation applies:

(a) on being informed that the consignee will not accept delivery of the consignment, the carrier shall inform the Department and seek instructions from the consignor;
(b) the consignor shall forthwith inform the carrier and the Department of his intentions as regards the consignment; and
(c) the carrier shall take all reasonable steps to ensure that the consignor's intentions are fulfilled.

(6) For the purposes of paragraph (5), the consignor may propose one of the following, namely:

(a) the delivery of the consignment to the premises from which it had been collected;
(b) the delivery of the consignment to the premises at which it had been produced;
(c) the delivery of the consignment to other specified premises subject to a permit necessary to authorise the receipt of the waste.

Requirement for a new consignment note (reg 32)

8.114 (1) This regulation applies where, in accordance with regulation 31(6), a consignor proposes that a consignment be delivered to other specified premises in respect of which there is held any waste management licence necessary to authorise receipt of the waste.

(2) Before the consignment is delivered to those premises:

(a) four copies of a new consignment note shall be prepared and

  (i) on each copy Parts A and B shall be completed and the relevant code entered, including the previous code;
  (ii) to each copy shall be attached a copy of any relevant previous carrier's schedule, annotated to show which consignment was not accepted;

(b) the carrier shall complete Part C on each of those copies;
(c) the consignor, subject to paragraph (3) below—

  (i) shall complete Part D on each of those copies;
  (ii) shall retain one copy (on which Parts A to D have been completed and the relevant codes entered); and
  (iii) shall give the three remaining copies (on which Parts A to D have been completed and the relevant codes entered) to the carrier.

(3) The carrier may, where he has received written instructions from the consignor to that effect, complete Part D of each of the copies of the consignment note on behalf of the consignor, and where he does so he shall send to the consignor, the consignor's copy (on which Parts A to D have been completed and the relevant codes entered).

(4) The carrier shall ensure that the three copies of the consignment note which he has received (or, if paragraph (3) applies, retained):

(a) travel with the consignment; and
(b) are given to the consignee on delivery of the consignment.

(5) Subject to regulation 9 (Specific waste to be treated as hazardous), on receiving the consignment the consignee shall—

(a) complete Part E on all copies of the consignment note given to him;
(b) retain one copy;
(c) give one copy to the carrier; and
(d) forthwith furnish one copy to the Department for the place to which the consignment has been transported.

(6) The carrier shall retain the copy of the consignment note given to him by the consignee.

Provisions as to furnishing of consignment notes (reg 33)

8.115 (1) Subject to paragraphs (2), (3) and (6), a copy of a consignment note required by regulation 23 (Standard procedure) or 27 (Carrier's rounds) to be furnished to the Department must be furnished not more than one month and not less than 72 hours before the removal of the consignment.

(2) Subject to paragraphs (3) and (6), a copy of a consignment note required to be furnished by regulation 27(2)(a)(ii) shall be furnished not less than 72 hours before the removal of the first consignment to which the consignment note relates.

(3) The copy of the consignment note mentioned in paragraphs (1) and (2) may be furnished to the Department within 72 hours before the removal where—

(a) the consignment is to be delivered to other specified premises pursuant to a proposal under regulation 31;
(b) the consignment can not lawfully remain where it is for 72 hours.

(4) The requirements of paragraphs (1) and (2) shall be treated as satisfied if:

(a) a facsimile of the copy is furnished to the Department by telephonic, electronic or other similar means of transmission in compliance with the time limits set out in those paragraphs; and
(b) the copy is furnished to the Department before or, in accordance with paragraph (5), forthwith upon removal of the consignment.

(5) A copy of a consignment note or a written explanation of reasons for refusing to accept delivery of any hazardous waste is furnished to the Department in accordance with this paragraph if it, and any document required to be furnished with it, is:

(a) delivered to the Department; or
(b) posted to the Department by pre-paid first class post;
(c) within one day of the receipt, removal or refusal to accept delivery of the hazardous waste in question, as the case may be.

(6) In reckoning any period of hours for the purposes of these Regulations, the hours of any:

(a) Saturday or Sunday;
(b) Good Friday or Christmas Day;
(c) day which is a bank holiday in Northern Ireland under the Banking and Financial Dealings Act 1971; or
(d) other public holiday,
(e) shall be disregarded.

Duty to deliver consignment within time limit (reg 34)

8.116 It is the duty of a carrier ordinarily to deliver the consignment to the consignee within one week of collection or in the case of a consignment under a carrier's round, in accordance with regulation 27 (carrier's rounds) below, one week from the time of the last collection under the round.

Where it is not reasonably practicable to deliver the consignment to the consignee in accordance with the time limit in paragraph (1) above, the carrier shall inform the Department and it is the duty of the carrier to deliver the consignment to the consignee without due delay.

Mutual recognition of consignment notes (reg 35)

8.117 Nothing in regulations 23 (Standard procedure) to 34 (Duty to deliver a consignment within a time limit) shall apply in relation to shipments of waste to which the provisions of Council Regulation (EEC) No. 259/93, other than Title III of that Regulation, apply.

Schedule 7 has effect in connection with the mutual recognition of consignment notes and the cross border movement of hazardous waste between Northern Ireland, England and Wales, Scotland and Gibraltar.

Where hazardous waste is produced in Northern Ireland, its transfer within the United Kingdom must be accompanied by a consignment note, in the form set out in Schedule 4 to these Regulations, and with a code provided by the Department.

Fees (reg 36)

8.118 (1) Subject to paragraph (2), in connection with the assignment or supply of a code for a consignment or a carrier's round in accordance with regulation 22(1), the Department shall require payment of a fee of:

(a) £10 in respect of a code relating to a consignment, or a round, which consists entirely of lead acid batteries;
(b) £24 in other cases.

(2) The Department shall not require payment of a fee where the code is assigned or supplied in connection with:

(a) a second or subsequent carrier's round in a succession of such rounds in which a single vehicle is used and in respect of which—

  (i) the carrier is also the consignee in relation to every consignment in all the rounds;
  (ii) no more than one consignment is collected from any consignor during the succession;
  (iii) the total weight of hazardous waste collected in each round does not exceed 400 kg; and

(iv) the time between the collection of the first consignment on the first round in the succession and the delivery of the last consignment to the place to which it is to be transported is no more than one week.

(b) the removal of a single consignment of hazardous waste for the purposes set out in regulation 24(1)(b) provided that the person to whom the product or material was supplied is satisfied that it fails to meet any specification which he expected it to meet; or

(c) the removal of hazardous waste from a ship in a harbour area—

(i) to a conveyance for transportation to a place outside that area;
(ii) to reception facilities provided within the same harbour area; or
(iii) by pipeline to reception facilities provided outside the harbour area.

(3) Where the Department assigns or supplies a code under regulation 22(1) without the fee required under this regulation having been paid to it, the person who requested the assignment or supply shall be required to pay the fee to the Department within the period of two months beginning with the date on which the request was made.

Part 6 Registers, records and Returns

8.119 Part 6 requires producers, holders, carriers, consignors and consignees to keep records. These must be kept for no less than three years. The Department of the Environment retains a right of inspection. The Regulations require consignees to provide producers, holders or consignors with a return as set out in Sch 8 or a copy of the consignment note within 30 days of accepting the consignment.

Registers of consignments (reg 37)

8.120 (1) The following provisions of this regulation apply in relation to registers of consignments and, as appropriate, to other registers required to be kept under regulations 37 (registers of consignments) to 40 (records of hazardous waste recovered or received by treatment facilities and transfer stations).

(2) Records must be made in a register kept for the purposes, and, insofar as is consistent with the provisions of regulations 37 (Registers of consignments) to 40 (records of hazardous waste recovered or received by treatment facilities and transfer stations) and paragraph (3) of this regulation, registers may be kept in any form.

(3) Producers, consignors, carriers, and consignees shall keep registers containing the information set out in paragraph (4).

(4) A person who is required to keep a register shall enter in the register each copy of:

(a) a consignment note (including where consignments are not accepted, the original note and the note prepared pursuant to regulation 32), and, where applicable, written explanation and schedule of carriers) completed to any extent by, or given to him, pursuant to these regulations;
(b) a consignee's return to the producer, holder or consignor received pursuant to regulation 41 (Consignee's return to the producer, holder or consignor); and
(c) a carrier's schedule given to him pursuant to regulation 26 (Schedule of carriers);

and, where a regulation requiring records to be kept is silent as to the length of time for which records are to be kept under these regulations, keep the copies on the register for not less than three years from the date on which the hazardous waste was first removed from the premises at which it was produced, or held, as the case may be.

(5) Any register kept or retained under this regulation shall be produced on demand by and given up for inspection to the Department.

(6) An establishment or undertaking to which hazardous waste is delivered for treatment or disposal, as the case may be, shall provide to the previous holder of the waste, or to the Department, on request, documentary evidence that the management operation concerned has been carried out, indicating the relevant operation where applicable in accordance with the numbering and description applicable in accordance with the list in Annex IIA or Annex IIB, as the case may be, of the Waste Directive.

(7) A person who is required to retain any record shall at any time during the period in which the record is required to be retained (and without prejudice to any duty to send the record to the Department at the end of that period), produce that record, in any form reasonably required, to the Department or emergency services on request.

(8) A producer, holder, consignor, carrier or consignee of hazardous waste shall supply to the Department on request such information as it may reasonably require for the purposes of performing its functions in connection with these Regulations and for the purposes of monitoring the production, movement, storage, recovery and disposal of hazardous waste.

Producers', holders' and consignors' records (reg 38)

8.121 (1) A producer or holder of hazardous waste, and where different from the producer, a consignor of hazardous waste, shall keep a record of the quantity, nature, origin and, where relevant, the destination, frequency of collection, mode of transport and treatment method of the waste.

(2) Where the waste is transported from the premises where it is produced or held by a person other than the producer or holder the duty in paragraph (1) includes a requirement to keep a record of particulars sufficient to identify that other person.

(3) The producer or holder, as the case may be, shall preserve the records to be made pursuant to this regulation whilst he remains the holder of the waste and for at least three years commencing on the date on which the waste is transferred to another person.

(4) The information to be recorded pursuant to the foregoing provisions of this regulation shall be recorded in a register kept by the producer or holder, as the case may be, for the purpose.

(5) Except where paragraph (6) applies, the register required to be kept and retained by a producer or holder under paragraph (4) of this regulation shall be kept at the premises at which the hazardous waste was produced.

(6) Where the period mentioned in paragraph (5) has not expired, for the remainder of that period the producer or holder shall keep the register at the producer's or holder's principal place of business where and only where:

(a) he ceases to have access to the premises in paragraph (5);
(b) his access to the premises in paragraph (5) is not and has never been regular access;

and in either case he shall notify the Department forthwith.

Any register kept and retained under this regulation shall be produced to and made available for inspection by the Department on request.

Site records (reg 39)

8.122 (1) Any person who makes a deposit of hazardous waste whether by way of storage or disposal in or on any land shall record, as soon as is reasonably practicable and in any event within 24 hours, the location and identity of the waste as required by paragraphs (2) to (4) below, and shall keep such records until his permit is surrendered or revoked and shall then send the records to the Department.

(2) Such records shall comprise either—

(a) a site plan marked with a grid, or
(b) a site plan with overlays on which deposits of the waste are shown in relation to the contours of the site.

(3) Deposits shall be identified by reference to the register of consignment notes kept under regulation 37, and shall contain a description of the composition of the waste including a reference to the six digit code in the List of Wastes save that where waste is disposed of:

(a) by pipeline, or
(b) within the curtilage of the premises at which it is produced,

the deposits shall be described by reference to a record of the composition of the waste and the date of its disposal.

Records of hazardous waste recovered or received by treatment facilities and transfer stations (reg 40)

8.123 (1) Any person who:

(a) recovers hazardous waste in or on land; or
(b) receives hazardous waste at a transfer station,

or both shall record and identify any hazardous waste received in accordance with the following requirements of this regulation and regulation 37 (Registers of consignments).

(2) The waste shall be identified in the record by reference to the relevant description and six-digit code in the List of Wastes, and the entry shall include a description of the composition of the waste.

(3) The record shall include:

(a) the quantity, nature and origin of any such waste;
(b) the relevant hazardous properties;
(c) where applicable, the recovery method in respect of the waste by reference to the numbering and description applicable in accordance with Annex IIB of the Waste Directive; and
(d) an inventory showing the specific location at which the waste is being held.

(4) The record shall be updated as soon as reasonably practicable and in any event no later than 24 hours after:

(a) each consignment of hazardous waste is received; and
(b) any recovery operations are carried out or any hazardous waste is placed in storage at the transfer station, as the case may be;
(c) any hazardous waste is removed from the premises.

(5) A person who is required to make or retain records pursuant to this regulation shall:

(a) keep the register on the site where the recovery operations take place or the transfer station is operated, as the case may be; or
(b) where the register cannot be kept on the site in accordance with paragraph (5)(a) above, at his principal place of business or such other address as agreed with the Department for that purpose; and
(c) retain the records—

    (i) where the hazardous waste is fully recovered, or remains at a transfer station, as the case may be, until it leaves the site and for three years thereafter;
    (ii) in any other case, until his waste permit pursuant to which the site is operated is surrendered or revoked, and shall then send those records to the Department.

Consignee's return to the producer, holder or consignor (reg 41)

8.124 (1) Without prejudice to any duty under these Regulations on the part of the consignee to send any document or copy thereof to the producer, holder or consignor, a consignee shall send:

(a) a producer or holder a return in a form corresponding to the form set out in Schedule 8 or a form substantially to like effect; or
(b) a copy of the consignment note together with a description of the method of disposal or recovery to the hazardous waste producer or holder identified in Part A of the consignment note, within 30 days of accepting a consignment of hazardous waste, or, where hazardous waste was delivered by pipeline in a case to which regulation 30 applies, of piping having ceased, or, where the piping is continuous and exceeds one quarter in duration, on the first day of each successive quarter.

(2) Where regulation 31 (duty of consignee not accepting delivery) applies, this regulation does not apply to the consignment or part concerned.

(3) Where the consignee is required to make a return in accordance with sub-paragraph (1) above, that return may be completed electronically or by other similar means of transmission.

Amendments to other legislation

8.125 Schedule 10 (Consequential Amendments), Part I omits references to special waste in the Waste and Contaminated Land (Northern Ireland) Order 1997 and replaces them with references to hazardous waste.

Schedule 10, Part II makes consequential amendments to secondary legislation so that references to special waste are omitted and references to hazardous waste are updated so that they are consistent with these Regulations.

Schedules

8.126 SCHEDULE 1 ANNEX I of the Hazardous Waste Directive

SCHEDULE 2 ANNEX II of the Hazardous Waste Directive

~~SCHEDULE 3 ANNEX III of the Hazardous Waste Directive~~[68]

SCHEDULE 4 Form of Consignment Note

SCHEDULE 5 Schedule of Carriers

SCHEDULE 6 Carriers' Schedule for Carrier's Round

SCHEDULE 7 Cross Boarder Movement of Hazardous Waste

SCHEDULE 8 Form of Consignee's return to producer or holder

SCHEDULE 9 Form of Fixed penalty notices

SCHEDULE 10 Consequential Amendments

68 NISR 2015/238, reg 5(7) revokes Sch 3.

SCOTLAND

8.127 In Scotland waste with hazardous properties, which may be harmful to human health or the environment is called special waste.

Special Waste Regulations 1996, SI 1996/972

8.128 The Special Waste Regulations replaced the Control of Pollution (Special Wastes) Regulations 1980, and were introduced after the EC Directive on Hazardous waste[69]. The purpose of the legislation is to control the movements of the most hazardous types of waste. The law refers in particular to a list of waste materials that are listed in the schedule of the Regulations, and also it refers to controlled wastes, which are prescribed medicinal products. A new consignment note has been introduced, a new system of fees is in operation and the mixing of special wastes is expressly prohibited.

Central to these regulations is that the general management of such wastes is covered by the Environmental Protection Act 1990, but that further legislation was required for the movement of such waste. In Scotland, the Special Waste Regulations 1996 (and their amendments) set out procedures to be followed when disposing of, carrying and receiving special waste. These regulations are the principal piece of legislation covering special waste arising in Scotland.

Classification and Assessment of Special Waste

8.129 The Landfill Waste Acceptance Criteria (WAC) analysis (specifically leachate test results) must not be used for waste classification and hazardous waste assessment purposes. This analysis is only applicable for landfill acceptance and does not give any indication as to whether a waste may be hazardous or non-hazardous.

The assessment and classification of waste is based on a combination of chemicals legislation and waste legislation. As of 1 June 2015 the way waste is classified and assessed changed.

The Dangerous Substances Directive (67/548/EC) ([1967] OJ 196 , 16/08/1967 p 0001–0098) and Dangerous Preparations Directive (1999/45/EC) ([1999] OJ L 200, 30.7.1999, p 1–68) were replaced by the Classification, Labelling and Packaging Regulation (CLP) (2008/1272/EC) ([2008] OJ L 353, 31.12.2008, p 1–1355), which introduced a new system of chemical classification based on hazard classes, categories and statement codes (rather than risk phrases and categories of danger).

Annex III of the Waste Directive (2008/98/EC) has been revised to:

69 EC Directive on Hazardous Waste 91/689/EEC.

- amend hazardous properties to align with the CLP;
- provide hazardous waste criteria based on hazard statement codes.

The List of Waste Decision (2000/532/EC) has been revised to:

- align with the CLP and Annex III;
- include criteria for persistent organic pollutants (POPs);
- make amendments arising from a review of the List itself.

The Environment Agencies for England and Wales, and the Northern Ireland Environment Agency, have updated guidance to align with these changes and created a revised guidance document – Waste Classification: Guidance on the classification and assessment of waste (WM3).

These changes were transposed into national legislation with the implementation of the Waste (Meaning of Hazardous Waste and European Waste Catalogue) (Miscellaneous Amendments) (Scotland) Regulations 2015, SSI 2015/188.

Waste (Scotland) Regulations 2012, SSI 2012/148

8.130 The Regulations ensure the Scottish Government meets its requirements under the revised Waste Framework Directive. Meeting the requirements of the WFD will require substantial changes to waste management in Scotland, including increasing recycling rates. The Zero Waste Plan is well aligned to the aims of the WFD, and the Regulations are intended to bring about the changes required under the WFD in a way that will maximise the economic opportunities afforded by a resource-centred approach to managing Scotland's waste.

It requires:

- a requirement to remove key recyclables (plastics and metals) from mixed waste prior to incineration (from 1 July 2012);
- a requirement for businesses to present dry recyclables (metals, plastics, paper, card and glass) and food waste of more than 50 kg/week for collection from end of 2013, with those producing less than 50kg/week exempt until end 2015;
- a requirement on local authorities to provide householders with a collection service for dry recyclables (end 2013) and food waste (end 2015);
- a ban on materials collected separately for recycling going to landfill or incineration (end 2013); and
- a ban on biodegradable municipal waste going to landfill, thus helping to substantially reduce harmful emissions of methane (end 2020).

Scotland has some sparsely populated areas within the UK and this provides challenges when collecting and managing waste. For instance significant distances need to be travelled which increases the financial cost and also the carbon emission from vehicles. These Regulations provide exemptions in 'rural areas' which also exempts businesses from presenting food waste for collection. The detail of what constitutes a rural area can be found in 'Defining Rural and Non Rural Areas to Support Zero Waste Policies'.

The Regulations also introduce measures to ensure that those producing or managing waste take steps to promote high quality recyclable materials, thereby increasing the likelihood that materials will be recycled in Scotland, supporting the local economy and insulating industry from global fluctuations and unpredictable future quality demands.

The Regulations correct an error in the Control of Volatile Organic Compounds (Petrol Vapour Recovery) (Scotland) Regulations 2011, SSI 2011/418, which inserted regulation 9G into the Pollution Prevention and Control (Scotland) Regulations 2000, SSI 2000/323 but did not make all the consequential amendments that were needed in that respect. They also remove a disproportionate requirement for newspaper advertising of applications relating to petrol vapour recovery installations.

Waste (Scotland) Regulations 2005, SSI 2005/22[70]

8.131 These Regulations amend the Environmental Protection Act 1990 to extend controlled waste to cover mine, quarry and agricultural waste. Categorises mine and quarry waste as industrial waste and agricultural waste as commercial.

Waste (Scotland) Regulations 2011, SSI 2011/226[71]

8.132 These Regulations enable the partial suspension, revocation, transfer and surrender of site licences, and allows licences held by the same person to be consolidated if they cover the same site or more than one mobile plant.

These Regulations and the Waste Management Licensing Regulations 2011 have four main drivers:

- To consolidate the Waste Management Licensing Regulations 1994 (as amended) (WMLR) with the aim of making Scottish waste legislation clearer and easier to understand for both regulators and operators.
- To amend certain provisions of the WMLR to provide a more flexible and efficient waste management licensing system, in line with the principles of Better Regulation.
- To transpose the revised Waste Framework Directive (2008/98/EC) fully into Scots law.
- To amend legislation relating to waste carriers to take account of a clarification of EU law by the European Court of Justice (ECJ).

70 Waste (Scotland) Regulations 2005, SSI 2005/22.
71 Waste (Scotland) Regulations 2011, SSI 2011/226.

Waste (Scotland) Regulations 2012, SSI 2012/148[72]

8.133 These Regulations set out a system to control pollution from any installation or mobile plant carrying out specified activities through permits, inspections and control of emissions. This covers the inclusion of best available techniques (BAT) and standard rules in permits. These Regulations revoke previous PPC legislation and the Waste Incineration (Scotland) Regulations 2003, SSI 2003/170.

Specific measures within the regulations include:

- a requirement to remove key recyclables (plastics and metals) from mixed waste prior to incineration (from 1 July 2012);
- a requirement for businesses to present dry recyclables (metals, plastics, paper, card and glass) and food waste of more than 50kg/week for collection from end of 2013, with those producing less than 50kg/week exempt until end 2015;
- a requirement on local authorities to provide householders with a collection service for dry recyclables (end 2013) and food waste (end 2015);
- a ban on materials collected separately for recycling going to landfill or incineration (end 2013); and
- a ban on biodegradable municipal waste going to landfill, thus helping to substantially reduce harmful emissions of methane (end 2020).

The Special Waste (Scotland) Regulations

Special Waste Regulations 1996, SI 1996/972[73]

8.134 Provides a definition of 'special waste' in Scotland, to cover all hazardous waste, and regulates waste carriers by requiring them to complete and keep consignment notes.

Special Waste (Amendment) Regulations 1996, SI 1996/2019[74]

8.135 Amends SI 1996/972 by defining special waste in Scotland, to cover all hazardous waste, and regulates waste carriers by requiring them to complete and keep consignment notes.

72 Waste (Scotland) Regulations 2012, SSI 2012/148.
73 Special Waste Regulations 1996, SI 1996/1972.
74 Special Waste (Amendment) Regulations 1996, SI 1996/2019.

Special Waste (Amendment) Regulations 1997, SI 1997/251[75]

8.136 Amends SI 1996/972 to revise the qualifications and experience required to deal with special waste.

Special Waste (Scotland) Regulations 1997, SI 1997/254[76]

8.137 Establishes that managers of specified waste facilities operated by local councils who meet certain criteria should be treated as technically competent for special waste facilities.

Special Waste Amendment (Scotland) Amendment Regulations 2004, SSI 2004/204[77]

8.138 Amends the Special Waste Amendment (Scotland) Regulations 2004, SSI 2004/112 2004 which in turn amend the Special Waste Regulations 1996, SI 1996/972 and defines special waste and sets out controls on labelling, packaging and separating it. Requires consignment notes to be used when special waste is transferred and producers to keep registers of documents for at least three years.

The majority of wastes that will be special waste under the Scottish Amendment Regulations are likely to have been special waste under the 1996 Regulations. However, there will be some new special wastes relevant to the FHE sector, such as end-of-life vehicles before they are de-polluted and fridges that contain CFC, HCFC or HFC. A recent revision of the list given in the EWC includes some waste streams that were not previously considered to be hazardous and that routinely arise in the HFE sector such as fluorescent tubes, TV sets, computer monitors and some batteries. These wastes are now therefore special waste in Scotland. Another significant difference is that not all prescription medicines will be special waste in Scotland.

Special waste is subject to the Duty of Care and the requirements of the Environmental Protection (Duty of Care) Regulations 1991, SI 1991/2839. Compliance with the Duty of Care, however, does not in any way discharge the need to comply with the Special Waste Regulations (see **Chapter 4** on Duty of Care).

Regulation 2(10) of the Amendment Regulations, which was made on 10 March 2004, inserts a new reg 15A into the principal Regulations, which requires

75 Special Waste (Amendment) Regulations 1997, SI 1997/251.
76 Special Waste (Scotland) Regulations 1997, SI 1997/257.
77 Special Waste Amendment (Scotland) Regulations 2004, SSI 2004/204.

producers of special waste to keep a register which details the documents required to be kept in implementation of Article 4(2) of the Hazardous Waste Directive. Regulation 15A comes into force on 1 July 2004.

These Regulations correct an error in the original Amendment Regulations, which requires a producer of special waste to enter into the register details of a producer return in relation to a consignment of special waste.

Special Waste Regulations 1996, SI 1996/972

8.139 These Regulations provide a new definition of special waste. They make provision for handling such waste and for implementing Council Directive 91/689/EEC on hazardous waste.

Meaning of special waste (reg 2)

8.140 Regulation 2 defines special waste, making reference to Ps I, II and III of Sch 2. This is to implement the definition of hazardous waste in the Directive and in particular in the List annexed to Council Decision 94/904/EC. The definition extends, in accordance with Article 4 of the Directive, to certain other waste considered by the United Kingdom to display particular hazardous properties. Household waste is excluded from the definition.

(1) Subject to paragraphs (5) and (6), any controlled waste

(a) to which a six-digit code is assigned in the list set out in Part I of Schedule 2 to these Regulations (which reproduces the list of hazardous waste annexed to Council Decision 94/904/EC[78] establishing a list of hazardous waste pursuant to Article 1(4) of the Hazardous Waste Directive); and
(b) which, subject to paragraph (3), displays any of the properties specified in Part II of that Schedule (which reproduces Annex III to the Hazardous Waste Directive),

is special waste.

(2) Subject to paragraph (6), any other controlled waste which:

78 ([1994] OJ No L 356, 31.12.1994, p.14).

(a) displays the property H3A [79](first indent), or subject to paragraphs (3) and (5), H4[80], H5[81], H6[82], H7[83] or H8[84] specified in Part II of Schedule 2, or

(b) is a medicinal product, as defined in section 130 of the Medicines Act 1968[85] (meaning of "medicinal product" etc.), of a description, or falling within a class, specified in an order under section 58 of that Act[86] (medicinal products on prescription only)

is special waste.

(3) For the purposes of paragraphs (1) and (2), waste displays a property mentioned in Part II of Schedule 2 as toxic, very toxic, harmful, corrosive, irritant or carcinogenic, if it is so classified or, being so classified, has a risk phrase assigned to it or is placed in a category by the approved classification and labelling guide, as the case may be:

(i) in the case of a substance which is both listed in the approved supply list and present at or above the concentration limit applicable under Part V of that List, on the basis of that list; or
(ii) in the case of any other substance, on the basis of the criteria laid down in the approved classification and labelling guide.

(4) Except in the case of a substance listed in the approved supply list and present at or above the concentration limit applicable under Part V of that List, the test methods to be used, for the purposes of deciding which (if any) of the properties mentioned in Part II of Schedule 2 to these Regulations are to be

79
- liquid substances and preparations having a flash point below 21°C (including extremely flammable liquids), or
- substances and preparations which may become hot and finally catch fire in contact with air at ambient temperature without any application of energy, or
- solid substances and preparations which may readily catch fire after brief contact with a source of ignition and which continue to burn or to be consumed after removal of the source of ignition, or
- gaseous substances and preparations which are flammable in air at normal pressure, or
- substances and preparations which, in contact with water or damp air, evolve highly flammable gases in dangerous quantities.

80 'Irritant': non-corrosive substances and preparations which, through immediate, prolonged or repeated contact with the skin or mucous membrane, can cause inflammation.

81 'Harmful': substances and preparations which, if they are inhaled or ingested or if they penetrate the skin, may involve limited health risks.

82 'Toxic': substances and preparations (including very toxic substances and preparations) which, if they are inhaled or ingested or if they penetrate the skin, may involve serious, acute or chronic health risks and even death.

83 'Carcinogenic': substances and preparations which, if they are inhaled or ingested or if they penetrate the skin, may induce cancer or increase its incidence.

84 'Corrosive': substances and preparations which may destroy living tissue on contact.

85 Medicines Act 1968, s 130 is amended by para 3(7)–(10) of Sch 1, and Sch 2, to the Animal Health and Welfare Act 1984 .

86 Section 58 is amended by s 1 of the Medicinal Products: Prescription by Nurses etc Act 1992.

assigned to a substance, are those described in Annex V to Council Directive 67/548/EEC, as amended[87].

(5) Controlled waste which, apart from this paragraph, would be special waste is not special waste if it displays any of the properties H4 to H8 below the threshold shown as applicable to that property in Part III of Schedule 2 (which reproduces those thresholds set out in Article 1 of Council Decision 94/904/EC in so far as they are relevant).

(6) Household waste is not special waste.

Application of Regulations (reg 2A)[88]

8.141 'Special Waste' means any waste, which is hazardous waste as defined by Article 1(4) of the Hazardous Waste Directive.

Domestic Asbestos Waste (reg 2B)[89]

8.142 (1) Subject to paragraphs (2) and (3), these Regulations apply to special waste, which is domestic asbestos waste, except in so far as they would, apart from this paragraph, impose obligations on a person to whom paragraph (2) applies.

(2) This paragraph applies to a person who is both the original producer of the domestic waste and who is either:

(a) a person who resides at the domestic premises at which the asbestos waste arises; or
(b) a person who is acting on behalf of such a person without reward.

(3) In the application of these Regulations to asbestos waste–

(a) not being domestic waste; and
(b) produced in the course of any of the activities of construction, modification, repair and maintenance (including structural works) or demolition of domestic premises or any part thereof,

these Regulations operate in relation to a domestic occupier who has engaged any person, other than a person mentioned in paragraph (2)(b), for the carrying out of any such activity so as to treat that person as the producer of the asbestos waste to the exclusion of the occupier.

87 ([1967] OJ No. 196, 16.8.1967, p1) as amended by Commission Directive 92/69/EEC ([1992] OJ No. L 383, 29.12.1992, p1).
88 Inserted by the Special Waste Amendment (Scotland) Regulations 2004, SSI 2004/112, reg 2(3).
89 Inserted by the Special Waste Amendment (Scotland) Regulations 2004, SSI 2004/112 reg 2(5).

Certain radioactive waste to be special waste (reg 3)

8.143 Section 62 (special provision with respect to certain dangerous and intractable waste) of the EPA 1990 Act[90] shall have effect, without modification, so as to empower the Secretary of State to make provision for waste which would be controlled waste but for the fact that it is radioactive waste within the meaning of the Radioactive Substances Act 1993[91]; and paragraphs (1) and (2) of Regulation 2 to 2B[92] shall apply to any such waste as if it were controlled waste.

Coding of consignments (reg 4)

8.144 Regulation 4 requires the Environment Agency (in relation to England and Wales) and the Scottish Environment Protection Agency (in relation to Scotland) ('the Agencies'), to give unique codes to be applied to consignments of waste or to carrier's rounds. Carrier's rounds consist of several consignments collected on the same journey and delivered to one place. The codes are to be shown, together with other required information, on consignment notes, which are to accompany the waste when transported. Regulations 5 to 10 and 12 and 13 and Sch 1 provide for the completion and handling of these notes and for pre-notification to the Agency of the consignment or round. Completion of such identification forms and their transport with waste are required by Article 5 of the Directive. The pre-notification provisions (regs 5(2)(b), 8(2)(a)(ii) and 12) are not implementing specific Community obligations.

(1) An Agency shall assign or supply forthwith to any person, on request, for the purpose of assigning to a consignment of special waste or, where the consignment is one in a carrier's round, to the consignments in that round, a code unique to that consignment or round, as the case may be.

(2) A code assigned or supplied in accordance with paragraph (1) may consist of letters, numbers or symbols, or any combination of letters, numbers and symbols, or a bar code which enables the consignment or carrier's round, as the case may be, to be identified electronically.

Packaging and Labelling of special waste (reg 4A)[93]

8.145 In the course of its collection, transport and temporary storage, the consignor shall ensure that special waste is properly packaged and labelled in

90 Section 62 is amended by para 80 of Sch 22 to the Environment Act 1995.

91 Radiocative Substances Act 1993. Section 78 of the Environmental Protection Act 1990 (which is amended by para 7 of Sch 4 to the Radioactive Substances Act 1993) provides that Pt II of the 1990 Act does not apply to radioactive waste as defined in the 1993 Act save to the extent that the Secretary of State so provides in regulations.

92 Inserted by the Special Waste (Amendment) (Scotland) Regulations 2004, SSI 2004/112, reg 2(6).

93 Inserted by the Special Waste Amendment (Scotland) Regulations 2004, SSI 2004/112, reg 2(7).

accordance with the relevant provisions of the Carriage of Dangerous Goods (Classification, Packaging and Labelling) and Use of Transportable Pressure Receptacles Regulations 1996, SI 1996/2092[94]".

Consignment notes: standard procedure (reg 5)

8.146 (1) Except in a case to which regulation 6 (Consignment notes: cases in which pre-notification is not required), 8 (Consignment notes: carriers rounds) or 9 (Consignment notes: removal of ships' waste to reception facilities) applies, this regulation applies where a consignment of special waste is to be removed from the premises at which it is being held.

(2) Before the consignment is removed—

(a) five copies of the consignment note shall be prepared, and, on each copy, Parts A and B shall be completed, and there shall be entered on the consignment note–

  (i) the relevant code;
  (ii) the six digit code assigned to the waste in the European Waste Catalogue; and

  (iii) the postcode of the producer of the waste.".[95]

(b) the consignor shall ensure that one of those copies (on which Parts A and B have been completed and the relevant code entered) is furnished to the Agency for the place to which the consignment is to be transported;
(c) the carrier shall complete Part C on each of the four remaining copies; and
(d) the consignor

  (i) shall complete Part D on each of those copies;
  (ii) shall retain one copy (on which Parts A to D have been completed and the relevant code entered); and
  (iii) shall give the three remaining copies (on which Parts A to D have been completed and the relevant code entered) to the carrier.

(3) The carrier shall ensure that the copies which he has received:

(a) travel with the consignment; and
(b) are given to the consignee on delivery of the consignment.

(4) Subject to regulation 10, on receiving the consignment the consignee shall

(a) complete Part E on the three copies of the consignment note given to him;
(b) retain one copy;
(c) give one copy to the carrier; and

94 SI 1996/2092 as amended by SI 1999/303.
95 The Special Waste Amendment (Scotland) Regulations 2004, SSI 2004/112, reg 2(8).

(d) forthwith furnish one copy to the Agency for the place to which the consignment has been transported.

(5) The carrier shall retain the copy of the consignment note given to him by the consignee.

Consignment notes: cases in which pre-notification is not required (reg 6)

8.147 (1) For the purposes of regulation 7 (consignment notes: procedure where pre-notification is not required), except in a case to which regulation 8 (consignments notes: carriers rounds) applies, this regulation applies:

(a) subject to paragraph (2)(a), to the removal, from the premises at which it is being held, of each of the second and any subsequent consignment of special waste in a succession of consignments of special waste,
(b) subject to paragraph (2)(b) to the removal as a consignment of special waste of a product or material for the purposes of the return by the person to whom the product or material had been supplied to the person who supplied it to him or who manufactured it
(c) subject to paragraph (2)(c) to the removal of a consignment of special waste where the consignor and the consignee are bodies corporate belonging to the same group,
(d) to the removal from a ship in a harbour area of a consignment of special waste to a conveyance for transportation to a place outside that area, and
(e) to the removal of a consignment of special waste which consists entirely of lead acid motor vehicle batteries.

(2) This regulation does not apply unless—

(a) in the case mentioned in paragraph (1)(a), in respect of each consignment—

(i) the waste is of the same description as the waste in the first of the consignments in the succession;
(ii) the consignor is the same person;
(iii) the consignee is the same person;
(iv) the premises from which the consignment is removed are the same;
(v) the place to which the consignment is transported is the same; and
(vi) the removal of the consignment takes place within one year of the removal of the first consignment in the succession.

(b) in the case mentioned in paragraph (1)(b), the person to whom the product or material was supplied is satisfied that, as supplied, the product or material fails to meet any specification which he expected it to meet;
(c) in the case mentioned in paragraph (1)(c), the removal is for the purposes of an operation within either paragraph 15 of Part III, or paragraph 13 of Part IV, of Schedule 4 to the 1994 Regulations, and the consignee either—

(i) is the holder of a waste management licence which authorises the relevant operation; or

(ii) carries on any activity to which section 33(1)(a) and (b) of the 1990 Act[96] does not apply by virtue of regulation 16 (site records) or 17 (restrictions on mixing special wastes) of the 1994 Regulations.

(3) In paragraph (1)(c) 'group', in relation to a body corporate, means that body corporate, any other body corporate which is its holding company or subsidiary and any other body corporate which is a subsidiary of that holding company; and for these purposes:

(a) 'body corporate' does not include a corporation sole or a Scottish partnership, but includes a company incorporated elsewhere than in Great Britain; and
(b) 'holding company' and 'subsidiary' have the meaning given by section 736 of the Companies Act 1985[97].

Consignment notes: procedure where pre-notification is not required (reg 7)

8.148 Paragraph (2) with the exception of sub-paragraph (b) and paragraphs (3) to (5) of regulation 5 shall apply in cases to which regulation 6 (consignment notes: cases in which pre-notification is not required) applies as if:

(a) 'four' were substituted for 'five' in sub-paragraph (a) of paragraph (2);
(b) references to the consignor were references—

   (i) in relation to the case mentioned in regulation 6(1)(b), to the person to whom the product or material was supplied; and
   (ii) in relation to the case mentioned in regulation 6(1)(d), to the master of the ship; and

(c) references to the consignee were references, in relation to the case mentioned in regulation 6(1)(b), to the person to whom the product or material is to be returned.

Consignment notes: carrier's rounds (reg 8)

8.149 (1) This regulation applies to a carrier's round or to a succession of such rounds by the same carrier starting and ending within a 12-month period in respect of which:

(a) every consignor is a person specified in the consignment note or in the schedule prepared in accordance with paragraph (2)(b)(iii) or whose particulars are notified in writing to the Agency not less than 72 hours before the removal of the first waste on the carrier's round;

96 Section 33(1)(a) and (b) is modified by para 9 of Pt I of Sch 4 to the Waste Management Licensing Regulations 1994.
97 Companies Act 1985, s 736 is substituted by s 144(1) of the Companies Act 1989.

(b) the premises from which the special waste is removed are:

 (i) specified in the consignment note or in the schedule prepared in accordance with paragraph (2)(b)(iii) or notified in writing to the Agency not less than 72 hours before the removal of the first waste on the carrier's round; and
 (ii) so located that the Agency for each of those premises is the same;

(c) the special waste is of a description specified in the consignment note; and
(d) in the case of a single round other than a round that satisfies the requirements of regulation 14(2)(a) the time between the collection of the first consignment and delivery to the consignee is no more than 24 hours.

(2) Before the first removal of waste, the carrier shall,

(a) on any carrier's round which is not in a succession or on the first round in such a succession, ensure that

 (i) Parts A and B of the consignment note are completed and that the relevant code is entered;
 (ii) one copy of the consignment note is furnished to the Agency for the place to which the special waste is to be transported;

(b) on every round—

 (i) prepare four copies of the consignment note in addition to one copy for each consignor from whom waste is to be collected during the round;
 (ii) complete on those copies Parts A and B, the carrier's particulars and particulars of transport in Part C, the code assigned or supplied under regulation 4 in respect of the round and, if it is a second or subsequent round, the code in respect of the first round; and
 (iii) ensure that four copies of a schedule are prepared in the form set out in Part II of Schedule 1 to these Regulations, or in a form substantially to the like effect, in addition to one consignor's copy for each site from which waste is to be collected during that round.

(3) The consignor shall, before the removal of waste from a site, complete on all the copies that part of the schedule indicated on it as for completion by them.

(4) The carrier shall ensure, before the removal of the waste, that:

(a) the part of the schedule indicated on it as for completion by him is completed on all the copies; and
(b) he has all copies of the schedule (on which the part to be completed by the consignor has been completed) except the copy to be retained by the consignor under paragraph (5).

(5) The consignor shall retain in respect of each site one copy of the consignment note and of that part of the schedule on which the parts to be completed by him and by the carrier have been completed.

(6) The carrier shall ensure that the copies of the consignment note and of the schedule which he has received:

(a) are completed with respect to that round;
(b) travel with the waste to which they refer;
(c) are given to the consignee on delivery of the waste.

(7) Subject to regulation 10, on receiving the waste collected on each round, the consignee shall:

(a) complete Part E on the three copies of the consignment note given to him;
(b) retain one copy of the consignment note and one copy of the schedule;
(c) give to the carrier a copy of the consignment note and a copy of the schedule; and
(d) forthwith furnish to the Agency for the place to which the consignment has been transported one copy of the consignment note and one copy of the schedule.

(8) The carrier shall retain the copies given to him in accordance with paragraph (7)(c).

Consignment notes: removal of ships' waste to reception facilities (reg 9)

8.150 (1) This regulation applies where special waste is removed from a ship in a harbour area to:

(a) reception facilities provided within that harbour area; or
(b) by pipeline to any such facilities provided outside a harbour area.

(2) Before the waste is removed from the ship:

(a) three copies of the consignment note shall be prepared and Parts A and B shall be completed and the relevant code entered on each of those copies;
(b) the operator of the facilities shall complete Part C on each of those copies; and
(c) the master of the ship:
   (i) shall ensure that Part D is completed on each of those copies;
   (ii) shall retain one copy (on which Parts A to D have been completed); and
   (iii) shall give the two remaining copies (on which Parts A to D have been completed) to the operator of the facilities.

(3) On receiving a consignment of special waste the operator of the facilities shall:

(a) complete Part E on the copies of the consignment note which he has received;
(b) retain one copy; and
(c) forthwith furnish the other copy to the Agency for the place where the facilities are situated.

Consignment notes etc: duty of consignee not accepting delivery of a consignment (reg 10)

8.151 (1) This regulation applies where the consignee does not accept delivery of a consignment of special waste.

(2) In a case to which this regulation applies the requirements of regulation 5(4) (including that paragraph as applied in cases to which regulation 6 applies) or 8(7), as the case may be, shall not apply to the consignee.

(3) If , in a case to which this regulation applies, copies of the consignment note have been given to the consignee he shall:

(a) indicate on Part E of each copy that he does not accept the consignment and the reasons why he does not accept the consignment;
(b) retain one copy;
(c) ensure that one copy, accompanied by one copy of any carrier's schedule given to him in accordance with regulation 8, are furnished forthwith to the Agency for the place to which the special waste has been transported; and
(d) ensure that the other copy is returned to the carrier forthwith.

(4) If, in a case to which this regulation applies, no copies of the consignment note have been given to the consignee he shall ensure that a written explanation of his reasons for not accepting delivery, including such details of the consignment and of the carrier as are known to him, is furnished forthwith to the Agency for the place to which the special waste has been transported

(5) In a case to which this regulation applies:

(a) on being informed that the consignee will not accept delivery of the consignment, the carrier shall inform the Agency and seek instructions from the consignor;
(b) the consignor shall forthwith inform the carrier and the Agency of his intentions as regards the consignment; and
(c) the carrier shall take all reasonable steps to ensure that the consignor's intentions are fulfilled.

(6) For the purposes of paragraph (5), the consignor may propose one of the following, namely:

(a) the delivery of the consignment to the premises from which it had been collected;
(b) the delivery of the consignment to the premises at which it had been produced;
(c) the delivery of the consignment to other specified premises in respect of which there is held any waste management licence necessary to authorise the receipt of the waste.

Consignment notes: duties of the Agencies (reg 11)

8.152 Regulation 11 requires the Agencies to provide certain information to one another, following notification, where waste is to be transported from

England and Wales to Scotland or vice versa. Regulation 14 requires the Agencies to charge fees on supplying a code under regulation 4. Rounds of low quantity fulfilling certain conditions attract no fees. Regulations 11 and 14 do not implement Community obligations.

(1) Subject to paragraph (2), where:

(a) an Agency ('the receiving Agency') has been furnished with a copy of a consignment note under regulation 5 (consignment notes: standard procedure), 7 (consignment notes: procedure where pre-notification is not required), 8 (Consignment notes: carriers rounds), 9 (consignment notes: removal of ships' waste to reception facilities) or 10 (consignment notes: duty of consignee not accepting delivery of a consignment) or with a copy of the explanation under regulation 10(4); and
(b) the other Agency is the Agency for the premises from which the special waste was removed,
(c) the receiving Agency shall, within two weeks of receipt, send to the other Agency one copy of the consignment note or explanation as the case may be.

(2) Where copies have been furnished:

(a) under regulation 7 (consignment notes: procedure where pre-notification is not required) in a case to which regulation 6 (consignment notes: cases in which pre-notification is not required) applies by virtue of paragraph (1) (d) of that regulation, or
(b) under regulation 9(3)(c),

paragraph (1) shall have effect as if the reference to the premises from which the special waste was removed were a reference to the harbour area in which the special waste was removed from the ship.

Consignment notes: provisions as to furnishing (reg 12)

8.153 (1) Subject to paragraphs (2), (3) and (6), a copy of a consignment note required by regulation 5 or 8 to be furnished to an Agency must be furnished not more than one month and not less than 72 hours before the removal of the consignment.

(2) Subject to paragraphs (3) and (6), a copy of a consignment note required to be furnished by paragraph 8(2)(a) shall be furnished not less than 72 hours before the removal of the first consignment to which the consignment note relates.

(3) The copy of the consignment note mentioned in paragraphs (1) and (2) may be furnished to the Agency within 72 hours before the removal where:

(a) the consignment is to be delivered to other specified premises pursuant to a proposal under regulation 10(6)(c);
(b) the consignment cannot lawfully remain where it is for 72 hours.

(4) The requirements of paragraphs (1) and (2) shall be treated as satisfied if:

(a) a facsimile of the copy is furnished to the Agency by telephonic, electronic or other similar means of transmission in compliance with the time limits set out in those paragraphs, and
(b) the copy is furnished to the Agency before or, in accordance with the paragraph below, forthwith upon removal of the consignment.

(5) A copy of a consignment note or a written explanation of reasons for refusing to accept delivery of any special waste is furnished to an Agency in accordance with this paragraph if it, and any document required to be furnished with it, is:

(a) delivered to the Agency, or
(b) posted to the Agency by pre-paid first class post,

within one day of the receipt, removal or refusal to accept delivery of the special waste in question, as the case may be.

(6) In reckoning any period of hours for the purposes of paragraphs (1), (2) and (3), the hours of any Saturday, Sunday, Good Friday, Christmas Day, bank holiday or other public holiday shall be disregarded.

Consignment notes: importers and exporters (reg 13)

8.154 (1) Subject to paragraphs (3) and (4), regulations 5 (consignment notes: standard procedure) to 12 (consignment notes: provisions as to furnishing) shall apply to special waste imported into Great Britain from Northern Ireland or Gibraltar as if:

(a) any reference to the consignor were a reference to the person importing the special waste;
(b) any reference to the premises at which the special waste is being held and from which it is removed were a reference to the place where it first enters Great Britain; and
(c) the special waste is removed from that place at the time when it first enters Great Britain.

(2) Subject to paragraph (4), these Regulations shall apply to special waste exported from Great Britain to Northern Ireland or Gibraltar as if:

(a) any reference to the consignee were a reference to the person exporting the waste; and
(b) the consignment of special waste is received by that person at the place where and the time when it leaves Great Britain.

(3) Paragraph (1) does not apply in a case to which either regulation 6(1)(d) or regulation 9 (consignment notes: removal of ships' waste to reception facilities) applies.

(4) Nothing in regulations 5 to 12 shall apply in relation to shipments of waste to which the provisions of Council Regulation (EEC) No. 259/93, other than Title III of that Regulation, apply.

Fees (reg 14)

8.155 (1) Subject to paragraph (2), when it assigns or supplies a code for a consignment or a carrier's round in accordance with regulation 4(1) an Agency shall require payment of a fee of:

(a) £10 in respect of a code relating to a consignment, or a round, which consists entirely of lead acid motor vehicle batteries;
(b) £15 in other cases.

(2) An Agency shall not require payment of a fee where the code is assigned or supplied in connection with:

(a) a second or subsequent carrier's round in a succession of such rounds, in respect of which:

  (i) the carrier is also the consignee in relation to every consignment in all the rounds;
  (ii) no more than one consignment is collected from any consignor during the succession;
  (iii) the total weight of special waste collected in each round does not exceed 400 kg; and
  (iv) the time between the collection of the first consignment on the first round in the succession and the delivery of the last consignment to the place to which it is to be transported is no more than one week.

(b) the removal of a single consignment of special waste for the purposes set out in regulation 6(1)(b) provided that the person to whom the product or material was supplied is satisfied that it fails to meet any specification which he expected it to meet; or
(c) the removal of special waste from a ship in a harbour area—

  (i) to a conveyance for transportation to a place outside that area;
  (ii) to reception facilities provided within the same harbour area; or
  (iii) by pipeline to reception facilities provided outside the harbour area.

Registers (reg 15)

8.156 Regulation 15 implements Article 4.3 of the Directive on the keeping of records by those consigning and carrying hazardous waste. They are both required to keep the documents for three years, although the Directive only requires carriers to keep records for at least 12 months. Together with reg 16, it also implements the requirements of Article 2.1 of the Directive on the keeping of records for sites where hazardous waste is deposited.

(1) At each site from which any consignment of special waste has been removed, the consignor shall keep a register containing:

(a) a copy of the consignment note; and

(b) where the consignment is one to which regulation 8 applies, a copy of that part of the carrier's schedule retained under regulation 8(5);
(c) applicable to each consignment removed from that site.

(2) Every carrier shall keep a register containing:

(a) a copy of the consignment note; and
(b) where the consignment is one to which regulation 8 applies, a copy of the carrier's schedule,
(c) applicable to each consignment which he has transported.

(3) At each site at which any consignment of special waste has been received, the consignee shall keep a register containing:

(a) a copy of the consignment note; and
(b) where the consignment is one to which regulation 8 applies, a copy of the carrier's schedule,

applicable to each consignment, other than a consignment to which regulation 10 applies, received at that site.

(4) A consignment note or carrier's schedule required by paragraphs (1) or (2) to be kept in a register shall be retained in the register for not less than three years from the date on which the waste to which it relates was removed from the premises at which it was being held.

(5) Subject to paragraphs (6) and (7), consignment notes and carrier's schedules required by paragraph (3) to be kept by a person shall be retained until his waste management licence for the site in question is surrendered or revoked entirely, at which time he shall send the register to the Agency for the site; and that Agency shall retain the register for not less than three years after its receipt.

(6) Where, by virtue of regulation 16(1)(a) or (b) of the 1994 Regulations, section 33(1)(a), (b) and (c) of the 1990 Act does not apply to any of the activities carried on at a site at which special waste is received, paragraph (5) shall have effect as if any reference to the surrender or revocation of a person's waste management licence were a reference to the surrender or revocation of his authorisation under Part I of the 1990 Act for the site in question.

(7) Where, in circumstances other than those mentioned in paragraph (6), section 33(1)(a) and (b) of the 1990 Act does not apply to any of the activities carried on at a site at which special waste is received, each consignment note and carrier's schedule required to be kept in a register shall be kept in that register for not less than three years from the date on which the consignment of special waste to which it relates was received at the site to which it was transported.

(8) Insofar as is consistent with the foregoing provisions of this regulation, registers under this regulation may be kept in any form.

(9)[98] It is the duty of the Scottish Environment Protection Agency to carry out inspections of the registers referred to within these Regulations at such intervals as it considers appropriate, having regard to the need to meet the requirements of Article 13 of the Waste Directive and Article 5(2) of the Hazardous Waste Directive.

Registers: special waste producers (reg 15A)[99]

8.157 (1) A special waste producer shall keep a record of the quantity, nature, origin and, where appropriate, the destination, frequency of collection and mode of transport of the special waste produced by that producer.

(2) Where special waste is transported from the premises where it was produced, by a person other than the producer of that waste, the requirement on the producer to record the destination of that waste includes a requirement to record particulars sufficient to identify that other person.

(3) A special waste producer shall preserve the records kept pursuant to this Regulation whilst that producer remains the holder of that waste and for at least three years commencing on the date upon which that waste is transferred to another person.

(4) The information required to be kept in accordance with paragraphs (1) and (2) shall be kept in a register maintained by the special waste producer for that purpose.

(5) A special waste producer shall also enter into the register a copy of each consignment note and, where applicable, carrier's schedule in respect of each consignment removed. , together with the producer return detailing that consignment[100].

(6) Except where paragraph (6) applies, the register required to be maintained under paragraph (4) shall be kept at the premises at which the special waste was produced.

(7) Where a special waste producer ceases to have access to the premises referred to at paragraph (6), but the period mentioned in paragraph (3) has not expired, for the remainder of that period, that producer shall keep the register at the producer's principal place of business and notify the Scottish Environment Protection Agency thereof forthwith.

(8) Any register kept and retained under this regulation shall be produced to, and made available for inspection by, the Scottish Environment Protection Agency on request.

98 Inserted by the Special Waste Amendment (Scotland) Regulations 2004, SSI 2004/112, reg 2(9).
99 Inserted by the Special Waste Amendment (Scotland) Regulations 2004, SSI 2004/112, reg 2(10).
100 The Special Waste Amendment (Scotland) Regulations 2004, SSI 2004/112, reg 2.

(9) Insofar as is consistent with the foregoing provisions of this regulation, registers required to be kept under this regulation may be kept in any form.

Site records (reg 16)

8.158 (1) Any person who makes a deposit of special waste in or on any land shall record the location of each such deposit, shall keep such records until his waste management licence is surrendered or revoked and shall then send the records to the Agency for the site.

(2) Such records shall comprise either—

(a) a site plan marked with a grid, or
(b) a site plan with overlays on which deposits are shown in relation to the contours of the site.

(3) Deposits shall be described in such records by reference to the register of consignment notes kept under regulation 15, save that where waste is disposed of:

(a) by pipeline, or
(b) within the curtilage of the premises at which it is produced,

the deposits shall be described by reference to a record of the quantity and composition of the waste and the date of its disposal.

(4) In the case of liquid wastes discharged without containers into underground strata or disused workings the record shall comprise only a written statement of the quantity and composition of special waste so discharged and the date of its disposal.

(5) Every record made pursuant to regulation 14 of the Control of Pollution (Special Waste) Regulations 1980 shall:

(a) be kept with the records referred to in paragraph (1) above for so long as is mentioned in that paragraph, and
(b) shall accompany those records when they are sent to the Agency in accordance with that paragraph.

Restrictions on mixing special waste (reg 17)

8.159 Regulation 17 prohibits the mixing of special waste with other waste or other categories of special waste except where this is authorised under, or exempted from the effect of, certain other waste management legislation. This regulation is to implement Articles 2.2 and 2.3 of the Directive.

(1) Subject to paragraph (2), an establishment or undertaking which carries out the disposal or recovery of special waste, or which collects or transports special waste, shall not:

(a) mix different categories of special waste; or
(b) mix special waste with waste which is not special waste.

(2) Paragraph (1) shall not apply if the mixing:

(a) is authorised by a waste management licence or under an authorisation granted under Part I of the 1990 Act; or
(b) is an activity to which, by virtue of regulation 17 of the 1994 Regulations, section 33(1)(a) and (b) of the 1990 Act does not apply.

Duty to separate mixed wastes (reg 17A)[101]

8.160 (1) An establishment or undertaking which carries out the disposal or recovery of special waste, or which collects or transports special waste, shall, where such waste is already mixed with other waste, substances or materials, effect separation as soon as reasonably practicable where–

(a) technically and economically feasible; and
(b) necessary in order to comply with the provisions of Article 4 of the Waste Directive set out in paragraph (2).

(2) The provisions referred to in paragraph (1)(b) are to ensure that waste is recovered or disposed of without endangering human health and without using processes or methods which could harm the environment and in particular without–

(a) risk to water, air, soil, plants or animals;
(b) causing a nuisance through noise or odours;
(c) adversely affecting the countryside or places of special interest.

Offences (reg 18)

8.161 Regulation 18 makes failure to comply with the Regulations a criminal offence except for an Agency member, officer or employee. There is a defence for those who take certain steps in cases of emergency or grave danger. The Agencies are made responsible by regulation 19 (responsibilities of the Agency) for supervising activities and persons subject to the Regulations. Insofar as they relate to provisions, which implement Community obligations these provisions are part of that implementation because their purpose is to make the implementation effective.

Responsibilities of the Agencies (reg 19)

8.162 The Agencies shall be responsible for supervising the persons and activities subject to any provision of these Regulations.

101 Inserted by the Special Waste Amendment (Scotland) Regulations 2004, SSI 2004/112, reg 2(11).

Mutual recognition of consignment notes (reg 19A)[102]

8.163 Where special waste is removed from premises situated outside Scotland, any consignment note that contains or purports to contain the same information as that required in the form set out in the Annex to Commission Decision 94/774 EC (concerning the standard consignment note referred to in Council Regulation (EEC) No 259/93 on the supervision and control of shipments of waste within, into and out of the European Community), that accompanies the special waste shall be treated for the purposes of these Regulations as if it was a consignment note raised in compliance or purported compliance with the provisions of these Regulations.

Transitional provisions for certificates of technical competence (reg 20)

8.164 Regulation 20 makes transitional provision for applications for certificates of technical competence under the Waste Management Licensing Regulations 1994, SI 1994/1056 as amended) where such applications were made before 1 March 1997 where the applicant was licensed to deal with special waste before the change in definition made by these Regulations. This provision does not implement a Community obligation.

Amendment of regulations relating to the assessment of environmental effects (reg 21)

8.165 Regulations 21 to 26 and Sch 3 make consequential amendments to, and revocations and saving in respect of, other legislation. Regulation 21 makes consequential amendments to regulations implementing Directive 85/337/EEC. Regulation 25 makes consequential amendments to the Waste Management Licensing Regulations 1994. Parts of those Regulations, in particular Schedule 4, implement Community obligations. Regulations 22, 23, 24 and 26 do not implement Community obligations.

In regulation 2(1) of the Town and Country Planning (Assessment of Environmental Effects) Regulations 1988, for the definition of 'special waste' there shall be substituted:

> '"special waste" means waste which is special waste for the purposes of the Special Waste Regulations 1996;'

In regulation 4(1) of the Environmental Assessment (Scotland) Regulations 1988, for the definition of 'special waste' there shall be substituted:

> '"special waste" means waste which is special waste for the purposes of the Special Waste Regulations 1996;'

102 Inserted by the Special Waste Amendment (Scotland) Regulations 2004, SSI 2004/ 112, reg 2(12).

Amendment of the Controlled Waste (Registration of Carriers and Seizure of Vehicles) Regulations 1991 (reg 22)

8.166 In Schedule 1 to the Controlled Waste (Registration of Carriers and Seizure of Vehicles) Regulations 1991 there shall be added at the end:

> 'the Special Waste Regulations 1996'.

Amendment of the Environmental Protection (Duty of Care) Regulations 1991 (reg 23)

8.167 In reg 2 of the Environmental Protection (Duty of Care) Regulations 1991:

(a) at the beginning of paragraph (1), (The transferor and the transferee shall, at the same time as the written description of the waste is transferred, ensure that such a document as is described in paragraph (2) ('a transfer note') is completed and signed on their behalf.) there shall be added "Subject to paragraph (3),"

(b) after paragraph (2), the following paragraph shall be added:

(3) Paragraph (1) shall not apply where the waste transferred is special waste within the meaning of the Special Waste Regulations 1996 and the consignment note and, where appropriate, schedule required by those Regulations are completed and dealt with in accordance with those Regulations.

Amendment of the Controlled Waste Regulations 1992 (reg 24)

8.168 In paragraph 18(2) of Schedule 3 to the Controlled Waste Regulations 1992, for the definition of 'tank washings', there shall be substituted:

> 'tank washings' has the same meaning as in paragraph 36 of Schedule 3 to the Waste Management Licensing Regulations 1994;

Amendment of the Waste Management Licensing Regulations 1994 (reg 25)

8.169 The 1994 Regulations shall be amended in accordance with Sch 3 to these Regulations.

Schedules

8.170 Schedule 1 Forms of consignment note and schedule[103];

Schedule 2 Special waste[104];

Schedule 3 Amendments to the Waste Management Licensing Regulations 1994.

103 The Special Waste Amendment (Scotland) Regulations 2004, SSI 2004/112, reg 2(13) updates the form in Pt 1 of Sch 1.

104 The Special Waste Amendment (Scotland) Regulations 2004, SSI 2004/112, reg 2(14). The Table (Hazardous Waste List) in Pt 1 of Sch 2 is omitted.

Chapter 9

Scrap Metal Dealers

SCRAP METAL DEALERS

9.1 Scrap metal theft has been on the increase since the rise in the price of metal. Between April 2012 and March 2013 there were 61,349[1] metal theft offences recorded by police in England and Wales, which corresponds to 2% of all police recorded crime in England and Wales for that period. Of the 61,349 metal theft offences recorded by police in England and Wales:

- 28,843 (47%) were infrastructure related offences;
- 25,869 (42%) were non-infrastructure related offences;
- 6,637 (11%) were not classified.

1 Crime in England and Wales. Year ending March 2013 Office for National Statistics available at: http://webarchive.nationalarchives.gov.uk/20160105160709/http://www.ons.gov.uk/ons/rel/crime-stats/crime-statistics/period-ending-march-2013/stb-crime--period-ending-march-2013.html.

Within the UK, the licensing of scrap metal dealers is dealt with through different legislation. In England and Wales the Scrap Metal Dealers Act 2013 (the Act) came into force and repealed the Scrap Metal Dealers Act 1964 (and linked legislation) and Part 1 of Vehicles (Crime) Act 2001. It creates a revised regulatory regime for the scrap metal recycling and vehicle dismantling industries and maintains the local authorities as the principal regulator. It also grants them the power to better regulate these industries by allowing them to refuse to grant a licence to 'unsuitable' applicants and the power to revoke licences if the dealer becomes 'unsuitable'. Suitability will be judged on the basis of a number of factors outlined in s 3 of the Act including any unspent relevant criminal convictions. The Act will also provide local authorities and police officers with appropriate powers of entry and inspection.

The Act also requires that the licence has a fee, which will be set locally by each local authority on a cost recovery basis. However, local authorities will have a duty to have regard to guidance issued by the Secretary of State which outlines the issues about what should be considered by local authorities when setting the fee and what activities the fee can cover. This fee provides local authorities with the funding they need to administer the regime and ensure compliance.

The Act also aims to raise trading standards across the scrap metal industry by requiring more detailed and accurate records of transactions to be kept. Scrap metal dealers will also be required to verify the identity of those selling metal to them.

Within the Act there is a separate regulatory scheme for motor salvage operators under the Vehicles (Crime) Act 2001. This replaces the current overlapping regimes for the vehicle salvage and scrap metal industries with a single regulatory scheme. The Act also revises the definition of 'scrap metal dealer' and 'scrap metal'.

Also repealed by the Act is the amendment to the Scrap Metal Dealers Act 1964 in s 146 of the Legal Aid, Sentencing and Punishment of Offenders Act 2012 which created the offence of buying scrap metal for cash. This offence came into force on 3 December 2012. The other measures within the Legal Aid, Sentencing and Punishment of Offenders Act 2012 which relate to tackling metal theft, a revision of police entry powers into unregistered scrap metal stores and an increase in the financial penalties for offences in the Scrap Metal Dealers Act 1964 have also been repealed on the grounds that these provisions are covered separately in the 2013 Act. Supplementary guidance is available for the England and Wales Act (the Scrap Metal Dealers Act 2013)[2].

In Scotland the Civic Government (Scotland) Act 1982 and the Air Weapons and Licensing (Scotland) Act 2015 make provision for the licensing of scrap metal dealers. Within the Air Weapons and Licensing (Scotland) Act 2015, new provisions were introduced that will prohibit the use of cash to buy/sell

2 Scrap Metal Dealers Act 2013 – Supplementary Guidance (last updated 2013) Home Office https://www.gov.uk/government/uploads/system/uploads/attachment_data/file/265871/SMDA_supplementary_ guidance.pdf.

metal and tighten and improve record keeping within scrap metal dealers. The relevant section within the Act provides similar powers to those found in the Scrap Metal Dealers Act 2013 for law enforcement and will help to create a level playing field for scrap metal businesses across Britain.

In Northern Ireland the statistics relating to burglary, robbery and theft offences, including those of scrap metal, rose from under 500 in 2007/08 to 1,426 in 2012/13. Northern Ireland found that it was out of step with the rest of the United Kingdom in dealing with scrap metal dealers. Concerned that lack of regulation could result in crime being displaced to Northern Ireland, the Bill for Scrap Metals Dealers[3] was introduced. In Northern Ireland the Scrap Metal Dealers Bill is still going through the Parliamentary process with the consideration stage in February 2016 and Royal Assent at some time after this. This Bill is closely related to the Scrap Metals Dealer Act 2013 which became law on 28 February 2013.

This chapter will look at the legislation that applies to scrap metal dealers in the UK and also look at the provisions in the Bill for Northern Ireland.

LICENSING OF SCRAP METAL DEALERS

England and Wales

9.2 In England and Wales it is a requirement that anyone who carries out the business of a scrap metal dealer will need a 'scrap metal licence'.

Carrying on a business as a scrap metal dealer is defined in s 21 of the Scrap Metal Dealers Act 2013 and scrap metal is defined in s 21(6). If anyone carries out their business without the required licence then they are liable for a fine not exceeding level 5.

It will be for a court to decide if the buying or selling of scrap metal forms the whole or part of a person's business, or alternatively, whether the buying or selling of scrap metal forms such a minimal part of their overall business dealings that the definition of scrap metal dealer in the Act is not made out. There are many factors a court may consider in reaching its judgment such as the proportion of the business related to scrap metal in terms of value or volume. There is an exemption for manufacturers selling scrap only as a by-product or as surplus materials not required for manufacturing.

Licence fee charges apply[4] and people required to hold a scrap metal dealer licence are:

3 Scrap Metal Dealers Bill 65/11-16 Available at: http://www.niassembly.gov.uk/assembly-business/legislation/ primary-legislation-current-bills/scrap-metal-dealers-bill/.

4 Scrap Metal Dealers Act 2013: guidance for licence fee charges available at: https://www.gov.uk/government/ publications/scrap-metal-dealer-act-2013-licence-fee-charges.

- Motor salvage operators – motor salvage operators will need to hold a scrap metal dealer's licence which replaces the need to separately register with the local council[5];
- Producers of ferrous and non-ferrous metals – producers of ferrous and non-ferrous metals purchase processed scrap metal as a raw material used in their manufacturing process. Whilst on a case by case basis this may meet the test of wholly or partly buying or selling scrap metal, the Act was never intended to extend beyond those who were expected to register under the Scrap Metal Dealers Act 1964 or the Vehicles (Crime) Act 2001 which this Act replaces.
- Agent, broker or trader – a person who buys or sells scrap metal on paper without actually operating a scrap metal site, carries on business as a scrap metal dealer and will require a licence. This person will carry out their business as a scrap metal dealer from premises and will need to hold a site licence as a site is any 'premises used in the course of carrying on business as a scrap metal dealer (whether or not metal is kept there)'[6] (Scrap Metal Dealers Act 2013, s 22(9)).
- Skip hire company – skip hire companies may need a scrap metal dealer's licence. It will be a question for a court to decide whether the buying or selling of scrap metal forms the whole or part of the person's business, or alternatively, whether the buying or selling of scrap metal forms such a minimal part of their overall business dealings that the definition of scrap metal dealer in the Act is not made out.
- Tradespersons – tradespersons will not require a scrap metal dealer licence if buying or selling scrap metal is an incidental function of their business (eg, being a plumber or electrician). It will be a question for a court to decide whether, under all the circumstances, the buying or selling of scrap metal forms the whole or part of the person's business, or alternatively whether the buying or selling of scrap metal forms such a minimal part of their overall business dealings that the definition of scrap metal dealer in the Act is not made out.
- Civic amenity sites – civic amenity sites run by councils, or contracted out to others, will not require a licence.
- Vehicle collection companies – if a scrap metal dealer (who is a motor salvage operator) holds a site licence and employs or sub-contracts a company to pick up cars on the company's behalf in the course of the business from that site, they will not need individual mobile collector's licences as this is not regularly engaging in collecting waste materials and old, broken, worn out or defaced articles by means of visits from door to door but pre-arranged appointments. If the vehicle is to be scrapped, the collection company will be committing an offence if they buy the vehicle for cash and do not verify the name and address of the supplier on behalf of the

5 The Act repeals Pt 1 of the Vehicles (Crime) Act 2001 which is the requirement to register as a vehicle salvage operator.

6 Scrap Metal Dealers Act 2013, s 22(9).

site licence holder. The scrap metal dealer will be responsible for ensuring that the vehicle collection company complies with these requirements.

Scotland

9.3 In Scotland, scrap metal dealers are required to hold a scrap metal dealer licence. The requirement to hold a scrap metal dealer licence is in the Civic Government (Scotland) Act 1982, s 28. The Air Weapons and Licensing (Scotland) Act 2015 (AWL(S)A 2015), s 72 means that a licence is required for those who carry on a business that '*wholly or substantially*' consists of buying or selling scrap. So people whose business involves a significant degree of dealing in scrap metal will require a licence but those whose involvement is peripheral will not. Each case may be looked at and the amount of scrap metal dealing which is carried out, considered. So a plumber who as part of their business has to dispose of some metal piping may not be required to be licensed – but a skip hirer who takes substantial amounts of metal from a building site, and that forms a substantial part of the hirer's business may require to be licensed.

Section 72 amends[7] the definition of both metal dealers and itinerant metal dealers to encompass those who buy or sell metal as opposed to those who both buy and sell metal (which was the previous position). The 1982 Act is amended to list the activities that are licensable and it is explicitly stated that a motor salvage operator, as defined in sub-section (3), is carrying out the business of a metal dealer and will require a licence to do so.

The Air Weapons and Licensing (Scotland) Act 2015, s 73 creates a new s 37A within the 1982 Act. The new section provides a regulation-making power that will allow the Scottish Ministers to set out circumstances where the metal dealer and itinerant metal dealer regime does not apply, thereby resulting in a licence not being required.

Penalties for Failure to Have Appropriate Licence or Comply with Conditions

9.4 AWL(S)A 2015, s 66 amends s 7 of the 1982 Act to increase the penalties for the offences of operating without a metal dealer licence or itinerant metal dealer licence (where a licence is required) to a maximum fine of £20,000 and/or six months' imprisonment on summary conviction. This section also increases the maximum penalties for the offence of failing to comply with a condition of a metal dealer or itinerant metal dealer licence to the same level, ie, a fine of £20,000 or six months' imprisonment, or both.

7 Amends the Civic Government (Scotland) Act 1982, s 37.

Exemption from Audit Removed

9.5 AWL(S)A 2015, s 67 has removed the exemption whereby dealers with an audited turnover in excess of £1m would be exempted from the licensing requirement. This now ensures that all dealers are subject to the licensing requirements.

Abolition of the Condition to Retain Metal for 48 Hours before Processing

9.6 AWL(S)A 2015, s 68 removes the mandatory requirement[8] that metal dealers should not process metal for 48 hours after receiving it. This would allow a dealer to process metal quickly (which may be required for the safe operation of the site).

Northern Ireland

9.7 The current regulation of scrap metal dealers falls under the Waste and Contaminated Land (Northern Ireland) Order 1997, SI 1997/2778, reg 6(1). This regulation makes provision for a waste management licence, defines controlled waste and makes provision for the treatment, keeping and disposal of waste. It also imposes on traders a duty of care in the transfer of waste.

In addition, the Waste Management Licensing Regulations (Northern Ireland) 2003, NISR 2003/493 and the Pollution Prevention and Control Regulations (Northern Ireland) 2003, NISR 2003/46 detail further the requirements for licensing either through a waste management licence or a pollution prevention control permit. The Controlled Waste (Duty of Care) Regulations (Northern Ireland) 2002, NISR 2002/271 require that identification documents are noted during the transfer process.

However, while the recovery of scrap metal may deal with controlled waste and fall within the existing licensing regime, many other individuals may sell metal on to scrap metal dealers without documentation and outside of any regulation.

Increase in metal theft has been shown to be closely related to metal prices. Crime figures indicate that a significant number of burglary, robbery and theft offences include property items listed as metal and there can be no doubt that the theft and illegal sale of scrap metal is a lucrative business. At the time of writing, the rise in metal theft, specifically aluminium and copper and the link to price has been reported by the Home Office as being a global trend.

8 Section 28 of the 1982 Act and repeals s 29 of the 1982 Act.

A voluntary code of conduct is in place in some areas in Northern Ireland under which each business displays at least one Police Service of Northern Ireland (PSNI) supplied poster showing that the firm is a 'metal theft aware premises'. The code asks that CCTV is used to record vehicles and customers, no metal is accepted without suitable identification and Vehicle Registration Mark (VRM) details are recorded. It has been noted in the consultation that the code is effective for those who use it.

The Police Service Northern Ireland[9], in its response to the Member's consultation on the Scrap Metal Dealers Bill, has stated that, 'The current absence of Scrap Metal Act 1964 powers and the Scrap Metal Dealers Act 2013 in Northern Ireland means we are not only more vulnerable to crimes related to any increase in metal prices, but exposed to additional displacement of thefts in the event of new legislation in the Republic of Ireland.'

This is supported by anecdotal evidence that Scotland had experienced an increase in scrap metal crime as a result of organised gang activity following the introduction of the Scrap Metal Dealers Act 2013 in England and Wales. There appears to be a displacement of criminal activity in England and Wales, which correlates with the introduction of new legislation covering those jurisdictions.

The Bill

9.8 Clause 1 introduces the requirement for a scrap metal dealer to hold a licence and a corresponding criminal offence and penalty for carrying on business as a scrap metal dealer without such a licence.

FORM AND EFFECT OF LICENCE

England and Wales

9.9 Section 2 of the Scrap Metal Dealers Act 2013 provides further detail in respect of the licence, including that there are two types of licence: one for a site and the other for a mobile collector (carrying on business otherwise than at a site). The licence authorises the licensee to carry on business as a scrap metal dealer at the sites listed in it (in the case of a site licence) or within the local authority area (in the case of a mobile collector's licence). A power is provided for the Secretary of State to prescribe the form and content of the licence in Regulations.

A licence will be issued by the local authority, in whose area a scrap metal site is situated, or (in respect of a mobile collector) in the area that the collector operates. A site licence will require all of the sites at which the licensee carries on the business as a scrap metal dealer within the local authority area to be

9 Scrap Metal Dealers Bill Explanatory and Financial Memorandum NIA Bill 65/11-16 EFM.

identified and a site manager to be named for each site. In doing so, they will be permitted to operate from those sites as a scrap metal dealer, including transporting scrap metal to and from those sites from any local authority area.

A collector's licence authorises the licensee to operate as a mobile collector in the area of the issuing local authority, permitting them to collect any scrap metal as appropriate. This includes commercial as well as domestic scrap metal. The licence does not permit the collector to collect from any other local authority area; a separate licence will need to be obtained from each local authority the individual wishes to collect in. A licence also does not authorise the licensee to carry on a business at a site within any area – should a collector wish to use a fixed site, they will need to obtain a site licence from the relevant local authority. There is no restriction as to the location where the collector can transport and sell their metals.

The form in which a licence is issued must enable it to be displayed in accordance with the Scrap Metal Dealers Act 2013, s 10.

(1) A scrap metal dealer who holds a site licence must display a copy of the licence at each site identified in the licence.
(2) The copy must be displayed in a prominent place in an area accessible to the public.
(3) A scrap metal dealer who holds a collector's licence must display a copy of the licence on any vehicle that is being used in the course of the dealer's business.
(4) The copy must be displayed in a manner which enables it to be easily read by a person outside the vehicle.
(5) A scrap metal dealer who fails to comply with this section is guilty of an offence and is liable on summary conviction to a fine not exceeding level 3 on the standard scale.

Scotland

9.10 Section 28 of the Civic Government (Scotland) Act 1982 provides that scrap metal dealers require a 'metal dealer licence'. Section 36 of the same Act defines the type of person requiring a licence.

Where a person carries on business:

- as a second-hand dealer;
- as a metal dealer;
- as an itinerant metal dealer;
- or as any two of these kinds of dealer,

he shall require the appropriate licence in respect of each activity.

Northern Ireland

9.11 Clause 2 of the Scrap Metal Dealers Bill provides further detail in respect of the licence, including that there are two types of licence, one for a site and the other for a mobile collector (carrying on business otherwise than at a site). It lists the information which must be detailed in the licence and establishes the Department of Environment as the issuing body.

The licence authorises the licensee to carry on business as a scrap metal dealer at the sites listed in it (in the case of a site licence) or within the Department area (in the case of a mobile collector's licence). A power is provided for the Department to prescribe the form and content of the licence in Regulations.

A site licence will require all of the sites at which the licensee carries on the business as a scrap metal dealer to be identified and a site manager to be named for each site. In doing so, they will be permitted to operate from those sites as a scrap metal dealer, including transporting scrap metal to and from those sites.

A collector's licence authorises the licensee to operate as a mobile collector, permitting them to collect any scrap metal as appropriate. This includes commercial as well as domestic scrap metal. A licence does not authorise the licensee to carry on a business at a site – should a collector wish to use a fixed site, they will need to obtain a site licence. There is no restriction as to the location where the collector can transport and sell their metals.

The form in which a licence is issued must enable it to be displayed in accordance with Clause 8.

(1) A scrap metal dealer who holds a site licence must display a copy of the licence at each site identified in the licence.
(2) The copy licence referred to in s 8(1) must be displayed in a prominent place in an area accessible to the public.
(3) A scrap metal dealer who holds a collector's licence must display a copy of the licence on any vehicle that is being used in the course of the dealer's business.
(4) The copy licence referred to in s 8(3) must be displayed in a manner, which enables it easily to be read by a person outside the vehicle.
(5) A scrap metal dealer who fails to comply with this section is guilty of an offence and is liable on summary conviction to a fine not exceeding level 3 on the standard scale.

ISSUE OF LICENCE

England and Wales

9.12 Section 3 of the Scrap Metal Dealers Act 2013 requires that the licensing authority is satisfied that an applicant is a suitable person to carry on business as a scrap metal dealer. In considering suitability, the local authority may look at any relevant information, including whether any relevant enforcement action has been taken against the applicant or whether the applicant has been convicted of a relevant offence.

The Secretary of State has a power to prescribe in regulations the meaning of relevant offence and relevant enforcement action. The intention is that this will be in line with the criteria used by the Environment Agency when issuing environmental permits under the Environmental Protection Act 1990. The authority must also have regard to any guidance on determining suitability, which is issued from time to time by the Secretary of State, and the authority may consult other organisations to assist in determining suitability.

Local Authorities can also include conditions in the licence if the licensee or site manager has been convicted of a relevant offence. Section 3(8)[10] specifies the two conditions that can be imposed by local authorities on a licence, namely a restriction on opening hours and imposing 'quarantine periods'.

Northern Ireland

The Bill

9.13 Clause 3 sets out the key criteria for determining whether to issue a licence, to ensure that a licensee is a suitable person. It requires the licensing authority to be satisfied that an applicant is a suitable person to carry on business as a scrap metal dealer. In considering suitability, the Department may have regard to any relevant information, including whether any relevant enforcement action has been taken against the applicant or whether the applicant has been convicted of a relevant offence. The Department has a power to prescribe in regulations the meaning of relevant offence and relevant enforcement action. It is intended that this will be in line with the criteria used by the Scrap Metal Dealers Act 2013 which bases its process on that used by the Environment Agency when issuing environmental permits under the Environmental Protection Act 1990.

This clause also allows the Department, when issuing a licence, to include conditions in it if the licensee or site manager has been convicted of a relevant offence. The intention is that these conditions represent a probationary arrangement when there was reason to believe that restrictions were appropriate but where non-issue (or revocation) of a licence was excessive. Clause 3(6) specifies the two conditions that can be imposed by local authorities on a licence, namely a restriction on opening hours and imposing 'quarantine periods'.

10 Scrap Metal Dealers Act 2013, s 3(8)
If the applicant or any site manager has been convicted of a relevant offence, the authority may include in the licence one or both of the following conditions—
(a) that the dealer must not receive scrap metal except between 9 am and 5 pm on any day;
(b) that all scrap metal received must be kept in the form in which it is received for a specified period, not exceeding 72 hours, beginning with the time when it is received.

REVOCATION OF LICENCE AND IMPOSITION OF CONDITIONS

England and Wales

9.14 The Scrap Metal Dealers Act 2013, s 4 provides the licensing authority with the discretion to revoke a licence on particular grounds, including where the local authority is no longer satisfied that the licensee is a suitable person to carry on the business as a scrap metal dealer. A licence can also be revoked if:

- the authority is satisfied that a site manager named in the licence does not act as site manager at any of the sites identified in the licence.
- the licensee or any site manager named in a licence is convicted of a relevant offence. The authority may vary the licence by adding one or both of the conditions set out in s 3(8)[11].
- A revocation or variation comes into effect when no appeal under para 9[12] of Sch 1 is possible in relation to the revocation or variation, or when any such appeal is finally determined or withdrawn.

But if the authority considers that the licence should not continue in force without conditions, it may by notice provide—

(a) that, until a revocation under this section comes into effect, the licence is subject to one or both of the conditions set out in s 3(8), or
(b) that a variation under this section comes into effect immediately.

11 (8) If the applicant or any site manager has been convicted of a relevant offence, the authority may include in the licence one or both of the following conditions—
  (a) that the dealer must not receive scrap metal except between 9 a.m. and 5 p.m. on any day;
  (b) that all scrap metal received must be kept in the form in which it is received for a specified period, not exceeding 72 hours, beginning with the time when it is received.

12 9 *Appeals*
  (1) An applicant may appeal to a magistrates' court against the refusal of an application made under paragraph 2 or 3.
  (2) A licensee may appeal to a magistrates' court against—
    (a) the inclusion in a licence of a condition under section 3(8), or
    (b) the revocation or variation of a licence under section 4.
  (3) An appeal under this paragraph is to be made within the period of 21 days beginning with the day on which notice of the decision to refuse the application, to include the condition, or to revoke or vary the licence under section 4, was given.
  (4) The procedure on an appeal under this paragraph is to be by way of complaint for an order and in accordance with the Magistrates' Courts Act 1980.
  (5) For the purposes of the time limit for making an appeal under this paragraph, the making of the complaint is to be treated as the making of the appeal.
  (6) On an appeal under this paragraph, the magistrates' court may—
  (such directions as it considers appropriate having regard to the provisions of this Act.
  (7) The authority must comply with any directions given by the magistrates' court under sub-paragraph (6).
  (8) But the authority need not comply with any such directions—
    (a) until the time for making an application under section 111 of the Magistrates' Courts Act 1980 (application by way of case stated) has passed, or
    (b) if such an application is made, until the application is finally determined or withdrawn.

Northern Ireland

The Bill

9.15 Clause 4 provides the Department with the discretion to revoke a licence in particular circumstances, including where the Department is no longer satisfied that the licensee is a suitable person to carry on business as a scrap metal dealer. A Department as the licensing authority can only carry out the revocation of a licence. The clause also allows the licensing authority to vary a licence, imposing the conditions stipulated in clause 3, if the licensee or a site manager is convicted of a relevant offence. This includes the power to impose conditions during the period in which an appeal may be made, or if made, is being heard.

FURTHER PROVISION ABOUT LICENCES

England and Wales

9.16 The Scrap Metal Dealers Act 2013, s 5 provides that Sch 1 has effect. Schedule 1 sets out procedural issues relating to the licence.

Term of the Licence

9.17 A licence will be issued for a period of three years from the date of issue. The Secretary of State will have the power under Sch 1, para 1(4) to alter the duration of the licence.

The local authority can also vary the licence where there is a change of circumstance relating to particular details contained in the licence. This means that there will not be a need to issue a new licence for a further three-year period, with a further fee being paid each time details change eg, the name of the licensee. The ability to vary a licence includes the ability to change it from a site licence to a collector's licence or vice versa.

Application

9.18 The application process for a licence is set out in Sch 1, para 2, including a requirement to provide information relating to relevant enforcement action and convictions for relevant offences. The Secretary of State has a power by order to alter the requirements as to what information must accompany the application.

Variation

9.19 The local authority can change the licence type and if the licensee has any change in circumstances they must make an application to change the licence. The power to change the name of the licensee does not mean that the licence can be transferred one to another.

Further Information

9.20 Schedule 1, para 4 allows the local authority to request (either when the application is made or later) further information from the applicant as the authority considers relevant for the purpose of considering the application. If an applicant fails to provide the information requested, the authority may decline to proceed with the application.

Offence of Making a False Statement

9.21 Schedule 1, para 5 states that an applicant who in an application or in response to a request makes a statement knowing it to be false in a material particular, or recklessly makes a statement which is false in a material particular, is guilty of an offence and is liable on summary conviction to a fine not exceeding level 3 on the standard scale.

Fee

9.22 Schedule 1, para 6 permits local authorities to charge a licence fee at cost recovery. This fee will be set locally. The Secretary of State will issue statutory guidance, which local authorities will have a duty to have regard to, which will outline the activities that can be funded by this licence fee.

Right to Make Representations

9.23 Schedule 1 also sets out how the applicant or licensee can make representations where the licensing authority proposes either to refuse to issue/ renew an application or to revoke a licence.

Notice of Decision

9.24 There will be a requirement for the licensing authority to provide the applicant with a notice to outline the decision that they propose to make about issuing/revoking a licence. The procedure for making representations will be outlined in the notice, giving at least a 14-day window, or further time as is reasonable for representations to be made.

Appeals

9.25 The applicant will also have the right to appeal to the magistrates' court against any decision made by the licensing authority to refuse to issue or renew a licence, to vary or revoke a licence or to impose conditions within a licence.

Northern Ireland

The Bill

9.26 Clause 5 provides that Schedule 1 of the Bill has effect. Schedule 1 sets out further and more detailed provisions about licences, including details as to the term of the licence, application procedures and fees, variations, requests for information and offences of giving false information, a representation and appeals process and a process for notification of decisions.

Term of the Licence

9.27 A licence expires at the end of the period of three years or such period, not being less than one year, as the Department may specify in the licence beginning with the day on which it is issued. If an application to renew a licence is received before the licence expires, the licence continues in effect and:

(a) if the application is refused, the licence expires when no appeal under paragraph 9 is possible in relation to the refusal or any such appeal is finally determined or withdrawn;
(b) if the licence is renewed, it expires at the end of the period of three years or such period, not being less than one year, as the Department may specify in the renewed licence beginning with the day on which it is renewed.

Sub-paragraphs 5(1) and 5(2) are subject to Clause 4 (revocation of licence).

Applications

9.28 The application process for a licence is set out in Sch 1, para 2, including a requirement to provide information relating to relevant enforcement action and convictions for relevant offences.

Variation of Licence

9.29 The Department can change the licence type and if the licensee has any change in circumstances they must make an application to change the licence. The power to change the name of the licensee does not mean that the licence can be transferred one to another.

Further Information

9.30 The Department may request (either when the application is made or later) that the applicant provide such further information as the Department considers relevant for the purpose of considering the application. If an applicant fails to provide information requested, the Department may decline to proceed with the application.

Offence of Making a False Statement

9.31 An applicant who in an application or in response to a request under Sch 1, para 4(a) makes a statement knowing it to be false in a material particular, or (b) recklessly makes a statement which is false in a material particular, is guilty of an offence and is liable on summary conviction to a fine not exceeding level 3 on the standard scale.

Fee

9.32 An application must be accompanied by a fee set by the Department. In setting a fee under this paragraph, the Department must have regard to any guidance issued from time to time by the Department for Finance and Personnel.

Right to Make Representations

9.33 If the Department proposes to refuse an application to revoke or vary a licence under Clause 4, the Department must give the applicant or licensee a notice which sets out what the Department proposes to do and the reasons for it.

Notice of Decision

9.34 If the Department refuses the application, or revokes or varies the licence under Clause 4, it must give a notice setting out the decision and the reasons for it. A notice must also state:

(a) an appeal can be made against the decision;
(b) the time within which such an appeal may be brought; and
(c) in the case of a revocation or variation under Clause 4, the date on which the revocation or variation is to take effect.

Appeals

9.35 An applicant may appeal to a magistrates' court against the refusal of an application made under Sch 1, para 2 or 3. An appeal is to be made within the period of 21 days beginning with the day on which notice of the decision to refuse the application, to include the condition, or to revoke or vary the licence under Clause 4, was given.

SUPPLY OF INFORMATION BY THE AUTHORITY (DEPARTMENT)

England and Wales

9.36 The Scrap Metal Dealers Act 2013, s 6 places a duty on the local authority to supply any such information as requested relating to a scrap metal licence to any other local authority in England and Wales, the Environment Agency, the Natural Resources Body for Wales and to police forces.

REGISTER OF LICENCES

England and Wales

9.37 Section 7 of the Scrap Metal Dealers Act 2013, requires that a Register of Licences issued under this Act should be maintained by the Environment Agency in England and the Natural Resources Body for Wales in Wales. Local authorities will provide the appropriate information on all licences issued in their geographic areas in order that this register can be updated regularly. The register should be made openly accessible to the public and will include: the name of the authority which issued the licence; the name of the licensee; any trading name; the type of licence; the site(s) covered by the licence and the expiry date of the licence.

Scotland

9.38 Section 71[13] creates a new s 35A in the Civic Government (Scotland) Act 1982allowing the Scottish Ministers to make regulations establishing a

13 Air Weapons and Licensing (Scotland) Act 2015.

register of metal dealers and itinerant metal dealers and make provision about its ongoing maintenance. The regulations may include details on things such as who is to keep and maintain the register, the duty to provide information, the information to be covered by the register, the form and publication of the register and any fees that may be relevant. Examples of information that may be included in the register might include the name (or trading names) of the person holding the licence, a contact address for the person and the person's place of business, which type (or types) of licence the person holds and the date on which the licence (or licences) will expire.

Northern Ireland

The Bill

9.39 There is currently no provision in the Bill to establish a register of scrap metal dealers.

NOTIFICATION REQUIREMENTS

England and Wales

9.40 Section 8 of the Scrap Metal Dealers Act 2013 imposes an obligation on an applicant to notify the relevant environmental body of any changes which materially affect the accuracy of any information, which the applicant has provided in connection with the application. It also imposes an obligation on a licensee to notify the licensing authority if the name under which the licensee carries on the business as a scrap metal dealer changes (if they trade under a different name to the name of the licensee). If there is a variation in other matters that are entered in the register, ie, those in s 7(3)(b) and (d), the licensee is required to make an application to vary the licence under paragraph 2 of Sch 1. The licensing authority has a duty to inform the relevant environmental body of any changes to the licence that affect the entry in the register in respect of that licence, including any licences that have been revoked, within 28 days of the change taking effect. The Environment Agency or the Natural Resources Body for Wales, whichever is appropriate, must then amend the register accordingly.

Scotland

9.41 Schedule 1, para 9 of the Civic Government (Scotland) Act 1982 details what a person must do if there are changes and alterations to their circumstances.

- Where there is a material change of circumstances affecting the holder of a licence, or the activity to which the licence relates, the holder of the licence shall, in accordance with this paragraph, notify the licensing authority of the change as soon as reasonably practicable after it has taken place.

- The holder of a licence which relates to an activity consisting of or including the use of premises, a vehicle or a vessel shall not, unless in accordance with a requirement imposed by or in pursuance of any enactment other than Parts I or II of this Act, make or cause to be made or permit there to be made, any material change in the premises or, as the case may be, the vehicle or vessel without the prior consent of the licensing authority.

This change is accompanied by a fee.

Northern Ireland

The Bill

9.42 Clause 6 imposes a duty on an applicant to notify the Department of any changes, which materially affect the accuracy of any information, which the applicant has provided in connection with the application. It also obligates a licensee to notify the Department if the name under which the licensee carries on the business as a scrap metal dealer changes (if they trade under a different name to the name of the licensee) or if the business ceases to operate. Failure to comply with these requirements is an offence.

CLOSURE OF UNLICENSED SITES

England and Wales

9.43 The Scrap Metal Dealers Act 2013, s 9 provides that Sch 2 has effect. Schedule 2 makes provision for the closure of sites at which a scrap metal business is carried on without a licence. The power is similar to the existing closure powers for unlicensed alcohol vendors under the Criminal Justice and Police Act 2001.

This closure power does not affect the ability of the police to bring charges for the offence in s 1.

Schedule 2 sets out that a police constable or an officer of the local authority can issue a closure notice specifying the reasons for such action and the steps that may be taken to ensure that the alleged use of the premises ceases. Once a closure notice has been issued, the local authority officer or constable can then make an application to a Justice of the Peace for a closure order to close that site. Applications cannot be made where a scrap metal dealer in the course of their business is no longer using the premises and there is no reasonable likelihood that the premises will be used in this way in the future. If the court is satisfied, the court can then make such order that it considers appropriate for the closure of the premises. The court can only make a closure order. Such a closure order can vary between requiring the immediate closure of the premises, immediately discontinuing its use by a scrap metal dealer, and requiring a sum to be retained by the court until the requirements are met.

A constable or local authority officer can issue a certificate at any time to terminate the closure order where they are satisfied that the need for the order has ceased. Likewise, the court can issue a discharge order at any time where they are satisfied that there is no longer a need for such an order. Appeals in relation to all stages of this process will be available to the crown court.

The offence of operating premises in contravention of a closure order is punishable on summary conviction with a fine not exceeding level 5 on the standard scale.

Scotland

9.44 It is currently unclear whether the police in Scotland have the power to close an unlicensed scrap metal dealer.

Northern Ireland

The Bill

9.45 Schedule 2 contains further and more detailed provisions relating to the closure of unlicensed sites, including a closure notice procedure, the provision for application to the court for a closure order, an appeals process and provisions for enforcement of a closure order.

It makes provision for the closure of sites at which a scrap metal business is carried on without a licence. This closure power does not affect the ability of the police to bring charges for the offence in Clause 1.

The Schedule sets out that a police constable or the Department can issue a closure notice specifying the reasons for such action and the steps that may be taken to ensure that the alleged use of the premises ceases. Once a closure notice has been issued, the Department or constable can then make an application to a magistrate for a closure order to close that site. Applications cannot be made where a scrap metal dealer in the course of their business is no longer using the premises and there is no reasonable likelihood that the premises will be used in this way in the future. If the court is satisfied, the court can then make such an order as it considers appropriate for the closure of the premises. A closure order can only be made by the court. Such a closure order can vary between requiring the immediate closure of the premises, immediately discontinuing its use by a scrap metal dealer, and requiring a sum to be retained by the court until the requirements are met.

A constable or the Department can issue a certificate at any time to terminate the closure order where they are satisfied that the need for the order has ceased.

Likewise, the court can issue a discharge order at any time where they are satisfied that there is no longer a need for such an order. Appeals in relation to all stages of this process will be available to the Crown Court.

The offence of operating premises in contravention of a closure order is punishable on summary conviction with a fine not exceeding level 5 on the standard scale.

CONDUCT OF BUSINESS AND DISPLAY OF LICENCE

England and Wales

9.46 Section 10 of the Scrap Metal Dealers Act 2013 requires that the licensee displays a copy of their licence. For site operators this must be in a prominent place in an area accessible to the public. For mobile collectors, it must be in a manner which enables the licence to be easily read by a person outside the vehicle. A criminal offence is committed by any scrap metal dealer who fails to fulfil this requirement. This offence is punishable on summary conviction with a fine not exceeding level 3 on the standard scale.

Scotland

9.47 The Civic Government (Scotland) Act 1982, Sch 1, para 2(2) is clear that under *applications for the grant and renewal of licences* a licence, then the application must be clearly displayed.

Where an application is for the grant or renewal of a licence in relation to an activity wholly or mainly to be carried on in premises, the applicant shall, for a period of 21 days beginning with the date on which the application was submitted to the licensing authority, display a notice complying with sub-paragraph (3) below at or near the premises so that it can conveniently be read by the public.

It is less clear what regulation there is for the display of the licence itself.

Northern Ireland

The Bill

9.48 Clause 8 requires that the licensee displays a copy of their licence. For site operators this must be in a prominent place in an area accessible to the public. For mobile collectors, it must be in a manner which enables the licence to be easily read by a person outside the vehicle. A criminal offence is committed by any scrap metal dealer who fails to fulfil this requirement. This offence is punishable on summary conviction with a fine not exceeding level 3 on the standard scale.

VERIFICATION OF SUPPLIER'S IDENTITY

England and Wales

9.49 Section 11 of the Scrap Metal Dealers Act 2013 places a requirement on scrap metal dealers, site managers and employees who have been delegated the responsibility to do so, to verify the identity of the person they are receiving metal from and the person's address. This verification must be done by reference to data, documents or other information obtained from a reliable or independent source, such as the Identity and Passport Service, the Driver and Vehicle Licensing Agency, a bank or utility company etc. The Secretary of State will prescribe in regulations the data or documents which are sufficient, or not sufficient as the case may be, for verifying identity.

It will be an offence not to obtain and verify the seller's identity, punishable by a fine not exceeding level 3 on the standard scale. The offence will apply to the scrap metal dealer, the site manager and any person, who under arrangements made by either the scrap metal dealer or the site manager, has responsibility for fulfilling this requirement on behalf of the business. A defence of making arrangements to ensure this requirement should be met and taking all reasonable steps to ensure those arrangements were complied with can be used by the scrap metal dealer or the site manager; this will cover situations where an employee has acted in a rogue manner or in breach of established working practices. In addition, a person who on delivering scrap metal to a scrap metal dealer gives a false name or address is liable on summary conviction to a fine not exceeding level 3 on the standard scale.

Scotland

9.50 According to the Civic Government (Scotland) Act 1982, s 34:

- Any metal dealer or itinerant metal dealer who disposes of metal to a person apparently under the age of 16 or who acquires metal from such a person, whether that person is acting on his own behalf or on behalf of another person, shall be guilty of an offence and liable, on summary conviction, to a fine not exceeding level 3 on the standard scale.
- In any proceedings for an offence under subsection (1) above, it shall be a defence to prove that the person to whom it is alleged the metal was disposed of or from whom it is alleged it was acquired was 16 years of age or over at the time.
- Any metal dealer or itinerant metal dealer who knowingly or recklessly furnishes false particulars under section 30 of this Act or, as the case may be, false information on any record or receipt which he is required by

section 33 of this Act to keep shall be guilty of an offence and liable, on summary conviction, to a fine not exceeding level 4 on the standard scale.
- Any person who, when selling metal to or purchasing metal from a metal dealer or itinerant metal dealer, gives the dealer a false name or address shall be guilty of an offence and liable, on summary conviction, to a fine not exceeding level 3 on the standard scale.

Under the Air Weapons and Licensing (Scotland) Act 2015, Pt 3, s 70, metal dealers are required to keep a record of the people they have dealt with and the method used to verify their identity.

Northern Ireland

The Bill

9.51 Clause 9 places a requirement on scrap metal dealers, site managers and employees who have been delegated the responsibility to do so, to verify the identity of the person (the supplier) they are receiving metal from and the person's address. This verification must be done by reference to specific documents which bear a photo of the supplier. Clause 9(3) details the acceptable identification documents. These are drawn from the Representation of the People (Northern Ireland) Regulations 2008, SI 2008/1741 which are used to verify identity when voting. Given the requirement for photographic identification, some of the specified documents are qualified to ensure that they meet the criteria (for example, Clause 9(3)(d)).

The Department may prescribe in regulations other documents which are acceptable or not, as the case may be, for verifying identity.

This clause also creates an offence of receiving scrap metal in breach of these requirements in failing to comply with this verification of identity requirement. Specifically, it will be an offence not to obtain and verify the seller's identity, punishable by a fine. The offence will apply to the scrap metal dealer, the site manager and any person who, under arrangements made by either the scrap metal dealer or the site manager, has responsibility for fulfilling this requirement on behalf of the business. The scrap metal dealer or the site manager can use a defence of making arrangements to ensure this requirement should be met and taking all reasonable steps to ensure those arrangements were complied with. However, this defence will not apply to the employee who has, notwithstanding those arrangements, breached the requirements of the section. This ensures there is always somebody liable if the requirements of this provision are not complied with – the intention is to provide a defence to the licensee/site manager (who might be relatively high up the managerial chain) where they had in good faith put in place procedures to meet these requirements and done what was reasonable to ensure these are carried out.

OFFENCE OF BUYING SCRAP METAL FOR CASH ETC

England and Wales

9.52 Section 12 of the Scrap Metal Dealers Act 2013, re-enacts with modifications the amendments made by s 146 of the Legal Aid, Sentencing and Punishment of Offenders Act 2012, which inserted s 3A into the Scrap Metal Dealers Act 1964, creating the offence of purchasing scrap metal for cash. The offence will apply to all scrap metal dealers.

This offence prohibits scrap metal dealers from paying for scrap metal other than by cheque or by electronic transfer – the Secretary of State can make an order to permit other methods of payment if considered appropriate under subsection (2).

As with s 11, the offence applies to the scrap metal dealer, the site manager and any person who makes the payment acting for the dealer. A defence of making arrangements to ensure this requirement is met, and taking all reasonable steps to ensure those arrangements were complied with, can be used by the scrap metal dealer or the site manager. Anyone convicted will be liable on summary conviction to a level 5 fine on the standard scale.

Scotland

9.53 The Air Weapons and Licensing (Scotland) Act 2015, s 69 creates a new s 33A in the Civic Government (Scotland) Act 1982. This specifies acceptable forms of payment that may be accepted by a metal dealer or itinerant metal dealer. The acceptable forms of payment are a cheque or electronic transfer into a bank or building society account. Cash is not an acceptable form of payment. A dealer who makes payment in a method not specified commits an offence. The offence extends to a person with day to day management responsibilities and the person who makes the payment. The metal dealer and manager are provided with a defence that they have made arrangements to ensure that payment is made by the specified methods and have taken all reasonable steps to ensure compliance. Subsection (7) gives Scottish Ministers the power by regulation to add or remove forms of payment that are acceptable and to make any consequential changes to ss 33B or 33C(3) in consequence of changes to the acceptable form of payment.

Section 69 also inserts a new s 33B into the 1982 Act, which provides a definition of what constitutes an acceptable bank or building society account for the purposes of the regime. Subsection (4) in particular stipulates types of account that are not acceptable, with reference to Pt 4A of the Financial Services and Markets Act 2000.

Northern Ireland

The Bill

9.54 Clause 10 makes it a criminal offence to purchase scrap metal except by means of an electronic transfer of funds or a non-transferrable cheque, ie, traceable methods. It is included to create a clearer trail of information in investigating theft of high value metals such as copper or aluminium. Recognising that small amounts of low value metal are sold, there is provision for cash payments of £100 or less (or payment in kind of the same value). This is included both to ensure the regulations for scrap metal dealers are not needlessly onerous and to not deter business. The safeguards against frequently recurring payments below the £100 threshold balance this.

There is provision for the Department to introduce by order other payment methods and amend, remove or set a time limit on the £100 threshold for cash payments.

The offence applies to the scrap metal dealer, the site manager and any person who makes the payment acting for the dealer. A defence of making arrangements to ensure this payment method requirement is met, and taking all reasonable steps to ensure those arrangements were complied with, can be used by the scrap metal dealer or the site manager. However, as with Clause 9, this defence will not apply to the employee who has breached the relevant requirements. Anyone convicted will be liable on summary conviction.

RECORDS: RECEIPT OF METAL

England and Wales

9.55 Section 13 of the Scrap Metal Dealers Act 2013 sets out the record-keeping requirements in respect of any scrap metal received by a scrap metal dealer in the course of their business. Information that is required to be recorded includes the type of metal being purchased; the time/date of the transaction; personal information on the seller; who is acting on behalf of the dealer and proof of the non-cash transaction. Failure to comply with the requirements of this section is an offence attracting a penalty up to level 5 on the standard scale (see s 15).

Scotland

9.56 The Air Weapons and Licensing (Scotland) Act 2015, s 70 makes it clear that records are required to be kept for all scrap metal that the dealers has. This requires the keeping of receipts and invoices for any metal whether it has value or not, the weight, date and time and the name of the person and the means by which their identification was verified.

Northern Ireland

The Bill

9.57 Clause 11 sets out the record-keeping requirements in respect of any scrap metal received by a scrap metal dealer in the course of their business. It is included as a means to discourage the sale of stolen goods (which in turn discourages the initial theft) and to create a robust audit trail to aid investigation. Information that is required to be recorded includes the type of metal being purchased; the time/date of the transaction; personal information on the seller; who is acting on behalf of the dealer and proof of the non-cash transaction. It also introduces obligations relating to the receipt of 'suspicious scrap metal' which has been included specifically to combat theft of high value metals. In this instance, the suspicious scrap metal is defined in detail in Clause 11(8) with a corresponding provision in Clause 11(9) for the Department to vary this description by order.

There is a related duty to inform the PSNI of any suspicious metal received (including all the relevant details that are required to be recorded) and to hold it available for inspection for no less than one week.

Failure to comply with the requirements of this clause is an offence.

RECORDS: DISPOSAL OF METAL

England and Wales

9.58 Section 14 of the Scrap Metal Dealers Act 2013 makes similar provision to the previous one but relates to records of metal being disposed of by a site licensee or collector. The same sanctions apply.

Scotland

9.59 The Air Weapons and Licensing (Scotland) Act 2015, Pt 3, s 70(3) makes it clear that records of disposal are to be kept.

Northern Ireland

The Bill

9.60 Clause 13 makes similar provision to the previous one but relates to records of metal being disposed of by a site licensee or collector. The same sanctions apply.

RECORDS: SUPPLEMENTARY

England and Wales

9.61 As with the Scrap Metal Dealers Act 2013, s 11, breach of the requirements of ss 13 or 14 is an offence. The offence applies to the scrap metal dealer, the site manager and any person who has responsibility for fulfilling the requirement under arrangements made by a dealer or seller. A defence of making arrangements to ensure this requirement should be met, and taking all reasonable steps to ensure those arrangements were complied with, can be used by the scrap metal dealer or the site manager. Anyone convicted will be liable on summary conviction to a level 5 fine on the standard scale.

Scotland

9.62 Section 70 of the Air Weapons and Licensing (Scotland) Act 2015 amends the record keeping requirements for metal dealers and itinerant metal dealers. A new s 33C is inserted into the Civic Government (Scotland) Act 1982 and provides the details that must be recorded by a dealer when metal is acquired or disposed of and supports the separate provisions stipulating acceptable forms of payment by requiring dealers to keep copies of documentation evidencing the form of payment used. Subsection (6) of s 33C provides the Scottish Ministers with the power by regulation to amend the record keeping requirement and to stipulate particular means that can be used for the purpose of establishing a person's name and address.

The section also inserts a new s 33D into the 1982 Act to stipulate how records should be stored and a new s 33E to require records to be kept for each place of business a dealer operates from.

The section also creates an offence in relation to a failure to comply with the new requirements in relation to record keeping and amends the existing offence in relation to providing false or misleading information.

Northern Ireland

The Bill

9.63 Clause 13 sets out the duties and offences in relation to the keeping of records and establishes that breach of the requirements is an offence.

The offence applies to the scrap metal dealer, the site manager and any person who has responsibility for fulfilling the requirement under arrangements made by a dealer or seller. A defence of making arrangements to ensure this requirement is met, and taking all reasonable steps to ensure those arrangements were complied with, can be used by the scrap metal dealer or the site manager.

However, as with Clause 9, this defence will not apply to the employee who has breached the relevant requirements.

Anyone convicted will be liable on summary conviction.

RIGHT TO ENTER AND INSPECT

England and Wales

9.64 Section 16 of the Scrap Metal Dealers Act 2013 provides police constables or officers of a local authority with a right to enter and inspect the premises of licensed and unlicensed scrap metal dealers.

Subsection (1) provides a right of entry and inspection to a constable or local authority officer to a licensed site at all reasonable times on notice to the site manager.

Subsection (2) sets out the circumstances in which a constable or local authority officer can exercise a right of entry and inspection, at a reasonable time, without giving notice to the site manager. Subsection (3) makes it clear these powers of entry in subsections (1) and (2) (which are exercisable without warrant) do not apply to residential premises and subsection (4) makes it clear they cannot be exercised with the use of force.

Subsection (5) provides a right of entry exercisable by warrant to secure compliance with the provisions of the Act and to ascertain whether the provisions of the Act are being complied with. This right of entry applies to premises, including residential premises, which are a licensed site or to premises that are not licensed but where there are reasonable grounds for believing that they are being used by a scrap metal dealer in the course of business. Reasonable force may be used in the exercise of a right of entry under warrant.

It is an offence punishable on summary conviction with a fine not exceeding level 3 on the standard scale to obstruct the right of entry or inspection or fail to produce the records that are required.

Northern Ireland

The Bill

9.65 Clause 14 provides the PSNI with a right to enter and inspect the premises of licensed and unlicensed scrap metal dealers. This is subject to the issue of a warrant in certain circumstances. This includes the right of a constable to enter and inspect a licensed site at all reasonable times on notice to the site manager as well as the circumstances where entry and inspection may be made without giving notice to the site manager and exempts residential premises from these rights of entry. It also establishes that force cannot be used to gain entry to premises in the aforementioned instances.

Clause 14(5)–(8) provides a right of entry exercisable by warrant to secure compliance with the provisions of the Act and to ascertain whether the provisions of the Act are being complied with. This right of entry applies to premises, including residential premises, which are a licensed site or to premises that are not licensed but where there are reasonable grounds for believing that they are being used by a scrap metal dealer in the course of business. Reasonable force may be used in the exercise of a right of entry under warrant.

This clause also establishes that the PSNI may ask for the production of any scrap metal or records kept in accordance with Clauses 10–13 as part of the right of entry and inspection.

This clause also makes it an offence to obstruct the right of entry or inspection or to fail to produce the records that are required. It is punishable on summary conviction with the maximum fine on the standard scale. The level 5 maximum fine is a departure from similar legislation in Great Britain where the penalty for preventing investigation was less severe than the penalties for breaching the requirements of the Act.

OFFENCES BY BODIES CORPORATE

England and Wales

9.66 The Scrap Metal Dealers Act 2013, s 17 makes provision for where an offence under the Act is committed by a body corporate to attribute liability to both an individual and the body corporate. This applies to offences that are committed with the consent or connivance of a director, manager, secretary or other similar officer.

Northern Ireland

The Bill

9.67 Clause 15 establishes that if an offence is committed by a body corporate, both individuals and the body corporate can be held liable. This applies to offences that are committed with the consent or connivance of a director, manager, secretary or other similar officer.

REVIEW OF ACT

England and Wales

9.68 Section 18 of the Scrap Metal Dealers Act 2013 places a duty on the Secretary of State to review the Act within five years of the licensing requirement coming into force. The purpose of this review is to assess whether the licence regime has achieved the objectives that it was set out to achieve and whether it is appropriate to retain the Act. The Secretary of State must publish a report of the conclusions of this review before both Houses.

Northern Ireland

The Bill

9.69 Clause 20 requires the Department to prepare and lay before the Assembly a report addressing the implementation of the Act, six months after Royal Assent and at least once every six months thereafter, until all provisions of the Act are in force.

CONSEQUENTIAL AMENDMENTS

England and Wales

9.70 Section 19 of the Scrap Metal Dealers Act 2013 makes the necessary consequential amendments to existing legislation to give effect to this Act including the Scrap Metal Dealers Act 1964 and all linked legislation; Pt 1 of the Vehicles (Crime) Act 2001 (which regulated motor salvage operators) and the relevant provisions of the Legal Aid, Sentencing and Punishment of Offenders Act 2012.

CASES

9.71 *R v Birch and other* [2012] EWCA Crim 1000

B, with another man, had stolen 200m of copper cable from British Telecom. They used two vehicles, heavy duty cable-cutting tools, and walkie-talkies. The value of the cable stolen was £9,588 and a further £3,500 was spent on the labour costs of engineers required to replace the cables.

Sentences of two-and-a-half years' imprisonment imposed following guilty pleas to theft of copper cable from British Telecom were upheld. The three appellants appealed against sentences of two years and six months' imprisonment imposed following their guilty pleas to theft.

B argued that the sentence was out of line with the authorities.

The appeals were dismissed.

R v Moss (Michael) [2010] EWCA Crim 1097, [2011] 1 Crim App Rep (S) 31

This case was a similar offence involving a significant length of copper wire and the court had concluded that the starting point in such a case was likely to be in the order of two to three years for a first-time offence after trial. However the Court of Appeal had not accepted the concerns of the trial judge as to the prevalence of such offending.

R v Mitchell (Billy Richard) [2011] EWCA Crim 1652, [2012] 1 Crim App Rep (S) 68

In this case the Court of Appeal had said that the offence was very prevalent; it was well known that cable theft had become a major problem. The theft had not only had financial consequences for BT, but also had the consequence that some 800 telephone lines were inoperative. Included among those would probably have been some people who relied on a panic alarm. Further, there had been not only domestic users but also shops whose card payment machines were inoperative. Such consequences were an important consideration. In *Mitchell* it had been suggested that a sentence of around three years was appropriate. The judge in the instant case had chosen a starting point of three years and nine months. The case had the aggravating feature of gang operation, and there had been two vehicles and equipment involved. The sentence was not manifestly excessive, *Mitchell* applied, *Moss* not followed.

Chapter 10

Powers of the Agencies

ENGLAND

10.1 The Environment Agency have published an Enforcement and Sanctions Statement – the latest version is Version 3[1]. This includes the application of civil sanctions and an outcome focused approach to enforcement. This outlines the general principles which are to be followed in relation to enforcement and sanctioning and should be seen in conjunction with the Enforcement and Sanctions Guidance[2] which contains more detail.

1 Enforcement and Sanctions Statement Policy 1429_10 (previously EAS/8001/1/1/1) Version 3. https://www.gov.uk/government/uploads/system/uploads/attachment_data/file/389348/LIT_5197.pdf.

2 Enforcement and Sanctions Guidance Version 4 https://www.gov.uk/government/uploads/system/uploads/ attachment_data/file/468315/LIT_5551.pdf.

Approach to Enforcement

10.2 Enforcement can range from providing advice and guidance through to serving notices to prosecute or may be a combination of these, which will achieve the desired outcome. Under the Environmental Damage (Prevention and Remediation) Regulations 2009, SI 2009/153, enforcement action may be required in the absence of any suspected offence, but in most cases an activity that leads to 'environmental damage' under the Regulations will also constitute an offence under environmental legislation.

When an offence has been committed but the delivery of the advice and guidance has not or will not achieve the desired outcomes then there are a number of sanctions available. These are:

- issuing a warning;
- statutory enforcement notices and works notices;
- prohibition notices;
- suspension or revocation of environmental permits;
- variation of permit conditions;
- injunctions;
- carrying out remedial works;
- civil sanctions;
- other civil and financial sanctions including Fixed Penalty Notices;
- issuing a formal caution;
- prosecution and orders ancillary to prosecution; and
- sanctions used in combination.

The Environment Agency also take into account the Macrory Penalty principles when considering what is the appropriate course of action to take.

These are:

- aim to change the behaviour of the offender;
- aim to eliminate any financial gain or benefit from non-compliance;
- be responsive and consider what is appropriate for the particular offender and regulatory issue, which can include punishment and the public stigma that should be associated with a criminal conviction;
- be proportionate to the nature of the offence and the harm caused;
- aim to restore the harm caused by regulatory non-compliance, where appropriate; and
- aim to deter future non-compliance.

Principles of Regulation

10.3 The Environment Agency takes account of the Regulators' Code when devising and implementing regulatory policies and systems.

The requirements of the Code do not apply in certain situations, for example:

- where we can demonstrate that immediate enforcement action is required to prevent or respond to a serious breach;

- where following the provisions would defeat the purpose of the proposed enforcement action.

The underlying principles to firm but fair regulation are:

- proportionality in the application of the law and in securing compliance;
- consistency of approach;
- transparency about how we operate and what those we regulate may expect from us;
- targeting of enforcement action; and
- accountability for the enforcement action we have taken.

Proportionality

10.4 The concept of proportionality is included in much of the regulatory system by balancing action to protect the environment against the risks and costs of such action. Some incidents or breaches of regulatory requirements cause, or have the potential to cause, serious environmental damage. Others may interfere with people's enjoyment or rights, or our ability to carry out our activities. The Environment Agency's first response will be to prevent harm to people and the environment from occurring or continuing. Any enforcement action will be proportionate to the risks posed to people and the environment and also to the seriousness of the breach of the law and its impact on legitimate business.

Consistency

10.5 This means taking a similar approach in similar circumstances to achieve similar ends. However, consistency does not mean simple uniformity. Officers will need to take account of many variables:

- the scale of environmental impact;
- the attitude and actions of individuals and managers of businesses;
- the history of previous incidents or breaches.

Decisions on enforcement action are a matter of professional judgment and officers need to be able to exercise discretion. The Environment Agency will continue to develop arrangements to promote consistency, including effective ways to liaise with other enforcing authorities.

Transparency

10.6 Transparency is important in maintaining public confidence in the Environment Agency's ability to regulate. It means helping regulated entities and others to understand what is expected of them and what they should expect from the Agency.

To aid transparency the Agency will:

- where remedial action is required, clearly explain why the action is necessary and when it must be carried out, making a distinction between best practice advice and legal requirements;
- provide the opportunity to discuss what is required to comply with the law before formal enforcement action is taken, unless urgent action is required, for example to protect the environment or to prevent evidence being destroyed;
- where an operator is required to take urgent action, then a written explanation will be provided with reasons as soon as possible after the event;
- a written explanation of any rights of appeal against formal enforcement action at the time the action is taken.

Targeting

10.7 Targeting makes sure that the Agency's regulatory effort is directed primarily towards those whose activities give rise to or create the risk of serious environmental damage, where the risks are least well controlled, or against deliberate or organised crime.

Regulated Industries

10.8 In the case of regulated industries, management actions are important. Repeated incidents or a series of related breaches of regulatory requirements may indicate an unwillingness to change behaviour, or an inability to achieve sufficient control. These may require a review of the regulatory requirements, the ability of the operator to run the site and/or additional investment. We recognise that a relatively low hazard site or a poorly-managed activity has potential for greater risk to the environment than a higher hazard site or activity where proper control measures are in place.

Accountability

10.9 Accountability means that the Agency takes responsibility for their decisions and will justify them where appropriate.

Enforcement

10.10 A sanction is defined as an enforcement requirement (such as a notice), a binding legal agreement or even a penalty applied by the Agency or a court.

To commence a prosecution the Agency must be satisfied that the case before them meets the test in the Code for Crown Prosecutors; namely that there is a realistic prospect of conviction.

For most sanctions issued under the Regulatory Enforcement and Sanctions Act 2008 (RES Act) there is a higher burden of proof, namely that the Agency must be satisfied beyond reasonable doubt that an offence has been committed before they can issue a sanction. In respect of enforcement undertakings received under that Act the Agency must have reasonable grounds to believe that an offence has been committed before they can accept the offer. In such cases the Agency will still want to be in a position to establish that an offence has been committed in case further enforcement action is necessary where an offender fails to comply with the undertaking.

Action Against the Crown

10.11 No contravention by the Crown of any provision will make the Crown criminally liable, nor can the Crown be made the subject of a RES Act civil penalty. However, the Environment Agency may apply to the High Court for a declaration that any act or omission of the Crown is unlawful. This is an unusual course to take and whether to do this will depend upon:

- the gravity of the incident or breach;
- whether liability is admitted; and
- the response of the body concerned.

Outcome Based Enforcement

10.12 Their use of sanctions to achieve environmental outcomes are divided into four basic principles.

Tools to Stop Offending

10.13 Injunctions

An injunction is an order of a court directing an individual (or company) to either:

- stop a particular activity (a prohibitory injunction); or
- carry out a particular activity (a mandatory injunction).

Failure to comply with an injunction is treated as a contempt of court and is punishable by an unlimited fine and/or up to two years' imprisonment.

Court Order under the Environmental Permitting (England and Wales) Regulations 2010, SI 2010/675 (EP Regulations 2010)

- This is similar to an injunction.

Notices

- Stop Notice;
- Suspension Notice;
- Groundwater Prohibition Notice;
- Anti-Pollution Works Notice.

Tools to Restore or Remediate

10.14 Notices

- Restoration Notice;
- Enforcement Notice;
- Notice to Remove Waste;
- Anti-Pollution Works Notice.

Powers

- Under the EP Regulations 2010, power to seek a regulation 44 remediation order;
- Acceptance of an Enforcement Undertaking or acceptance of a Third Party Undertaking. This can include restitution to adversely affected third parties and communities.

Tools to Bring Under Regulatory Control

10.15 Notices

- Compliance Notice;
- Enforcement Notice;
- Suspension Notice;
- Anti-Pollution Works Notice;
- Revocation of an Environmental Permit; and
- Variation of Permit Conditions.

Powers

- Acceptance of an Enforcement Undertaking.

Tools to Deter and/or Punish

10.16

- Criminal Sanctions;
- Variable Monetary Penalties, Fixed Monetary Penalties.

Advice and Guidance – Compliance Assistance

10.17 In the context of enforcement, the Agency will normally provide advice and guidance after an offence is committed or where we consider that an offence is likely to be committed.

Warnings

10.18 A warning is a written notification issued when it is believed that an offence has been committed. The notification can be either:

- a warning letter; or
- a site warning that is normally issued on-site or otherwise as a result of a compliance visit to a permitted site or activity.

It will be recorded and may, in the event of further non-compliance, influence subsequent choice of sanction. Warnings can be used to achieve any of the four types of environmental outcome sought.

Non-RES Notices, Powers and Orders

10.19 Many of the regimes already contain powers to serve specific enforcement notices, which require the recipient to stop offending, to restore or remediate an affected environment or to comply with regulatory requirements. These existing provisions may be used wherever appropriate.

Examples of these include:

Environmental Permitting (England and Wales) Regulations 2010, SI 2010/675 (EP Regulations 2010):

- Notices: Variation notice (reg 20), Enforcement notice (reg 36), Suspension notice (reg 37), Revocation notice (reg 22), Groundwater prohibition notice (Sch 22, para 9);
- Court order: Enforcement by the High Court (reg 42);
- Remediation powers: Power to prevent or remedy pollution and recover costs (reg 57), Power to dispose of radioactive waste (Sch 23, Pt 3, para 4), Orphan sources (Sch 23, Pt 4, para 8).

Environmental Protection Act 1990:

- Notice to remove waste (ss 59 and 59Z).

Water Resources Act 1991:

- Anti-pollution works notice (ss 161A–161D).

The RES Act 2008 has created notice powers to complement the existing notices, particularly to provide powers to require an offender to stop the unlawful activity, to bring the activity into regulatory compliance and to restore damage that has been done. These new notices, namely Stop, Compliance and Restoration Notices, have been made available in order to address situations not capable of being dealt with by the notices described above.

Criminal Sanctions

Fixed Penalty Notices (FPNs)

10.20 A Fixed Penalty Notice (FPN) is a financial penalty for an offence, imposed by the regulator, which if unpaid can be dealt with by way of prosecution in the criminal courts. FPNs are available for a limited number of offences.

Where a Fixed Penalty Notice is served, payment of the penalty discharges the liability. Where this happens a record of the payment of the FPN will be kept and treated in the same way as a record arising from a warning.

Where a Fixed Penalty Notice is imposed but not paid, the recipient will normally be prosecuted for the original offence.

Formal Caution

10.21 A formal caution is the written acceptance by an offender that they have committed an offence and may only be used where a prosecution could properly have been brought. To this extent it differs from a warning as described above, which is simply a record and warning about an offence that has been or may be committed. The formal caution is a formal recorded criminal sanction, which will be produced in court if there is further offending. It differs from the imposition of a civil sanction as the circumstances which led to the offence have been considered to be appropriate for a prosecution and, indeed, a repetition of similar offending would be likely to lead to such a response. Formal cautions are intended to be a specific deterrent to an offender and are suitable for cases where, although a prosecution could be initiated, other factors mitigate against this. We must however consider the test under the Code for Crown Prosecutors, namely whether there is sufficient evidence and also the public interest factors involved.

Where a formal caution is not accepted we will normally prosecute for the original offence.

Prosecution

10.22 The sanction of prosecution is available for all criminal offences by law. The legislation which establishes the penalty provisions, gives the courts considerable scope to punish offenders and to deter others. In some cases imprisonment and unlimited fines may be imposed. Where it is decided that a criminal sanction is appropriate the case will be assessed in accordance with the requirements of the Code for Crown Prosecutors before commencing a prosecution.

Orders Imposed by the Court Ancillary to Prosecution

10.23 The Code for Crown Prosecutors requires the Agency to apply for compensation and ancillary orders such as anti-social behaviour orders and confiscation orders, in all appropriate cases. Listed below are the ancillary orders that a court may make following a conviction:

- disqualification of directors;
- confiscation of assets – Proceeds of Crime Act 2002;
- anti-social behaviour orders;
- forfeiture of equipment used to commit the offence;
- disqualification from driving;
- compensation;
- vehicle seizure;
- remediation – under the Environmental Permitting Regulations 2010.

Consideration may also be given to a person's competence to be the holder of some Environmental Permitting Regulation permits or the desirability of a person continuing to be a registered waste carrier or broker.

RES Act Civil Sanctions

10.24 In some cases the outcome may be achieved by a civil sanction made available under the RES Act. Civil sanctions are not available for all offences. The following illustrates the sorts of circumstances where civil sanctions might be used or accepted:

Fixed Monetary Penalties (FMPs)

10.25 These are fixed penalties which will be most suitable for offences with minor or no direct environmental impact, such as paperwork and administration offences. They will be most appropriate where advice and guidance has failed to secure the necessary improvements.

Fixed Monetary Penalties are set at £300 for businesses and £100 for individuals, with discounts for early payment. They may be used for offences where a low-level monetary penalty is more likely to change the offender's behaviour and encourage future compliance, for example because advice and guidance has failed. They can be issued for minor offences that need some kind of enforcement action but which, depending on the public interest factors, may not be serious enough to warrant a prosecution.

Variable Monetary Penalties (VMPs)

10.26 These are monetary penalties, which can be imposed directly for more serious offences. They follow the published methodology[3] to set the level of the VMP. Variable Monetary Penalties may be used instead of criminal sanctions for offences where imposing a financial penalty may change offender behaviour and deter others and/or lead to faster resolution. A VMP does not carry the stigma of a criminal conviction. It may also provide a saving for the recipient in avoiding the need to pay for representation, and enable the recipient to offer a Third Party Undertaking to make restitution to adversely affected third parties, including local communities.

Compliance Notices

10.27 These require the offender to come back into compliance. They may be used in a case where the offender has previously been in compliance with a requirement such as regularly submitting returns, but is currently not fulfilling their obligations. The notice should ensure that the offender takes action to stop the non-compliance, addresses the underlying causes and comes back into compliance. They may be used where previous advice or guidance to encourage compliance has not been followed and a formal notice has become necessary to ensure compliance. They can be combined with a Variable Monetary Penalty and a Restoration Notice.

Restoration Notices

10.28 These require the offender to take steps to put right any damage caused as a result of the non-compliance and address any harm. These notices can be used where damage has been caused to the environment and the action and work needed to address the damage can be identified and carried out by the offender. They may be used where restoration has not been undertaken voluntarily or it is considered that a formal notice is necessary to ensure it is undertaken. They may be used where there is no other suitable enforcement notice available and where previous advice or guidance to encourage restoration has not been followed. They can be combined with a Variable Monetary Penalty and a Compliance Notice.

Stop Notices

10.29 A Stop Notice can be used in two sets of circumstances:

3 Enforcement and Sanctions Guidance Annex 1: Variable Monetary Penalty methodology.

- to immediately stop an activity that is causing, or presents a significant risk of causing, serious harm to human health or the environment and where a specified offence is being, or is likely to be, committed;
- to immediately stop an activity that is likely to be carried on that will cause, or will present a significant risk of causing, serious harm to human health or the environment, and the activity likely to be carried on involves or will be likely to involve a specified offence being committed.

Enforcement Undertakings (EUs)

10.30 These are legally binding voluntary agreements offered by those who may have committed an offence and accepted by us. An Enforcement Undertaking can be accepted from a person where there are reasonable grounds to suspect that a specified offence has been committed. They will only be accepted where there is sufficient confidence that the terms of the EU will be delivered. They are more likely to be accepted when they are offered early. An EU is unlikely to be accepted where it has already been decided that a prosecution is required in a particular case. Equally it is unlikely that an EU will be accepted when discussions are already underway as to the level of a Variable Monetary Penalty in a particular case.

The terms of the EU will normally contain an element of restoration as well as steps to ensure future compliance such as long-term investment in environmental management systems. EUs should encourage legitimate business operators to make amends, come into compliance and prevent recurrence.

Other Civil Sanctions

10.31 The Climate Change Regimes (CRC Energy Efficiency Scheme, the EU Emissions Trading System, the Climate Change Agreements Scheme and the Energy Savings Opportunity Scheme) have a different civil sanctions framework[4].

In relation to the Landfill Allowance scheme, the legislation requires suspected non- compliance to be reported to the Secretary of State.

Choosing the Appropriate Response

10.32 RES Act civil sanctions are not available for all offences and where they are available, not all of the civil sanctions can be applied. Similarly non-RES Act notices, powers and orders are not available for all offences. A list of

4 Enforcement and Sanctions Guidance Annex 4: Climate Change Regimes – Civil Sanctions.

all available sanctions for each offence is contained within the Environment Agency document Offence Response Options (ORO)[5].

The consideration of a sanction will also involve considering whether a regulatory position statement has been published for that offence. Regulatory position statements, where available, are set out in the ORO.

Consideration is firstly given to the outcome that is being sought in line with the Environment Agency's commitment to outcome-focused regulation. The Agency considers what immediate action is needed as a response to protect the environment, such as the need to immediately serve a notice, and whether further action is then required to secure the desired outcomes.

Where the Agency considers that advice and guidance has not or will not achieve the necessary outcome and that some form of sanction (either criminal or civil) is required to secure that outcome, then they will consider the facts against public interest factors (see the Enforcement and Sanctions Guidance LIT 5551, Version 4) in order to decide what type of sanction to impose.

Where the circumstances leading to the offence could reasonably have been foreseen, and adequate avoiding and/or preventative measures were not taken, the response will normally give rise to a sanction beyond advice and guidance or issuing a warning.

Public Interest Factors

10.33 The importance of each public interest factor may vary on a case-by-case basis. Deciding on the public interest is not simply a matter of adding up the number of factors in favour of and against applying a sanction. The Agency will decide how important each factor is in the circumstances of each case and make an overall judgment.

Consideration of these factors will normally indicate the appropriate sanction(s) likely to produce the desired outcome. These outcomes may include lasting compliance with the law, redress for environmental harm and obtaining a good and lasting benefit for the environment, affected local communities and ensuring a level playing field for businesses and others.

Intent

10.34 Offences that are committed deliberately, recklessly or with gross negligence are more likely to result in prosecution. Where an offence was committed as a result of an accident or a genuine mistake this is more likely to result in the use of advice and guidance, warning or an available civil sanction.

5 Environment Agency – Offence Response Options https://www.gov.uk/government/uploads/system/uploads/ attachment_data/file/514087/LIT_9052.pdf.

Foreseeability

10.35 Where the circumstances leading to the offence could reasonably have been foreseen and no avoiding and/or preventative measures were taken, the response will normally be to impose a sanction beyond advice and guidance or the issuing of a warning.

Environmental Effect

10.36 Any response will need to address the potential and actual harm to people and the environment. Normally where an offence is classified under the Common Incident Classification Scheme (CICS) or Compliance Classification Scheme (CCS) (or the equivalent under the Reservoirs Dam Risk Categorisation Scheme) as Category 1 or 2 the Agency would consider a prosecution, formal caution or a Variable Monetary Penalty. Certain offences are of a type designed to prevent environmental harm, for example particular abstraction offences. Where such an offence could give rise to either a Category 1 or 2 CCS or CICS incident the Agency would normally consider a prosecution, formal caution or a Variable Monetary Penalty.

Nature of the Offence

10.37 Where offending impacts on the efficiency and effectiveness of the Regulator, for example where staff are obstructed in the conduct of their duties; where the Agency is targeting a particular type of offending or where the Agency has been provided with false or misleading information, then a prosecution would normally follow.

Financial Implications

10.38 Where legitimate business is undercut, or where profits are made or costs are avoided, such as costs saved by not obtaining a permit, this will normally lead to the imposition of a Variable Monetary Penalty or a prosecution. This will include offences motivated by financial gain.

Deterrent Effect

10.39 When choosing a sanction the Agency will consider the deterrent effect, both on the offender and others. Prosecutions, which have a greater level of stigma if a conviction is secured, may be appropriate even for minor non-compliances where they might contribute to a greater level of overall deterrence. Where the use of a sanction is likely to reduce future self-reporting of offences or non-compliance, a different sanction may be appropriate.

Previous History

10.40 The degree of offending and/or non-compliance (including site-specific offending or generic failures by the offender) will be taken into account. Enforcement response would normally be escalated where previous sanctions have failed to achieve the desired outcome. For example, where a formal caution has previously been used to deal with offending but has failed to encourage behaviour change and prevent a repeat of the offending, it is likely that the offender will be prosecuted or, where available, a Variable Monetary Penalty will be served.

Attitude of the Offender

10.41 Where the offender has a poor attitude towards the offence and/or is unco-operative with the investigation or remediation, this will normally mean a prosecution or a Variable Monetary Penalty. Conversely, where the offender provides details of an offence voluntarily or through a self-reporting mechanism, this will be taken into account when deciding on a sanction or whether advice and guidance will suffice.

Personal Circumstances

10.42 Personal circumstances of the offender will be considered (for example if the offender is suffering from a serious illness). A first offence by a juvenile will not normally result in prosecution. Where a sanction incorporates a financial penalty or a requirement to perform costly remediation then the offender's ability to pay will be considered. Where that consideration indicates the penalty is beyond the offender's ability to pay it will be modified to place it within the reach of the offender.

Other Public Interest Considerations

Serious Offences

10.43 When dealing with overt criminality, gross negligence or reckless behaviour or where the seriousness of an offence requires that it be heard in a public forum, then the choice for prosecution will normally be made.

Subject to the consideration of public interest factors, a prosecution would normally follow where:

- the offending has been intentional, reckless or grossly negligent or involves outright criminal activity;
- the offending has created serious harm (or has the potential to cause such harm) to the environment or to people;

- there has been large-scale and protracted non-compliance with regulatory provisions;
- Agency staff have been subject to harassment, alarm, distress or fear of violence;
- intentionally, recklessly or wilfully making a false or misleading statement or record;
- the Agency has been obstructed in their duties and this obstruction has prevented the investigation of potentially criminal activity or an offender has impersonated an Environment Agency officer;
- where an offender has failed to comply with a Stop Notice.

Minor Breaches

10.44 When the Agency is dealing with minor breaches of a permit condition where there is no environmental impact, they will normally choose to provide advice and guidance to help the business back into compliance.

Repeat Offending

10.45 Where offending has continued or been repeated despite the Agency taking enforcement action they will normally increase the level of enforcement response and impose a more serious or severe sanction.

Failure to Comply with a Notice

10.46 Where a recipient fails to comply with a notice the Agency will seek a sanction that is likely to punish and/or deter.

Operating Without a Permit, Licence or Other Authority

10.47 Where a person has failed to obtain the necessary authorisation, then the Agency are more likely to impose a sanction that is likely to punish and/or deter.

Multiple Site Operations

10.48 The Environment Agency will always have regard to the compliance history of an offender and will not normally take account of offences committed on another of an operator's sites unless they are of a similar type or demonstrate an overall management failure on the part of the operator.

Offences by Bodies Corporate

10.49 Where a company is involved, enforcement will usually be against the company where the offence resulted from the company's activities. However, where an offence has been committed by a body corporate and is attributable to the consent, connivance or neglect of any director, manager, secretary or other officer, that person can be guilty of an offence and is liable to be the subject of enforcement action for that offence.

In such circumstances action will be taken against the most appropriate body (corporate and/or individual) or in combination, using the full range of the enforcement tools at the Agency's disposal. In appropriate cases, the Agency will consider seeking disqualification of directors under the Company Directors' Disqualification Act 1986.

Combining Sanctions

10.50 Other than where specifically allowed by the legislation it is not normally possible to combine criminal and civil sanctions for the same offence.

Combinations of sanctions will in some circumstances be possible but careful regard must be had to the relevant legislative provisions. For instance, the RES Act 2008 allows the imposition of Variable Monetary Penalties, Restoration and Compliance Notices for the same offence.

Where it appears desirable to use a combination of sanctions consideration of the extent to which this is lawful and appropriate will be given. For regimes which involve annual registration and compliance such as those established by Producer Responsibility obligations, where the choice is one of a civil sanction, the Agency will endeavour to reflect all compliance failures by choosing one civil sanction option, which most fairly characterises the offences arising, rather than issuing a number of sanctions for each compliance period.

Other Notices, Powers and Orders

10.51 Many of the regimes the Agency enforces already contain powers to serve specific enforcement notices, which require the recipient to stop offending, to restore or remediate an affected environment or to come into compliance with regulatory requirements. These existing provisions may be used wherever appropriate.

Procedures for RES Act Civil Sanctions

10.52 The RES Act sets out highly prescriptive requirements and procedures for regulators wishing to use civil sanctions. These are expanded by the:

- Environmental Civil Sanctions (England) Order 2010, SI 2010/1157 (ECSO)[6]; and

6 SI 2010/1157.

- Environmental Civil Sanctions Miscellaneous Amendments (England) Regulations 2010, SI 2010/1159 (ECSR)[7].

The legislation does not set out all of the considerations which arise in relation to proceedings, which are alternatives to criminal sanctions. In carrying out an investigation as to whether or not an offence has been committed, or is likely to be committed, the Agency apply the safeguards laid down by the Police and Criminal Evidence Act 1984.

Enforcement Cost Recovery Notices (ECRNs)

10.53 Enforcement Cost Recovery Notices can be served requiring payment of the Agency's costs up to the time of imposition of most civil sanctions. An Enforcement Cost Recovery Notice may not be served where a Fixed Monetary Penalty is imposed as the legislation does not allow this. Where an Enforcement Undertaking is offered the Agency is not entitled to require payment of their costs but will look more favourably upon Enforcement Undertakings, which offer a contribution towards the costs, associated with an alleged offence and the acceptance of the undertaking through to its discharge under a completion certificate.

Whilst it may be possible to reduce the amount of a civil sanction by reason of an offender's inability to pay the amount due, it is unlikely that a reduction will be made in the Enforcement Cost Recovery Notice. This is because the Enforcement Cost Recovery Notice should reflect the actual costs associated with the offending and the sanction imposed and because of the Agency's duty to protect the public purse.

There is a right of appeal against the imposition of an Enforcement Cost Recovery Notice as provided for by the relevant Statutory Instruments.

Non-Compliance Penalties (NCPs)

10.54 Where an offender fails to comply with a compliance notice then a Non-Compliance Notice, Restoration Notice or Third Party Undertaking can be imposed.

For Compliance Notices and Restoration Notices the Agency can prosecute for non-compliance with a notice or serve a Non-Compliance Penalty. If the requirements of the original sanction or Third Party Undertaking are complied with before the time set for payment of the Non-Compliance Penalty, the penalty falls.

7 SI 2010/1159.

Appeals

10.55 Enforcement action (specifically the imposition of a sanction) can normally be appealed either through the criminal court process or as a result of specific appeal provisions. The notices set out the rights of appeal, which apply in the specific circumstances of each sanction or provision. When considering any type of appeal against enforcement and sanctioning action it will usually be appropriate for the recipient to obtain independent legal advice.

Appeals Against RES Act Sanctions

10.56 RES Act sanctions cannot be imposed in the following circumstances:

- where that offence is not specified as having a RES Act sanction available;
- where that offence does not have that particular RES Act sanction available;
- where no offence has been committed;
- where it is not possible for the offence to be proved beyond reasonable doubt;
- where representations indicate that a defence is available.

References to specified offences mean those that are specified under the Environmental Civil Sanctions (England) Order 2010, SI 2010/1157 (ECSO) and the Environmental Civil Sanctions Miscellaneous Amendments (England) Regulations 2010, SI 2010/1159 (ECSR).

Where a RES Act sanction can be imposed, the notices will set out rights of appeal. The First Tier Tribunal has issued forms and guidance in respect of such appeals[8].

Generally, the grounds of appeal against the imposition of a RES Act sanction are as follows[9]:

- that the decision was based on an error of fact;
- that the decision was wrong in law;
- that the decision was unreasonable;
- the amount is unreasonable;
- any other reason.

8 http://hmctsformfinder.justice.gov.uk/HMCTS/GetForms.do?court_forms_category=Tax%20 and%20Chancery %20Chamber%20(Upper%20Tribunal).

9 Enforcement and Sanction Guidance Version 4 Annex 3: Grounds for Appeal for RES Act sanctions.

Compensation Provisions – Stop Notices

10.57 Where a Stop Notice has been served the Agency must compensate a person if, as a result of the service of the notice or the refusal of a completion certificate, that person has suffered loss, and:

- the Agency withdraw or amend the Stop Notice because the decision to serve it was unreasonable or any step specified in the notice was unreasonable;
- the Agency are in breach of their statutory obligations;
- the person successfully appeals against the Stop Notice and the Tribunal finds that their service of the notice was unreasonable;
- the person successfully appeals against the refusal of a Completion Certificate and the Tribunal finds that the Agency's refusal was unreasonable.

In these circumstances compensation should cover loss including exceptional costs resulting from the service of the Stop Notice (for example, legal or expert advice).

A person who wishes to claim compensation must do so in writing to the address which appears on the Stop Notice with which they have been served.

Any dispute over whether compensation should be paid, or over the amount of any such compensation, will be referred to the arbitration of a single arbitrator appointed by agreement between the parties to the dispute or, in default of agreement, a person will have the right of appeal to the First Tier Tribunal against the Agency's decision not to provide compensation or against the amount of compensation. The compensation scheme is without prejudice to a person's rights to seek redress through the Ombudsman, judicial review or civil litigation.

WALES

Legislation

10.58 From 1 April 2013 the Natural Resources Body for Wales took the combined functions of the Countryside Council for Wales, and the Welsh devolved functions of the Environment Agency in Wales and the Forestry Commission in Wales.

This policy also applies to ongoing investigations, which were commenced by the Environment Agency in Wales, the Forestry Commission in Wales and the Countryside Council for Wales prior to 1 April 2013.

The Natural Resources Body for Wales was formed under the Natural Resources for Wales (Establishment) Order 2012, SI 2012/1903.

This Order was made under the Public Bodies Act 2011 ('the Act') which established a new statutory body, the Natural Resources Body for Wales ('the

Body'), and provides for its form, purpose, membership, procedure, financial governance and initial functions.

The Body's principal function is to prepare to assume substantive regulatory and other functions, relating to the environment and natural resources of Wales, at a later stage. This preparatory work will include setting up the internal structures of the Body and making ready for the transfers of those functions and of staff, property and other rights and liabilities, in subsequent legislation.

The Body is set up in this way because the Welsh Ministers are still in the process of finalising their proposals, under the Act, as to which functions to transfer to the Body from existing organisations, and whether any of those functions should be modified.

Article 6 gives the Body its initial, preparatory, functions. Paragraph (1) of art 6 identifies the categories of Welsh Minister proposals to which this preparatory function relates. Paragraph (2) makes it plain that if a proposal requires the approval of the National Assembly for Wales (or any other body) in order to be implemented, then nothing in this Order removes the need for that approval.

The Order also gives the Body other powers it may need in order to carry out its preparatory functions: for example, the power to enter into agreements (art 9), to borrow money (art 14) and to employ staff (para 13(4) of the Schedule). The Order places certain conditions on the exercise of the Body's functions (see arts 7 and 8).

The Body is not given any substantive regulatory or other functions in relation to the environment or natural resources of Wales in this Order. Unless or until those functions are transferred to the Body, they will remain with the bodies or office-holders in which they are currently vested.

The Order vests a number of functions regarding the Body in the Welsh Ministers, including the power to appoint and remove a number of its members (paras 2, 4 and 7 of the Schedule); powers in relation to its corporate and financial governance (paras 10–15, 18, 19 and 21–24 of the Schedule) and powers to give the Body guidance (art 5) and directions (art 11).

The Body

10.59 There is to be a body corporate to be known as the Natural Resources Body for Wales or Corff Adnoddau Naturiol Cymru (referred to in this Order as 'the Body'). The Schedule contains further provisions about the Body

Purpose of the Body

10.60 The purpose of the Body is to ensure that the environment and natural resources of Wales are:

- sustainably maintained;

- sustainably enhanced; and
- sustainably used.

'sustainably' (*yn gynaliadwy*) means:

- with a view to benefitting, and
- in a manner designed to benefit,

the people, environment and economy of Wales in the present and in the future;

'environment' ('*amgylchedd*') includes, without limitation, living organisms and ecosystems.

Wherever the Body exercises any function in relation to, or that affects, the Welsh zone (as defined in s 158(1) of the Government of Wales Act 2006)[10], both the references to 'Wales' in para (1) are to be interpreted as including references to the Welsh zone.

Guidance with Respect to the Body's Purpose

10.61 The Welsh Ministers may give guidance to the Body with respect to the manner in which it should exercise its functions so as to give effect to its purpose. In preparing any guidance the Welsh Ministers must have regard to the Body's responsibilities and resources.

When discharging its functions, the Body must have regard to guidance given under this article.

Before any guidance is given to the Body under this article, the Welsh Ministers must consult the Body and such other bodies or persons as the Welsh Ministers consider appropriate.

The Welsh Ministers must publish any guidance given under this article as soon as is reasonably practicable after giving the guidance.

The power to give guidance under this article also includes power to vary or revoke it.

Initial Functions of the Body

10.62 The Body has the following functions:

(a) the function of facilitating the implementation of any proposal of the Welsh Ministers for the transfer (with or without modification) to the Body of—

    (i) any function of the Countryside Council for Wales;

10 'Welsh zone' means the sea adjacent to Wales which is—
(a) within British fishery limits (that is, the limits set by or under s 1 of the Fishery Limits Act 1976), and
(b) specified in an Order in Council under s 58 or an order under subsection (3).

(ii) any Welsh devolved function of the Environment Agency or the Forestry Commissioners;
(iii) any function of a Welsh Flood and Coastal Committee;
(iv) any of their own functions relating to the environment; or
(v) any Welsh environmental function of any person.

(b) the function of facilitating the implementation of any other proposal of the Welsh Ministers made in association with any proposals falling within sub-paragraph (a) which—

(i) relates to the subject-matter of those proposals, or
(ii) is consequential on those proposals, or supplementary or incidental to them, or relates to transitional matters.

This applies to a proposal of the Welsh Ministers regardless of whether the National Assembly for Wales or any other person or body has given any consent or approval on which, by law, implementation of that proposal depends, but does not obviate the need for any such consent or approval to be obtained before the proposal can be implemented.

The Body must carry out its functions under art 6(1) according to the criteria set out in the following paragraphs.

The first criterion is that the Body must ensure, so far as possible without compromising the achievement of its functions under art 6(1), that there is effective co-operation in relation to the implementation of any proposal between itself, the Welsh Ministers, and any other person or body which is—

(a) referred to in art 6(1)(a), and
(b) affected by the relevant proposal.

The second criterion is that the Body must not interfere with the effective carrying out by any of the persons or bodies mentioned in art 6(1)(a) of any of their functions.

General Duty of the Body to have Regard to Costs and Benefits in Exercising Powers

10.63 In considering whether or not to exercise any power conferred upon it by or under any enactment, the Body must take into account the likely costs and benefits of the exercise or non-exercise of that power. The body must also take into account the likely costs and benefits of its exercise in the manner in question.

These duties apply unless, or to the extent that, it is unreasonable for the Body to be subject to them in view of the nature or purpose of the power or in the circumstances of the particular case.

These duties do not affect the Body's obligation to discharge any duties, comply with any requirements, or pursue any objectives, imposed upon or given to it by any enactment other than this article.

General Incidental Function of the Body

10.64 The Body may do anything that appears to it to be conducive or incidental to the discharge of its functions.

In particular, the Body may:

(a) enter into agreements;
(b) acquire or dispose of property and carry out such engineering or building operations as it considers appropriate;
(c) subject to the approval of the Welsh Ministers, form bodies corporate or acquire or dispose of interests in bodies corporate;
(d) form charitable trusts;
(e) accept gifts;
(f) invest money.

'engineering or building operations' ('*gwaith peirianyddol neu waith adeiladu*'), without prejudice to the generality of that expression, includes:

(a) the construction, alteration, improvement, maintenance or demolition of any building or structure or of any reservoir, watercourse, dam, weir, well, borehole or other works, and
(b) the installation, modification or removal of any machinery or apparatus.

Advice and Assistance to Welsh Ministers

10.65 The Body must provide the Welsh Ministers with such advice and assistance as they may request.

Directions

10.66 The Welsh Ministers may give the Body general or specific directions as to the exercise of its functions.

The Welsh Ministers must publish any directions given under this article as soon as is reasonably practicable after giving the directions.

The power to give directions under this article includes power to vary or revoke the directions.

The Body must comply with any directions given under this article.

Grants

10.67 The Welsh Ministers may make grants to the Body. A grant under this article may be made subject to conditions.

Borrowing Powers

10.68 The Body may borrow in accordance with the following provisions of this article, but not otherwise.

The Body may borrow such sums in sterling as it may require for meeting its obligations and carrying out its functions.

The Body may borrow:

- from the Welsh Ministers, or
- from persons other than the Welsh Ministers, but only with the consent of the Welsh Ministers.

Welsh Ministers' Guarantees of the Body's Borrowing

10.69 The Welsh Ministers may guarantee, in such manner and on such conditions as they think fit, the repayment of the principal of, the payment of interest on, and the discharge of any other financial obligation in connection with, any sum, which the Body borrows from any person.

If any sums are paid out in fulfilment of a guarantee under this article, the Body must make to the Welsh Ministers, at such times and in such manner as they from time to time direct:

(a) payments of such amounts as they direct in or towards repayment of the sums so paid out, and
(b) payments of interest, at such rate as they direct, on what is outstanding for the time being in respect of sums so paid out.

The aim of the Natural Resources of Wales is to deliver the best possible environmental, social and economic outcomes for people and communities in Wales. They do this by education, by providing advice and by regulating activities. Their main approach is to provide clear advice and guidance and by using enforcement powers including civil sanctions and prosecution.

The Natural Resources for Wales have extensive functions including the management of Wales' forests and woodlands, pollution control, waste regulation, the management of water resources, flood and coastal risk management, fisheries, navigation and the safeguarding of protected sites and species. They also regulate other activities from recreation to the control of plant diseases and the protection of rare species and emissions form industrial processes.

Approach to Enforcement

10.70 Enforcement is any action taken where it is suspected that an offence has occurred or in some cases is about to occur. This action can range from providing advice and guidance, serving notices through to prosecution, or any combination that best achieves the desired outcome.

Under the Environmental Damage (Prevention and Remediation) Regulations 2009[11], enforcement action may be required in the absence of any suspected offence, but in most cases an activity that leads to 'environmental damage' under the Regulations will also constitute an offence under environmental legislation.

Sanctions

10.71 Where an offence has been committed and the delivery of advice and guidance has not or will not achieve the necessary outcome, then the issuing of a form of sanction is considered as well as taking any other preventative or remedial action necessary to protect the environment or people.

Enforcement Options

10.72 This includes:

- providing advice and guidance;
- issuing a warning;
- statutory enforcement notices and works notices;
- prohibition notices;
- suspension or revocation of environmental permits;
- variation of permit conditions;
- injunctions;
- carrying out remedial works;
- civil sanctions;
- other civil and financial sanctions including Fixed Penalty Notices;
- issuing a formal caution;
- prosecution and orders ancillary to prosecution;
- sanctions used in combination.

Cost Recovery

10.73 Where the law allows then Natural Resources Wales will always seek to recover the costs of investigation and enforcement proceedings. Also where remedial works have been carried out then they will seek to recover the full costs incurred from those responsible under the 'polluter pays' principle.

11 SI 2009/995 (W 81).

Penalty Principles

10.74 When the course of action is considered, this follows the Macrory Penalty Principles set out in the regulators' Compliance Code[12]. A sanction should:

- aim to change the behaviour of the offender;
- aim to eliminate any financial gain or benefit from compliance;
- be responsive and consider what is appropriate for the particular offender and the regulatory issue, which can include punishment and the public stigma that should be associated with a criminal conviction;
- be proportionate to the nature of the offence and the harm caused;
- aim to restore the harm caused by regulatory non-compliance, where appropriate.

Regulators Compliance Code

10.75 They take account of the Regulators' Compliance Code when devising and implementing regulatory policies and systems. However those provisions do not apply directly to individual cases. The requirements of the Code do not apply in certain situations, for example:

- where it is necessary to respond immediately to either prevent or respond to a serious breach or incident;
- where following the provisions would defeat the purpose of the proposed enforcement action;
- where there are reasonable grounds to suspect that a serious criminal offence has been committed, in particular where such offending would have a damaging effect on legitimate business and desirable regulatory outcomes.

Proportionality

10.76 Some incidents or breaches of regulatory requirements cause, or have the potential to cause, serious environmental damage. Others may interfere with people's enjoyment or rights, or the ability of Natural Resources Wales to carry out its activities. The first response will be to prevent harm to people and the environment from occurring or continuing. Any enforcement action taken by Natural Resources Wales will be proportionate to the risks posed to people and the environment and also to the seriousness of the breach of the law and its impact on legitimate business.

12 Regulatory Justice: Making Sanctions Effective Final report November 2006 http://webarchive.nationalarchives.gov.uk/20121212135622/http:/www.bis.gov.uk/files/file44593.pdf.

Consistency

10.77 This means taking a similar approach in similar circumstances to achieve similar ends.

Discretion

10.78 Officers will need to take account of many variables:

- the scale of environmental impact;
- the attitude and actions of individuals and managers of businesses;
- the history of previous incidents or breaches.

Decisions on enforcement action are a matter of professional judgment and the officers of Natural Resources Wales need to be able to exercise discretion. Natural Resources Wales will continue to develop arrangements to promote consistency, including effective ways to liaise with other enforcing authorities.

Transparency

10.79 Transparency means helping regulated entities and others to understand what is expected of them and what they should expect from Natural Resources Wales. It also means making clear why an officer intends to take, or has taken, enforcement action.

Targeting

10.80 By targeting regulatory effort it means that it is directed primarily towards those whose activities give rise to or create the risk of serious environmental damage, where the risks are least well controlled, or against deliberate or organised crime.

Risk Based Approach

10.81 The risk-based approach of Natural Resources Wales allows them to prioritise their inspection activity. This includes categorisation of non-compliances at permitted sites and categorisation of other incidents based on an assessment of risk to and actual impact on the environment. They will give high priority to investigations that involve organised crime, overtly criminal activity, substantial profit, threats of violence or other aggravating factors.

Regulated Industry

10.82 Where regulated industries are involved then management actions are important. Repeated incidents or a series of related breaches of regulatory requirements may indicate an unwillingness to change behaviour, or an inability to achieve sufficient control. These may require a review of the regulatory requirements, the ability of the operator to run the site and/or additional investment. Natural Resources Wales recognise that a poorly managed relatively low hazard site or activity, can present a greater risk to the environment than a higher hazard site or activity, where proper control measures are in place.

SCOTLAND

10.83 In Scotland, the Scottish Environment Protection Agency purpose is set out in the Regulatory Reform (Scotland) Act 2014 and is to ensure that Scotland's environment is protected and improved and to ensure that the natural resources are managed in a sustainable way. SEPA's main statutory functions, duties and powers are set out in the Environment Act 1995. They have been given additional functions under legislation, which has been enacted since SEPA was formed.

The main statutory functions are:

- the regulation of activities or processes that may pollute water, air or land;
- the regulation of waste storage, transport, treatment and disposal;
- the regulation of keeping and disposing of radioactive materials;
- running Scotland's flood warning systems;
- participating as a statutory consultee to the planning system.

SEPA also has other principal responsibilities including:

- monitoring, analysing and reporting on the state of Scotland's environment helping implement the Zero Waste Strategy;
- controlling, in conjunction with the Health and Safety Executive, the risk of major accidents at industrial sites;
- operating the Scottish component of the Radioactive Incident Monitoring Network;
- providing advice to the SG in a variety of policy areas.

Regulatory Reform (Scotland) Act 2014

10.84 Part 3, Chapter 5 defines the general purpose of SEPA.

(1) SEPA is to carry out the functions conferred on it by or under this Act or any other enactment for the purpose of protecting and improving the environment (including managing natural resources in a sustainable way).
(2) In carrying out its functions for that purpose SEPA must, except to the extent that it would be inconsistent with subsection (1) to do so, contribute to—

 (a) improving the health and well being of people in Scotland, and
 (b) achieving sustainable economic growth.

(3) In subsection (1), 'enactment' includes an enactment comprised in, or in an instrument made under, an Act of the Scottish Parliament.

The Environmental Regulation (Enforcement Measures) (Scotland) Order 2015, SSI 2015/383

10.85 The instrument is made by the Scottish Ministers in exercise of the powers conferred by ss 20 to 32, 53, 58(1), and 59 of the Regulatory Reform (Scotland) Act 2014 ('The Act') and s 107 of the Courts Reform (Scotland) Act 2014 and all other powers enabling them to do so. The instrument is subject to the affirmative procedure.

Policy Objectives

10.86 The Act's primary aim is to introduce various measures to improve the way regulation is developed and applied, creating more favourable business conditions in Scotland and delivering benefits for the environment.

Part 3 of the Act provides a simpler legislative framework so that the Scottish Environment Protection Agency ('SEPA') can be more transparent, accountable, proportionate, consistent and targeted in carrying out its regulatory functions. This will enable SEPA to better identify and focus most effort on, the most important environmental risks and harms, ensuring more effective and efficient protection of the environment and reducing the regulatory burden on business.

The Order provides SEPA with the power to impose fixed monetary penalties (FMPs) and variable monetary penalties (VMPs) and to accept enforcement undertakings to better tackle environmental non-compliance, to better protect communities and to help create a level playing field for business. These enforcement measures are designed to offer SEPA a more proportionate alternative set of tools for dealing with 'relevant offences'.

In particular, the Order enables SEPA to impose FMPs or VMPs where SEPA is satisfied on the balance of probabilities that the person has committed an offence, which is a 'relevant offence' for those purposes.

There are a variety of penalties and undertakings that may be imposed or accepted under the Order where SEPA determines that an offence has been committed. The Order sets out how the penalties, undertakings and criminal proceedings interact, as required by the Act. The general policy aim is that when a decision is made by SEPA to use a particular enforcement measure for an offence, then criminal proceedings cannot be taken and SEPA cannot subsequently use another enforcement measure.

If criminal proceedings are taken for an offence, SEPA cannot subsequently use an enforcement measure. Where a different or ongoing offence is committed, however, criminal proceedings may be taken in respect of that separate offence.

If more than one person is involved in the act or omission giving rise to the offence then criminal proceedings can be taken, or enforcement measures issued, against each person separately.

The measures will be supported by new enforcement guidance, which is required to be issued under the Order, and a revised enforcement policy from SEPA. They will also be underpinned by guidelines from the Lord Advocate issued in accordance with s 31 of the Act, which will ensure that the new enforcement measures are applied consistently and proportionately as part of the whole range of sanctions that are available (including prosecution).

Fixed Monetary Penalties

10.87 FMPs are primarily intended to be used for lower level offences, particularly those where the circumstances of the offence are relatively consistent (eg, there is little scope for different levels of harm caused to be taken into account), there is little or no actual environmental impact and where a low level penalty is likely to be sufficient to change the offender's behaviour.

The Order sets the amount of an FMP for each 'relevant offence' at £300, £600 or £1,000, as the case may be, in Sch 1 and Sch 4. The Order also provides for a late payment penalty of 40%, which will be imposed if the penalty is not paid on time, to incentivise timely payment. These penalty amounts and the late payment penalty, were consulted on in 2014 and there was majority support for the three proposed penalty levels.

Variable Monetary Penalties

10.88 VMPs are intended for use in relation to offences where the circumstances of the offence are variable (eg, the type and level of harm caused, the behaviour of the offender prior to and after the offence or whether a financial advantage has been gained by the offender as a result of the offence) and need to be taken into account in delivering a fair and proportionate sanction.

The Order sets the maximum amount of a VMP in Sch 2 at either the maximum fine that may be imposed on summary conviction (for offences triable summarily (whether or not also triable on indictment) and punishable on summary conviction with a fine) for the relevant offence, or £40,000 (in any other case). The Order also provides for late payment interest of 8% per annum to apply if the penalty is not paid on time, to incentivise timely payment.

Schedule 2 also enables SEPA to require a person who has been given a variable monetary penalty to pay the costs it has incurred in imposing the penalty. This is in line with the 'polluter pays' principle. Where someone has committed an offence and SEPA has imposed a variable penalty, then the policy objective is to ensure that the offender, rather than the public purse, contributes to the costs associated with the investigation.

Schedule 2 also provides that SEPA may accept a VMP undertaking. This allows a person to discharge criminal liability for a 'relevant offence' voluntarily without a final penalty notice being imposed by SEPA. Such undertakings involve a person agreeing to take specified actions to restore the position to what it would have been if an offence had not been committed, to benefit the environment where harm has been caused or to ensure that no financial benefit arises from the commission of an offence, or all of the above.

If a person fails to comply with a VMP undertaking, the Order allows SEPA to impose a non-compliance penalty of 140% of the VMP amount which would have been imposed unless there are exceptional circumstances justifying a lower amount (paras 14 to 16 of Sch 2) to incentivise compliance with the undertaking. The proposed uplift of 140% was consulted on in 2014 and the majority of respondents considered that to be a sufficient penalty for non-compliance with a VMP undertaking.

Enforcement Undertakings

10.89 Schedule 3 provides that SEPA may accept an enforcement undertaking. These are voluntary undertakings to take certain types of action, for example to secure that an offence does not recur, or to restore, benefit or improve the environment.

Where SEPA accepts an undertaking, this allows a person to discharge criminal liability for a 'relevant offence' voluntarily, unless the person subsequently fails to comply with the undertaking and SEPA issues a certificate of non-compliance.

Relevant Offences

10.90 Schedule 4 lists the offences, which are to be considered 'relevant offences' in relation to which SEPA may use fixed monetary penalties, variable monetary penalties (and related measures) and enforcement undertakings. It also sets out the amount of a fixed monetary penalty in relation to each offence.

Appeals

10.91 In response to stakeholder desire for an appeals route independent of the Scottish Ministers, the Order provides that appeals against enforcement measures are to the Scottish Land Court (SLC).

To provide access to justice for appellants, the Order allows the SLC to determine appeals without a formal hearing where appropriate or where both parties agree. This provision is designed to encourage a 'light touch' appeals process, which does not require the engagement of legal representation on either side and which, as a consequence, is both quicker and less costly to both appellants and SEPA. An applicant may still request a hearing, although the SLC is not obliged to grant one if it considers that it can justly determine the matter without a hearing. If granted, however, a daily fee applies to each hearing day. The Order exempts appeals under the Order from the fee to lodge an appeal usually applied by the SLC.

The Order also provides that expenses may not be awarded in relation to an appeal except in relation to any court fees paid, or where the court considers that a party has acted unreasonably. This is principally to ensure that potential appeals (particularly those from small businesses or sole traders) are not discouraged by appellants being held liable for expenses incurred by SEPA should they be unsuccessful in relation to an appeal.

SEPA's Powers of Enforcement

Fixed Monetary Penalties

10.92 SEPA may impose a fixed monetary penalty on a person in relation to a relevant offence.

The fixed monetary penalty:

- may be imposed on a person only where SEPA is satisfied on the balance of probabilities that the person has committed the offence to which the penalty relates;
- is to be imposed by notice; and
- may not be imposed on a person in relation to an offence constituted by an act or omission if a fixed monetary penalty has already been imposed on that person in respect of the same offence constituted by the same act or omission.

The 'fixed monetary penalty' is a requirement to pay SEPA a penalty amount specified in an order made and imposed by SEPA for a relevant offence.

The maximum amount of such a penalty that may be specified in relation to a particular offence is an amount equivalent to level 4 on the standard scale.

'The standard scale' has the meaning given by s 225(1) of the Criminal Procedure (Scotland) Act 1995.

Fixed Monetary Penalties: Procedure

10.93 Where a fixed monetary penalty is issued that:

- must secure 'the mandatory results';
- may secure 'the optional result'.

The mandatory results are that:

- where SEPA proposes to impose a fixed monetary penalty on a person, it must serve on the person a notice of what is proposed (a 'notice of intent');

- except where the person has discharged liability the person may make written representations to SEPA in relation to the proposed imposition of the fixed monetary penalty (including that the person would not, by reason of any defence, be liable to be convicted of the offence to which the penalty relates);
- SEPA must, after the end of the period for making representations, decide whether to impose the fixed monetary penalty;
- SEPA must, in so deciding, have regard to any representations;
- where SEPA decides to impose the fixed monetary penalty, the notice imposing it ('the final notice') complies with subsection (5); and
- the person on whom a fixed monetary penalty is imposed may appeal against the decision to impose it.

The optional result is that the notice of intent also offers the person the opportunity to discharge their liability for the fixed monetary penalty by payment of a sum specified in the notice of intent (which sum must be less than or equal to the amount of the penalty).

The notice of intent must include information as to:

- the grounds for the proposal to impose the fixed monetary penalty;
- the right to make written representations;
- the period within which representations may be made;
- where provision is made as an optional result:
    - how payment to discharge the liability for the fixed monetary payment may be made,
    - the period within which liability for the fixed monetary penalty may be discharged, and
    - the effect of payment of the sum referred to in subsection (3).

The final notice must include information as to:

- the grounds for imposing the penalty;
- how payment may be made;
- the period within which payment must be made;
- any early payment discounts or late payment penalties;
- rights of appeal; and
- the consequences of non-payment.

A person may appeal against a decision of SEPA.

They should:

- include the grounds that:
    - the decision was based on an error of fact,
    - the decision was wrong in law, and
    - the decision was unreasonable, but
- do not include the ground that SEPA failed to comply with guidance issued to it by the Lord Advocate under s 31(1).

Fixed Monetary Penalties: Effect on Criminal Proceedings etc

10.94 Provision under s 20 (Fixed Monetary Penalties) must ensure that when a fixed penalty notice has been served on a person:

- no criminal proceedings may be commenced against the person in respect of that offence constituted by that act or omission:
- before the end of any period in which the person may discharge liability for the fixed monetary penalty pursuant to s 21(3), or
- if the person so discharges liability, and
- the period as mentioned in subsection (2) is not to be counted in calculating any period within which criminal proceedings in respect of that offence constituted by that act or omission must be commenced.

The period is that beginning with the day on which the notice of intent is served and ending with the day, which is the final day on which written representations may be made in relation to the notice.

Provision under s 20 must also secure that, in a case where a fixed monetary penalty is imposed on a person in respect of an offence constituted by an act or omission:

- no criminal proceedings may be commenced against the person in respect of that offence constituted by that act or omission.
  - before the end of any period in which the person may discharge liability for the fixed monetary penalty pursuant to s 21(3), or
  - if the person so discharges liability, and
- the period as mentioned below is not to be counted in calculating any period within which criminal proceedings in respect of that offence constituted by that act or omission must be commenced.

The period is that beginning with the day on which the notice of intent is served and ending with the day which is the final day on which written representations may be made in relation to the notice.

Provision under s 20 must also secure that, in a case where a fixed monetary penalty is imposed on a person in respect of an offence constituted by an act or omission, no criminal proceedings may be commenced against the person in respect of that offence constituted by that act or omission.

The references to criminal proceedings being commenced are to be read as if they included references to:

(a) a warning being given by the procurator fiscal;
(b) a conditional offer (within the meaning of s 302 of the Criminal Procedure (Scotland) Act 1995) being sent;
(c) a compensation offer under s 302A of that Act being sent;
(d) a combined offer under s 302B of that Act being sent; and
(e) a work order under s 302ZA of that Act being made.

Variable Monetary Penalties

10.95 The Scottish Ministers may by Order make provision for or about the imposition by SEPA of a variable monetary penalty on a person in relation to a relevant offence.

Provision must provide that a variable monetary penalty:

(a) may be imposed on a person only where SEPA is satisfied on the balance of probabilities that the person has committed the offence to which the penalty relates;
(b) is to be imposed by notice; and
(c) may not be imposed on a person in relation to an offence constituted by an act or omission if a variable monetary penalty has already been imposed on that person in respect of the same offence constituted by the same act or omission.

For the purposes of this Chapter, a 'variable monetary penalty' is, subject to the subsection below, a requirement to pay SEPA a penalty of such amount as SEPA may in each case determine.

SEPA may not in any case impose a variable monetary penalty that exceeds the maximum amount specified in an order made under subsection (1) in relation to that case.

The maximum amount that may be so specified is:

(a) in the case mentioned in subsection (6), the maximum amount of the fine that may be imposed on summary conviction in such a case;
(b) in any other case, £40,000.

The case is one where the offence in respect of which the variable monetary penalty is imposed:

(a) is triable summarily (whether or not it is also triable on indictment); and
(b) is punishable on summary conviction by a fine (whether or not it is also punishable by a term of imprisonment).

The Scottish Ministers may by order substitute another sum for the one for the time being mentioned as £40,000.

Variable Monetary Penalties: Procedure

10.96 (1) Provision under s 23 must secure the results in subsection (2).

(2) The results are that—

(a) where SEPA proposes to impose a variable monetary penalty on a person, it must serve on the person a notice (a 'notice of intent') which complies with subsection (3);
(b) the person may make written representations to SEPA in relation to the proposed imposition of the variable monetary penalty (including that the

person would not, by reason of any defence, be liable to be convicted of the offence to which the penalty relates);
(c) SEPA must, after the end of the period for making such representations, decide whether to impose a variable monetary penalty and, if so, the amount of the penalty;
(d) SEPA must, in so deciding, have regard to any representations;
(e) where SEPA decides to impose a variable monetary penalty, the notice imposing it (the 'final notice') complies with subsection (4); and
(f) the person on whom a variable monetary penalty is imposed may appeal against the decision as to the imposition or amount of the penalty.

(3) To comply with this subsection the notice of intent must include information as to—

(a) the grounds for the proposal to impose the variable monetary penalty;
(b) the right to make written representations; and
(c) the period within which representations may be made.

(4) To comply with this subsection the final notice must include information as to:

(a) the grounds for imposing the penalty;
(b) how payment may be made;
(c) the period within which the payment must be made;
(d) any early payment discounts or late payment penalties;
(e) rights of appeal; and
(f) the consequences of non-payment.

(5) Provision to secure the result must include provision for:

(a) the person on whom the notice of intent is served to be able to offer an undertaking as to action to be taken by that person, within such period as may be specified in the undertaking, for all or any of the following purposes—

    (i) to secure that the position is, so far as possible, restored to what it would have been if the offence had not been committed;
    (ii) to benefit the environment to the extent that the commission of the offence has harmed the environment;
    (iii) to secure that no financial benefit arising from the commission of the offence accrues to the person,

(b) SEPA to be able to accept or reject such an undertaking; and
(c) SEPA to take any undertaking so accepted into account in its decision.

(6) Provision to secure the result that the grounds on which a person may appeal against a decision of SEPA:

(a) include the grounds that:

    (i) the decision was based on an error of fact;
    (ii) the decision was wrong in law;
    (iii) the amount of the penalty is unreasonable; and
    (iv) the decision was unreasonable for any other reason; but

(b) do not include the ground that SEPA failed to comply with guidance issued to it by the Lord Advocate under s 31(1).

Variable Monetary Penalties: Effect on Criminal Proceedings etc

10.97 Provision under s 23 must secure the result in a case where:

(a) either—

  (i) a variable monetary penalty is imposed on a person; or
  (ii) an undertaking referred to in s 24(5) is accepted from a person; or

(b) both such a penalty is imposed on, and such an undertaking is accepted from, a person.

The result is that no criminal proceedings may be commenced against the person for an offence constituted by an act or omission if the variable monetary penalty or, as the case may be, the undertaking related to that offence constituted by that act or omission.

Provision under s 23 must provide that the period mentioned in subsection (4) is not to be counted in calculating any period within which criminal proceedings in respect of an act or omission in relation to which a notice of intent under s 24(2)(a) is served must be commenced.

The period is that beginning with the day on which the notice of intent is served and ending with the day which is the final day on which written representations may be made in relation to the notice.

The reference to criminal proceedings being commenced is to be read as if it included a reference to:

(a) a warning being given by the procurator fiscal;
(b) a conditional offer (within the meaning of s 302 of the Criminal Procedure (Scotland) Act 1995) being sent;
(c) a compensation offer under s 302A of that Act being sent;
(d) a combined offer under s 302B of that Act being sent; and
(e) a work order under s 302ZA of that Act being made.

Undertakings under section 24: non-compliance penalties

10.98 Provision under s 23 may include provision for a person to pay a monetary penalty (in this Part, a 'non-compliance penalty') to SEPA if the person fails to comply with an undertaking referred to in s 24(5) which is accepted from the person.

Where such provision is included, it may also:

(a) specify the amount of the non-compliance penalty;
(b) provide for the amount to be calculated by reference to criteria specified by order by the Scottish Ministers;

(c) provide for the amount to be determined by SEPA (subject to any maximum amount set out in the provision);
(d) provide for the amount to be determined in any other way.

Where provision is included as mentioned, it must provide that the maximum amount of the non-compliance penalty that may be imposed in any case is not to exceed the maximum amount of the variable monetary penalty to which the non-compliance penalty relates in such a case.

Where provision is included as mentioned in subsection (1), it must secure that:

(a) the non-compliance penalty is imposed by notice served by SEPA; and
(b) the person on whom it is imposed may appeal against the notice.

Provision must secure that the grounds on which a person may appeal against a notice referred to in that subsection include that—

(a) the decision to serve the notice was based on an error of fact;
(b) the decision was wrong in law;
(c) the decision was unreasonable for any reason (including, in a case where the amount of the non-compliance penalty was determined by SEPA, that the amount is unreasonable).

Enforcement Undertakings

10.99 The Scottish Ministers may by order make provision:

(a) for or about enabling SEPA to accept an enforcement undertaking from a person in a case where SEPA has reasonable grounds to suspect that the person has committed a relevant offence; and
(b) for the acceptance of the undertaking to have the consequences in subsection (4).

For the purposes of this Chapter, an 'enforcement undertaking' is an undertaking to take action of a type mentioned in subsection (3) and specified in the undertaking within such period as may be so specified.

The types of action are:

(a) action to secure that the offence does not continue or recur;
(b) action to secure that the position is, so far as possible, restored to what it would have been if the offence had not been committed;
(c) action of a description specified by order by the Scottish Ministers.

The consequences are that, unless SEPA has certified under provision made in pursuance of subsection (5)(g) that the person from whom the enforcement undertaking is accepted has not complied with the undertaking or any part of it:

(a) no criminal proceedings may be commenced against the person from whom the enforcement undertaking is accepted in respect of an offence constituted by an act or omission if the undertaking relates to that offence constituted by that act or omission;

(b) SEPA may not impose on the person a fixed monetary penalty which it would otherwise have power to impose by virtue of s 20 in respect of the act or omission; and
(c) SEPA may not impose on the person a variable monetary penalty which it would otherwise have power to impose by virtue of s 23 in respect of the act or omission.

An order under this section may in particular include provision:

(a) as to the procedure for entering into an enforcement undertaking;
(b) as to the terms of an enforcement undertaking;
(c) as to publication of an enforcement undertaking by SEPA;
(d) as to variation of an enforcement undertaking;
(e) as to circumstances in which a person may be regarded as having complied with an enforcement undertaking;
(f) as to monitoring by SEPA of compliance with an enforcement undertaking;
(g) as to certification by SEPA that an enforcement undertaking or any part of it has not been complied with;
(h) for appeals against such certification;
(i) in a case where a person has given inaccurate, misleading or incomplete information in relation to an enforcement undertaking, for that person to be regarded as not having complied with it;
(j) in a case where a person has complied partly but not fully with an enforcement undertaking, for that partial compliance to be taken into account in the imposition of any criminal or other sanction on the person;
(k) for the purpose of enabling criminal proceedings in respect of an act or omission in relation to which SEPA has accepted an enforcement undertaking to be commenced against a person who has not complied with the undertaking or any part of it, for the period mentioned in subsection (6) not to be counted in calculating any period within which such proceedings must be commenced.

The period is that beginning with the day on which the enforcement undertaking is accepted and ending with:

(a) the day on which SEPA certifies, under provision made in pursuance of subsection (5)(g), that the undertaking or any part of it has not been complied with; or
(b) where an appeal against such a certification is taken, the day on which the appeal is finally determined.

The reference in subsection (4)(a) to criminal proceedings being commenced is to be read as if it included a reference to:

(a) a warning being given by the procurator fiscal;
(b) a conditional offer (within the meaning of s 302 of the Criminal Procedure (Scotland) Act 1995) being sent;
(c) a compensation offer under s 302A of that Act being sent;
(d) a combined offer under s 302B of that Act being sent; and
(e) a work order under s 302ZA of that Act being made.

References in this section to taking action specified in an enforcement undertaking include references to refraining from taking such action.

Combination of Sanctions

10.100 Provision may not be made by order under s 20 and s 23 conferring powers on SEPA in relation to the same offence unless it secures that:

(a) SEPA may not serve a notice of intent referred to in s 21(2)(a) on a person in relation to an act or omission where a variable monetary penalty has been imposed on that person in relation to the act or omission; and
(b) SEPA may not serve a notice of intent referred to in s 24(2)(a) on a person in relation to any act or omission where—

  (i) a fixed monetary penalty has been imposed on the person in relation to the act or omission; or
  (ii) the person has discharged liability for a fixed monetary penalty in relation to that act or omission pursuant to s 21(3).

Provision under s 20 must secure that in a case where a notice of intent referred to in s 21(2)(a) is served on a person—

(a) SEPA may not, before the end of any period in which the person may discharge liability to the fixed monetary penalty pursuant to s 21(3), impose a variable monetary penalty on the person in respect of the act or omission to which the notice relates; and
(b) SEPA may not, if the person so discharges liability, impose a variable monetary penalty on the person in respect of that act or omission.

Provision under s 20 must also secure that in a case where a fixed monetary penalty is imposed on a person, SEPA may not impose a variable monetary penalty on the person in respect of the act or omission giving rise to the penalty.

Provision under s 20 must also secure the result that a fixed monetary penalty in respect of an offence constituted by an act or omission may not be imposed on a person if, in respect of that offence as constituted by that act or omission:

(a) criminal proceedings have been commenced against the person
(b) the person has been given a warning by the procurator fiscal;
(c) the person has been sent a conditional offer (within the meaning of s 302 of the Criminal Procedure (Scotland) Act 1995 (fixed penalty: conditional offer by procurator fiscal));
(d) the person has accepted, or is deemed to have accepted, a compensation offer issued under s 302A of that Act (compensation offer by procurator fiscal);
(e) the person has accepted, or is deemed to have accepted, a combined offer issued under s 302B of that Act; or
(f) a work order has been made against the person under s 303ZA of that Act (work orders).

Provision under s 23 must also secure the result that a variable monetary penalty in respect of an offence constituted by an act or omission may not be imposed on a person if, in respect of that offence as constituted by that act or omission:

(a) criminal proceedings have been commenced against the person;
(b) the person has been given a warning by a procurator fiscal;
(c) the person has been sent a conditional offer (within the meaning of s 302 of the Criminal Procedure (Scotland) Act 1995 (fixed penalty: conditional offer by procurator fiscal));
(d) the person has accepted, or is deemed to have accepted, a compensation offer issued under s 302A of that Act (compensation offer by procurator fiscal);
(e) the person has accepted, or is deemed to have accepted, a combined offer issued under s 302B of that Act; or
(f) a work order has been made against the person under s 303ZA of that Act (work orders).

Monetary Penalties

10.101 An order under this Chapter which confers power on SEPA to require a person to pay a fixed monetary penalty, a variable monetary penalty or a non-compliance penalty may include provision for:

(a) early payment discounts;
(b) the payment of interest or other financial penalties for late payment of the penalty (such interest or other financial penalties not in total to exceed the amount of the penalty);
(c) enforcement of the penalty.

Where such provision is included, it may also provide for:

(a) SEPA to recover the penalty, and any interest or other financial penalty for late payment, as a civil debt;
(b) the penalty and any interest or other financial penalty for late payment, to be recoverable as if it were payable under an extract registered decree arbitral bearing a warrant for execution issued by a sheriff of any sheriffdom.

Costs Recovery

10.102 Provision under s 23 may include provision for SEPA to require a person on whom a variable monetary penalty is imposed to pay the costs incurred by SEPA in relation to the imposition of the penalty up to the time of its imposition.

Where such provision is included, it must secure that:

(a) a requirement to pay the costs is imposed by notice;

(b) the notice specifies the amount required to be paid;
(c) SEPA may be required to provide a detailed breakdown of the amount;
(d) the person required to pay costs may appeal against—
    (i) the decision of SEPA to impose the requirement;
    (ii) the decision of SEPA as to the amount of the costs (including that some or all of the costs were unnecessarily incurred);
(e) SEPA is required to publish guidance about how it will exercise the power conferred by the provision.

In subsection (1), the references to costs include in particular—

(a) investigation costs;
(b) administration costs;
(c) costs of obtaining expert advice (including legal advice).

Subsections (1)(b) and (c) and (2) of s 29 apply to costs required to be paid by virtue of subsection (1) of this section as they apply to a fixed monetary penalty, a variable monetary penalty or a non-compliance penalty.

Guidance as to Use of Enforcement Measures

10.103 The Lord Advocate may issue and from time to time revise, guidance to SEPA on the exercise of its functions relating to enforcement measures.

SEPA must comply with such guidance or revised guidance in exercising those functions.

In this section, an 'enforcement measure' means a fixed monetary penalty, variable monetary penalty or enforcement undertaking (and any references in this Chapter to the imposition of an enforcement measure include acceptance of an enforcement undertaking).

Where power is conferred on SEPA by virtue of this Chapter to impose an enforcement measure in relation to an offence, the provision conferring the power must secure the results in the subsection below.

The results are that—

(a) SEPA must publish guidance about—
    (i) how the offence is enforced;
    (ii) the sanctions (including criminal sanctions) to which a person who commits the offence may be liable;
    (iii) the action which SEPA may take to enforce the offence, whether by virtue of this Chapter or otherwise;
    (iv) the circumstances in which SEPA is likely to take any such action;
    (v) SEPA's use of the enforcement measure.
(b) in the case of guidance relating to a fixed monetary penalty or variable monetary penalty, the guidance must contain the relevant information; and

(c) SEPA must have regard to the guidance in exercising its functions.

In the case of guidance relating to a fixed monetary penalty, the relevant information referred to in subsection (5)(b) is information as to—

(a) the circumstances in which the penalty is likely to be imposed;
(b) the circumstances in which it may not be imposed;
(c) the amount of the penalty;
(d) how liability for the penalty may be discharged and the effect of discharge; and
(e) rights to make representations and rights of appeal.

In the case of guidance relating to a variable monetary penalty, the relevant information referred to in subsection (5)(b) is information as to—

(a) the circumstances in which the penalty is likely to be imposed;
(b) the circumstances in which it may not be imposed;
(c) the matters likely to be taken into account by SEPA in determining the amount of the penalty (including, where relevant, any discounts for voluntary reporting of non-compliance); and
(d) rights to make representations and rights of appeal.

SEPA may from time to time revise guidance published by it by virtue of the subsection above.

The references in subsections (5) to (7) to guidance include references to any revised guidance

Before publishing any guidance or revised guidance by virtue of this section, SEPA must consult:

(a) the Lord Advocate; and
(b) such other persons as it considers appropriate.

Publication of Enforcement Action

10.104 Subsection (2) applies where the Scottish Ministers make provision by order under:

(a) section 20 as to the imposition by SEPA of a fixed monetary penalty;
(b) section 23 as to the imposition by SEPA of a variable monetary penalty; or
(c) section 27 as to the acceptance by SEPA of an enforcement undertaking.

The order may require SEPA to publish such information as may be specified in the order as regards cases in which it has done what the order permits it to do.

Interpretation of Chapter 2

10.105 In this Chapter:

'early payment discounts' means early payment discounts included in an order under this Chapter by virtue of s 29(1);

'enforcement undertaking' has the meaning given in s 27;

'fixed monetary penalty' has the meaning given in s 20;

'late payment penalties' means a requirement to pay interest or other financial penalties for late payment of a fixed monetary penalty, a variable monetary penalty or a non-compliance penalty included in an order under this Chapter by virtue of s 29(1);

'non-compliance penalty' has the meaning given in s 26(1);

'variable monetary penalty' has the meaning given in s 23.

NORTHERN IRELAND

10.106 The Northern Ireland Environment Agency (NIEA) was formed in 1996 as an executive agency of the DoE. It has a wide range of responsibilities including protecting and improving the natural environment and built heritage.

The Department of the Environment comprises a number of central policy divisions and three Executive Agencies – The Environment and Heritage Service (EHS), the Planning Service (PS) and the Driver and Vehicle Agency (DVA). The work of the department is monitored by the Department Board involving senior DoE civil servants and the Chief Executives and Agencies.

The separation of operational enforcement from policy is done by maintaining a policy function within the Department. It is the DoE, which takes the lead in developing and implementing new policies. The passing of legislation has provided new responsibilities and powers meaning that the Department and its' agencies are now interfacing more directly with the main criminal justice organisations.

A review of the Northern Ireland Environment Agency's Environmental Crime Unit, was carried out in 2015 by the Chief Inspector of the Criminal Justice System in Northern Ireland.

The Environmental Crime Unit was established in late 2008 to deal with waste crime and has 37.4 full-time equivalent staff which represents some 5.4% of overall staffing and an operating budget of £1.7m. The report considered the effectiveness of the Environmental Crime Unit within the Northern Ireland Environment Agency and the wider context of regulation and enforcement.

This review takes account of previous wider inspections which touched on the work of the NIEA, which were published by CJI in 2007 and 2011. A further DoE commissioned review, which covered matters of enforcement and the wider strategies of the NIEA was published in 2014 and included a group of recommendations in this area. This review sought to build upon the findings and recommendations of these inspections and reviews, whilst focusing in particular on the operational aspects of the ECU.

Waste Crime

10.107 Waste disposal is currently subject to stringent regulation and often involves substantial costs in the disposal of both general and specialised waste. This creates a market for the illegal disposal at lower prices and without any regulation. The high profits which can arise, have attracted Organised Crime Groups (OCGs). The nature of waste crime is adequately demonstrated by the discovery in 2012 of an estimated total of 516,000 tonnes of macerated waste in an area adjacent to the River Faughan in the townland of Mobuoy near Derry/Londonderry. It was deposited in an area stretching to almost 1.4km. This was a sophisticated operation, which had been carried out over a number of years. It continues to be the subject of a criminal investigation by the ECU. A further visible manifestation of the involvement of OCGs in waste crime in Northern Ireland concerns the by-product of fuel laundering in the form of hazardous waste; which is often dumped illegally with consequent environmental damage and clean-up costs for tax/rate payers.

The NIdirect website summarises the local position in Northern Ireland and states:

> 'Waste crime is the illegal management of waste. It is becoming an increasing problem in Northern Ireland and in particular, the dumping of fuel laundering waste, is a concern.
>
> Fuel laundering is an illegal process whereby marker dyes are removed from red and green diesel, which is then sold to motorists or fuel retailers. This process produces waste. The waste is typically disposed of in industrial bulk containers, which are big plastic cubes with metal cages around them. The laundered waste looks like a dark reddish brown sludge with dark coloured water on top and is very hazardous.
>
> Criminals are involved in illegal waste activities – from waste collection and the provision of skips to illegal landfilling and burning of the waste to get rid of it. Their motive is making money. They usually charge the legal going rate to collect and transport waste and then dispose of it illegally and pocket the profits.'

Illustrating the issue of clean-up costs, it is currently estimated that the costs of remediating the Mobuoy site is £800k and could rise to well over £100m. These are costs to the Northern Ireland tax payer.

The report found that there was a wider problem concerning the acceptance of non-compliance with the laws surrounding waste. This risks aiding a culture of contravention both for these and other laws.

This non-compliance culture also impacts on how strategy and policy for regulation and enforcement in this area is framed. While accepting that discretion is an important tool in regulation and enforcement, the underlying principle of full compliance and respect for the law should be clearly stated and emphasised by the DoE and the NIEA.

Specifically in a section on enforcement the Waste Management Strategy states:

> 'In adopting the principles of Better Regulation and taking a risk-based approach to how we regulate waste activities, a certain degree of responsibility is being placed upon the waste industry to do the right thing. When the regulator takes a step back from the lower risk activities, allowing businesses to "get on with the job", it is essential that unscrupulous operators do not undercut legitimate businesses and that the regulator has an effective suite of penalties and sanctions to be able to take swift and appropriate action. At the other end of the spectrum, serious criminality involving waste management has far reaching implications for the environment and human health; for the health of the legitimate waste industry; the economy and on Northern Ireland's susceptibility to incurring hefty EU infraction fines. A number of specific work strands are being taken forward to ensure that the mechanisms are in place to tackle the various levels of illegal activity.'

The Report found that the DoE has recognised the risks associated with illegal waste and noted:

> 'As a result of ineffective regulation of the waste sector and/or poor working relationships with councils, there is an increased risk of the dumping of illegal waste which will result in damage to the environment and the wider economy, and additional liabilities falling to public expenditure.'

The report found that one of the key actions was a strategy for combatting waste crime. Another was the development of effective working relationships with councils. Significantly, both were still in development at the time of inspection. The assessment of risk, particularly in the context of serious waste crime, could be strengthened by a comprehensive strategic analysis of waste offending.

The report fond that there was little recognition in the documents provided to Inspectors of an assessment of the risks surrounding closed landfill sites and other areas, for example, in the disposal of gypsum. The latter is potentially significant given that there is only one licensed recycling facility for Northern Ireland based in Downpatrick. There is a risk of unauthorised waste activity in this sector. The absence of a strategic assessment could itself be a risk, which leads to a weakened assessment of threat.

The report considered alternative enforcement models.

Waste Crime Enforcement Should be the Responsibility of Core Law Enforcement Agencies

10.108 This would see the transfer of all crime investigations from the NIEA to the Police Service of Northern Ireland (PSNI) and/or the National Crime Agency (NCA). The reports say that the main advantage would be the application of core enforcement methodologies where investigative skills, confidence and enforcement infrastructures could be brought to bear. However, this option is not considered feasible for a number of reasons, principally:

- the importance of waste crime enforcement would inevitably become diluted in competition with other law enforcement/policing priorities and resource pressures;
- the difficulties of the transfer of cases between silos within the DoE and NIEA are already a feature of inefficiencies and would be exacerbated by the introduction of a further agency with its own different priorities; and
- waste legislation can be complex and requires specialist knowledge.

The DoE Should Create a Larger Department Wide Enforcement Team

10.109 This option would see the creation of a much larger law enforcement directorate across the DoE and could incorporate a range of enforcement areas such as in the transport sector, with some staff specialising in bespoke areas of enforcement. The reports sees the advantagcs would bc that all criminal enforcement could be dealt with more efficiently in one division with a more consistent approach (subject to thresholds being consistently applied in the passage of cases for enforcement). Over a period of time such a unit could build the expertise required in investigation and enforcement.

Strategy and Governance

10.110 This option is ruled out for the present given resource constraints and again the difficulties inherent in the transfer of cases between agencies/ departments and business units. In addition, this model would dilute the skills of regulatory staff to escalate their work, where appropriate, to enforcement while maintaining continuity.

Maintain the Status Quo of an ECU

10.111 This is seen as the optimum operating model for the present time. The ECU is already well established and while there is room for improvement, its current positioning means that it is best placed to deliver future success. With some revisions to existing processes and wider alignments within the NIEA, overall efficiency and effectiveness can be maintained and further enhanced.

The report makes a number of strategic recommendations:

The ECU Strategy should be reviewed to reflect a balanced focus on strategic objectives, which are aligned to departmental, and agency strategies, and are performance focused and SMART.

The ECU should develop guidelines for levels of enforcement action, which should complement the Enforcement and Prosecution Policy and support staff in decision making in this regard.

Any suggestion that the law is negotiable should be avoided, as it was a concern that the existing Enforcement Policy itself provided areas of ambiguity. The Policy outlines that:

> 'We recognise that the initiation of a prosecution is a serious matter which should only be taken after appropriate consideration of the implications and consequences, one of which may be imprisonment.'

Strategic Recommendation 1

10.112 The ECU Strategy should be reviewed to reflect a balanced focus on strategic objectives, which are aligned to departmental, and agency strategies, and are performance focused and SMART.

Operational Recommendation 1:

10.113 The ECU should develop guidelines for levels of enforcement action, which should complement the Enforcement and Prosecution Policy and support staff in decision making in this regard.

The report found that the Policy outlines that:

> 'We recognise that the initiation of a prosecution is a serious matter which should only be taken after appropriate consideration of the implications and consequences, one of which may be imprisonment.'

They felt that this left the impression that prosecution would be the last resort. It also highlighted that the Policy also stated that cases will be referred to the Public Prosecution Service for Northern Ireland (PPS) where 'there are excessive or persistent breaches of regulatory requirements'. It was also concerned that the Policy included matters that would be taken into consideration which are stated to include:

> 'the attitude of the offender;
>
> the deterrent effect of prosecution, on the offender and others; and
>
> the personal circumstances of the offender.'

The report went on to say:

> 'Accepting that discretion is a valid tool in enforcement, this does not mean that the Enforcement Policy should be open to interpretation, by NIEA staff or by the regulated sector.'

Operational Recommendation 2

10.114 In its revision of the Enforcement and Prosecution Policy, the NIEA should ensure that there is clarity of approach and consistent messaging which removes ambiguity.

Operational Recommendation 3

10.115 Clear guidance should be provided on the rationale for prioritisation of investigations, their inclusion and/or linking to priority operations/investigations and their subsequent management.

Strategic Recommendation 2

10.116 The NIEA should conduct a strategic assessment of the waste sector in Northern Ireland, including illegality in the regulated and non-regulated sectors. This should incorporate the risk of transfer of waste from and to Northern Ireland.

Operational Recommendation 4

10.117 The ECU should seek to fill vacancies from candidates with specific core investigative skills. If these skills are not available within the Northern Ireland Civil Service pool, external recruitment should be considered.

Operational Recommendation 5

10.118 The NIEA should examine the feasibility of a single waste/pollution reporting mechanism.

Operational Recommendation 6

10.119 The weekly 'intelligence' meeting should be renamed to become the TTCG.

Operational Recommendation 7

10.120 A fundamental review of the continuing need for the LUGUS system should be undertaken. This review should examine:

- how the system could be retained and its value maximised by extending its use and access (subject to access level controls) across the NIEA; and
- whether there is continuing need to retain the system balanced against the costs and the need for the protection of data.

Operational Recommendation 8

10.121 A single incident and enforcement database should be developed.

Operational Recommendation 9

10.122 In addition a Central Evaluation Unit should be established and act as the central nervous system of the RED providing one of the principle governance mechanisms.

Operational Recommendation 10

10.123 The ECU should introduce improved systems in the following areas:

- Case management, which includes the call-up of incident reports and enforcement, files; and
- a formalised system of the investigative review of all enforcement files by a competent Manager at intervals to be agreed.

All these recommendations would seem to bring the regulatory enforcement process in Northern Ireland closer to the rest of the UK – time will tell how this will work.

Chapter 11

Public Registers

INTRODUCTION

11.1 The Environment Agency and the Scottish Protection Agency (SEPA) have public registers that provide environmental information. SEPA have an online register for 2014 onwards. Any requests for information from before that date should be emailed to: producer.responsibility@sepa.org.uk.

These registers contain information on applications made for licences by businesses and individuals who carry out activities that have the potential to harm the environment. The registers show how such licences are granted, any conditions that may be attached to that licence and how they are regulated and monitored. It also included details of any breaches of the licence conditions and enforcement action taken by the Environment Agency or SEPA against the licence holder. Natural Resources for Wales also hold environmental registers.

The main public register covers:

- environmental permits (industrial installations);
- environmental permits (waste operations);
- environmental permits (radioactive substances);
- environmental permits (water discharge);
- waste carriers and dealers (registrations before 27 March 2015);
- waste brokers and dealers (registrations before 27 March 2015);
- hazardous waste producers;

- waste exemptions;
- water discharge exemptions;
- enforcement actions;
- scrap metal dealers.

The Agency also holds registers for WEEE:

- EEE producers;
- WEEE approved producer compliance schemes;
- WEEE approved authorised treatment facilities and approved exporters;
- End of Life vehicle authorised treatment facilities;
- Packaging and batteries – held on the National Packaging Waste database;
- Large raised reservoirs – held on register of large raised reservoirs.

Types of Information

11.2 The types of information which are classified as environmental information under the Environmental Information Regulations 2004[1] or the Environmental Information (Scotland) Regulations 2004[2] include:

(a) Information on the state of elements of the environment.
This includes air and atmosphere, water, soil, land, landscape and natural sites including wetlands, coastal and marine areas, biological diversity and its components, including genetically modified organisms, and the interaction between these elements.

(b) Information on factors which affect the elements of the environment.
These factors include substances such as energy, noise, radiation or waste, including radioactive waste, emissions, discharges and other releases into the environment, affecting or likely to affect the elements of the environment referred to in (a).

(c) Information on measures or activities affecting the environment or factors affecting it.
These measures (including administrative measures), such as policies, legislation, plans, programmes, environmental agreements, and activities affecting or likely to affect the elements and factors referred to in (a) and (b) as well as measures or activities designed to protect those elements.

(d) Reports on the implementation of environmental legislation.

(e) Associate cost benefit legislation.
This is specifically cost-benefit and other economic analyses and assumptions used within the framework of the measures and activities referred to in (c).

(f) Information on the state of human health and safety.

1 Environmental Information Regulations 2004, SI 2004/3991, Pt I, reg 2(1)(a)–(f).
2 The Environmental Information (Scotland) Regulations 2004 , SI 2004/520.

This includes the contamination of the food chain, where relevant, conditions of human life, cultural sites and built structures inasmuch as they are or may be affected by the state of the elements of the environment referred to in (a) or, through those elements, by any of the matters referred to in (b) and (c).

Registers and Access to Information

11.3 The Freedom of Information Act 2000[3] and the Environmental Information Regulations 2004, SI 2004/3991 give the public in England, Wales and Northern Ireland the right to access environmental information held by public authorities. There is no requirement to prove any interest or give a reason why the information is required.

In Scotland there is legislation with similar provisions[4,5].

Duty to Hold Information

11.4 The duty to keep a public register is found at reg 46 and Sch 24 of the Environmental Permitting (England and Wales) Regulations 2010, SI 2010/675.

What the Register Must Hold

11.5 The Environmental Permitting (England and Wales) Regulations 2010, SI 2010/675, Sch 24 states that:

A public register must contain a copy of—

(a) every application—

- for the grant of an environmental permit;
- to vary an environmental permit;
- to transfer an environmental permit in whole or in part or;
- to surrender an environmental permit in whole or in part.

(b) every notice requesting further information under para 4(1) of Sch 5[6].

3 Freedom of Information Act 2000, Pt 1, Sch 1.
4 Freedom of Information (Scotland) Act 2002 as amended.
5 Environmental Information (Scotland) Regulations 2004, SI 2004/520 as amended.
6 **4.**—(1) If the regulator considers that it requires further information to determine a duly-made application, it may serve a notice on the applicant specifying the further information and the period within which it must be provided.

(c) all representations made in respect of an application for an environmental permit or to vary an environmental permit.
(d) every environmental permit, variation, transfer in whole or in part, or surrender in whole or in part granted or made by the regulator.
(e) every determination or decision notified under para 17(2)(a) of Sch 5[7].
(f) every prohibition notice, enforcement notice, revocation notice, suspension notice, landfill closure notice or notice withdrawing such a notice served by the regulator.
(g) in relation to an appeal to an appropriate authority, every—

- notice of appeal;
- document relating to the appeal;
- representation made in respect of the appeal; and
- determination of the authority, including any report accompanying that determination.

(h) all monitoring information obtained by the regulator—

- as a result of its own monitoring;
- by virtue of any environmental permit condition; or
- under reg 60[8] the power to require the provision of information.

(i) all other information given to the regulator in compliance with—

- an environmental permit condition;
- an enforcement notice;
- a suspension notice;

7 (2) The regulator must—
(a) notify the applicant or, for a regulator-initiated variation, the operator of—
(i) its determination or decision,
(ii) the rights of appeal the applicant or operator has under regulation 31, and
(iii) the requirements relating to the exercise of those rights in paragraphs 2 and 3 of Schedule 6; and
(b) if paragraph 10 applies, notify the appropriate authority of the determination or decision.

8 **Power to require the provision of information**
60.—(1) For the purposes of discharging its functions under these Regulations, an appropriate authority, regulator or exemption registration authority, by notice served on any person, may require that person to provide such information in such form and within such period as is specified in the notice.
(2) A notice under paragraph (1) may require a person to provide any information on emissions where that requirement is reasonable, including the provision of information—
(a) not in the person's possession; and
(b) which would not usually come into the person's possession.
(3) For the purposes of this regulation the discharge by the appropriate authority of—
(a) an obligation of the United Kingdom under the Community Treaties; or
(b) an international obligation of the United Kingdom,
must be treated as a function of the authority under these Regulations.
(4) For the purposes of this regulation the compilation of an inventory of emissions (whether or not from a regulated facility) must be treated as a function of the regulator under these Regulations.

- a landfill closure notice;
- mining waste facility closure notice; or
- regulation 60 – the power to require the provision of information.

(j) every report published by the regulator relating to an assessment of the environmental consequences of the operation of an installation.

(k) every direction given to the regulator or the exemption registration authority by an appropriate authority under these Regulations, other than a direction given under regulation 47[9] (Exclusions from public registers of information affecting national security) or paragraph 9 of Schedule 2 (Exclusions from the register of information affecting national security)[10].

9 **Exclusion from public registers of information affecting national security**

47.—(1) The appropriate authority may direct the regulator that in the interests of national security specified information or information of a specified description must be excluded from a public register.

(2) The regulator must notify the appropriate authority of any information (other than information relating to a radioactive substances activity) that it excludes from a public register pursuant to such a direction.

(3) The appropriate authority may direct the regulator that in the interests of national security information of a specified description—

(a) must be referred to the authority for its determination as to whether or not the information may be included on a public register; and

(b) must not be included on a public register unless the appropriate authority determines that it may be included.

(4) A person may give a notice to the appropriate authority stating that, in the person's opinion, the inclusion of information on a public register would be contrary to the interests of national security.

(5) A notice under paragraph (4) must specify the information and indicate its apparent nature.

(6) A person giving a notice under paragraph (4) must at the same time notify the regulator.

(7) The regulator must not include information notified under paragraph (4) on a public register unless the appropriate authority determines that it may be included.

10 **Exclusion from the register of information affecting national security**

9.—(1) The appropriate authority may direct the exemption registration authority that, in the interests of national security, specified information or information of a specified description must be excluded from the register.

(2) The exemption registration authority must notify the appropriate authority of any information it excludes from the register pursuant to such a direction.

(3) The appropriate authority may direct the exemption registration authority that, in the interests of national security, before information of a specified description is included on the register, the information must be referred to the appropriate authority for determination as to whether or not it should be excluded from the register.

(4) A determination by the appropriate authority under sub-paragraph (3) to exclude information from the register must be given effect by a direction under sub-paragraph (1).

(5) A person may give a notice to the appropriate authority stating that, in the person's opinion, the inclusion of information on the register would be contrary to the interests of national security.

(6) A notice under sub-paragraph (5) must specify the information and indicate its apparent nature.

(7) A person giving a notice under sub-paragraph (5) must at the same time notify the exemption registration authority.

(8) The exemption registration authority must not include information notified under sub-paragraph (5) on the register unless the appropriate authority determines that it may be included.

(9) A determination by the appropriate authority to exclude information notified under sub-paragraph (5) from the register must be given effect by a direction under sub-paragraph (1).

A Register Must Also Include

11.6

(a) details of any conviction or formal caution for an offence under reg 38[11](Offences) in respect of an environmental permit granted by the regulator or a failure to apply to the regulator for an environmental permit;

---

(10) In relation to an exempt facility that is the subject of a direction or notice given under this paragraph, the requirement in paragraph 3(1)(b), 4(b) or 5(b)—
  (a) does not apply where a direction or notice requires the exclusion of all relevant particulars from the register; and
  (b) is satisfied where—
    (i) a direction under sub-paragraph (1) requires the exclusion of some, but not all, relevant particulars from the register,
    (ii) a direction under sub-paragraph (3) requires the exclusion of relevant particulars from the register, or
    (iii) a notice under sub-paragraph (5) which is pending a determination under sub-paragraph (8) requires the exclusion of relevant particulars from the register,

and where any relevant particulars which are not subject to a direction or notice are included on the register.

11 **Offences**

38.—(1) It is an offence for a person to—
  (a) contravene regulation 12(1); or
  (b) knowingly cause or knowingly permit the contravention of regulation 12(1)(a).

(2) It is an offence for a person to fail to comply with or to contravene an environmental permit condition.

(3) It is an offence for a person to fail to comply with the requirements of an enforcement notice or of a prohibition notice, suspension notice, landfill closure notice or mining waste facility closure notice.

(4) It is an offence for a person—
  (a) to fail to comply with a notice under regulation 60(1) requiring the provision of information, without reasonable excuse;
  (b) to make a statement which the person knows to be false or misleading in a material particular, or recklessly to make a statement which is false or misleading in a material particular, where the statement is made—
    (i) in purported compliance with a requirement to provide information imposed by or under a provision of these Regulations,
    (ii) for the purpose of obtaining the grant of an environmental permit to any person, or the variation, transfer in whole or in part, or surrender in whole or in part of an environmental permit, or
    (iii) for the purpose of obtaining, renewing or amending the registration of an exempt facility;
  (c) intentionally to make a false entry in a record required to be kept under an environmental permit condition;
  (d) with intent to deceive—
    (i) to forge or use a document issued or authorised to be issued or required for any purpose under an environmental permit condition, or
    (ii) to make or have in the person's possession a document so closely resembling such a document as to be likely to deceive.

(5) It is an offence for an establishment or undertaking to—
  (a) fail to comply with paragraph 14(3) or (4) of Schedule 2; or
  (b) intentionally make a false entry in a record required to be kept under that paragraph.

(6) If an offence committed by a person under this regulation is due to the act or default of some other person, that other person is also guilty of the offence and liable to be proceeded against and punished accordingly, whether or not proceedings for the offence are taken against the first-mentioned person.

(b) an inventory of closed mining waste facilities as required under Article 20 of the Mining Waste Directive[12]; [attributes of the inventory are not specified]
(c) a list identifying all waste incineration installations which—

- have a capacity of less than 2 tonnes per hour; and
- are the subject of an environmental permit containing conditions which give effect to the Waste Incineration Directive; and

(d) details of—

- all fees and charges paid to the local authority pursuant to a scheme under reg 65[13], (Fees and charges in relation to the exercise of regulator's functions by local authorities); and

12 The Directive was published in the Official Journal of the European Union on 11 April 2006 (L102/15).
Article 20 **Inventory of closed waste facilities** Member States shall ensure that an inventory of closed waste facilities, including abandoned waste facilities, located on their territory which cause serious negative environmental impacts or have the potential of becoming in the medium or short term a serious threat to human health or the environment is drawn up and periodically updated. Such an inventory, to be made available to the public, shall be carried out by 1 May 2012, taking into account the methodologies as referred to in Article 21, if available.

13 **Fees and charges in relation to the exercise of regulator's functions by local authorities**
65.—(1) An appropriate authority may make, and from time to time revise, a scheme prescribing—
(a) fees payable to a regulator in respect of applications—
(i) for the grant of an environmental permit,
(ii) for the variation of an environmental permit,
(iii) for the transfer of an environmental permit in whole or in part,
(iv) for the surrender of an environmental permit in whole or in part;
(b) fees payable to a regulator in respect of a regulator-initiated variation; and
(c) charges payable to a regulator in respect of the subsistence of an environmental permit.
(2) A scheme may in particular—
(a) prescribe specific fees and charges or the methods by which they are to be calculated;
(b) make different provision for different cases, including different provision in relation to different persons, circumstances or localities;
(c) subject to the requirements of these Regulations, provide for the time when, and the manner in which, payments required by the scheme are to be made; and
(d) make such incidental, supplementary and transitional provision as appears necessary or expedient to the appropriate authority.
(3) In making or revising a scheme, so far as practicable the appropriate authority must ensure that the fees and charges payable are sufficient to cover expenditure by a regulator—
(a) in exercising its functions under these Regulations;
(b) in making payment to any person who prepares guidance in relation to an installation or mobile plant that is—
(i) mentioned in regulation 32(2), or
(ii) specified in a direction under regulation 33;
(c) in making payment to the Agency in relation to the exercise of the Agency's functions under regulation 58.
(4) A scheme must provide for the payment of sums by the regulator to the Agency where those sums are related to expenditure by the Agency under regulation 58 or in preparing guidance referred to in paragraph (3)(b).

- the total expenditure of the authority in exercising its functions under these Regulations in respect of permits granted by the authority.

Environmental Monitoring Information

11.7 Under Sch 24, para1(1)(h) the information also includes information obtained by the regulator:

(i) as a result of its own monitoring;
(ii) as a result of monitoring required under an environmental permit condition; or
(iii) under reg 60 in relation to monitoring.

Omission of a Representation

11.8 The regulator is also able to omit a representation referred to in sub-paragraph (1) from its public register at the request of the person making the representation, however they must then include in the public register a statement that a representation was made and was the subject of such a request.

Omission of Monitoring Information

11.9 The regulator can omit from its public register any representation, which substantially duplicates a representation already included in the public register, however the regulator must then include on the register a statement of the number of representations that have been omitted on this basis[14].

If the regulator decides to omit monitoring information referred to in sub-paragraph (1) from its public register on the grounds that it is commercially or industrially confidential[15], the regulator must include in the public register

(5) If a regulator considers that an operator has failed to pay a charge specified in a scheme in respect of the subsistence of the operator's permit, the regulator may revoke or suspend the permit.
(6) A revocation or suspension must be by way of notice served under regulation 22(3) or regulation 37.
(7) In this regulation, "regulator" means a local authority on which functions are conferred by regulation 32 or by a direction under regulation 33.

14 (4) The regulator may omit from its public register any representation which substantially duplicates a representation already included in the public register, but the regulator must then include on the register a statement of the number of representations that have been omitted on this basis.

15 (5) If the regulator omits from its public register information referred to in sub-paragraph (1) (h) on the grounds that it is commercially or industrially confidential, the regulator must include in the public register a statement indicating whether or not there has been compliance with any environmental permit condition related to that information and requiring compliance with emission limit values.

a statement indicating whether or not there has been compliance with any environmental permit condition related to that monitoring information and requiring compliance with emission limit values.

In this paragraph, 'waste incineration installation' has the meaning given in Sch 13[16].

Commercially Confidential Information

11.10 Under Sch 48 confidential information can be excluded form the register. The information must be excluded if the regulator considers the information to be confidential information or they receive notice that states the information is confidential and reasons are given as to why this is the case[17].

16 **Interpretation**

2.—(1) In this Schedule, "waste incineration installation" means that part of an installation or Part A mobile plant in which any of the following activities is carried on—

(a) the incineration of waste falling within the following provisions of Section 5.1 of Part 2 of Schedule 1—

(i) paragraphs (a) to (c) of Part A(1), or

(ii) paragraph (a) or (b) of Part A(2); or

(b) any other activity falling within Part 2 of Schedule 1 which is carried on in a co-incineration plant (as that term is defined in Section 5.1 of Part 2 of Schedule 1).

(2) When interpreting the Waste Incineration Directive for the purposes of this Schedule—

(a) an expression that is defined in Section 5.1 of Part 2 of Schedule 1 has the meaning given in that Section;

(b) except where also defined in Section 5.1 of Part 2 of Schedule 1, an expression that is defined in Part 1 of these Regulations has the meaning given in that Part;

(c) "permit" means environmental permit; and

(d) the competent authority is the regulator.

17 **Exclusion from public registers of confidential information**

48.—(1) The regulator must exclude information from a public register, unless a condition in paragraph (2) is met, if it—

(a) considers that the information may be confidential information; or

(b) receives notice from the information subject which—

(i) states that the information subject considers the information is confidential information, and

(ii) gives reasons for that view.

(2) The conditions are that—

(a) in relation to paragraph (1)(a), the regulator has given a notice under regulation 49(1) and the information subject has given notice of consent under regulation 49(2)(a);

(b) in relation to paragraph (1)(a) or (b)—

(i) a final confidentiality decision that the information should be included on the register has been made, or

(ii) the appropriate authority has given a direction under regulation 56(1) which requires the information to be included on the register.

Exclusions from the Register of Information, Material Affecting National Security

11.11 The appropriate authority may direct the exemption registration authority that, in the interests of national security, specified information or information of a specified description must be excluded from the register. This is covered under Sch 2, para 9[18].

The exemption registration authority must notify the appropriate authority of any information it excludes from the register after such a direction.

The appropriate authority may direct the exemption registration authority that, in the interests of national security, before information of a specified description is included on the register, the information must be referred to

18 **Exclusion from the register of information affecting national security**

9.—(1) The appropriate authority may direct the exemption registration authority that, in the interests of national security, specified information or information of a specified description must be excluded from the register.

(2) The exemption registration authority must notify the appropriate authority of any information it excludes from the register pursuant to such a direction.

(3) The appropriate authority may direct the exemption registration authority that, in the interests of national security, before information of a specified description is included on the register, the information must be referred to the appropriate authority for determination as to whether or not it should be excluded from the register.

(4) A determination by the appropriate authority under sub-paragraph (3) to exclude information from the register must be given effect by a direction under sub-paragraph (1).

(5) A person may give a notice to the appropriate authority stating that, in the person's opinion, the inclusion of information on the register would be contrary to the interests of national security.

(6) A notice under sub-paragraph (5) must specify the information and indicate its apparent nature.

(7) A person giving a notice under sub-paragraph (5) must at the same time notify the exemption registration authority.

(8) The exemption registration authority must not include information notified under sub-paragraph (5) on the register unless the appropriate authority determines that it may be included.

(9) A determination by the appropriate authority to exclude information notified under sub-paragraph (5) from the register must be given effect by a direction under sub-paragraph (1).

(10) In relation to an exempt facility that is the subject of a direction or notice given under this paragraph, the requirement in paragraph 3(1)(b), 4(b) or 5(b)—

- (a) does not apply where a direction or notice requires the exclusion of all relevant particulars from the register; and
- (b) is satisfied where—
  - (i) a direction under sub-paragraph (1) requires the exclusion of some, but not all, relevant particulars from the register,
  - (ii) a direction under sub-paragraph (3) requires the exclusion of relevant particulars from the register, or
  - (iii) a notice under sub-paragraph (5) which is pending a determination under sub-paragraph (8) requires the exclusion of relevant particulars from the register,

and where any relevant particulars, which are not subject to a direction or notice, are included on the register.

the appropriate authority for determination as to whether or not it should be excluded from the register.

Information That is no Longer Relevant for Public Participation

11.12 A regulator is not required to keep in its public register, information which is no longer relevant for the purposes of public participation required under these Regulations.

Formal Cautions

11.13 A regulator must remove details of any formal caution from its public register, five years after the caution was given.

Licence Holder's Convictions

11.14 The public register must contain under Sch 24, para 1(2)(a) details of any conviction or formal caution for an offence under reg 38 in respect of an environmental permit granted by the regulator or a failure to apply to the regulator for the grant of an environmental permit.

Duty to Provide Information

11.15 All public authorities holding environmental information have a duty under the Environmental Information Regulations 2004, SI 2004/3391 (EIRs) or the Environmental Information (Scotland) Regulations 2004, SI 2004/520 to provide information to any member of the public who asks for it unless they can claim an exemption (reg 5 of both sets of Regulations). (

How to Request Information

11.16 Information can be requested verbally or in writing. The Environment Agency has information online where a request can be made. See http://epr.environment-agency.gov.uk/ePRInternet/searchregisters.aspx

Charges

11.17 Public Authorities are allowed to charge for the supply of information under the Environmental Information Regulations 2004, SI 2004/3391, reg 8 and the Environmental (Scotland) Regulations 2004, SI 2004/520, reg 8.

It is possible for a public authority to set their own charges but they must have published the details of their charges, when they are to be made and the charges must be reasonable.

Public Authorities in England, Wales and Northern Ireland mostly follow the rates that are provided under the Freedom of Information and Data Protection (Appropriate Limit and Fees) Regulations 2004, SI 2004/3244. These Regulations state that there is no charge for a request below £600 for Central Government or below £450 for other public authorities. Where the cost goes above this amount then the authorities may work with the applicant to change the request to fall below the limits.

Refusal to Grant Information

11.18 Under the Environmental Information Regulations 2004[19] and Environmental Information (Scotland) Regulations 2004[20] public authorities can refuse to disclose environmental information if one or more of the following exemptions applies:

- it does not hold the information at the time of the request. If the authority knows or believes another authority has the information then it has a duty to transfer the request to the other authority;
- the request is too vague or general and the applicant is unable to or refuses to make the request clearer;
- the information requested is not yet complete. In this case the authority should estimate the time when it will be available;
- the request is unreasonable (this is similar to the exemption under Freedom of Information legislation for vexatious requests[21]);
- meeting the request requires the disclosure of internal communications;
- the request relates to personal data that does not belong to the applicant (this is covered by the Data Protection Act).

Making a Complaint

Public Authorities' Internal Complaints Procedure

11.19 Public Authorities are required by the EIRs to set up internal complaints procedures to deal with complaints from the public in their handling of environmental information requests. It is important that these producers are followed before any other forms of redress are sought. Many public authorities have the details of their complaints process on their websites. They have 40 days

19 Environmental Information Regulations 2004, SI 2004/3391, reg 14.
20 Environmental Information (Scotland) Regulations 2004, SI 2004/520, reg 13.
21 Freedom of Information Act 2000, Pt I, s 14.

to deal with a complaint and some organisations aim to deal with complaints quicker then this (ie, Defra aim for 15 days).

Complaints to the Information Commissioner

11.20 If a person is not satisfied with the response from the public authority then a complaint can be made to the Information Commissioner. All complaints must be in writing and must be sent as soon as possible after the outcome of the public authority's internal review has been given. There is no time limit in the EIRs for the Information Commissioner's Office to review complaints; it is expected to act in a timely and efficient manner. The public authority would be expected to comply with the Information Commissioner's decision although it has the right to appeal to the Information Tribunal.

Right of Appeal to the First Tier Tribunal (Information Rights)

11.21 If a person is not satisfied with the decision from the Information Commissioner, then an appeal can be brought to the First Tier Tribunal. This is dealt with by the Tribunal's General Regulatory Chamber. The procedure for appeals is set out in the Tribunal Procedure (First-tier Tribunal) (General Regulatory Chamber) Rules 2009[22].

Court Action

11.22 If a person is not satisfied with the tribunal's decision, then redress can be sought through the courts.

There is guidance on the complaints and appeals procedure available on the Information Commissioner's Office websites: https://ico.org.uk.

AVAILABLE INFORMATION

11.23 The Environmental Permitting (England and Wales) Regulations 2007, SI 2007/3538 (EPR) came into force on 6 April 2008. These combined the Pollution Prevention and Control (PPC) and Waste Management Licensing (WML) Regulations. All existing PPC permits or Waste Management Licences automatically became Environmental Permits from 6 April 2008. The regulations were amended in April 2010 such that all Water Discharge consents,

22 The Tribunal Procedure (First-tier Tribunal) (General Regulatory Chamber) Rules 2009, SI 2009/1976.

Groundwater authorisations and Radioactive Substances registrations and authorisations automatically became Environmental Permits.

Permits are now issued under the Environmental Permitting (England and Wales) Regulations 2010, SI 2010/675 (EPR) which have replaced (or partly replaced) the following regulations:

- Pollution Prevention and Control (England and Wales) Regulations 2000, SI 2000/1973 (PPC Regulations);
- Environmental Protection Act 1990, s 20 and the Environmental Protection (Applications, Appeals and Registers) (Amendment No 2) Regulations 1996, SI 1996/979;
- Waste Management Licensing Regulations 1994, SI 1994/1056 and amendments;
- End-of-Life Vehicles Regulations 2003, SI 2003/2635 and the End-of-Life Vehicles (Producer Responsibility) Regulations 2005, SI 2005/263 (Authorised Treatment Facilities);
- Water Resources Act 1991, s 190 and the Control of Pollution (Applications, Appeals and Registers) Regulations 1996, SI 1996/2971;
- Groundwater (England and Wales) Regulations 2009,SI 2009/2902; Groundwater Regulations (Northern Ireland) 2009, NISR 2009/254; Water Environment (Groundwater and Priority Substances) (Scotland) Regulations 2009, SSI 2009/420;
- Radioactive Substances Act 1993, s 39;
- Mining Waste Directive 2006/21/EC.

WASTE

11.24

- The Register of Carriers of Controlled Waste;
- The Register of Brokers of Controlled Waste;
- The Register of Dealers in Controlled Waste;
- The Register of Exempt Activities;
- The Producer Responsibility Register;
- Waste Electrical and Electronic Equipment Regulations 2006, SI 2006/3289;
- Waste Batteries and Accumulators Regulations 2009, SI 2009/890;
- Waste and Emissions Trading Act 2003;
- Landfill Allowances Scheme (Wales) Regulations 2004, SI 2004/1490;
- Landfill Allowances and Trading Scheme (England) Regulations 2004, SI 2004/3212.

Register of Carriers of Controlled Waste

11.25 A public register maintained under s 2(2)(b) of the Control of Pollution (Amendment) Act 1989 and the Controlled Waste (Registration of Carriers and Seizure of Vehicles) Regulations 1991, SI 1991/1624, reg 3 and prescribed by SI 1991/1624, reg 6 (since 14 October 1991) (registration of a carrier).

Regulation 6 prescribes the following to be held on a public register:

- entries showing persons as registered carriers of controlled waste with their allocated registration number, including any letter;
- the date the registration takes effect and the date of expiry;
- registered person's business name, address of his principal business office and telephone, telex or fax numbers and if an individual, date of birth;
- for bodies corporate, the names of each director, manager, secretary or other similar officers and their dates of birth;
- for companies registered in Great Britain, the registered number, otherwise the country in which it was incorporated;
- relevant conviction details including person's name, details of the offence, date of conviction, penalty imposed, the name of the Court and, if an individual, their date of birth;
- details of waste management licences held.

Schedule 3 sets out the details required for a certificate of registration under reg 6:

- of the public to readily trace information on the register;
- register of waste carriers and dealers (registrations before 27 March 2015).

The Register of Brokers of Controlled Waste

11.26 A public register held under the Waste (England and Wales) Regulations 2011, SI 2011/988, reg 28 includes:

- broker's registration number;
- date the registration takes effect and date of its expiry;
- broker's business name and address of the principal place of business including telephone number, telex or fax numbers and if an individual, their date of birth;
- for bodies corporate, the names of each director, manager or similar position and their dates of birth;
- for companies registered in Great Britain, the registered number, otherwise the country in which it was incorporated;
- relevant conviction details including person's name, details of the offence, date of conviction, penalty imposed, the name of the court and if an individual, their date of birth;
- details of waste management licences held;
- date of renewal and revised expiry date;
- any other changes to registered details and the date of amendment.

The Register of Dealers in Controlled Waste

11.27 The Waste (England and Wales) Regulations 2011, SI 2011/988, reg 28 states that a register of professional dealers must be maintained, although details are not specified. Under the previous legislation the information to be held was:

- name and address of the establishment or undertaking;
- address of its principal place of business or address of any place at, or from which, it carries on its business.

The Producer Responsibility Register

11.28 A public register maintained under the Producer Responsibility Obligations (Packaging Waste) Regulations 2010, SI 2010/871 reg 33 (as amended) (as detailed in Sch 7) must include:

The name and address of the registered office or principal place of business of:

(a) each registered producer;
(b) each registered operator of a scheme and each member of the scheme for which he is the operator; and
(c) each accredited reprocessor and accredited exporter.

In relation to accredited reprocessors and exporters:

(a) each material type accepted;
(b) in relation to any recovery operation to be used, the appropriate classification of the applicable operation provided for in Annex II of the Waste Directive;
(c) in relation to the incineration at waste incineration plants with energy recovery, the appropriate classification of the applicable operation under Annex I or II of the Waste Directive;
(d) whether the reprocessor or exporter is accredited to issue PRNs or PERNs for either 400 tonnes or less, or more than 400 tonnes of packaging waste;
(e) the trading name, address and telephone number of the reprocessor or exporter;
(f) the reprocessing site address for an accredited reprocessor;
(g) the reference number supplied by the appropriate Agency;
(h) whether or not the quarterly returns and annual returns have been provided in accordance with the conditions set out in paragraph 1(n) and (o) of Schedule 5;
(i) whether the appropriate Agency has served a notice to suspend or cancel accreditation; and
(j) whether accreditation has been suspended or cancelled.

A statement in relation to each registered producer as to whether a satisfactory certificate of compliance has been furnished.

A statement in relation to each registered operator of a scheme as to whether he has complied with his obligations under regulation 12(1)[23].

For all registrations:

- a note of any amendment made to any register entry and the date of amendment;
- an index to allow members of the public to readily trace information on the register.

Waste Electrical and Electronic Equipment Regulations 2006, SI 2006/3289

11.29 For producers, Sch 10 of the Regulations sets out the information to be held on a public register. This must include:

(1) The producer's name.
(2) The address of the registered office or, where the producer is not a body corporate registered in the United Kingdom, the address of the principal place of business in the United Kingdom of that producer.
(3) The producer's EEE producer registration number.
(4) The name of the scheme that the producer has joined indicating whether that producer has joined the scheme for the purposes of discharging his obligations under regs 8(1) or 9(1) or both.
(5) The name of the operator of that scheme and the address of his registered office if he is a body corporate registered in the United Kingdom or, where he is not a body corporate registered in the United Kingdom, the address of his principal place of business in the United Kingdom.
(6) A description of the producer identification mark used on that producer's products in compliance with regulation 16[24].

For schemes, reg 62[25] states —

23 **Schemes: general provisions**
12.—(1) The operator of a scheme shall carry out the recovery and recycling obligations and where applicable, consumer information obligations, that every producer who is a member of the scheme that he operates would have had, but for their membership of that scheme.

24 **Marking EEE with a producer identification mark and a date mark**
16.—(1) A producer shall mark EEE that he puts on the market in such a manner that—
(a) he can be easily identified by that mark as the producer of the equipment ("the producer identification mark"); and
(b) the equipment can be easily identified as having been put on the market after 13th August 2005 ("the date mark").
(2) The producer identification mark and the date mark shall be affixed in a visible, legible and indelible form to each item of equipment.

25 **Approval of schemes**
62.—(1) The appropriate authority shall maintain and publish a list of—
(a) all schemes that it has approved under regulation 41; and
(b) the operators of the schemes referred to in sub-paragraph (a).

(1) The appropriate authority shall maintain and publish a list of—

(a) all schemes that it has approved under regulation 41; and
(b) the operators of the schemes referred to in sub-paragraph (a).

For Approved Authorised Treatment Facilities and Approved Exporters, reg 64[26] provides that —

(1) The appropriate authority shall be under a duty to maintain and publish a list of all AATFs, operators of AATFs and approved exporters.

Waste Batteries and Accumulators Regulations 2009, SI 2009/890

11.30 For producers, reg 76 (Registration of Producers) provides that—

(1) The appropriate authority must maintain and make available in accordance with this regulation a register relating to—

(a) those producers who are registered with it in accordance with regulation 26[27]; and

(b) those producers who are registered with the Secretary of State in accordance with reg 42[28] and whose details have been provided to the appropriate authority by the Secretary of State under regulation 73(1)[29], and containing the information specified in Schedule 6[30].

Schedule 6 specifies:

---

(2) A list maintained by the appropriate authority in accordance with paragraph (1) shall not be required to contain any information that has been superseded by later information after four years have elapsed from that later information being entered in the list.
(3) The appropriate authority shall issue an invoice for payment of the annual producer charge referred to in regulation 45 to each operator of a scheme that it has approved under regulation 41.
(4) The appropriate authority shall publish the format in which—
(a) the information referred to in Part 1 of Schedule 7 shall be submitted to it in an application for approval made under regulation 41;
(b) the information referred to in regulation 27 shall be submitted to it in accordance with that regulation; and
(c) the information referred to in regulation 28 shall be submitted to it in accordance with that regulation.

26 Also: (2) The appropriate authority shall publish the format in which the information referred to in Part 1 of Schedule 8 shall be submitted to it in an application for approval made under regulation 47 or in an application for an extension of a grant of approval made under regulation 48 – **Application for extension of approval of an exporter to an additional site.**
27 Regulation 26: Registration of producers of portable batteries.
28 Regulation 42: Registration of producers: Industrial batteries or automotive batteries.
29 Regulation 71(1): Information for appropriate authorities in relation to producers of industrial or automotive batteries.
30 Schedule 6: Register of producers.

(1) The name of the producer and any brand name under which the producer operates in the United Kingdom.
(2) The following contact details for the producer—

  (a) the full postal address (including post code) and telephone number of—

    (i) the producer's registered office; or
    (ii) if the producer is not a company registered in the United Kingdom, the producer's principal place of business in the United Kingdom;

  (b) a website address; and
  (c) where available, a fax number and e-mail address.

(3) An address for service of notices on the producer if different from the addresses mentioned in paragraph 2.
(4) An indication of the categories of battery placed on the market by the producer.
(5) Information as to—

  (a) whether the producer meets its responsibilities under these Regulations individually or collectively; and
  (b) if collectively—

    (i) the name of the battery compliance scheme;
    (ii) the name of the scheme operator; and
    (iii) the address and telephone number of the registered office of the scheme operator or, if not a company registered in the United Kingdom, the principal place of business and telephone number of the scheme operator in the United Kingdom.

(6) The producer's battery producer registration number.
(7) The date of the application for registration.

For schemes, reg 79[31] provides that:

(1) The appropriate authority must maintain and publish a list of—

  (a) all battery compliance schemes that it has approved under regulation 49[32];

  and

  (b) the scheme operators.

For approved battery treatment operators and approved exporters reg 80[33] provides that —

(1) The appropriate authority must maintain and publish a list of all approved battery treatment operators and approved battery exporters.

31 Regulation 79: General duties in relation to battery compliance schemes.
32 Regulation 49: Approval of proposed schemes.
33 Regulation 80: General duties in relation to approved battery treatment operators and approved exporters.

Waste and Emissions Trading Act 2003

11.31 Section 15[34] requires that the monitoring authority keeps a register and s 16[35] requires that it is made accessible to the public under reg 10[36] of the Landfill Allowances Scheme (Wales) Regulations 2004, SI 2004/1490 and reg 16[37] of the Landfill Allowances and Trading Scheme (England) Regulations 2004, SI 2004/3212.

Landfill Allowances Scheme (Wales) Regulations 2004, SI 2004/1490

11.32 The Environment Agency is the monitoring authority for the purposes of these regulations.

Regulation 10 provides that:

The monitoring authority must establish and maintain a landfill allowances register which, in relation to each waste disposal authority for each scheme year, contains—

(a) the allowance allocated under section 4 of the Act;
(b) any alteration of the allowance referred to in paragraph (a) under section 5 of the Act;
(c) the amount of biodegradable municipal waste sent to landfills by that waste disposal authority; and
(d) the balance of the following:
    (i) the allowance registered under paragraph (a), as altered by any alteration registered under paragraph (b), (the 'total allowance'); and
    (ii) the amount of biodegradable municipal waste sent to landfills by that waste disposal authority as registered under paragraph (c).

Registration of the Waste Collection Authority and Other Waste Carriers

11.33 The Environment Agency Public Register only contains waste carriers and brokers registration before 27 March 2015. For registration after that date then the register is held at: https://www.wastecarriersregistration.service.gov.uk/registrations/search

For **Welsh** based businesses and organisations, Natural Resources Wales are now providing their own public register for:

34 Section 15: Monitoring information: registers.
35 Section 16: Registers: Public Access.
36 Regulation 10: Landfill allowances register.
37 Regulation 16: Landfill allowances register.

Hazardous Waste registrations;
Enforcement Actions;
Scrap Metal Dealers;
Waste Carrier, Dealer and Broker registrations;
Waste and Water Exemptions registrations made after 20 October 2014;
Environmental Permits (Industrial Installations);
Environmental Permits (Waste Operations);
Environmental Permits (Radioactive Substances);

and can be found on: http://naturalresources.wales/how-we-regulate-you/find-out-if-a-site-has-a-permit-licence-or-exemption/?lang=en.

EU EMISSIONS TRADING

Stationary Installations

11.34 Under the Greenhouse Gas Emissions Trading Scheme Regulations 2005, SI 2005/925 (reg 36[38]), the regulator is required to publish the names of operators who are liable to the civil penalty for failing to surrender sufficient allowances to cover their reportable emissions. There are no other specific public register provisions but please see the Registry section below at para **11.36**.

Aviation

11.35 Under the Aviation Greenhouse Gas Emissions Trading Scheme Regulations 2010, SI 2010/1996 (reg 49[39]), the regulator is required to publish, by 30 June each year, a list of aircraft operators that are liable to the civil penalty for failing to surrender sufficient allowances to cover their aviation emissions. There are no other public register provisions but please see the Registry section below at para **11.36**. Regulation 55[40] covers confidentiality of information provided by the aircraft operators and states that the regulator must not disclose or publish any information provided to it under the Regulations unless: (a) disclosure is required in the Regulations or otherwise by law, necessary for the regulator to perform its functions or made with the consent of the person who provided the information, or (b) disclosure is between the regulator and the Secretary of State.

38 Greenhouse Gas Emissions Trading Scheme Regulations 2005, SI 2005/925: Regulation 36 Publication of a list of operators subject to penalties.

39 Aviation Greenhouse Gas Emissions Trading Scheme Regulations 2010, SI 2010/1996: Regulation 49 Naming of operators.

40 Aviation Greenhouse Gas Emissions Trading Scheme Regulations 2010, SI 2010/1996: Regulation 55 Confidentiality.

Registry

11.36 The Environment Agency also maintains a Registry and acts as 'registry administrator' under the EU ETS. Operators, aircraft operators and traders may hold accounts in the Registry. Article 74 (which refers to information contained in Annex XIII) of Commission Regulation (EU) No 920/2010 ([2010] OJ L 270, 14.10.2010, p 1–52) requires the registry administrator to publish certain information contained in the Registry on a publicly accessible website. Article 75 of the Regulation contains provisions about confidentiality of other information within the Registry.

HAZARDOUS WASTE REGULATIONS

11.37 The duty to provide information to the public is detailed under reg 17 of the Control of Major Accident Hazards Regulations 2015[41]:

- notifications to the Competent Authority under reg 6[42];
- notifications under reg 16(2);
- communications under reg 17(1)(a);
- COMAH improvement notices served under s 21 of the Health and Safety at Work etc Act 1974;
- COMAH prohibition notices served under reg 18(3).

The Environment Agency, the Health and Safety Executive (HSE) and the Scottish Environment Protection Agency (SEPA) work together to implement the COMAH regulations. Public register details of all the COMAH establishments can be obtained from HSE at http://www.hse.gov.uk/comah/public-register.htm

HSE will provide, on request, an up to date list containing the name and address of the operator and whether the site is top or lower tier.

WATER QUALITY

11.38 The public register for Water Quality Discharge Consents (now Environmental Permits) is now held on the Public Register under the Environmental Permitting (England and Wales) Regulations 2010, SI 2010/675 (see above at para **11.36**).

41 Control of Major Accident Hazards Regulations 2015, SI 2015/483, Pt 5, reg 17 Provision of information to the public.

42 Regulation 6 Notifications.

Pollution Control Registers

11.39 The information about water discharge consents and groundwater consents previously held under the Water Resources Act 1991 (WRA 1991), s190 and the Control of Pollution (Applications, Appeals and Registers) Regulations 1996, SI 1996/2971 and the Groundwater (England and Wales) Regulations 2009, SI 2009/2902 is now mostly held under the Environmental Permitting (England and Wales) Regulations 2010, SI 2010/675, reg 46[43] and Sch 24[44].

The following elements of the WRA 1991 pollution control registers remain in place:

- WRA 1991, s 190(1)(a) – a public register of notices of water quality objectives or other notices served as required under s 83 of WRA 1991.
- WRA 1991, s 190(1)(e) – a register of samples of water or effluent taken by the Agency for water pollution provisions, analyses; samples and analyses taken by others and acquired by the Agency for water pollution provisions and the steps taken in consequence of any such information (ie, information relating to non-permitted discharges).
- WRA 1991, s 190(1)(l) – directions given by the Secretary of State in relation to the Agency's functions under the water pollution provisions of this Act.
- WRA 1991, s 190 (1)(o) – anti-pollution works notices served under s 161A of this Act.
- WRA 1991, s 190(1)(p) – appeals under s 161C against anti-pollution works notices.

ENVIRONMENTAL PERMITS

11.40 Permits are now issued under the Environmental Permitting (England and Wales) Regulations 2010, SI 2010/675 (EPR 2010) which have replaced (or partly replaced) the following Regulations:

- Pollution Prevention and Control (England and Wales) Regulations 2000, SI 2000/1973 (PPC Regulations) and the Pollution Prevention and Control (Scotland) Regulations 2000, SSI 2000/323;
- Environmental Protection Act 1990, s 20 and the Environmental Protection (Applications, Appeals and Registers) (Amendment No 2) Regulations 1996, SI 1996/979;
- Waste Management Licensing Regulations 1994, SI 1994/1056 and amendments;

43 Environmental Permitting (England and Wales) Regulations 2010, SI 2010/675: Regulation 46 Duty of the regulator to maintain a public register.

44 Environmental Permitting (England and Wales) Regulations 2010, SI 2010/675: Application for the surrender of an environmental permit.

- End-of-Life Vehicles Regulations 2003, SI 2003/2635 and End-of-Life Vehicles (Producer Responsibility) Regulations 2005, SI 2005/263 (Authorised Treatment Facilities);
- Water Resources Act 1991, s 190 and the Control of Pollution (Applications, Appeals and Registers) Regulations 1996, SI 1996/2971;
- Groundwater (England and Wales) Regulations 2009,SI 2009/2902; Groundwater Regulations (Northern Ireland) 2009, NISR 2009/254; Water Environment (Groundwater and Priority Substances) (Scotland) Regulations 2009, SSI 2009/420;
- Radioactive Substances Act 1993, s 39.

CONTAMINATED LAND

England

11.41 Part 2A of the Environmental Protection Act 1990 (as amended and modified) sets out the provisions for identifying, remediating and regulating contaminated land. The local authority is the primary regulator, but some contaminated land will also be designated as 'special sites' (pursuant to the following regulations) which are regulated by the Environment Agency. The Agency's public register is only required to contain information about these 'special sites'. The local authorities maintain the Public Registers for the ordinary contaminated land in their area.

In addition to regulating the significant harm and pollution of controlled waters caused by contaminated land, the regime was modified to enable it to also address harm from radioactivity – all such sites are 'special sites'. They are regulated by the Environment Agency and subject to the same public register requirements.

The Contaminated Land (England) Regulations 2006, SI 2006/1380 set out provisions relating to the identification and remediation of contaminated land. The public register provisions are in reg 13, which refers to s 78R(1) of the Environmental Protection Act 1990 (EPA 1990). The particulars to be included on the register are set out in Sch 3 of the Contaminated Land Regulations:

- remediation notices served by the Environment Agency – including:
  - the name and address of the person upon whom the notice is served;
  - the location and extent of the contaminated land;
  - the significant harm or pollution of controlled waters which resulted in it being identified as contaminated;
  - the contaminating substances;
  - the current land use;
  - what action the notice requires;
  - the date of the notice.
- any appeal against a remediation notice served by the Environment Agency;

- any decision on such an appeal;
- remediation declarations (including similar information as for notices as set out above);
- in relation to a remediation notice, the location and extent of the land and the nature of the contamination (including similar information as for notices as set out above);
- remediation statements (including similar information as for notices as set out above);
- in relation to a remediation statement, the location and extent of the land and the nature of the contamination;
- any appeal against a charging notice;
- any decision on such an appeal;
- designation of Special Sites, and any termination of all or part of that designation;
- notification of remediation steps claimed to have been taken;
- convictions under EPA 1990, s 78M;
- the date of any guidance issued under EPA 1990, s 78V(1);
- where the Environment Agency is precluded from serving a remediation notice, the location and extent of land identified as contaminated land under Pt 11A of the EPA 1990 but which is subsequently dealt with under other environmental controls;
- where the Environment Agency is precluded from serving a remediation notice, the location and extent of land identified as contaminated by the deposit of waste but which is dealt with by another authority;
- where the Environment Agency is precluded by the existence of a water discharge activity in a permit from specifying remediation action, a copy of the permit, the location and extent of the land and the nature of the contamination.

In relation to regs 14, 15 and 16, the register requires the inclusion of:

- location and extent of land;
- the significant harm or pollution of controlled waters which resulted in it being identified as contaminated;
- the contaminating substances;
- the current land use;
- steps taken under the regime to address the contamination.

The register contains a note if any information has been excluded for reasons of national security under EPA 1990, s 78S or for reasons of commercial confidence under EPA 1990, s 78T.

Wales

11.42 The Contaminated Land (Wales) Regulations 2006, SI 2006/2989 set out further provisions for Wales for the identification and remediation of contaminated land. The Regulations add land contaminated by radioactive substances as a 'special site'. The register provisions are in reg 13[45], which refers to s 78R(1) of the Environmental Protection Act 1990. The particulars to be included on the register are set out in Sch 3 of these Regulations, as listed above.

WATER RESOURCES

The Water Abstraction and Impounding Register

11.43 A register maintained under the Water Resources Act 1991, s 189[46] and prescribed by the Water Resources (Abstraction and Impounding) Regulations 2006, SI 2006/641, reg 34.

45 **The Contaminated Land (Wales) Regulations 2006, SI 2006/2989**

13. Registers—(1) For the purpose of section 78R(1) (registers), the particulars that must be contained in a register maintained under that subsection are specified in Schedule 3.

(2) The following descriptions of information are prescribed for the purposes of section 78R(2) as information to be contained in notifications for the purposes of section 78R(1)(h) and (j)—

(a) the location and extent of the land in sufficient detail to enable it to be identified;

(b) the name and address of the person who it is claimed has done each of the things by way of remediation;

(c) a description of any thing which it is claimed has been done by way of remediation; and

(d) the period within which it is claimed each such thing was done.

(3) The following places are prescribed for the purposes of section 78R(8) as places at which any registers or facilities for obtaining copies must be available or afforded to the public in pursuance of paragraph (a) or (b) of that subsection—

(a) where the enforcing authority is the local authority, its principal office; and

(b) where the enforcing authority is the Environment Agency, its office for the area in which the contaminated land in question is situated.

46 **189. Register of abstraction and impounding licences**

(1) The Agency shall keep, in such manner as may be prescribed, registers containing such information as may be prescribed with respect—

(a) to applications made for the grant, revocation or variation of licences under Chapter II of Part II of this Act, including information as to the way in which such applications have been dealt with; and

(b) to persons becoming the holders of such licences by virtue of section 49 above of this Act or regulations made under section 50 above.

(2) Every register kept by the [**F1**Agency] under this section shall also contain such information as may be prescribed with respect—

(a) to applications made in accordance with regulations under section 64 above; and

(b) to licences granted or deemed to be granted, and licences revoked or varied, in accordance with regulations made under that section.

(3) Subject to any regulations under this section, the information which the [**F1**Agency] is required to keep in registers under this section shall continue to include the information which immediately before 1st September 1989 was contained in a register kept by a water authority under section 53 of the Water Resources Act 1963.

(4) The contents of every register kept under this section shall be available, at such place as may be prescribed, for inspection by the public at all reasonable hours.

In summary, details include:

- name and address of the applicant, date of the application and brief details of its proposal;
- the Agency's decision (including refusal), date of decision and brief details of any licence granted or revocation or variation;
- date (if any) on which a licence expires;
- in relation to an appeal or application made directly to him, the Minister's decision, date of decision and brief details of any licence granted or revocation or variation;
- name and address of a person who becomes the holder of a licence following its transfer or apportionment and details of the transfer notice or licence granted including: the date of the notice, date of effect and any date of expiry;
- name and address of a person in whom a licence has vested following the death or bankruptcy of the holder and the date of vesting.

Water Resources (Environmental Impact Assessment) (England and Wales) Regulations 2003, SI 2003/164 (as amended by SI 2006/3124)

11.44 The Environment Agency is required to keep a Register of environmental statements and further information.

Regulation 16 states:

The Agency shall keep in the register maintained under section 189 of the 1991 Act (register of abstraction and impounding licences)—

(a) any determination under regulation 4 [whether a project is a relevant project] or 14 [any variation of a licence to abstract or impound];
(b) any environmental statement or further information furnished under regulation 6 [the statement and opinions of the Agency on its contents];
(c) any documents, other than the environmental statement prepared in accordance with regulations 5 and 6, which have been provided to the Agency and are relevant to the determination under regulation 14; and
(d) any further information which is relevant to the determination under regulation 14 and which becomes available to the Agency after the time that the public was given notice in accordance with section 52(4) of the 1991 Act [variation or revocation at the instance of the Agency or Secretary of State].

Provision of Information about Water Flow etc

11.45 A register held under s 197[47] of the Water Resources Act 1991 contains information about the flow, level or volume of any inland waters or any water contained in underground strata, about rainfall or any fall of snow, hail or sleet or about the evaporation of any water.

Large Raised Reservoirs

11.46 Since the Water Act 2003 came into force, the duties in respect of this public register transferred from the local authorities to the Environment Agency. The register is held by the Environment Agency at the Exeter Regional

47 **197 Provision of information about water flow etc.**

(1) It shall be the duty of the [**F1**Agency]—

(a) to provide a water undertaker with all such information to which this section applies as is in the possession of the [**F1**Agency] and is reasonably requested by the undertaker for purposes connected with the carrying out of its functions; and

(b) to provide reasonable facilities to all persons—

(i) for the inspection of the contents of any records kept by the [**F1**Agency] and containing information to which this section applies; and

(ii) for the taking of copies of, or of extracts from, any such records.

(2) It shall be the duty of every water undertaker to provide the [**F1**Agency] with all such information to which this section applies as is in the possession of the undertaker and is reasonably requested by the [**F1**Agency] for purposes connected with the carrying out of any of its functions.

(3) Where records of the flow, level or volume of any inland waters, other than discrete waters, are kept by a person other than a water undertaker, the [**F1**Agency] shall have the right at all reasonable times—

(a) to inspect the contents of any of those records; and

(b) to take copies of, or of extracts from, the contents of any of those records;

and any person who, without reasonable excuse, refuses or fails to permit the [**F1**Agency] to exercise its right under this subsection shall be guilty of an offence and liable, on summary conviction, to a fine not exceeding level 1 on the standard scale.

(4) Information provided to a water undertaker or to the [**F1**Agency] under subsection (1) or (2) above shall be provided in such form and in such manner and at such times as the undertaker or, as the case may be, the [**F1**Agency] may reasonably require; and the duties of the [**F1**Agency] under subsection (1) above shall extend to information provided to or obtained by the [**F1**Agency] under subsection (2) or (3) above.

(5) Information or facilities provided under subsection (1) or (2) above to the [**F1**Agency], to a water undertaker, to a local authority or joint planning board, or to an internal drainage board, shall be provided free of charge; and facilities provided under subsection (1) above to other persons may be provided on terms requiring the payment by persons making use of the facilities of such reasonable charges as they may determine.

(6) The duties of a water undertaker under subsection (2) above shall be enforceable under section 18 of the Water Industry Act 1991 by the Secretary of State.

(7) This section applies to information about the flow, level or volume of any inland waters or any water contained in underground strata, about rainfall or any fall of snow, hail or sleet or about the evaporation of any water.

Office. If information is requested, it can be provided on application to any office.

The Reservoirs Act 1975 (as amended by the Water Act 2003) states at s 2 that:

It shall be the duty of [each local authority] to establish and maintain for their area a register showing the large raised reservoirs situated wholly or partly in the area, and giving the prescribed information about each of them; and—

(a) the register maintained by [a local authority] under this subsection and copies of it or a prescribed part of it shall be kept at such place or places as may be prescribed; and
(b) the register so maintained, and any copy required under sub-paragraph (a) above to be kept at any place, shall be available for inspection at all reasonable times by any person.

Registration of the Waste Collection Authority and other Waste Carriers

11.47 The Environment Agency public register only contains waste carriers and brokers' registration before 27 March 2015. For registration after that date then the register is held at: https://www.wastecarriersregistration.service.gov.uk/registrations/search

For **Welsh** based businesses and organisations, Natural Resources Wales are now providing their own public register for:

- Hazardous Waste registrations;
- Enforcement Actions;
- Scrap Metal Dealers;
- Waste Carrier, Dealer and Broker registrations;
- Waste and Water Exemptions registrations made after 20 October 2014;
- Environmental Permits (Industrial Installations);
- Environmental Permits (Waste Operations);
- Environmental Permits (Radioactive Substances).

These can be found on: http://naturalresources.wales/how-we-regulate-you/find-out-if-a-site-has-a-permit-licence-or-exemption/?lang=en

THE REGISTER OF EXEMPT ACTIVITIES

11.48 A public register held under and prescribed by Sch 2, para 7[48] of the Environmental Permitting (England and Wales) Regulations 2010, SI 2010/675.

48 7. **Register of exempt facilities**—(1) Every exemption registration authority must establish and maintain a register of exempt facilities in relation to which it is the exemption registration authority.

For a waste operation:

- name and address of the exempt operation, establishment or undertaking and for a water discharge activity or a groundwater activity, the occupier or operator;
- description of the exempt activity;
- place where the activity occurs.

If the waste operation is a WEEE operation:

- type and quantity of waste subject to the operation;

Periodic Inspections of Establishments and Undertakings (Exempt)

11.49 Under Sch 2, Pt 15, every exemption registration authority must carry out appropriate periodic inspections of establishments and undertakings carrying on exempt waste operations in respect of which it is the exemption registration authority. If an exempt waste operation involves a WEEE operation, the exemption registration authority must discharge the duty in sub-paragraph (1) by carrying out an inspection that complies with the third and fourth paragraphs of Article 6(2) of the WEEE Directive.

---

(2) Subject to sub-paragraphs (3), (4) and (5), the exemption registration authority must ensure the register contains the relevant particulars—
  (a) for an exempt waste operation, within 5 working days after the date that it receives—
    (i) notification of the relevant particulars and the information specified in paragraph 6(4), and
    (ii) for a WEEE operation, payment of the applicable fee in accordance with paragraph 6(6); and
  (b) for an exempt water discharge activity or exempt groundwater activity, within 15 working days after the date that it receives notification of the relevant particulars.

(3) The exemption registration authority must ensure the register is updated to reflect any changes notified under paragraph 13(1) of this Schedule or under Part 2 or 3 of Schedule 3—
  (a) for exempt waste operations, within 5 working days after the date that it receives the notification, or
  (b) for exempt water discharge activities and exempt groundwater activities, within 15 working days after that date.

(4) The exemption registration authority must not enter the relevant particulars on the register in relation to a WEEE operation until it has carried out an inspection in relation to the operation that complies with the second and third paragraphs of Article 6(2) of the WEEE Directive.

(5) Every exemption registration authority must—
  (a) ensure that its register is open to inspection by the public free of charge at all reasonable hours; and
  (b) provide reasonable facilities to the public for obtaining a copy of an entry on payment of a reasonable charge.

(6) A register may be kept in any form.

Duty to Remove Entries from the Register

11.50 (1) The duty to maintain a register in paragraph 7 (see para **11.9**) also includes a duty to remove an entry if:

(a) the exemption registration authority becomes aware that the exempt facility is no longer in operation at the place stated in the relevant particulars;
(b) the facility ceases to be an exempt facility; or
(c) in the case of a WEEE operation, the exemption registration authority—

  (i) has carried out an inspection in accordance with paragraph 15(2), and
  (ii) is not satisfied as to the particulars to be verified pursuant to the third paragraph of Article 6(2) of the WEEE Directive.

(2) If the exemption registration authority removes an entry from the register under sub-paragraph (1), it must notify without delay the occupier, operator or other person registered in relation to the exempt facility.

(3) Sub-paragraph (2) does not apply if the exemption registration authority was notified by the person registered in relation to the facility that the facility is no longer in operation at the place stated in the relevant particulars.

Restrictions on Registering Exempt Waste Operations Carried Out in the Same Place

11.51 Within Sch 2, para 10 of the Environmental Permitting (England and Wales) Regulations 2010:

(1) An establishment or undertaking may not be registered more than once during a valid registration period in relation to any one waste operation described in Part 1 of Schedule 3 that is carried on or is to be carried on at the same place.

(2) If an establishment or undertaking is registered more than once in contravention of sub-paragraph (1), the second and any subsequent registration is invalid.

(3) Subject to sub-paragraph (7), not more than one establishment or undertaking may be registered at the same time in relation to the same waste operation described in Part 1 of Schedule 3 if the operation is carried on or is to be carried on at the same place.

(4) If more than one establishment or undertaking is registered in contravention of sub-paragraph (3), only the registration specified in sub-paragraph (5) is valid.

(5) The registration in this sub-paragraph is the registration of the establishment or undertaking in respect of which the relevant particulars first appeared on the register in the relevant period.

(6) In this paragraph, "relevant period" means the period during which the relevant particulars appear on the register.

(7) Sub-paragraph (3) does not apply in relation to a waste operation falling within a description in paragraph

Fee for Registration Relating to WEE Operations

11.52 Within Sch 2, para 11 of the Environmental Permitting (England and Wales) Regulations 2010:Although there is the power to make a charging scheme under s 41 of the 1995 Act, the Agency may impose a fee until—

(a) the fee is superseded by such a charging scheme; or
(b) 1 April 2013,

whichever is the earlier.

The fee referred to is £840 in respect of the registration of an establishment or undertaking in relation to a WEEE operation.

Record Keeping of Exempt Waste Operations

11.53 Within Sch 2, para 12 of the Environmental Permitting (England and Wales) Regulations 2010:(1) This applies to every exempt waste operation where the waste operation—

(a) is a WEEE operation; or
(b) falls within a description in—
    (i) subject to sub-paragraph (2), paragraph U10 or U11,
    (ii) paragraph T9,
    (iii) paragraph T3 or T7 where the operation is carried on by waste mobile plant.

(2) But it does not apply in relation to an exempt waste operation where—

(a) the waste operation falls within a description in paragraph U10 or U11; and
(b) the establishment or undertaking is required to keep records in relation to the operation and the waste that is subject to that operation under the Nitrate Pollution Prevention Regulations 2008 or the Nitrate Pollution Prevention (Wales) Regulations 2008.

(3) An establishment or undertaking which carries on an exempt waste operation to which this paragraph applies must—

(a) keep records of the quantity, nature, origin and, where relevant, the destination and treatment method of all waste disposed of or recovered in the course of that operation; and

(b) where the waste operation falls within a description in paragraph T3 or T7 and is carried on by waste mobile plant, keep records of the places where the operation is carried on.

(4) An establishment or undertaking which carries on an exempt waste operation to which this paragraph applies must—

(a) retain any records that it is required to keep under sub-paragraph (3) for a period of 2 years; and
(b) during that period make those records available to the exemption registration authority on request.

Index

[*all references are to paragraph number*]